Unification of Elementary Forces and Gauge Theories

Unification of Elementary Forces and Gauge Theories

Papers presented at the Ben Lee Memorial
International Conference on Parity Nonconservation,
Weak Neutral Currents and Gauge Theories,
Fermi National Accelerator Laboratory,
October 20-22, 1977

Editors:

David B. Cline
Fermi National Accelerator Laboratory
and University of Wisconsin

Frederick E. Mills
Fermi National Accelerator Laboratory

harwood academic publishers
LONDON – CHUR

Published under license by:

Harwood Academic Publishers Ltd.
Chansitor House
37/38, Chancery Lane
London WC2A 7EL

and

Harwood Academic Publishers GmbH
Poststrasse 22
7000 Chur

Editorial Office for the United States of America:
Post Office Box 786
Cooper Station
New York, New York 10003

British Library Cataloguing in Publication Data

Ben Lee Memorial International Conference
on Parity Nonconservation, Weak Neutral
Currents and Gauge Theories, *Fermi National Accelerator Laboratory, Batavia, Ill., 1977*
Unification of elementary forces and gauge
theories.
1. Particles (Nuclear physics) – Congresses
I. Title II. Cline, David B
III. Mills, Frederick E
539.7'54 QC793.9

ISBN 0-906346-00-2

Printed in the United States of America

TABLE OF CONTENTS

4. Parity Nonconservation in Nuclei

5. Search for New Particles Beyond Charm

6. Weak Interactions in Astrophysics

7. Theory

PREFACE

David B. Cline

Fermilab and
University of Wisconsin

The Ben Lee Conference was organized to investigate the status of the experimental evidence for the proposed unification of weak and electromagnetic interactions. The most promising theory for unification is a gauge theory. The meeting was distinguished by excellent state-of-the-art experimental talks and some rather lucid theoretical talks. The broad nature of the subject lead to talks from several separate fields of physics, (particle physics, astrophysics, nuclear, and atomic physics). Thus the conference was perhaps the first "unity of science" conference in many years.

Some highlights of the conference were the excellent introductory talk by C. N. Yang, laying the framework for the gauge theory concepts from classical physics to field theory. All of the talks were excellent. The summary talk of S. Weinberg was particularly enlightening. A notable talk was given by J. D. Bjorken in which he surveyed the experimental and theoretical evidence for the gauge theory unification of weak and electromagnetic interaction. The reader is urged to study this paper and draw his own conclusion.

This conference will not be the last such endeavor, and perhaps is really the first in a series of meetings where the common ground in many areas of physics are discussed. In this sense it will be a lasting tribute to Ben Lee and his own broadminded vision of physics. Thus it is appropriate to record here some personal recollections of Ben's work.

The HPWF neutrino detector at Fermilab received a neutrino beam in the early spring of 1973. The detector had been outfitted with a trigger mode to select "muonless events", which simply required an energy deposition in the calorimeter and no obvious muon signal in the iron magnetic spectrometer. Of course, when the first neutrino data were recorded the "muonless trigger" lit up--there were many such events, so many that the experimenters were very confused as to the origin of these events. They were not unaware of similar events being seen in the GGM experiment at CERN at the same time (the CERN group was also informed of the observations at Fermilab in July, 1973). The only question was how to conclusively prove that the "apparent" muonless events were real muonless events. Both groups submitted papers for journals and for conferences in the summer of 1973.

In the fall of 1973, the HPWF detector was reconstructed for the purpose of uniquely establishing the muonless effect, and as might be expected in such a drama, the initial results from the new detector were incorrect and appeared to limit the size of the neutral current effect to about 10% of the charged current signal. The confusion in the mind of the experiments was overwhelming. However, Ben Lee never doubted the existence of neutral currents and gave a talk at Fermilab in the fall of 1973 about the beauty of the unified theory and the necessary existence of weak neutral currents, in his opinion.

In retrospect I believe the HPWF group finally resolved the confusion in their results by a better understanding of their detector and beam because Ben Lee and others were so certain the new results were wrong. And Ben was never too shy to say that the experimenters may be wrong and not the theory. I believe that this was the first close encounter that Ben Lee had with the confused world of experimental high energy physics. Before that he had worked on more abstract ways of mathematical physics.

The next experimental problem Ben Lee helped resolve was the existence of naked charm. He, Rosner and Gaillard wrote a classic paper on the properties of charmed particles in the spring of 1974. By 1975 there was strong indirect evidence for charm from several sources (ψ/J, dileptons in neutrino interactions, a $\Delta S = \Delta Q$ event at BNL) and yet the SLAC-LBL group at SPEAR published a paper with a low limit on the existence of charmed particles produced in e^+e^- collisions. Ben didn't believe this result, as he told me several times. He also told the SLAC-LBL group the same over the years. In the summer of 1976 the situation clarified. Once again, Ben was correct. Over the past few years Ben showed an intense interest in the experiments at Fermilab, including the dirty details of the detector and corrections for detection efficiency, etc. He even served on a committee at PEP and made important suggestions for the first round of experiments there.

It is therefore appropriate that the conference whose proceedings are published here is devoted to Ben Lee. His contribution to the subject is seen at many levels including helping the resolution of key experimental questions.

The conference was called at this time because a number of experimental and theoretical issues seemed to be on the verge of resolution. Although Ben Lee did not originate the conference he was very excited about the concept and with his usual enthusiasm joined the organizing committee early on. I only wish he could have been at the conference to enjoy the extremely stimulating intellectual atmosphere. He would have been overjoyed.

I wish to thank Judy Ward, Helen Peterson, Anne Burwell and Rene Donaldson for their excellent help in organizing the conference and the proceedings. Also, I would like to thank the other members of the Organizing Committee, Gertrude Scharff-Goldhaber, Ernest Henley, Fred Mills and David Schramm for their contributions toward the success of the conference.

The proceedings of the conference printed here follow in order of the talks. I would like to thank all of the speakers and chairpersons for their assistance in the successful operation of the conference.

N A L 3 1 4

WU INFOMASTER 1-001542A075 03/16/78
ICS IPMVEIC MVN
07044 (0705) (1-075673G075)PD 03-16 0707
TWX 9102303233 N A L 3 1 4
NOVOSIBIRSK URSS 40 16 1030
PROF V KADYSHEVSKY
FERMILAB
BATAVIA IL

BARKOV ZOLOTORYOV OBSERVED OPTICAL ACTIVITY OF BISMUTH VAPOR
THE EFFECT IS FOUR STANDARD DEVIATIONS ABOVE ZERO AND
COMPATIBLE WITH WEINBERG SALAM MODEL THE RESULT IS PUBLISHED
IN PISMA ZHETF 27 379 1978 PRIVETOM
SKRINSKY

0839 EST

N A L 3 1 4

SPEECH AT THE BENJAMIN W. LEE MEMORIAL SESSION

Conference on Parity Nonconservation, Weak Neutral Current and Gauge Theories
October 20-22, 1977

C. N. Yang

Benjamin W. Lee was born in 1935 and died earlier this year at the age of 42. He had published more than one hundred research papers in theoretical physics in his lifetime.

The discipline of theoretical physics has as its principal aim the formulation of theoretical descriptions of the physical world which are concise and comprehensive. It has produced, for understanding the subtle, complex and often confusing manifestations of nature, thoretical descriptions of unimagined accuracy. It utilizes, and helps to originate and to develop, mathematical concepts of the greatest beauty and depth.

Because nature is subtle and complex, the pursuit of theoretical physics requires bold and enthusiastic ventures into the muddy waters at the frontiers of newly discovered phenomena. Because the concepts used are beautiful and deep, the pursuit of theoretical physics requires appreciation of and insight into the structural aspects of the theoretical apparatus.

Ben Lee's work was characterized by his ability to excel in both of these requirements. His paper with Gaillard and Rosner in the Reviews of Modern Physics, completed before the discovery of J/ψ, was remarkable to read at the time and even more remarkable to read now, after the experimental discovery of charm. His work on the renormalizability of gauge theories is among the very important works on the fundamental structure of theoretical physics in recent years. We know that few contributions in theoretical physics remain noticable after ten years. Fewer after twenty. I venture to guess that the renormalizability of gauge fields will remain important fifty years from now.

We at Stony Brook were fortunate to have had Ben in our midst for many years. He was stimulating as a physicist, cooperative and generous as a colleague and friend. He had the admirable quality of always balancing enthusiasm with good judgment and restraint. Many of us, and I in particular, had benefited from many, many enjoyable discussions with him.

Ben's death occurred at the peak of his career as a physicist. He was exuberant and full of ideas about what is in store for our field. Let me quote from a speech he made last January at the Chicago Meeting of the American Physical Society, when Steve Weinberg was awarded the Heineman prize:

> " Do we understand, or hope to understand, weak interactions as well as, say electrodynamics, in the present framework ? Perhaps. We have yet to come to grips with CP violation and ultrahigh energy behaviors of weak interactions, on which subjects I have a few remarks to make. But I am more optimistic than ever that we are on the right track, and I can say that Steve has earned the honor bestowed on him today. "

Ben's untimely death was a great loss to his family and his friends. It was a great loss to the Fermilab. It was a great loss to the science of physics.

DEDICATION

I.

Ben Lee was to have been with us at this conference. But four months earlier we learned of his sudden death at the age of 42. A tragic accident took him from us while still at the peak of his powers, with so much that he had initiated and planned still taking form.

In his all too brief life he earned international recognition as a leading researcher and teacher in the field of particle physics. He played a crucial role in creating the theoretical group at Fermilab and in the laboratory as a whole.

Several of Ben's colleagues recall:

"I first met Ben Lee in the late fifties, when Ben was a graduate student at the University of Pittsburgh, and I was an assistant professor. It was a wonderful experience to have a student so outstanding. Before Ben was finished getting his master's degree, he told me he wanted to do a Ph.D. thesis in field theory. Neither I nor anyone else at Pitt was equipped to guide him; so I contacted Abe Klein and Wilbur Ufford at the University of Pennsylvania. I told them that Ben was the smartest student that they would ever see, and that they should give him a job. Fortunately they listened to my tout and the rest is history.

"Over the years, Ben and I remained friends. I felt close to him; I knew that I was always welcome at his laboratory and at his home. His advice and interest were only a phone call away, and I often availed myself of them..."

"Ben Lee went to the University of Pennsylvania graduate school in 1958. He received his Ph.D. in particle theory with Professor A. Klein in 1960, and was appointed as assistant professor at Pennsylvania in 1961. He moved directly to a tenured position in 1963, and to the rank of professor in 1965. This rapid rise simply followed from his accomplishments. As a young graduate student he passed the Ph.D. qualifying examination with one of the highest grades ever compiled at Pennsylvania, and demonstrated a grasp of physics far beyond his years. His thesis was concerned with a dispersion representation of kaon-nucleon scattering, but he was already publishing prolifically while bringing his thesis work to conclusion.

"It was a pleasure to participate as a member of the committee to hear Ben's thesis defense. This was a discussion simultaneously on a high theoretical level and also concerned with basic empirical questions. It presaged things to come because much of Ben's later contribution stemmed from his ability to theorize on the highest conceptual and mathematical level, while at the same time keeping firmly in mind the hints and constraints of experimental data.

"... in one respect, perhaps more than others, Ben as child was father to Ben the man. The joy that Ben took from doing physics, his delight in every feature of that intellectual process, was not attentuated as he matured and gained increasing recognition. To talk and think with Ben, for those of us who had that good fortune, was to be uplifted by this feeling of joy in what we were doing. It was a source of strength to all of us. And it was a measure of Ben that, the more he gave of himself in that way, the more he had to give."

"It was an education to talk physics with him. Almost always, at the start, some tough skepticism. Then, if your ideas were badly muddled, a gentle lowering of the tone with patient explanations couched in tactful phrases, e.g., 'what you're really saying is...' When the physics was less unpromising the response was less tactful and more vigorous: hard objections, probing reformulation, better ideas from Ben, phone calls, quick estimates and often at the end, an inadequate, brief footnote of acknowledgement to Ben in a publication whose best parts were his. In some cases, full collaboration, joyous and exhausting.

"I experienced both kinds of interaction with Ben. Mostly the muddled kind. Once, however, my own confusion was so outrageous as to produce an outcome in a class all by itself: a paper written by Ben with my blessings. Its contents and conclusions diametrically contradicted some erroneous ideas I had raised with him. In his paper I was warmly thanked for stimulating discussions: characteristically, though, not a word that this was the stimulation of raw foolishness. He was a great physicist and a gentleman."

"Late in 1973 he asked: 'Do you really think people have looked hard enough for charmed particles?' In that instance and in many others, he knew precisely the right question to ask."

"Ben came up to me one day and suggested that I work with him on a problem which he had considered previously. It was an exhilarating experience to work with such a brilliant and intense physicist. After this problem had been resolved, and even before we had finished writing the brief note reporting our results, Ben came into my office and urgently asked me if I wanted to work with him on estimating charm production. Again in this work I was struck by his infectious enthusiasm, the wonderful power of his mind, both in its speculative and analytical abilities and his admirable concern for confronting the new unified theories with experimental data."

"Ben had excellent taste without ever having to be deprecating about the work of others. In part this must have reflected his innate humility and gentleness. But physics also was so much a source of joy to him that one cannot imagine his spoiling that delight by uttering harsh or condescending words. Ben's example was a shining reminder of one's best reasons for doing physics -- an innate curiosity, a love of nature, and a boundless pleasure in sharing the fruits of one's quest with one's fellow man."

II.

It was one of the aspects of Ben Lee's preeminence as a physicist that he was not only gifted in discovering new results but also, and equally so, in explaining these results clearly and logically. This is amply attested to by his many monographs, reviews, and summer school lectures. His pedagogical works are characterized by his masterful command of the most difficult aspects of field theory, combined with his concern for extracting and testing the experimental predictions of the new theories. This range of Ben's interests and accomplishments is also clearly evident in the invited talks he gave at the NAL Conference in 1972, the SLAC conference in 1975, and the Coral Gables and Aachen Conferences in 1976.

III.

He had high moral standards and was concerned about social justice and injustice, both in the nearby physics community and in the world. His remarks in these areas, like his physics, showed imagination, individual thinking and analysis, and somewhat deeper understanding of the issues that existed in the community at large. At lunch and afterwards at coffee, he took great pleasure in conversing on a wide range of nonphysics topics with both theorists and experimentalists. Those who participated in these conversations remember well how articulately and precisely Ben would make his points and how broad was his range of interests. They remember too his candor, his warm sense of humor, and his understanding and appreciation of other people.

IV.

It is entirely appropriate that this conference on Parity Nonconservation, Weak Neutral Currents and Gauge Theories be dedicated to the memory of Ben Lee. He played an essential role in the organization of the conference in its early stages. His long term interest in weak interactions, his pioneering role in the modern developments of gauge theories and his broad interests in many areas of physics made him a natural supporter of an "interdisciplinary" conference of this type. He welcomed the opportunity to bring together physicists working in particle physics, nuclear physics, atomic physics and astrophysics whose work all had bearing on the investigations of the properties of weak interactions. He would have participated enthusiastically in all discussions, and his absence was keenly felt.

The conference has proved to be extremely interesting and stimulating. Its dedication to Ben Lee is a fitting tribute to his memory.

H.J. Lipkin, A.K. Mann, S. Meshkov,

C. Quigg, J. Rosner, R. Shrock, S. Treiman

Introduction

Weak Neutral Currents in Particle Physics

Parity Nonconservation in Atomic Processes

Parity Nonconservation in Nuclei

Search for New Particles Beyond Charm

Weak Interactions in Astrophysics

Theory

Introduction

SYMMETRIES IN PHYSICS

Chen Ning Yang
Institute for Theoretical Physics
State University of New York
Stony Brook, New York 11794

1. Earlier Roles of Symmetry Principles

Symmetry principles and invariance principles have played important roles in the development of physics. Twenty years ago[1] I discussed these roles and used as illustration the following diagram:

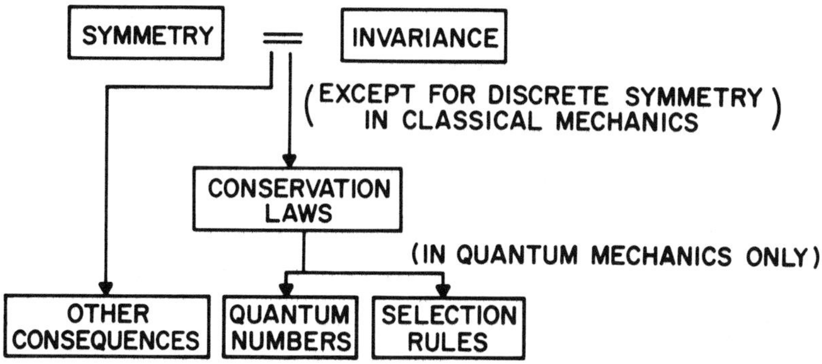

Figure I

While symmetry principles were useful in classical mechanics, it was only in quantum mechanics that symmetry principles assumed an essential role in the very language of physics. Quantum numbers and selection rules are, for example, basic in the vocabulary of quantum mechanics. What is the reason for this increased role? The answer is[2] quite simple: in quantum mechanics the state of a system is described by state vectors that satisfy the superposition principle. Because of the superposition principle, elliptical orbits as well as circular orbits of an electron in a hydrogen atom possess symmetry properties. Also because of the superposition principle, these symmetry properties can be discussed using the representation theory of the SO_3 group, leading to elegant and far reaching mathematical and physical results. As physicists proceeded from atomic to nuclear and then to elementary particle physics, from quantum mechanics to quantum field theory, the representation theory of groups became a powerful and indispensible tool in the description of nature.

2. Expanded Role of Symmetry Principles: Gauge Fields

In recent years a greatly expanded role of symmetry principles in physics came about through the concept of gauge fields. The history of this concept has been described elsewhere[3]. We need only emphasize here that one arrives naturally at this concept if one endeavors to make an internal symmetry local in nature. Another point that is worth emphasizing is that using the gauge principle, one writes down essentially unique interactions between the gauge field and different particles possessing conserved quantities which are sources of the gauge field. This is of course analoguous to the way one writes down the interaction between the electromagnetic field and charged particles. One might say that through the concept of gauge fields, symmetry dictates the interaction, whereas without the gauge principle it was the interaction that dictates the symmetry. [Einstein's general relativity also uses symmetry to dictate the form of interaction. This is quite natural because general relativity is in fact a kind of gauge theory.]

Two important developments about which we shall hear a lot during this conference are the possible unification of weak and electromagnetic interactions[4] and the related and important question of the renormalizability of non-Abelian gauge fields[5]. Ben Lee has contributed in essential ways to both of these two developments, and we shall miss him.

A most exciting lesson that we have learned is that gauge symmetry is good for renormalizability. The full implication of this statement, in my opinion, is yet to be completely understood.

3. Further Expansion Through Nontrivial Bundles: Dirac's Monopole as an Example

It was recognized in recent years that the physical concept of gauge fields is identical to some concepts in the mainstream of modern mathematics. A table of translation[6], reproduced here as Table 1, has been given for the terminology used by physicists and mathematicians. The last two entries introduce the mathematical concept of a trivial or nontrivial bundle. To explain the basic idea of a nontrivial bundle, let us start with an analysis of Dirac's magnetic monopole.

Gauge Field Terminology	Bundle Terminology
gauge (or global gauge)	principal coordinate bundle
gauge type	principal fiber bundle
gauge potential b^k_μ	connection on a principle fiber bundle
S Eq. (8)	transition function
phase factor ϕ_{QP}	parallel displacement
field strength $f^k_{\mu\nu}$	curvature
source[a] J^k_μ	?
electromagnetism	connection on a U_1 bundle
isotopic spin gauge field	connection on a SU_2 bundle
Dirac's monopole quantization	classification of U_1 bundle according to first Chern class
electromagnetism without monopole	connection on a trivial U_1 bundle
electromagnetism with monopole	connection on a nontrivial U_1 bundle

[a]I.e., electric source

Table 1. Translation of Terminology

Magnetic Monopole and Need to Introduce Two Regions

The magnetic monopole is the magnetic change. While the idea of magnetic monopoles must have been discussed in classical physics early in the history of electricity and magnetism, modern discussions date back to 1931 in the important paper of Dirac[7] in which he pointed out that magnetic monopoles in quantum mechanics exhibit some extra and subtle features. In particular, with the existence of a magnetic monopole of strength g, electric charges and magnetic charges must necessarily be quantized, in quantum mechanics. We shall give a new derivation of this result as follows:

If one wants to describe the wave function of an electron in the field of a magnetic monopole, it is necessary to find the vector potential $\vec{A}$ around the monopole. Dirac chose a vector potential which has a string of singularities. The necessity of such a string of singularities is obvious if we prove the following theorem[6].

Theorem. Consider a magnetic monopole of strength $g \neq 0$ at the origin and consider a sphere of radius R around the origin. There does not exist a vector potential $\vec{A}$ for the monopole magnetic field which is singularity free on the sphere. This theorem can be easily proved in the following way. If there were a singularity free $\vec{A}$ we consider the loop integral

$$\oint A_\mu dx^\mu$$

around a parallel on the sphere as indicated in Figure 2. By Stoke's theorem

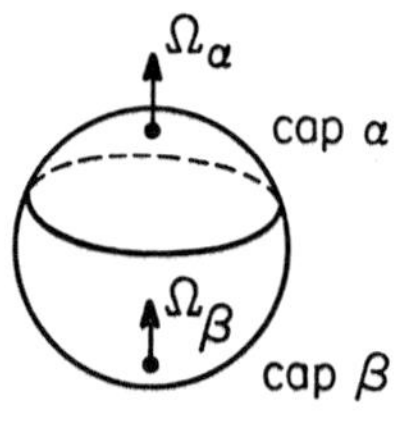

Figure 2. A sphere of radius R with a magnetic monopole at its center. The parallel divides the sphere into two caps α and β.

this loop integral is equal to the total magnetic flux through the cap α

$$\oint A_\mu dx^\mu = \Omega_\alpha . \tag{1}$$

Similarly we can apply Stoke's theorem to cap β obtaining

$$\oint A_\mu dx^\mu = \Omega_\beta . \tag{2}$$

Here Ω_α and Ω_β are the total upward magnetic flux through the caps α and β, both of which are bordered by the parallel. Subtracting these two equations we obtain

$$0 = \Omega_A - \Omega_B , \tag{3}$$

which is equal to the total flux out of the sphere, which in turn is equal to $4\pi g \neq 0$. We have thus reached a contradiction.

Having proved this theorem, we observe that R is arbitrary. Thus one concludes that there must be a string of singularities or strings of singularities in the vector potential to describe the monopole field. Yet we know that the magnetic field around the monopole is singularity free. This suggests that the string of singularities is not a real physical difficulty. Indeed the situation is reminiscent of the problem that one faces when one wants to find a parametrization of the surface of the globe. The coordinate system that we usually use, the latitude and the longitude, is not singularity free. It has singularities at the north pole and at the south pole. Yet the surface of the globe is evidently without singularities. We deal with this situation usually in something like the way illustrated in Figure 3. We consider a rubber sheet with nicely defined coordinates and stretch and wrap it down onto the globe so that it covers more than the northern hemisphere. Similarly, we consider another rubber sheet with nicely defined coordinates and stretch and wrap it upwards so that it covers more than the southern hemisphere. We now have a double system of coordinates to describe the points on the globe. The description is now analytic in the domain covered by each sheet, if we had done no violence in the stretching and wrapping. In the overlapping region covered by both sheets, one has two coordinate systems which are transformable into each other by an analytic non-vanishing Jacobian. This double coordinate system is an entirely satisfactory way to parametrize the globe.

Following this idea we shall now try to exorcise[6] the string of singularities in the monopole problem by dividing space into two regions. We shall call the points outside of the origin, above the lower cone in Figure 4, region R_a. Similarly, we shall call the points outside of the origin, under the upper cone, R_b.

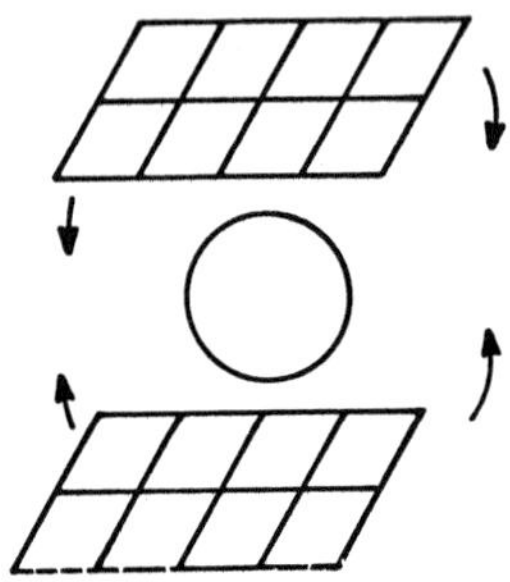

Figure 3. Method of parametrizing the globe.

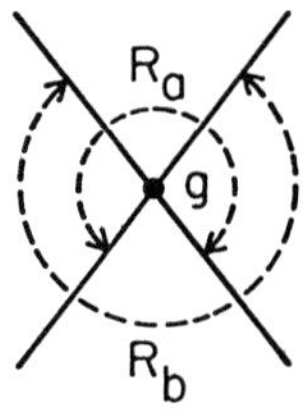

Figure 4. Division of space outside of monopole g into overlapping regions R_a and R_b.

The union of these two regions gives all points outside of the origin. In R_a we shall choose a vector potential for which there is only one non-vanishing component of A, the azimuthal component:

$$(A_r)_a = (A_\theta)_a = 0, \; (A_\phi)_a = \frac{g}{r \sin\theta}(1 - \cos\theta), \tag{4}$$

It is important to notice that this vector potential has no singularities anywhere in R_a. Similarly in R_b we choose the vector potential

$$(A_r)_b = (A_\theta)_b = 0, \quad (A_\phi)_b = \frac{-g}{r \sin\theta}(1 + \cos\theta), \tag{5}$$

which has no singularities in R_b. It is simple to prove that the curl of either of these two potentials give correctly the magnetic field of the monopole.

In the region of overlap, since both of the two sets of vector potentials share the same curl, the difference between them must be curlless and therefore must be a gradient. Indeed a simple calculation shows

$$(A_\mu)_a - (A_\mu)_b = \partial_\mu\alpha, \; \alpha = 2g\phi, \tag{6}$$

where ϕ is the azimuthal angle. The Schrödinger equation for an electron in the monopole field is thus

$$\frac{1}{2m}(p-eA_a)^2\psi_a + V\psi_a = E\psi_a, \qquad \text{in } R_a,$$

$$\frac{1}{2m}(p-eA_b)^2\psi_b + V\psi_b = E\psi_b, \qquad \text{in } R_b,$$

where ψ_a and ψ_b are respectively the wave functions in the two regions. The fact that the two vector potentials in these two equations are different by a gradient tells us, by the well known gauge principle, that ψ_a and ψ_b are related by a phase factor transformation

$$\psi_a = [\exp(2iq\phi)]\psi_b, \; q = eg, \tag{7}$$

or

$$\psi_a = S\psi_b, \; S = \text{transition function} = \exp(2ieg\phi). \tag{8}$$

Around the equator which is entirely in R_a, ψ_a is single valued. Similarly, since the equator is also entirely in R_b, ψ_b is single valued around the equator. Therefore, S must return to its original value when one goes around the equator. That[6] implies Dirac's quantization condition:

$$2q = \text{integer}. \tag{9}$$

<u>Hilbert Space of Sections</u>

Two ψ's, ψ_a and ψ_b, in R_a and R_b respectively, that satisfy the <u>condition of transition</u> (8) in the overlap region, is called a <u>section</u> by the mathematicians. We see that around a monopole the <u>electron wave function is a section and not an ordinary function</u>. We shall call these wave sections.

Different wave sections (belonging to different energies, for example), clearly satisfy the same condition of transition (8) with the same q. This suggests the concept[8] of a Hilbert space of sections. Thus, we define the scalar product of two sections ξ, η (for the <u>same</u> q) by

$$(\eta, \xi) \equiv \int \eta^* \xi d^3 r. \tag{10}$$

(The question of convergence at r = 0 and r = ∞ is ignored here). Notice that in the overlap

$$(\eta_a)^* \xi_a = (\eta_b)^* \xi_b \tag{11}$$

so that (10) is well defined.

It is clear that if ξ is a section, then $x\xi$ is also a section, since

$$x \xi_a = S(x \xi_b).$$

Thus x is an <u>operator</u> in the Hilbert space of sections. Similarly we prove that the components of $(\vec{p}-e\vec{A})$ are operators, but those of p are not. Furthermore $\vec{x}$ and $\vec{p}-e\vec{A}$ are both Hermitian.

Following Fierz[9] we shall now try to construct angular momentum operators. Define

$$\vec{L} = \vec{r} \times (\vec{p} - e\vec{A}) - \frac{q\vec{r}}{r}. \tag{12}$$

It is clear that L_x, L_y, L_z are Hermitian operators on the Hilbert space of sections. The following commutation rules can be easily verified:

$$[L_x, x] = 0, \quad [L_x, y] = iz, \quad [L_x, z] = -iy,$$

$$[L_x, p_x - eA_x] = 0, \ [L_x, p_y - eA_y] = i(p_z - eA_z),$$

$$[L_x, p_z - eA_z] = -\, i(p_y - eA_y). \tag{13}$$

It follows from these that

$$[L_x, L_y] = iL_z, \text{ etc.} \tag{14}$$

Eq. (13), together with its consequence (14), show that L_x, L_y, L_z are the angular momentum operators. We emphasize that neither the Hilbert space, nor these operators, possess any "singularities". (The singularities of A_a and A_b are not real singularities because they occur outside of R_a and R_b, respectively.)

Monopole Harmonics $Y_{q,\ell,m}$

Since $[r^2, \vec{L}] = 0$, we can diagonalize r^2 and study operator $\vec{L}$ for fixed r^2. I.e. we shall study sections of the form

$$\delta(r^2 - r_0^2)\xi,$$

where ξ is a section dependent only on angular coordinates θ and ϕ. $\vec{L}$ operates then on "angular sections".

Eq. (14) shows that $[L^2, L_z] = 0$. Simultaneous diagonalization produces the familiar multiplets with eigenvalues $\ell(\ell + 1)$ and m

$$L^2 Y_{q,\ell,m} = \ell(\ell+1) Y_{q,\ell,m}; L_z Y_{q,\ell,m} = m Y_{q,\ell,m} \tag{15}$$

where $\ell = 0, \frac{1}{2}, 1, \ldots$ and, for each value of ℓ, m ranges from $-\ell$ to $+\ell$ in integral steps of increment. The $Y_{q,\ell,m}$ are eigensections[8] which are called monopole harmonics. The allowed values of ℓ and m are

$$\begin{aligned} \ell &= |q|, \ |q| + 1, \ |q| + 2, \ldots, \\ m &= -\ell, \ -\ell + 1, \ldots, \ell, \end{aligned} \tag{16}$$

and each of these ℓ,m combinations occur exactly once. We shall choose each Y normalized so that

$$\int_0^{\pi} \sin\theta \, d\theta \int_0^{2\pi} |Y_{q,\ell,m}|^2 d\phi = 1. \tag{17}$$

The set of all $Y_{q,\ell,m}$ for a given $q = eg$, then forms a complete orthonormal set of eigensections. For the special case of $eg = 0$, they reduce to the usual spherical harmonics.

The monopole harmonics $Y_{q,\ell,m}$ do not have any singularities. That means $Y_{q,\ell,m}^{(a)}$ has no singularities in R_a and $Y_{q,\ell,m}^{(b)}$ has no singularities in R_b.

Many of the formulae for spherical harmonics can be generalized[10] to monopole harmonics.

Using monopole harmonics the scattering[11] and bound state[12] problems of a Dirac electron in the field of a magnetic monopole have recently been studied.

Remarks

(A) It is important to realize that the above-described way of using $(A)_a$ and $(A)_b$ together to describe the magnetic field of a monopole has an additional advantage: It gives the magnetic field H correctly everywhere. In older papers one oftentimes took a single A with a string of singularities. Since by definition

$$\nabla\cdot(\nabla \times A) = 0,$$

the magnetic field described by $\nabla \times A$ must have continuous flux lines. Thus its flux lines consist of the dotted lines of Figure 5, plus the bundle of lines described by the solid line, so as to make the net flux at the origin zero. Thus, $\nabla \times A$ does not correctly describe the magnetic field of the monopole, a point already emphasized by Wentzel.[13]

(B) It is instructive to go back to the reasoning concerning Figure 2 and try to repeat the steps for the combined A_a, A_b description of the magnetic field. Choose the parallel to be the equator. Then

$$\oint (A_\mu)_a dx^\mu = \Omega_\alpha$$

$$\oint (A_\mu)_b dx^b = \Omega_\beta .$$

Thus $$4\pi g = \Omega_a - \Omega_\beta = \oint [(A_\mu)_\alpha - (A_\mu)_\beta] \, dx^\mu$$

which is, by (6), equal to the increment of α around the equator i.e. $2g(2\pi) = 4\pi g$.

We have arrived at an identity. My reason for going through this simple argument is that it embodies exactly the gist of the proof of the famous Gauss-Bonnet-Allendoerfer-Weil-Chern theorem and the later Chern-Weil theorem which play seminal roles in contemporary mathematics.

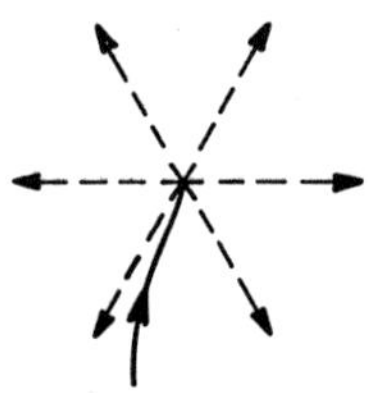

Figure 5. Magnetic Flux lines due to A. Since $\nabla\cdot(\nabla \times A) = 0$, flux lines are everywhere continuous. Hence there is "return flux" along solid line.

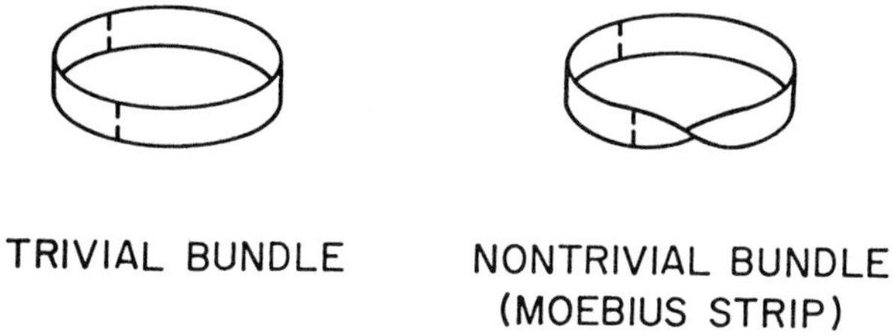

Figure 6. Examples of a trivial and nontrivial fiber bundle.

Magnetic Monopole and "Nontrivial Bundle"

We can now understand the basic reason why the monopole is related to a nontrivial bundle.

Consider Figure 6 which illustrates (i) a trivial and (ii) a nontrivial bundle. If they are cut along the dotted lines, each would break into two pieces. Looking at the resultant pieces we cannot differentiate between cases (i) and (ii). The two cases are different only in the way the "resultant pieces" are put together. In case (ii), a twist of one of the "resultant pieces" is necessary. Thus case (i) corresponds to

$$\psi_a = S\psi_b, \text{ where } S = 1, \text{ (no twist)};$$

and case (ii) corresponds to ($q \neq 0$),

$$\psi_a = S\psi_b, \text{ where } S = \exp(2iq\phi), \text{ (twist)}.$$

A bundle where the transition function S is necessarily different from 1 is called nontrivial, because a twist is needed. Hence electromagnetism with a magnetic monopole is nontrivial [cf. (7) and (8)].

4. Other Nontrivial Bundles in Physics

We have seen that the Dirac monopole requires a mathematical description in terms of a nontrivial bundle. Are there other nontrivial bundles which may be of interest to physicists?

About ten years ago Wu and I looked[14] for classical solutions of the sourceless SU_2 gauge field and for a second quantized theory around such solutions. The motivation for this search can be expressed in two different but equivalent ways. (A) A second quantized field theory does not necessarily define just one Hilbert space. Every classical solution of the field equations is a candidate for defining a Hilbert space. (B) In the Feynman path integral formalism, a classical solution of the field equations is a stationary point of the integrand. Each stationary point is a candidate for a meaningful definition of the path integral.

Wu and I did find some classical solutions, but they were not entirely satisfactory. In 1975 Belavin, Polyakov, Schwartz, and Tyukin[15] found a solution for Euclidean space which they called "pseudoparticle solution". It was later referred to as an "instanton" solution, a name that I think is misleading. The solution was later generalized, and all these solutions represent connections on nontrivial bundles.

The physical implication of these solutions is an interesting but difficult problem about which there has been much recent work[16-21]. If one believes that non-Abelian gauge fields do exist, then this difficult problem must be tackled.

Footnotes

[1]Chen Ning Yang, Science, 127, 565 (1958).

[2]Chen Ning Yang, Lecture at Symposium on Symmetries, Bryn Mawr College 75th Anniversary Celebration, 1958.

[3]Chen Ning Yang, Annals of N. Y. Acad. Science 294, 86 (1977).

[4]S. Weinberg, Phys. Rev. Lett. 19, 1264 (1967); A. Salam in Elementary Particle Theory, edited by N. Svartholm (Almquist and Forlag, Stockholm, (1968).

[5]G. t'Hooft, Nuclear Physics B35, 167 (1971); E. S. Abers and B. W. Lee, Physics Reports, 9C, 1 (1973). This last paper contains also a review of earlier works of R. P. Feynman, B. S. DeWitt, P. W. Higgs, L. D. Faddeev, V. N. Popov, and M. T. Veltman.

[6]Tai Tsun Wu and Chen Ning Yang, Phys. Rev. D12, 3845 (1975).

[7]P. A. M. Dirac, Proc. Roy. Soc. A133, 60 (1931).

[8]Tai Tsun Wu and Chen Ning Yang, Nuclear Phys. B107, 365 (1976).

[9]M. Fierz, Helv. Phys. Acta 17, 27 (1944).

[10]Tai Tsun Wu and Chen Ning Yang, Phys. Rev. D16, 1018 (1977).

[11]Yoichi Kazama, Chen Ning Yang and Alfred S. Goldhaber, Phys. Rev. D15, 2287 (1977).

[12]Yoichi Kazama and Chen Ning Yang, D15, 2300 (1977).

[13]G. Wentzel, Prog. Theor. Phys. Suppl. 37-38, 163 (1966).

[14]Tai Tsun Wu and Chen Ning Yang, in Properties of Matter under Unusual Conditions, Edited by H. Mark and S. Fernbach, (Interscience, 1969). See especially Section III where fluctuations near the classical solution were discussed.

[15]A. Belavin, A. Polyakov, A. Schwartz, and Y. Tyukin, Phys. Lett. 59B, 85 (1975). The "pseudoparticle" solution given in this paper is, in a 5-dimensional view, a generalization of the Dirac monopole of §3 above. See R. Jackiw and C. Rebbi, Phys. Rev. D14, 517 (1976). Chen Ning Yang, to appear in J. Math. Phys.

[16]A. M. Polyakov, Phys. Lett. 59B, 82 (1975).

[17] G. 't Hooft, Phys. Rev. Letters 37, 8 (1976).

[18] R. Jackiw and C. Rebbi, Phys. Rev. Letters 37, 172 (1976).

[19] C. G. Callan, R. F. Dashen, and D.J. Gross, Phys. Letters 63B, 334 (1976).

[20] J. L. Gervais and B. Sakita, "Collective Coordinates in SU(2) Yang-Mills Theory" CCNY-HEP-76/11.

[21] K. M. Bitar and Shan-jin Chang, Fermilab preprint 77/67-THY.

PRESENT STATUS OF INELASTIC AND ELASTIC HADRONIC WEAK NEUTRAL CURRENTS

A. K. Mann

Department of Physics
University of Pennsylvania
Philadelphia, Pennsylvania 19104

ABSTRACT

Since the first observations of weak neutral currents in the reactions $\nu_\mu(\bar{\nu}_\mu)$ + nucleon $\rightarrow$ $\nu_\mu(\bar{\nu}_\mu)$ + hadrons in 1973, there have been a number of additional measurements of those reactions, and also of the elastic reactions $\nu_\mu(\bar{\nu}_\mu) + p \rightarrow \nu_\mu(\bar{\nu}_\mu) + p$. In this paper the methods utilized are briefly compared and the results obtained are summarized. Inferences are drawn from these results concerning the space-time structure and scale invariance of the hadronic weak neutral current.

Historical Introduction

Let me begin with a few minutes of history. Almost exactly five years ago in this same lecture hall the material shown in Table I was presented by Professor D. Perkins in his review[1] of experiments on neutrino interactions. One sees in Table I the experimental limits available near the end of 1972 on purely leptonic weak neutral current (WNC) neutrino scattering reactions and also on certain $\Delta S = 0$ WNC semileptonic neutrino scattering processes. These limits, taken in conjunction with the much lower experimental upper limits on the rates of $|\Delta S| = 1$ WNC decays of K mesons shown in Table II, had led to the widely held conclusion that weak neutral current processes were probably absent in nature.

For many years we have visualized charged lepton scattering reactions, elastic and inelastic, as occurring through an electromagnetic neutral current as in the Feynman diagrams of Figs. 1a and 1b. On the other hand, on the basis of the evidence in Table I, it was assumed in the phenomenological theory of the weak interaction (prior to present gauge theories) that neutrino scattering reactions occurred only through charged current weak interactions as illustrated in Figs. 1c and 1d, i.e., through the exchange of a (real or virtual) charged vector boson (W^+), and never through neutral vector boson (W^o) exchange as in Figs. 2a and 2b. Consistency with experiment (Table II) also required that the diagram of Fig. 2c be forbidden.

Failure to observe WNC processes was clearly a stumbling block for theories seeking a unifying connection between the weak interaction and electrodynamics.[2] The absence of a WNC was also a serious problem for the phenomenological weak interaction theory with only weak charged currents because WNC processes should appear in higher order in a charged current theory[3] at a rate relative to charged current processes determined by the magnitude of the mass of the propagator (W^+) in Figs. 1c and 1d; for $M_W \geqslant G_F^{-\frac{1}{2}}$, the WNC rate should be of the same order of magnitude as the WCC rate.

Less than one year after the summary in Table I was presented at Fermilab, two experiments (in one of which the author of Table I was a participant) were in fact observing neutrino-induced reactions without a charged lepton in the final state.

One of these experiments was done at CERN in the large volume ($11.5\,m^3$) heavy liquid freon bubble chamber Gargamelle, with neutrinos (ν_μ) and antineutrinos ($\bar{\nu}_\mu$) of average energy about 1 GeV. One of the inelastic reactions observed in Gargamelle is illustrated in Fig. 3 which shows the tracing of an actual event observed in the chamber. In that event all of the tracks can be identified as those of strongly interacting particles, from which it was concluded that no charged lepton emerged from the neutrino-nucleon interaction.

The other experiment was carried out at Fermilab by a group of physicists from Harvard, Pennsylvania, Wisconsin and Fermilab (HPWF) using a scintillation counter-spark chamber apparatus and incident neutrinos and antineutrinos with an average energy of about 20 GeV. Here the target-detector of the neutrino interactions was a large ionization calorimeter consisting of optically separated segments, each of which was $(3 \times 3)m^2$ in area and 0.46 m long and filled with liquid scintillator of density 0.87 gm/cm^3. Each segment was looked at by 12 photomultiplier tubes, the summed output of which measured the ionization energy deposited by the products of a neutrino (antineutrino) interaction. Because of the large volume and mass of the target-detector, most of the energy in the hadronic-electromagnetic cascade from a neutrino interaction was contained within the ionization calorimeter.

Interspersed between every four segments of the calorimeter were optical spark chambers which gave a visual display of the secondary particle cascade. Immediately downstream of the ionization calorimeter was a muon detector, consisting of large area scintillation counters and optical spark chambers, to identify muons from WCC reactions and to measure the momentum of a subsample of the muons. The experiment was triggered by the deposition of energy in the target detector greater than a preset minimum of a few GeV. A diagram of a somewhat later version of the HPWF apparatus is shown in Fig. 4 which also includes the reproduction of a WNC event. Another such event - centrally located in the target-detector - is shown in Fig. 5.

These experiments had progressed to the extent that by August, 1973, a review of the subject of weak neutral currents by G. Myatt[4] contained the material shown in Table III and later published separately by the two experimental teams.[5,6] It was particularly impressive that experiments using experimental methods so dissimilar and mean neutrino energies so different should independently obtain a positive result of approximately the same magnitude for the semileptonic inclusive weak neutral current reactions, $\nu_\mu(\bar{\nu}_\mu)$ + nucleon $\rightarrow$ $\nu_\mu(\bar{\nu}_\mu)$ + hadrons. It is fair to say that a new era in the study of weak interactions began then.

What has happened in the experimental sector in the four years since the initial experiments? A partial list of accomplishments is:

(i) Much more data have been acquired on the semileptonic inclusive reactions

$$\nu_\mu(\bar{\nu}_\mu) + N \rightarrow \nu_\mu(\bar{\nu}_\mu) + X \qquad (1)$$

where N is a nucleon and X is the sum of hadronic final states.

(ii) The elastic semileptonic neutral current processes

$$\nu_\mu(\bar{\nu}_\mu) + p \rightarrow \nu_\mu(\bar{\nu}_\mu) + p \qquad (2)$$

have been observed and measured roughly.

(iii) Measurements have been made of exclusive reactions, particularly single pion production, that bear directly on the isospin properties of the hadronic WNC.

(iv) Preliminary data on purely leptonic WNC reactions have been obtained.

(v) Ambitious programs to search for WNC induced parity nonconservation effects in atoms and nuclei, as well as in charged lepton scattering processes, have been started.

In this talk I shall concentrate, at the request of the organizing committee, on topics (i) and (ii), although there is significant overlap with certain of the other topics. These will, however, be covered extensively by other speakers and, indeed, item (v) is the subject of two entire sessions.

Semileptonic Inclusive Neutral Current Measurements

Deep inelastic neutral current scattering of neutrinos and antineutrinos has been observed in many experiments using different techniques in the last four years. For example, in Fig. 6 is shown the method of distinguishing neutral current events from charged current events first employed in a CalTech-FNAL experiment which used a sampling type hadron calorimeter of alternate layers of relatively thick iron plates and scintillator.[7] The penetration of particles makes possible a clear separation of the two event types. This method has also been adopted by the CERN-Dortmund-Heidelberg-Saclay (CDHS) experiment[8] in which the thick Fe plates of the hadron calorimeter are also magnetized to provide an integrated function detector. The CDHS data are shown in Fig. 7. A similar method using a different variable, the transverse momentum, $p_\perp$, of charged particles relative to the direction of the hadron cascade, yielded the data in Fig. 8

taken in the large (15') bubble chamber at FNAL.[9] Again, there is a clear separation (at low $p_\perp$) of the neutral current from the charged current events.

Refinements of technique by the Gargamelle collaboration have led to better measurement of the neutron background in that experiment and significantly improved data. Corresponding refinements in the HPWF experiment have led to better understanding of the systematic effects in that experiment as shown, for example, in Fig. 9a where the raw results for the ratio

$$R^{\nu}_{INC} = \frac{\sigma(\nu_\mu + \text{nucleon} \to \nu_\mu + \text{hadrons})}{\sigma(\nu_\mu + \text{nucleon} \to \mu^- + \text{hadrons})}$$

are given for the two muon detectors in Fig. 4. These detectors have very different properties: detector 1 subtends a larger solid angle for muons than does detector 2 and hence the correction for escaping muons is less for detector 1 than for detector 2; on the other hand, there is a correction of about -10% on average to the charged particles observed in detector 1 due to hadron penetration of the iron shield in front of detector 1, while that correction does not exist for detector 2. The raw results for R^{ν}_{INC} of Fig. 9a are therefore quite different for the two detectors, but the corrected results of Fig. 9b are essentially the same, except for a possible small residual systematic difference, demonstrating that the corrections applied to these data are moderately well understood.

It seems safe to conclude that the semileptonic inclusive reactions of eq. (1) have in fact been observed using different methods in a number of experiments. A summary of the most recent results on R^{ν}_{INC} and $R^{\bar{\nu}}_{INC}$ is given in Table IV which is derived in part from a recent review of this subject by P. Musset.[10] The agreement among the experiments is remarkable, even though the hadronic energy (E_H) cut-off is not the same for all the experiments. A measure of that effect in the HPWF data is shown by the two entries for $E_H > 0$ GeV and $E_H > 4$ GeV given for that experiment. The effect on $R^{\bar{\nu}}_{INC}$ will be smaller for most of the other high energy experiments which used a harder $\bar{\nu}$ spectrum than did HPWF.

The data in Table IV bear directly on the space-time structure of the hadronic WNC. They may be used in the first instance to obtain the cross section ratio $\sigma^{\bar{\nu}}_{NC}/\sigma^{\nu}_{NC}$, since

$$\sigma^{\bar{\nu}}_{NC}/\sigma^{\nu}_{NC} = \left[R^{\bar{\nu}}(E_H > 0)/R^{\nu}(E_H > 0)\right]\left(\sigma^{\bar{\nu}}_{CC}/\sigma^{\nu}_{CC}\right) .$$

To make the correction from $R^{\bar{\nu}}(E_H > E_C)/R^{\nu}(E_H > E_C)$, where E_C is the experimental cutoff in hadronic energy, to $R^{\bar{\nu}}(E_H > 0)/R^{\nu}(E_H > 0)$, we use the simple quark-parton model (QPM) with the measured values of $\sigma^{\bar{\nu}}_{CC}/\sigma^{\nu}_{CC}$, and assume various couplings for the WNC. This procedure[11] leads to corrected experimental values of $\sigma^{\bar{\nu}}_{NC}/\sigma^{\nu}_{NC}$ for each assumed WNC coupling, which may be directly compared with the value of $\sigma^{\bar{\nu}}_{NC}/\sigma^{\nu}_{NC}$ predicted for that type of coupling.

Thus, in the QPM, with $y = E_H/E_\nu$, the cross sections for the reactions in eq. (1) are

$$\frac{d\sigma^{\nu}}{dy} = \frac{G^2 ME_\nu}{\pi}\left[a + b(1-y)^2 + cy^2\right]$$

$$\frac{d\sigma^{\bar{\nu}}}{dy} = \frac{G^2 ME_{\bar{\nu}}}{\pi}\left[b + a(1-y)^2 + cy^2\right] \qquad (3)$$

where, for example,

(i) any V, A combination implies $c = 0$

(ii) pure V or pure A coupling implies $a = b$

(iii) pure S or pure T coupling implies $a = b = 0$

(iv) pure T coupling implies $a = b = c$

The total cross sections for any assumptions concerning a, b and c are obtained by integrating over y.

A comparison of the corrected measured values of $\sigma_{NC}^{\bar{\nu}}/\sigma_{NC}^{\nu}$ from the HPWF experiment[11] with the expected (QPM) values is shown in Table V. Observe that the couplings (V+A), pure S or pure P, or (S,P) or (S,P)T mixtures are appreciably less likely than the other coupling forms in Table V. Also pure V (or A) and pure T are less probable than (V-A) by about 3 standard deviations.

Similar conclusions are obtained from the distributions in E_H measured in the HPWF experiment[11] when compared with the distributions expected for different couplings. There are shown in Fig. 10 comparisons which confirm the lower probability of S, P and T combinations and of (V+A).

If, somewhat arbitrarily, we restrict the WNC amplitudes to those involving only V and A and isospins 0 and 1, we can write in place of eq. (3), providing scale invariance violating effects are neglected,

$$\frac{d\sigma_{NC}^{\nu}}{dy} = \frac{G_{NC}^2 ME_{\nu}}{\pi} \; [g_L + (1-y)^2 \; g_R]$$

and $$\tag{4}$$

$$\frac{d\sigma_{NC}^{\bar{\nu}}}{dy} = \frac{G_{NC}^2 ME_{\bar{\nu}}}{\pi} \; [g_R + (1-y)^2 \; g_L]$$

with

$$g_L \sim |\epsilon_L|^2 + \frac{\bar{Q}}{Q} |\epsilon_R|^2$$

$$\tag{5}$$

$$g_R \sim |\epsilon_R|^2 + \frac{\bar{Q}}{Q} |\epsilon_L|^2$$

and

$$|\epsilon_L|^2 = |\epsilon_L(u)|^2 + |\epsilon_L(d)|^2$$

$$= |\tfrac{1}{4} (\alpha + \beta + \gamma + \delta)|^2 + |\tfrac{1}{4} (-\alpha - \beta + \gamma + \delta)|^2$$

$$|\epsilon_R|^2 = |\epsilon_R(u)|^2 + |\epsilon_R(d)|^2 \tag{6}$$

$$= |\tfrac{1}{4} (\alpha - \beta + \gamma - \delta)|^2 + |\tfrac{1}{4} (-\alpha + \beta + \gamma - \delta)|^2$$

where $\epsilon_R(u)$, $\epsilon_R(d)$, etc. are the strengths of the weak neutral coupling to up (u) and down (d) quarks, respectively, $\bar{Q}$ is the antiquark ($\bar{u}$ and $\bar{d}$) content and Q the quark (u and d) content of the nucleon, and α, β, γ, δ are coupling coefficients involving the space-time and isospin structures: α, vector-isovector; β, axial-isovector; γ, vector-isoscalar; δ, axial-isoscalar. Detailed analyses have been carried out in several papers[12] to extract the combination of α, β, γ and δ that best fits the WNC data, including single pion production data and data on the elastic scattering reactions in eq. (2).

For our purpose here we limit the discussion to a determination of the values of g_L and g_R from which we can draw directly some very general conclusions regarding the WNC coupling. We show in Table VI the experimental values of g_L and g_R (normalized so that $g_L + g_R = 1$). The weight of this evidence strongly indicates that pure right-handed ($g_R = 1$, $g_L = 0$; V+A) coupling is ruled out, and also that pure V or pure A coupling ($g_L = g_R = 0.5$) is ruled out. Furthermore, these data suggest that pure left-handed coupling ($g_L = 1$, $g_R = 0$; V-A) is disfavored, but the strength of this conclusion is somewhat less in view of the possibility of scale breaking corrections even at low energy.

In Fig. 11 is shown the plot of $R^{\bar{\nu}}_{INC}$ against R^{ν}_{INC} for the data in Table IV and also the prediction of the SU(2)xU(1) gauge theory model of Weinberg[13] and Salam[14] with $\bar{Q}/Q = 0.1$ and $\bar{Q}/Q = 0.14$. The agreement of the data with this model is seen to be excellent. The value of the single parameter of the model, $\sin^2\theta_W$, is determined from Fig. 10 to be in the interval

$$0.25 \leq \sin^2\theta_W \leq 0.30.$$

In view of small systematic differences among the experiments, it seems better to present limits on $\sin^2\theta_W$ in this way rather than to give in the usual way a value with an assigned error.

Finally, we present in Fig. 12 the ratio R^{ν}_{INC} as a function of average neutrino energy $< E_{\nu} >$ of the beams in the several experiments of Table IV. We see that R^{ν}_{INC} is essentially independent of $< E_{\nu} >$, indicating that the cross section for inelastic neutral current neutrino scattering tracks the approximate linear dependence on laboratory neutrino energy of inelastic charged current neutrino scattering. Without attempting here a detailed fit to the data of Fig. 12, it is clear that a limit on the mass of the weak neutral current propagator (the Z^o in the SU(2)xU(1) model) can be determined roughly and is greater than about 10 GeV/c^2. It is therefore unlikely that the structure observed at 9.4 GeV/c^2 in the invariant mass distribution of dimuons from proton-nucleus collisions[13] is due to the production of a Z^o.

Semileptonic Elastic Scattering Measurements

We turn now to measurements of the elastic scattering of neutrinos and antineutrinos by protons, the reactions of eq. (2). Three experiments, utilizing ν_{μ} and $\bar{\nu}_{\mu}$ in the 0.5 to 10 GeV energy region, have yielded data on these reactions: the Columbia-Illinois-Rockefeller (CIR) experiment[14] and the Harvard-Pennsylvania-Wisconsin (HPW) experiment,[15] both at BNL, and, very recently, the Aachen-Padova experiment[16] at CERN.

There are several advantages to the study of elastic scattering. Theoretically, the reaction is simple enough to hold the promise that a model independent analysis of elastic scattering data may be made. Even if that promise is not fully realized, the predictions of current models are unambiguous. Experimentally, the differential cross section with respect to the square of the momentum transferred to the proton, $d\sigma/dq^2$, may be obtained directly from measurement of the kinetic energy T of the recoiling proton, since $|q^2| = 2\ m_p T$. Furthermore, comparison of the ν and $\bar{\nu}$ total cross sections leads to the nature of the hadronic weak neutral current coupling;

for example, equality of the cross sections implies pure V (or A) coupling, while any difference of the cross sections means that polar vector-axial vector interference is present.

It is interesting to compare the apparatus of the CIR experiment, shown in Fig. 13, with that of the HPW experiment, outlined in Fig. 14.

The former arrangement utilizes thick plate optical spark chambers and liquid scintillator counters as the neutrino target. The spark chambers provide good visual identification of an event and good angular resolution. The liquid scintillation counters allow precision timing of an event relative to the radiofrequency structure of the extracted proton beam from the BNL synchrotron, and also measurement of dE/dx, but no measurement of the total energy of a final state charged particle is available. The CIR apparatus is unshielded, i.e., no massive shielding is nearby, and relies on time-of-flight to discriminate against neutron-induced background. The apparatus is approximately 1.8 m x 1.8 m in transverse dimensions and 28 tons in total mass, with 8 tons of fiducial volume. The Aachen-Padova apparatus is similar to that of CIR; it is described in detail in Professor Faissner's talk at this conference.

The HPW apparatus, in contrast, is a completely active liquid scintillator ionization calorimeter of total mass 33 tons and fiducial mass 10 tons. The transverse dimensions are 2.6 m x 2.7 m. It is divided into 192 optically isolated cells in which dE/dx is measured along the particle path, the total energy is obtained from summation of dE/dx along the path, and the angle of the trajectory relative to the incident beam direction is roughly determined. To verify the method of angle measurements using only the cell structure of the calorimeter, the downstream section of the calorimeter includes four drift chambers as shown in Fig. 14, which are utilized as illustrated in Fig. 15. Also illustrated in Fig. 15 is the method of particle identification by means of the dE/dx measurement applied to an actual event. The basic idea behind this ionization calorimeter is that the fiducial (inner) volume is actively shielded by the exterior liquid scintillator elements as shown in Fig. 15. Because of its sensitivity to neutrons and low energy charged particles, this apparatus is protected by a massive, passive shield, and relies on time-of-flight as well. Furthermore, much of the out-of-time neutron background and a part of the in-time neutron background (arising from neutrino interactions in the passive shield) is eliminated by observing a second interaction of the neutron as it traverses the large volume of the detector and the active shield; an energy deposition in the detector or shield larger than 2 MeV serves to veto (off-line) an event if it is spatially separated from the single charged particle trajectory that constitutes a useful event.

The time-of-flight spectrum of events identified as $\nu_\mu + p \rightarrow \nu_\mu + p$ in the HPW experiment is shown in Fig. 16a. This is to be compared with the corresponding spectrum for the charged current elastic events, i.e., from $\nu_\mu + n \rightarrow \mu^- + p$, shown in Fig. 16b. The spatial distributions (in x, y and z) of the WNC events are shown in Fig. 17; the uniformity of these distributions is evidence that most events are neutrino-induced rather than neutron-induced.

Backgrounds in the CIR, Aachen-Padova and HPW experiments due to other neutral current reactions, e.g., $\nu_\mu + n \rightarrow \nu_\mu + n$ with $n + p \rightarrow n + p$, $\nu_\mu + p \rightarrow \nu_\mu + p + \pi^o$, etc., are substantial [($\approx$ 25 to 35)% of the total observed rate], but are different in detail for the different experiments. For example, there is a net correction of roughly -10% in the HPW and CIR experiments and -40% in the Aachen-Padova experiment due to the reaction chains $\nu_\mu + n \rightarrow \nu_\mu + n$ with $n + p \rightarrow n + p$ (signal increase) and $\nu_\mu + p \rightarrow \nu_\mu + p$ with $p + n \rightarrow p + n$ (signal decrease). The $\nu_\mu + p \rightarrow \nu_\mu + p + \pi^0$ channel, however, contributes between 5 and 10% to the HPW total signal but less to the CIR and Aachen-Padova total signal because photons from π^0 that convert close to the event origin are difficult to recognize in the coarser grained HPW apparatus.

Despite differences in apparatus, signal identification, and backgrounds, the CIR and HPW experiments have obtained values of $R^{\nu}_{el} = \sigma(\nu_\mu + p \rightarrow \nu_\mu + p)/\sigma(\nu_\mu + n \rightarrow \mu^- + p)$ that are in good agreement within experimental errors, as shown in Table VII. A preliminary value of R^{ν}_{EL} from the Aachen-Padova experiment, presented at this conference, is included in Table VII. An independent experiment by HPW has also determined the value of $R^{\bar{\nu}}_{EL}$ given in Table VII. It should be emphasized that corrections to the data of Table VII, particularly background subtractions, are as yet very crude.

Plots of $d\sigma/dq^2$ for $\nu_\mu(\bar{\nu}_\mu) + p \rightarrow \nu_\mu(\bar{\nu}_\mu) + p$ obtained by HPW are shown in Figs. 18a and 18b, which also give the q^2 dependence for the charged elastic processes $\nu_\mu + n \rightarrow \mu^- + p$ and $\bar{\nu}_\mu + p \rightarrow \mu^+ + n$. Although the data in Figs. 18a and 18b are rough, they do suggest that larger data samples acquired by this experimental method might provide direct, independent determinations of $\sin^2\theta_W$ from the shapes of the q^2 distributions, and perhaps make possible experimental confirmation that $d\sigma/dq^2$ is equal for ν and $\bar{\nu}$ at $q^2 = 0$, which must hold in any model if charge symmetry invariance is valid.

A useful comparison of experiment and theory[17] is given in Fig. 19 where the ratio of WNC cross sections $\sigma(\bar{\nu}_\mu + p \rightarrow \bar{\nu}_\mu + p)/\sigma(\nu_\mu + p \rightarrow \nu_\mu + p)$ is plotted against $\sin^2\theta_W$. For $0.25 \leq \sin^2\theta_W \leq 0.30$, as suggested by the

R^{ν}_{INC} and $R^{\bar{\nu}}_{INC}$ data, the elastic data are consistent with the prediction of the Weinberg-Salam SU(2)xU(1) model (allowing for reasonable uncertainties in nuclear form factors) and approximately 3 standard deviations from pure V coupling.

Summary and Prospects

We can summarize the conclusions to be drawn from the material presented here as follows:

(i) The space-time structure of the hadronic weak neutral current appears to be an admixture of polar vector (V) and axial vector (A), as is the structure of the hadronic weak charged current. Although pure scalar (S), pseudoscalar (P) and tensor (T) couplings are eliminated for both types of weak currents, as are mixtures of only S and P, it is not possible to rule out unequivocally a mixture of S, P and T that would reproduce a given (V,A) mixture. Nor is it possible to rule out small contributions of S, P or T to a dominant (V,A) mixture.

(ii) Assuming that only V and A are present, the data indicate that

(a) pure (V+A) is ruled out with very high probability

(b) pure V, pure A are ruled out with high probability

(c) pure (V-A) is unlikely.

(iii) In any theory of weak interactions with no more than one real neutral vector boson, the presence of both V and A couplings directly implies parity non-conservation in the hadronic weak neutral current.

(iv) The (V,A) admixture determined by experiment is consistent with that of the SU(2)xU(1) model with

$$0.25 \le \sin^2\theta_W \le 0.30,$$

specified largely by the inelastic data.

(v) The dependence on energy of the cross section for the inclusive reactions

$$\nu_\mu + N \to \nu_\mu + X$$

is roughly linear up to about 150 GeV.

It may also be useful to mention the immediate prospects for improvements in the data we have discussed.

(i) We can expect soon to have more precise semileptonic elastic and inelastic differential and total cross section data that will lead directly to

(a) a unique determination of the left and right-handed weak couplings to u and d quarks

(b) precise values of $\sin^2\theta_W$ from both inelastic and elastic neutrino scattering data

(c) detailed and precise tests of scale invariance of weak neutral current phenomena.

(ii) We may hope also to have more intensive searches for flavor changing in weak neutral current processes.

REFERENCES

* Supported in part by the U.S. Energy Research and Development Administration.

1. D. H. Perkins, Proc. XVI Int'l. Conf. on High Energy Physics, Chicago-Batavia, Vol. 4, p. 189, 1972 (edited by J. D. Jackson and A. Roberts).

2. S. Weinberg, Phys. Rev. Lett. 19, 1264 (1967) and 27, 1688 (1972); Phys. Rev. D5, 1412 (1972); A. Salam and J. C. Ward, Phys. Lett. 13, 168 (1964).

3. See, for example, the talk by J. D. Bjorken at this conference.

4. G. Myatt, Proc. Sixth Int'l. Symposium on Electron and Photon Interactions at High Energies, Bonn, 1973, edited by H. Rollnik and W. Pfell (North-Holland, Amsterdam, 1974), p. 389.

5. F. J. Hasert et al, Phys. Lett. 46B, 138 (1973).

6. A. Benvenuti et al, Phys. Rev. Lett. 32, 800 (1974).

7. B. C. Barish et al, Phys. Rev. Lett. 34, 538 (1975).

8. M. Holder et al, CERN preprint, August (1977).

9. Data from the ANL-Carnegie Mellon-Purdue collaboration presented by A. F. Garfinkel at the 1977 Int'l. Symposium on Lepton and Photon Interactions at High Energies, Hamburg (1977).

10. P. Musset, Review talk given at the 1977 Int'l. Symposium on Lepton and Photon Interactions at High Energies, Hamburg (1977).

11. A. Benvenuti et al, Phys. Rev. Lett. 37, 1039 (1976); P. Wanderer et al, HPWF preprint 77/1 (1977), submitted to Phys. Rev. D.

12. See, for example, J. D. Bjorken, Proc. of Summer Institute on Particle Physics, SLAC Report No. 198, November, 1976 (edited by Martha C. Zipf); L. M. Sehgal, Aachen preprint (1977); P. Q. Hung and J. J. Sakurai, preprint UCLA/77/TEP/17 (1977).

13. W. R. Innes et al, Phys. Rev. Lett. 39, 1240 (1977).

14. W. Lee et al, Phys. Rev. Lett. 37, 186 (1976).

15. D. Cline et al, Phys. Rev. Lett. 37, 252, 648 (1976).

16. Data presented to this conference.

17. C. H. Albright, C. Quigg, R. E. Schrock and J. Smith, Phys. Rev. D14, 1780 (1976).

Table I. Limits on Weak Neutral Current Couplings from D. H. Perkins, Proc. XVI Int'l. Conf. on High Energy Physics, Vol. 4, p. 189, Chicago-Batavia, Sept. 1972.

Cross Section	Author	90% CL Upper Limit
$\sigma(\bar{\nu}_e + e^- \rightarrow \bar{\nu}_e + e^-)$	Gurr, Reines, Sobel PRL 28 1406 (1972)	$< 3.0\ \sigma_{V-A}$
$\sigma(\nu_\mu + e^- \rightarrow \nu_\mu + e^-)$	CERN Gargamelle (paper #785, this Proc.)	$< 0.44\ \sigma_{V-A}$
$\sigma(\bar{\nu}_\mu + e^- \rightarrow \bar{\nu}_\mu + e^-)$	CERN Gargamelle (#785)	$< 0.62\ \sigma_{V-A}$
$\dfrac{\sigma(\nu_\mu n \rightarrow \nu_\mu n\pi^o + \nu_\mu p \rightarrow \nu_\mu p\pi^o)}{2\sigma(\nu_\mu n \rightarrow \mu p \pi^o)}$	W. Lee (#239) CERN Gargamelle (#785)	$< 0.14^*$ $< 0.11^*$
$\dfrac{\sigma(\nu_\mu p \rightarrow \nu_\mu \Delta^+)}{\sigma(\nu_\mu p \rightarrow \mu^- \Delta^{++})}$	Cundy et al** PL 31B 478 (1970) Cho et al (#473)	< 0.46 < 0.31
$\dfrac{\sigma(\nu_\mu p \rightarrow \sigma_\mu p)}{\sigma(\nu_\mu n \rightarrow \mu^- p)}$	Cundy et al (ibid)	< 0.22
$\dfrac{\sigma(\nu_\mu N \rightarrow \nu_\mu + \text{anything})}{\sigma(\nu_\mu N \rightarrow \mu^- + \text{anything})}$	CERN 1.2 m HLBC (unpublished)	≤ 0.2

* See comments in text

** The experiment gave $\dfrac{\sigma(\nu_\mu p \rightarrow \nu_\mu n\pi^+)}{\sigma(\nu_\mu p \rightarrow \nu_\mu p\pi^+)} \leq 0.08 \pm 0.04.$

Table II. Limits on Weak Neutral Current Couplings in $|\Delta S| = 1$ Transitions. Data are from Rev. Mod. Phys. 45, No. 2, Part II (1973).

Decay Mode	Upper Limit
$\dfrac{\Gamma(K^+ \to \pi^+ + e^+ + e^-)}{\Gamma(K^+ \to \pi^o + e^+ + \nu_e)}$	0.8×10^{-4}
$\dfrac{\Gamma(K_L^o \to \mu^+ + \mu^-)}{\Gamma(K^+ \to \mu^+ + \nu_\mu)}$	0.8×10^{-9}
$\dfrac{\Gamma(K^+ \to \pi^+ + \nu + \bar{\nu})}{\Gamma(K^+ \to \pi^o + e^+ + \nu_e)}$	0.3×10^{-4}

Table III. Extract of material from a talk entitled "Neutral Currents" by G. Myatt in Proc. Int'l. Symposium on Electron and Photon Interactions at High Energies, p. 389, Bonn, 1973.

2.2 Inclusive Semileptonic Neutrino Reactions

2.2.1 Gargamelle Experiment at CERN

$$(NC/CC)_\nu = 0.23 \pm 0.03$$

$$(NC/CC)_{\bar{\nu}} = 0.46 \pm 0.09$$

2.2.2 A New Result From NAL

$$\frac{\text{Events With No Muon}}{\text{Events With Muon}} = 0.26 \pm 0.07$$

for a "natural" mixture of ν and $\bar{\nu}$ from an unfocussed beam.

Table IV. Experimental Values of R^{ν}_{INC} and $R^{\bar{\nu}}_{INC}$.

Expt.	R^{ν}_{INC}	$R^{\bar{\nu}}_{INC}$	E_H / GeV
(~ Isoscalar Targets)			
D_2BC (7'B)	0.25 ± 0.05	-	
GGM (C)	0.26 ± 0.04	0.39 ± 0.06	> 0
HPWF (F)	0.30 ± 0.04	0.33 ± 0.09	> 0
	0.29 ± 0.04	0.39 ± 0.10	> 4
CITFR(F)	0.28 ± 0.03	0.35 ± 0.11	> 12
CDHS (C)	0.29 ± 0.01	0.35 ± 0.03	> 12
BEBC (C)	0.33 ± 0.04	0.35 ± 0.09	> 15 (ν)
			> 30 ($\bar{\nu}$)
(Proton Target)			
FBHM (15'F)	0.48 ± 0.17		> 0
AC-MP(15'F)		0.49 ± 0.14	

Table V. Measured and Expected Values of $\sigma_{NC}^{\bar{\nu}}/\sigma_{NC}^{\nu}$ from the HPWF Experiment.

Form of WNC	Corrected Experimental Value	Expected Value
V-A	0.61 $\pm$ 0.25	0.38
V or A	0.40 $\pm$ 0.17	1.00
V+A	0.37 $\pm$ 0.16	2.65
S or P*	0.26 $\pm$ 0.09	1.00
(S,P)T	0.31 $\pm$ 0.11	1.00
T	0.45 $\pm$ 0.15	1.00

* or (S,P) mixture

Table VI. Experimental Values of g_L and g_R.

Expt.	g_L	g_R
GGM	0.77 ± 0.16	0.23 ± 0.07
HPWF	0.85 ± 0.12	0.15 ± 0.12
CITRF	0.64 ± 0.09	0.36 ± 0.09
CDHS	0.85 ± 0.03	0.15 ± 0.03

Table VII. Measured values of R_{EL}^{ν} and $R_{EL}^{\bar{\nu}}$ for $0.3 \leq |q^2| \leq 0.9\ (GeV/c)^2$.

Experiment	Quantity	Value	Reference
CIR	R_{EL}^{ν}	0.23 ± 0.09	14
HPW	R_{EL}^{ν}	0.17 ± 0.05	15
Aa-Pa	R_{EL}^{ν}	0.17 ± 0.04	16
HPW	$R_{EL}^{\bar{\nu}}$	0.2 ± 0.1	15

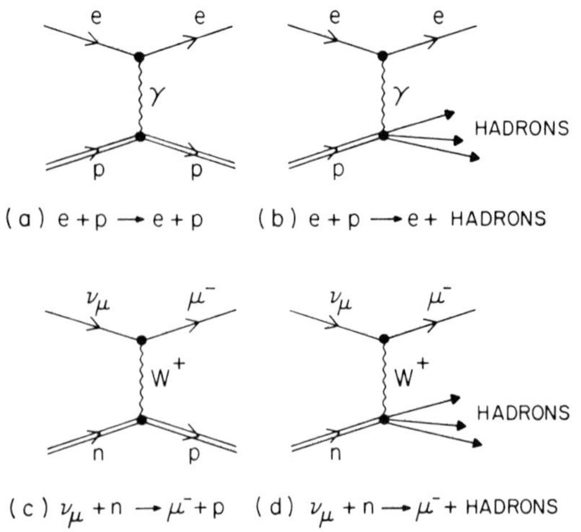

Fig. 1. Feynman diagrams for (a) elastic scattering and (b) inelastic scattering of charged leptons by protons and the corresponding diagrams (c) and (d) for weak charged current (WCC) scattering of neutrinos by neutrons.

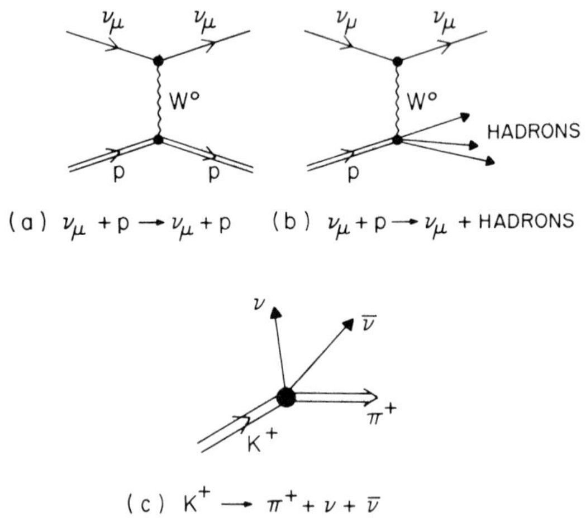

Fig. 2. Diagrams for weak neutral current (WNC) (a) elastic and (b) inelastic neutrino scattering by protons and (c) WNC ($\Delta S = 1$) decay of a charmed K meson.

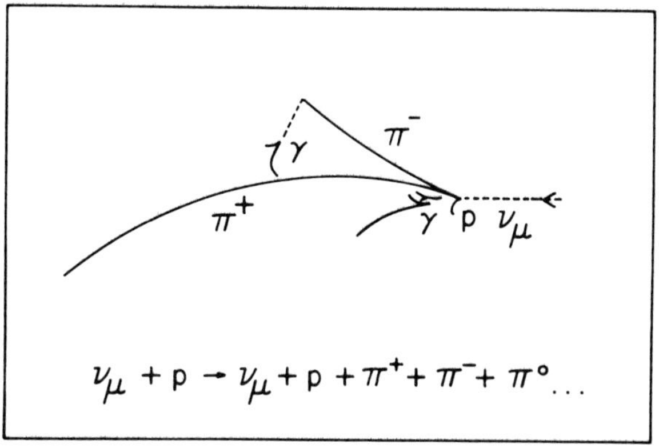

Fig. 3. Tracing of an example of a candidate for an interaction involving a neutral current in the CERN experiment. The neutrino enters from the right and no charged lepton emerges from its interaction with a proton. All the observed tracks can be identified as strongly interacting particles by measuring their curvatures, observing their range in the heavy liquid and their subsequent interactions. CERN Courier 13, No. 10 (October, 1973).

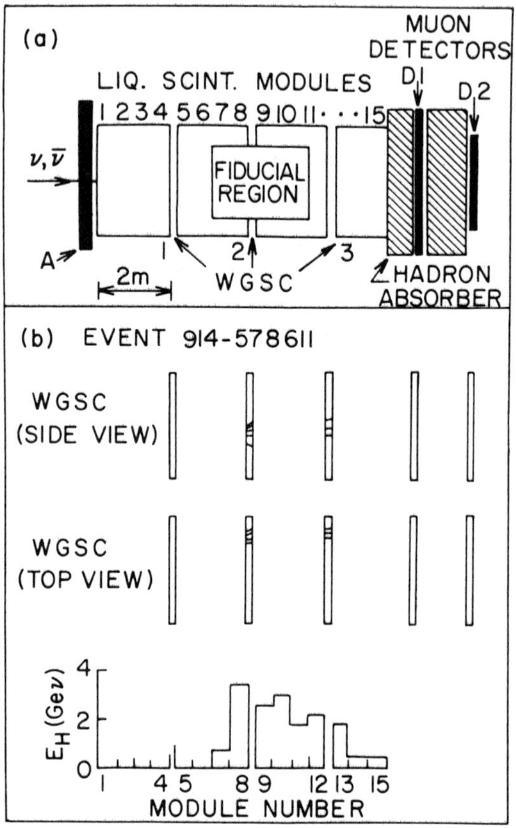

Fig. 4. (a) Schematic view of that portion of the target-detector used to identify neutral current and charged current neutrino interactions in the HPWF experiment. Not shown are the remaining three sections of iron toroidal magnet. (b) Display of the calorimeter and spark chamber information for a neutral current event. The side view of the wide gap spark chambers (WGSC) is from one of the 15° stereo views; the top view of the WGSC is from the 90° stereo view. The pulse heights in the calorimeter modules illustrate the development and absorption of the hadron shower, which had $E_H = 16.5$ GeV. (The gaps in the pulse height distribution show the location of the WGSC.) No particles appear in D1 or D2 behind the hadron absorber.

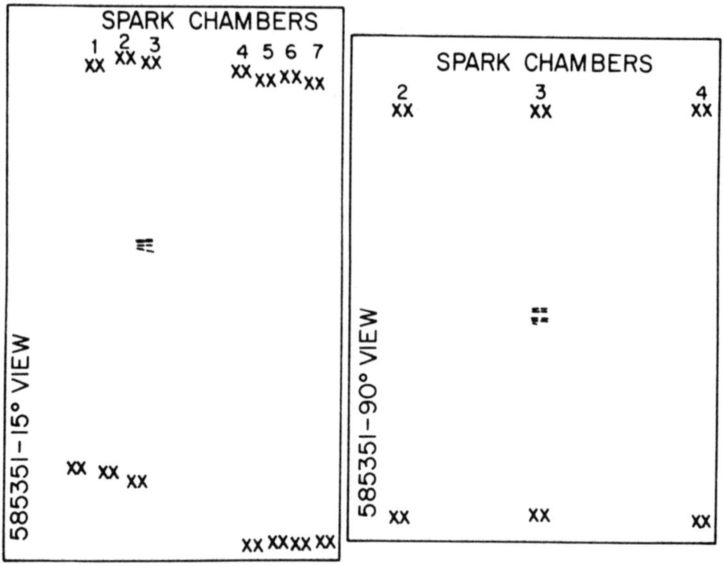

Fig. 5. Stereoscopic views of a WNC event centrally located in the HPWF apparatus. The X-marks are fiducial marks on the wide gap spark chambers. The actual locations of WGSC 1-5 are shown in Fig. 4.

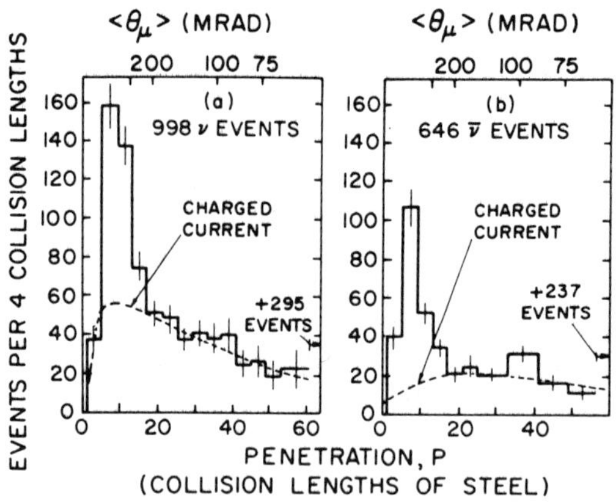

Fig. 6. Penetration distributions of events with no particles exiting through the end of the apparatus for (a) ν, and (b) $\bar{\nu}$ from the CalTech-FNAL experiment (reference 7). The smooth curves show the expected shape from CC events alone (wide-band contribution is included). The typical CC muon angle is shown on the top scale.

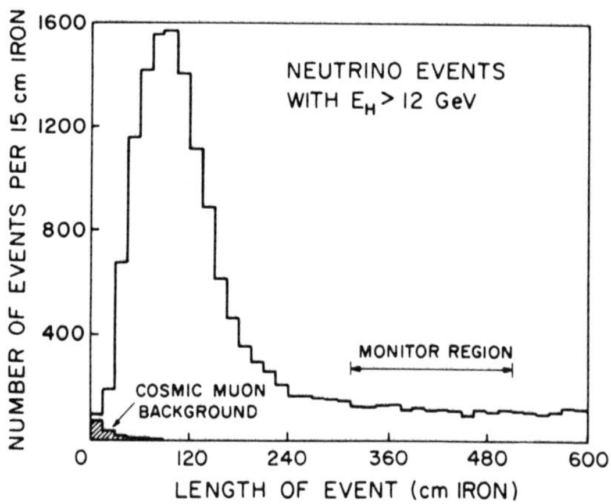

Fig. 7. Distribution in event length from the CDHS experiment (reference 8). Note the similarity to Fig. 6a.

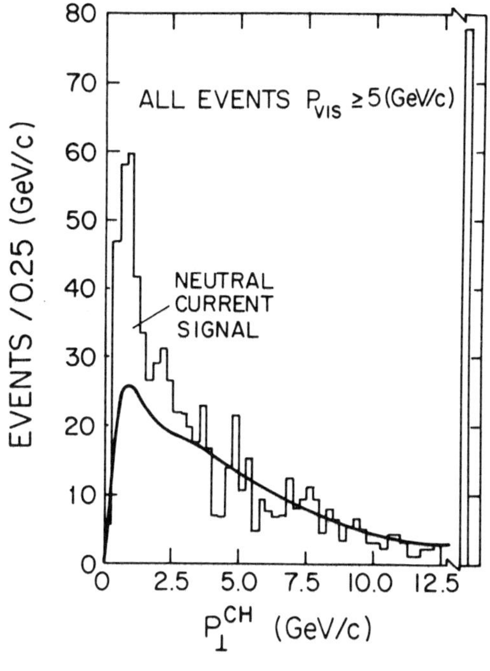

Fig. 8. Experimental distribution in transverse momentum of the charged particles in anti-neutrino events observed in the 15' FNAL bubble chamber (reference 9). The curve is the expected distribution for charged current events.

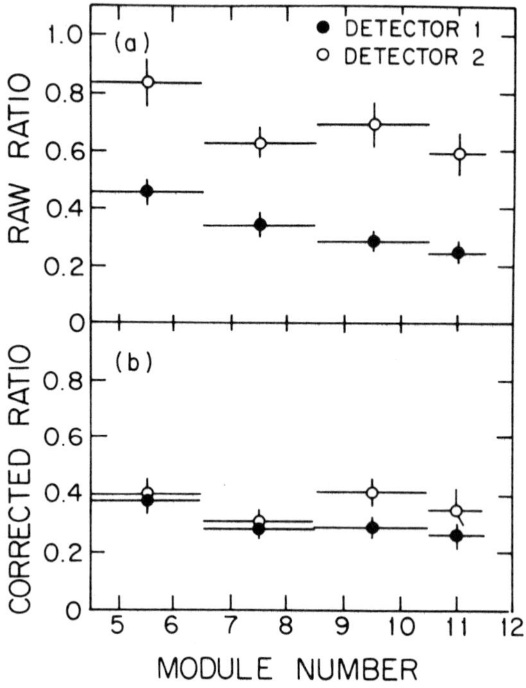

Fig. 9. The ratios of neutral current events to charged current events in the HPWF data (reference 11) plotted against longitudinal position (calorimeter module number). (a) Directly observed data and (b) after corrections for muon detection efficiency and hadron penetration.

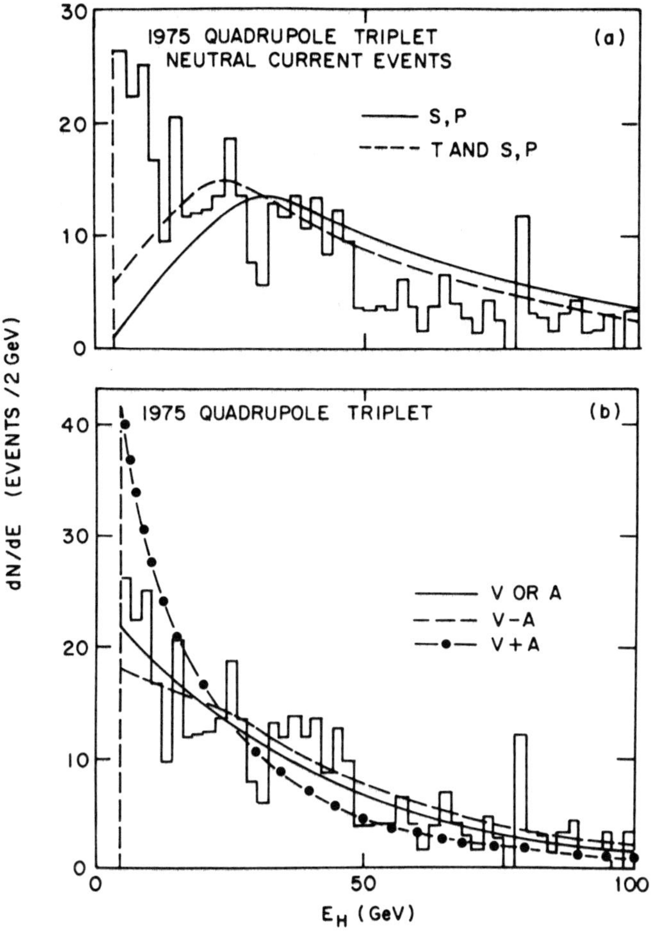

Fig. 10. Plots of the distributions in hadron energy (E_H) of the WNC events observed in the HPWF experiment (reference 11) compared with the distributions expected for different types of coupling.

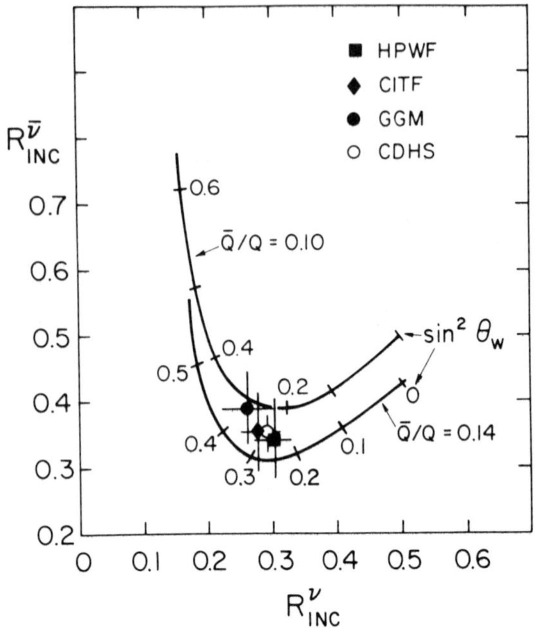

Fig. 11. Plot of $R^{\bar{\nu}}_{INC}$ against R^{ν}_{INC} for much of the data of Table IV. The solid curves are calculated for the SU(2)xU(1) model of Weinberg and Salam for antiquark to quark ratios of 0.1 and 0.14.

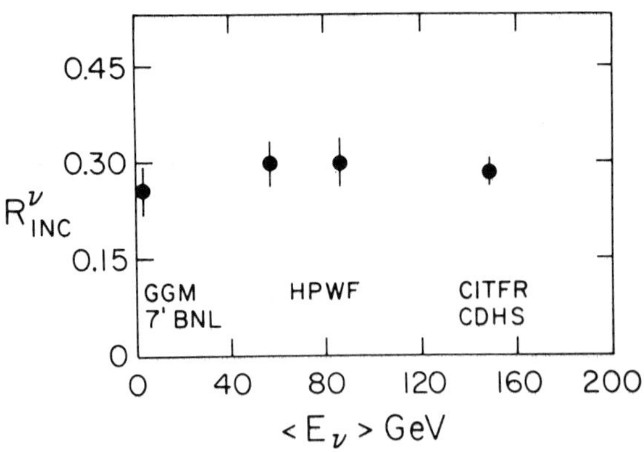

Fig. 12. Plot of R^{ν}_{INC} for different values of $<E_{\nu}>$.

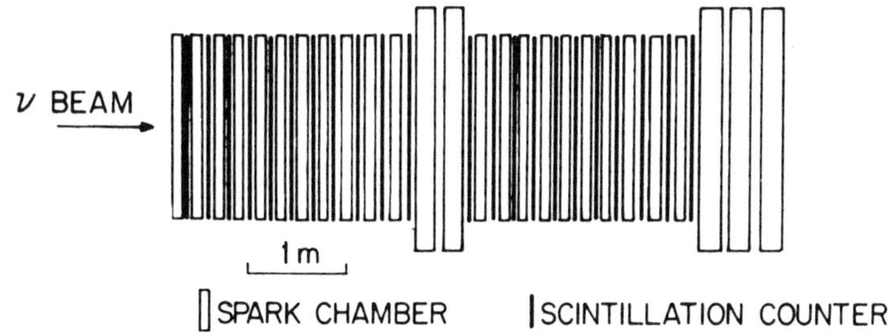

Fig. 13. Schematic representation of the Columbia-Illinois-Rockefeller detector.

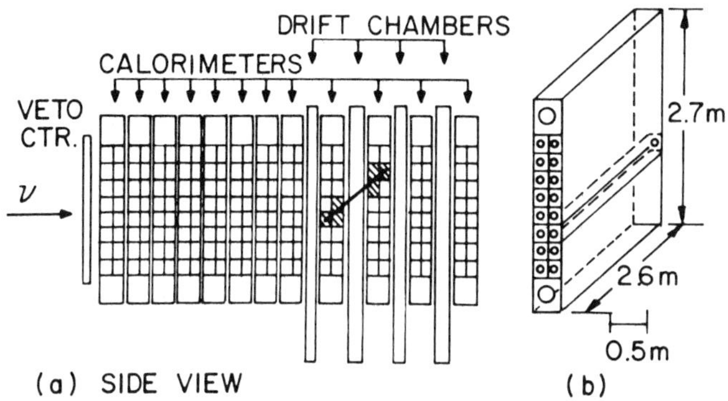

Fig. 14. (a) Side view of the Harvard-Pennsylvania-Wisconsin detector showing a typical recoiling proton. (b) Details of a single calorimeter module.

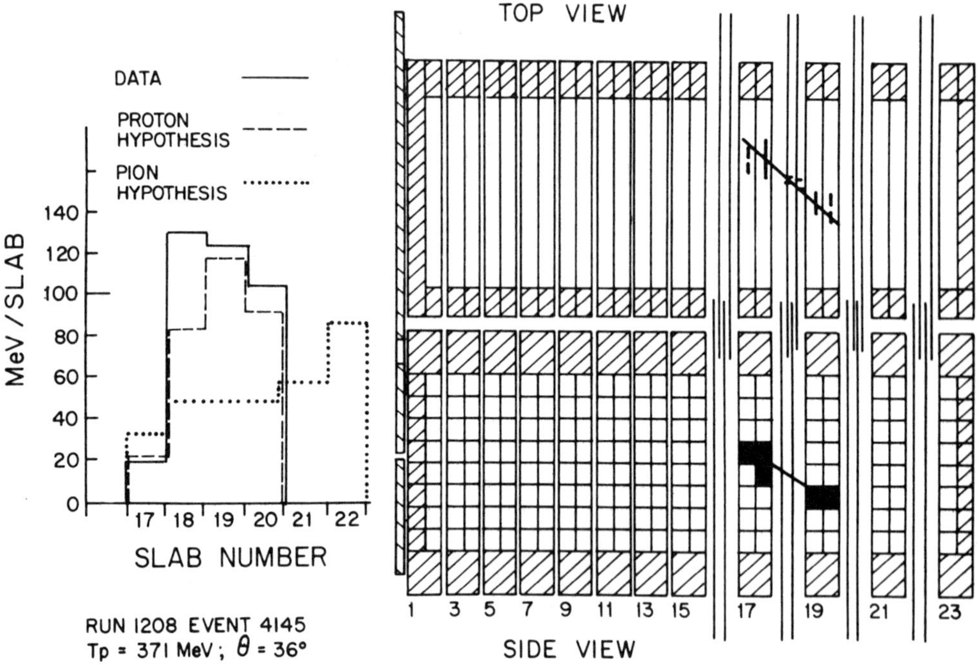

Fig. 15. Illustration of the method of measurement of proton angle and energy in the HPW experiment. Shaded areas are the active shield of the fiducial volume. Trajectory location in the top view is done by both pulse height and time comparison. Also shown is the method of particle identification for an actual event.

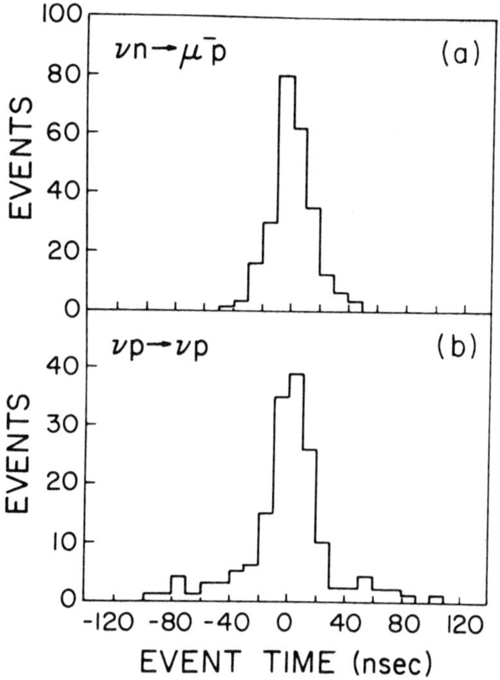

Fig. 16. Distribution of event times for (a) neutral current and (b) charged current elastic events in the HPW experiment. Event times are measured with respect to the 12 radio-frequency bunches of the AGS beam; each bunch is 50 ns fwhm, and bunch spacing is 220 ns.

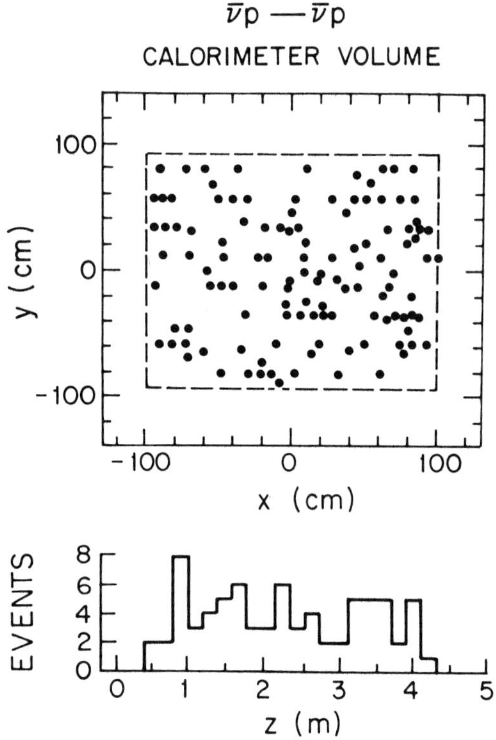

Fig. 17. Distributions of $\nu p \rightarrow \nu p$ event origins in the HPW data. Coordinate Z is along the incident beam direction. The dashed line indicates the fiducial volume.

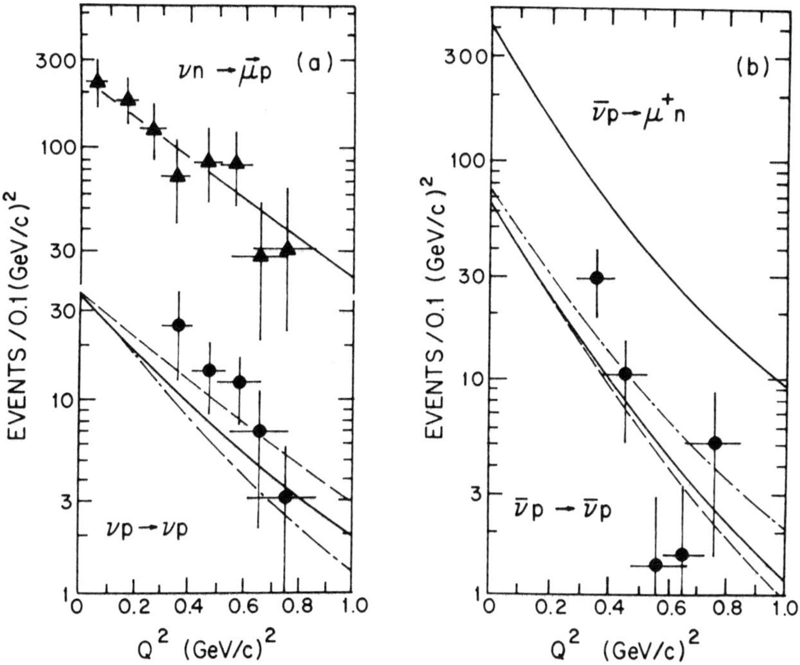

Fig. 18. (a) Observed q^2-distributions for charged and neutral current elastic scattering of neutrinos from the HPW experiment (reference 15). The data have been corrected for geometric acceptance and the $\nu p \rightarrow \nu p$ data have had an uniform background of 23% subtracted. The solid line for $\nu n \rightarrow \mu^- p$ is calculated for $m_A = 0.9$ GeV/c^2 and $m_V = 0.84$ GeV/c^2. Predictions of the Weinberg-Salam model for $\nu p \rightarrow \nu p$ correspond to $\sin^2\theta_W = 0.2$ (dashed line), 0.3 (solid line), and 0.4 (dash-dot line). (b) Observed distributions for antineutrino scattering. A background of 35%, uniform in q^2, has been subtracted. Weinberg-Salam model predictions are for $\sin^2\theta_W = 0.2$ (dashed line), 0.3 (solid line), and 0.4 (dash-dot line).

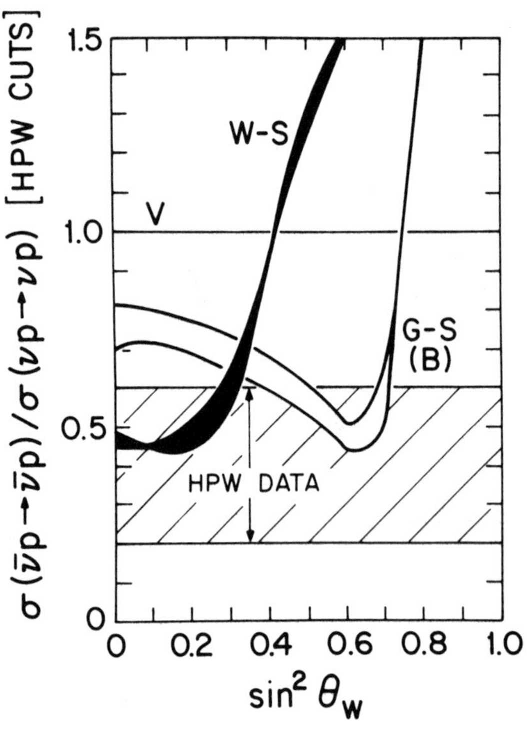

Fig. 19. Comparison of the ratio $\sigma(\bar{\nu}p \to \bar{\nu}p)/\sigma(\nu p \to \nu p)$ measured in the HPW experiment with the ratio predicted in the Weinberg-Salam model and the Gursey-Sikivie model. Graph is from reference 17.

Weak Neutral Currents in Particle Physics

LEPTONIC NEUTRAL WEAK CURRENTS

A. ROUSSET

DGRST (Delegation Generale a la
Recherche), Paris

1 - INTRODUCTION

The neutrino scatterings on electron targets are very simple processes. Very clean informations are expected on the fondamental properties of the weak interaction. Experimental results are presently available on three different reactions :

$$\nu_\mu + e^- \longrightarrow \nu_\mu + e^- \qquad (1)$$

$$\bar{\nu}_\mu + e^- \longrightarrow \bar{\nu}_\mu + e^- \qquad (2)$$

$$\bar{\nu}_e + e^- \longrightarrow \bar{\nu}_e + e^- \qquad (3)$$

The two first reactions (1) and (2) can only be produced via the neutral weak currents. Their detections are direct proof of the existence of these currents. The reaction (3) can be produced via the neutral and charged weak currents.

The experimental data are very limited in statistics. Therefore the only significant measurements are the total cross-sections. The comparaison with theory (1) (2) can only be done on a small number of parameters. The leptonic current is assumed to be a linear combination of vector (V) and axial vector (A) currents, with no contribution of scalar(S) , pseudoscalar (P) or tensor (T) currents. The three total cross-sections can be expressed with the two parameters g_V and g_A, vector and axial vector coupling constants of the neutral weak currents.

$$\sigma_1 = \frac{G^2 mEe}{2\pi}\left[(g_V + g_A)^2 + \frac{1}{3}(g_V - g_A)^2\right] \qquad (E_e >> m)$$

$$\sigma_2 = \frac{G^2 mEe}{2\pi}\left[\frac{1}{3}(g_V + g_A)^2 + (g_V - g_A)^2\right] \qquad (Ee >> m)$$

$$\sigma_3 = \frac{G^2 mEe}{2\pi}\left[\frac{1}{3}(2 + g_V + g_A)^2 + (g_V - g_A)^2 - \frac{m}{2Ee}(g_V - g_A)(2 + g_V + g_A)\right]$$

$$(Ee \sim m)$$

The third term, proportionnal to $\frac{m}{Ee}$ is neglected in σ_1 and $\sigma_2$because the available experimental data are at high energy (CERN beam, $E_\nu \sim E_{\bar{\nu}} \sim 2$ Gev). This term is not negligible in σ_3 because the experimental results are coming from $\bar{\nu}e$ interaction at low energy reactor flux ($E\bar{\nu}_e \sim 1$ Mev).

At a given total cross section σ_1, the g_V and g_A are constrained to be on an ellipse (Fig. 1). The ν_μ and $\bar{\nu}_\mu$ ellipses are centered on the origin ; the $\bar{\nu}e$ ellipse center corresponds to $g_A = g_V = -1$ because the charged current contribution. The large axis of the two $\bar{\nu}_\mu$ and $\bar{\nu}_e$ ellipses are parallel to the first bissectrix, the ν_μ large axis is parallel to the second bissectrix. If the theoretical assumption of limited V and A contributions is correct,these 3 ellipses have at least one common point. In the Salam-Weinberg model this point is expected to be on the segment.

$$g_A = -\frac{1}{2}$$

$$g_V = -\frac{1}{2} + 2\,\mathrm{Sin}^2\theta w \qquad -\frac{1}{2} < g_V < \frac{3}{2}$$

A determination of the Weinberg angle $Sin^2\theta w = \frac{e^2}{g^2}$ can be extracted from the experimental result.

Assuming the Salam-Weinberg model, the 3 cross-sections σ_1, σ_2 and σ_3 can be expressed in fonction of $Sin^2\theta w$ and are shown in the fig.2. From the semi-leptonic neutral currents the parameter $Sin^2\theta w$ is determined between 0.25 - 0.35. In this range the two cross-sections σ_1 and σ_2 are expected to be very near to the minimum values, of the order of 10^{-42} cm^2 at 1 Gev. The detection of these processes at these very small cross-sections will be very difficult. Another unpleasant consequence of this $Sin^2\theta w$ value is the fact that the prediction of the Saclam-Weinberg model is very near to the prediction of a pure V-A model without neutral current. So the separation between the two models from this experiment will be very difficult.

2 - THE ν_μ AND $\bar{\nu}_\mu$ SCATTERINGS ON ELECTRONS

The experimental results come from two experiments done at CERN on the same neutrino beam. The first detector is the heavy liquid bubble chamber Gargamelle (GGM) ; the second detector is the aluminium spark chambers of the Aachen-Padova experiment (NUE).

The ν_μ and $\bar{\nu}_\mu$ spectra are presented on fig.3. The mean value of the neutrino energy is about 2 Gev. The expected electron energy after scattering is expected mainly around 1 Gev. The figure 3 shows also the νe flux. This spectra is extended to higher energy (4-5 Gev), due to the K decay origin. These νe interact with neutrons and give background events :

$$\nu_e + n \rightarrow e^- + (p)$$

when the recoiling proton is not detected. The resulting background is strongly reduced when the high energy events are removed. The same cut off is applied at 2 Gev on the events in the GGM and NUE experiments.

The mass of fiducial volume of GGM (3), (4), is 10 tons of heavy freon (CF_3B_1). The radiation length is 11 cm. The electron is identified by spiralisation of the trajectory in the magnetic field and bremstrahlung γ (fig.4). The vertex is visible within 1 mm. The sign of the electron is determined . The angle error is typically 1°, the relative energy error is about 25%.

The mass of the NUE (5), (6), detector is 19 tons of aluminium . The thickness of the plates and their spacing are equal to 1cm. The radiation length is 22 cm. An electron is identified as an electromagnetic shower with very narrow development in the first gaps (fig.5). Therefore the electrons cannot be separated from γ. The electron angle error is also about 1° and the relative energy error 22%.

A large background coming from γ'_s is expected at low energy. Cuts off at 0.3.Gev and 0.2 Gev are respectively applied in the GGM and NUE experiments to the electron energy.

Another important cut off is applied on the angle between the neutrino and the electron trajectory. From the kinematic of the elastic reaction of a neutrino hitting a light target - the electron- the following equation is derived :

$$\theta_e^2 = 2m \left(\frac{1}{Ee} - \frac{1}{E\nu}\right) \qquad (4)$$

and consequently the inequality

$$\theta e < \frac{2m}{Ee}$$

The electron angle is expected to be small. In the GGM experiment a cut off at 5° was applied, independently of the electron energy.

In the NUE first analysis (in 1976) the used cut off on the angle was :

$$\theta e < 2 \frac{m}{Ee}$$

In the final analysis (in 1977) the relation (4) was used to determine the neutrino energy from the measured quantities Ee and θe and a cut off was applied on this energy :

$E_\nu > 0.8$ Gev

because very few events are expected at lower energy. This additional cut off is partly responsible of the important change in the results between the 76 and the 77 analysis. It has to be noticed that the error on θe, introduces a very large error of E_ν, and decreases the efficiency of this cut on the events. Table gives also the estimation of three different sources of background :

- The ν_e ($\bar{\nu}_e$) scattering on electron. Calculations on this background are made on the basis of the Weinberg-Salam model. The relative low flux of ν_e ($\bar{\nu}_e$) in comparison with ν_μ ($\bar{\nu}_\mu$) reduces the level of this contribution.

- The ν_e interaction on neutron with an invisible proton. $\nu_e + \eta \rightarrow e^- + (p)$. This background is reduced by the application of the cut off on the electron energy at 2 Gev. The contribution is estimated using the similar reaction $\nu_\mu + \eta \rightarrow \mu^- + (p)$ to determine the probality for proton to escape the detection, and using the events from the reaction $\nu_e + \eta \rightarrow e^-$ + visible protons for the normalisation.

- The background coming from γ is estimated from the scanning of isolated γ. The contribution to the background in GGM is strongly reduced because the small probability for a γ to give only a negative electron ($3.4 \pm 0.7\%$).

The contribution in the NUE experiment is higher, and is reduced by the application of more severe cut on the electron angle θe. The contribution to the background is determined by independent methods using the neutral currents events ($\nu N \rightarrow \nu N\pi^\circ$, $\nu P \rightarrow \nu P\pi^\circ$) and charged currents events ($\nu N \rightarrow \mu^- P\pi^\circ$, $\bar{\nu}P \rightarrow \mu^+ N\pi^\circ$), and also a simple extrapollation of the γ density in θ^2,Ee plot to the selection area. The figures 6 and 7 show the θ^2, Ee plot of the e^- candidates in the two experiments.

On the basis of a small number of e^- candidates in the GGM experiment (1 in ν_μ exposure, 3 in $\bar{\nu}_\mu$ exposure) , the cross-sections are estimated to be

$$\sigma_{\nu\mu e^-} < 3.0 . E\nu \; 10^{-42} \; cm^2 \qquad E_\nu \text{ in Gev}$$

$$\sigma_{\bar{\nu}\mu e^-} = 1.0 \,^{+2.1}_{-0.9} \, . \, E_\nu \; 10^{-42} \; cm^2$$

On the basis of larger number of events, the NUE results are :

$$\sigma_{\nu_\mu e^-} = 2.4 \pm 1.2.\ E_\nu\ 10^{-42}\ cm^2 \quad \text{(in 76)}$$
$$= 1.2 \pm 0.6.\ E_\nu\ 10^{-42} cm^2 \quad \text{(in 77)}$$

$$\sigma_{\bar{\nu}_\mu e^-} = 5.4 \pm 1.7 \quad E_\nu\ 10^{-42}\ cm^2 \quad \text{(in 76)}$$
$$= 2.4 \pm 1.1 \quad E_\nu\ 10^{-42}\ cm^2 \quad \text{(in 77)}$$

The 77 NUE results are compatible with the GGM results, and will be discussed in the following paper by Professor Faissner.

The figure 8 represents the GGM results on the $g_V\ g_A$ plot. The ellipses at one standard deviation cut the Salam-Weinberg segment and allow the constraint on $Sin^2\theta w$.

$$0.1 < Sin^2\theta w < 0.4$$

3 - THE $\bar{\nu}e$ SCATTERINGS ON ELECTRONS

The $\bar{\nu}e$ flux ($2 10^{13}\ \bar{\nu}e\ /cm^2/sec$) going out from the Savannah River reactor(1800 Megawatts) is used to detect the scattering on electron (7). The detector (fig.9) is 15.9 kg of scintillator plastic segmented into 16 optically isolated elements. 82% of $\bar{\nu}_e e^-$ events (Ee < 4.5 Mev) give a single pulse in one element ; 18% give multiple pulses. This characteristic will be used to eliminate background coming from γ. The rest of the apparatus is veto counters and shielding around central part where the active plastic is located. Successively there is 300 kg of sodium iodide to detect the γ rays, 10 cm of lead against the γ, a cadmium sheet against the neutrons, 2200 liters of scintillator liquid. Any events in the plastic in coincidence with one veto counter pulse is rejected.

The tables gives the results condensed into two energy bins. The signal results from a small difference between two counting rates, reactor on and reactor off. The two signals are significant at the level of 4.1 and 4.8 standard deviations. To attribute this effect to $\bar{\nu}_e e^-$ events, all other processes have to be eliminated, and it is the goal of the discussion on possible backgrounds.

A first source of background is the reaction $\bar{\nu}_e$ +p η + e^+ which is expected to give 30 more events than the reaction of $\bar{\nu}$e on electrons. Fortunately the neutron and the positron can be detected in a large fraction of events from the delayed γ after capture of the neutron and from the two γ of the position annihilation. From the events associated to a neutron, with 0,1 or 2 prompt γ, the γ detection is measured to be 0.96 $\pm$ 0.04. From the events associated to 2 prompts γ (e^+ signature) and detection or not of the neutron, the probability to detect the neutron is measured to be 0.76$\pm$ 0.07. The probability not to recognize a $\bar{\nu}$ep event is therefore very small and the resulting background is smaller than 3%.

The γ and neutron background can be eliminated on the basis of two observations : first the ratio $\frac{M}{S}$, multiple pulses to single pulse in the plastic, for the signal is founded to be 0.18, very close to the expected one $\frac{0.18}{0.82}$ = 0.22. γ or neutrons sources give much higher ratio, respectively 6.3 and 1.3. Secondly, the ratio plastic/NaI is 4.3 10^{-2} for the signal, and much smaller for γ and η sources, respectively 1.4 10^{-4} and 1.4 10^{-3}.

The signal, not explained by γ, η, $\bar{\nu}$ep background, is attributed to $\bar{\nu}_e$e reactions. It would be interesting to have a better understanding of the reactor off counting rate, and particularly of the contribution of cosmic rays.

The fig.1 shows the final result on the $g_V g_A$ plot. The two elliptic surfaces are compatible with a value of $\sin^2\theta_w$ equal to 0.29 $\pm$ 0.05. But, as expected, a pure V-A charged current is not excluded.

4 - CONCLUSIONS

The three experimental results on pure leptonic neutral weak currents represent a considerable effort of the experimentalists because the cross-sections of these processes are very small. The detection of $\nu_\mu e^-$ and $\bar{\nu}_\mu e^-$ scatterings are clean evidence of leptonic currents, and the first determinations of cross-sections are in agreement with the Salam-Weinberg model. The value of $\sin^2\theta_w$ is in agreement with the determination of this parameter using the semi-leptonic interactions (fig.11).

Nevertheless, these first results have to be improved. The events in GGM are very well identified but the statistics are very poor. The events in NUE are more numerous but the γ background is not easy to be separated. The reactor experiment gives a complementary result but the signal appears only a small difference between two large counting rates.

Significant improvements are expected from the next generation of experiments, particularly with neutrino beams from 400 Gev accelerators. The signal will be increased because the cross-sections are expected to be proportional to the energy (factor $\frac{400}{25} = 16$), and the backgrounds from $\nu_e n \rightarrow e^- (p)$ and from $\nu N \rightarrow \nu N \pi^\circ$ are expected to remain constant.

About 100 events are expected from the three large bubble chambers, the 15', BEBC and GGM. Already 3 good candidates have been found in the 15' bubble chamber (8).

One counter experiment is presently in progress at Fermi-Lab, with aluminium plates and proportional chambers. Another experiment with marble plates will be installed in 1978 at CERN. If the isolated γ rays background is negligible, these experiments, with massive detectors, will be able to produce events in the narrow band beams. The $y = \frac{Ee}{E\nu}$ distribution will be determined, and an independent measurement of g_V/g_A will be deduced.

The missing channel $\nu_e e^-$ could be investigated at the LAMPF accelerator, using νe from the decay of $\mu+$ ($\mu^+ \rightarrow e^+ + \nu_e + \bar{\nu}_\mu$). Experiments are in preparation.

Finally, it has to be noticed that informations on pure leptonic neutral weak currents will be extracted from the collisions e^+e^- at high energy (PETRAM PEP). The interference between the electromagnetic current and the neutral weak current will produce a measurable forward-backward asymmetry in the pure leptonic process $e^+e^- \rightarrow \mu^+\mu^-$. A muon longitudinal polarization is also expected. These effect will be more important at higher energy, in the range of 100-200 GW in the center of mass.

REFERENCES

1) G. t'Hooft - Predictions for neutrino-electron cross-sections in Weinberg's model of weak interactions. Physics letters 37 B, 2 p.395.

2) M. Gourdin - Pure Leptonic Currents. Proceedings of the International Neutrino Conference Aachen 1976, p.234.

3) J. Blietschau et al. Evidence for the leptonic neutral current reaction $\bar{\nu}\mu e^- \longrightarrow \bar{\nu}\mu e^-$. Nuclear Physics B 114 (76) p.189.

4) J. Blietschau et al. Upper limit to the cross-section for the process $\nu\mu e^- \longrightarrow \nu\mu e^-$ (to be published)

5) H. Faissner et al. Measurement of muon-neutrino and antineutrino scattering off electrons. Proceeding of the International Neutrino Conference Aachen 1976, p.223.

6) H. Reithler. $\nu\mu e^- \longrightarrow \nu\mu e^-$ at CERN. Aachen Padova Experiment. 1977 International Symposium on Lepton and Photon Interaction at High Energies, Hamburg, and H. Faissner et al. Muon-Neutrino and antineutrino scattering off electrons and the structure of the weak leptonic neutral current, Caucasus conference on neutrino physics (1977).

7) F. Reines, H.S. Gurr and H.W. Sobel. Detection of $\bar{\nu}ee^-$ scattering. Physical Review Letters 37 - 6. p.215.

8) C. Baltay, Private Communication.

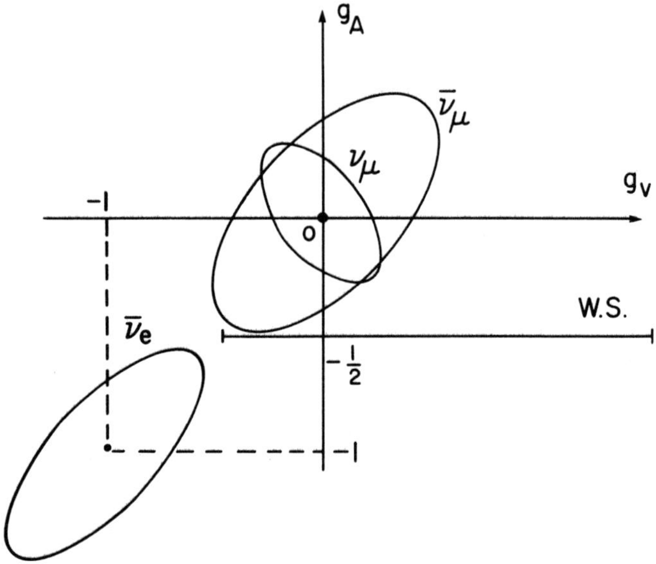

1- The g_V/g_A plot. The notation on g_V and g_A is such that these parameters are the same in the 3 expressions of the 3 total cross-sections $\nu\mu e^-$ (1) , $\bar{\nu}\mu e^-$ (2) and $\bar{\nu}_e e^-$ (3). The theoretical predictions with V and A currents are ellipses at a given ratio T/E_ν. In the Salam-Weinberg model $g_A = -\frac{1}{2}$ and $g_V = -\frac{1}{2} + 2\,\mathrm{Sin}^2\theta w$.

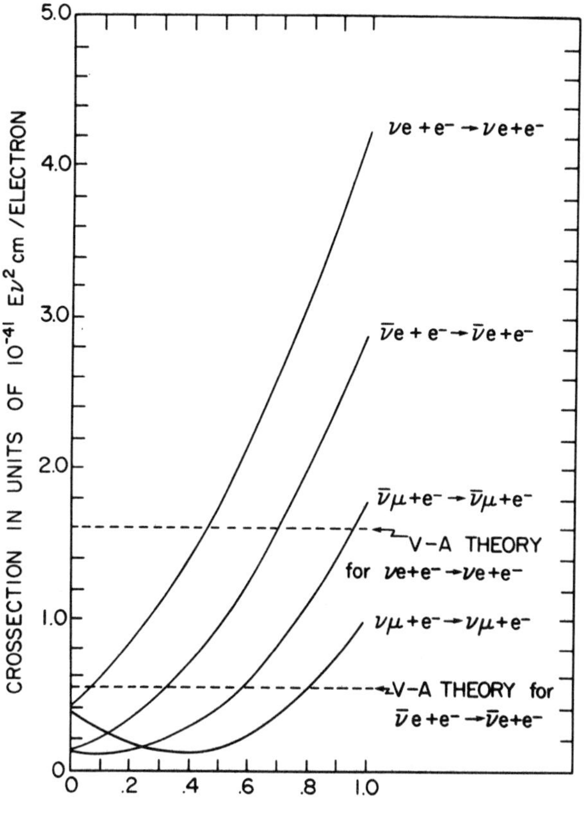

2 - The leptonic cross-sections in the Salam-Weinberg model and in the pure V-A charged currents theory.

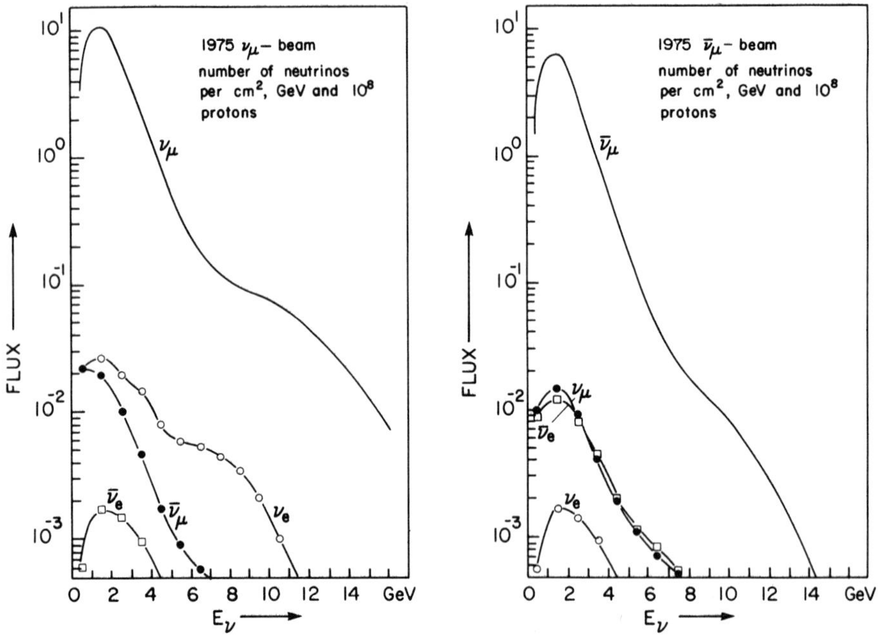

3 - The neutrino and antineutrino flux at CERN. The ν_e flux is an important information for the background. It has to be noticed that the relative ν_e flux is much smaller in the $\bar{\nu}\mu$ beam that in the $\nu\mu$ beam.

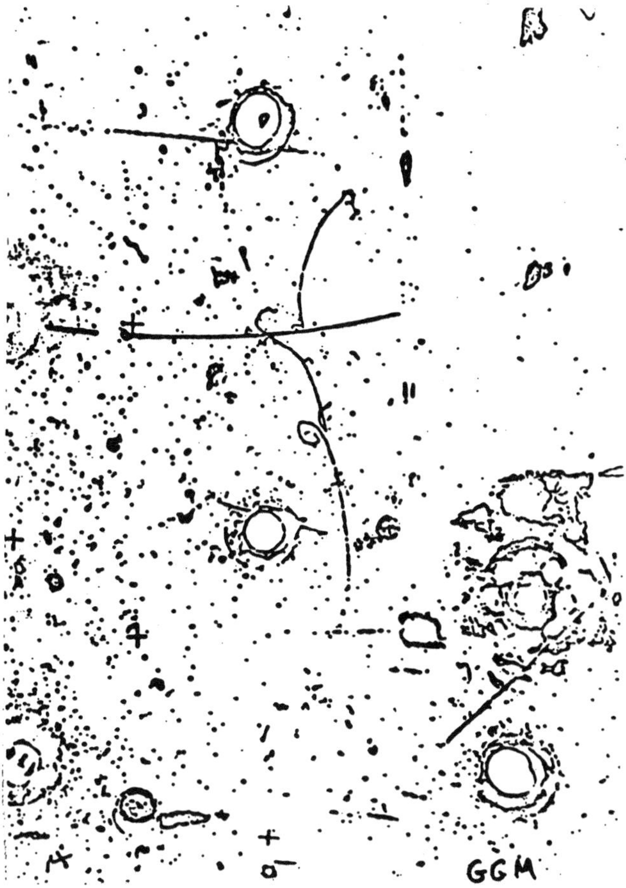

4 - One $\bar{\nu}_\mu e^- \rightarrow \bar{\nu}_\mu e^-$ candidate in GGM. The electron trajectory spiralise in the magnetic field, bremstrahlung are detected. The curvature of the trajectory is visible. The vertex is clean.

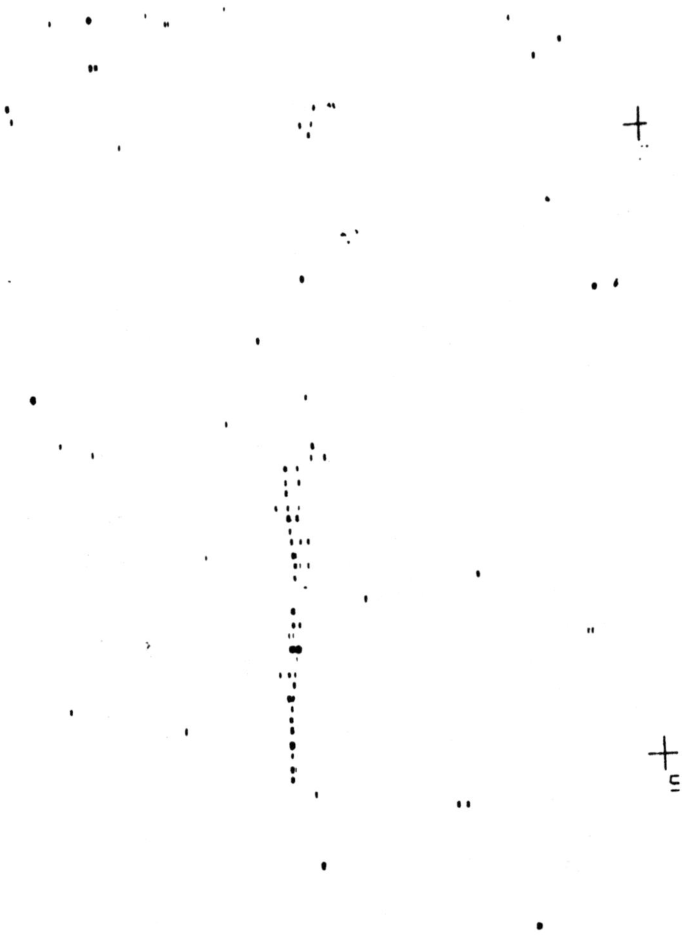

5 - One $\nu_\mu e^- \longrightarrow \nu_\mu e^-$ candidate in the NUE detector. Single sparks are visible in the first seven gaps. The shower developpement is characteristically of the electromagnetic nature.

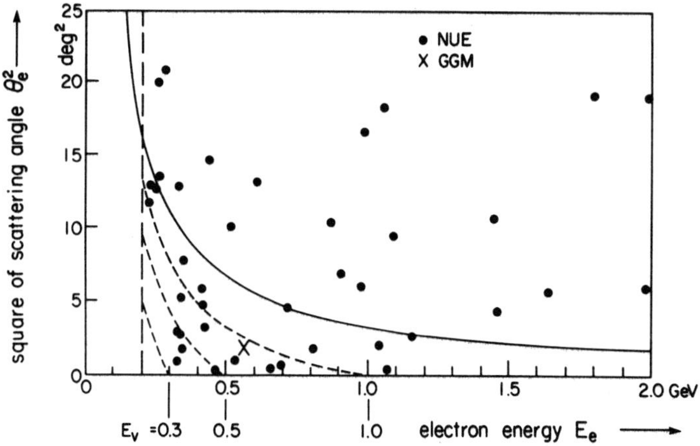

6 - The $\theta^2 E_e$ plot with the $\nu\mu e^-$ candidates. The GGM event is the unique candidate in the rectangle ($\theta e < 5°$, $0,3 < Ev < 2,0$ Gev). The NUE events above the $\theta^2 < 2 \frac{m}{E_e}$ curve and below $E_\nu < 0.8$ Gev have been eliminated.

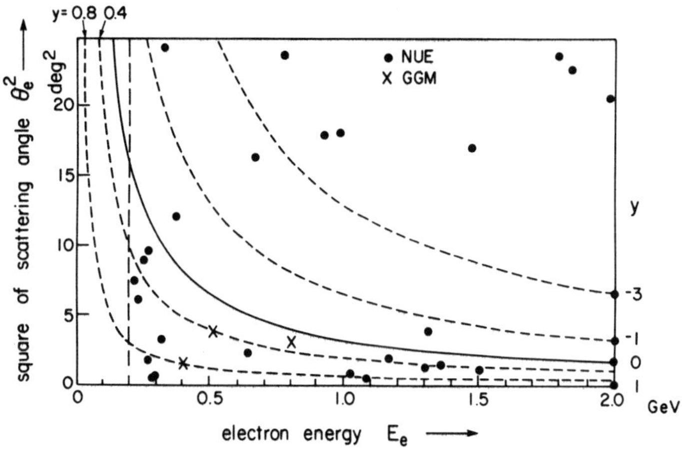

7 - The θ^2, Ee plot with the $\bar{\nu}_\mu$ e^- candidates. In the first analysis (76) of the NUE experiment, the events above the curve y = 1 ($\theta^2 = \frac{4m}{Ee}$) have been eliminated.

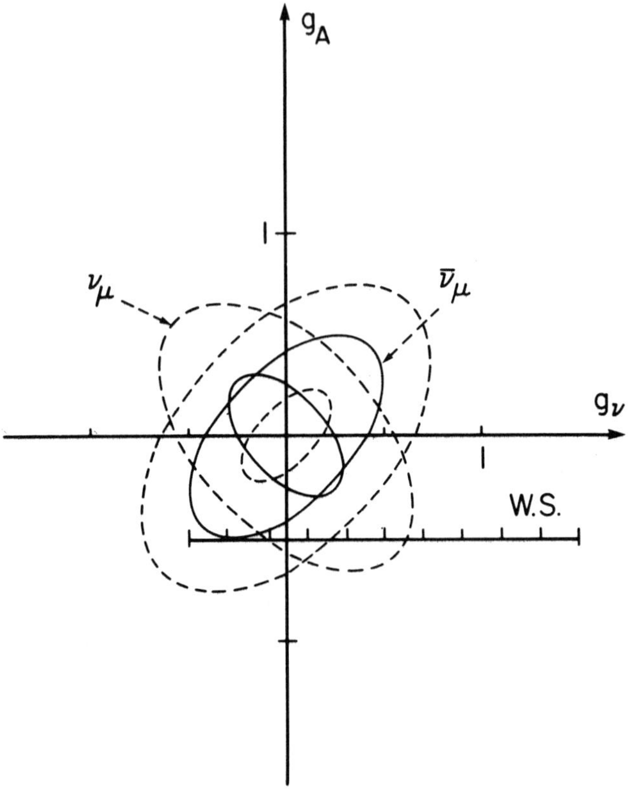

8 - The $g_V g_A$ plot with the GGM results. The overlapping of the ellipses at one standard deviation and the Salam-Weinberg prediction, show that a possible range for $Sin^2\theta w$ is $0.1. < Sin^2\theta w < 0.4$.

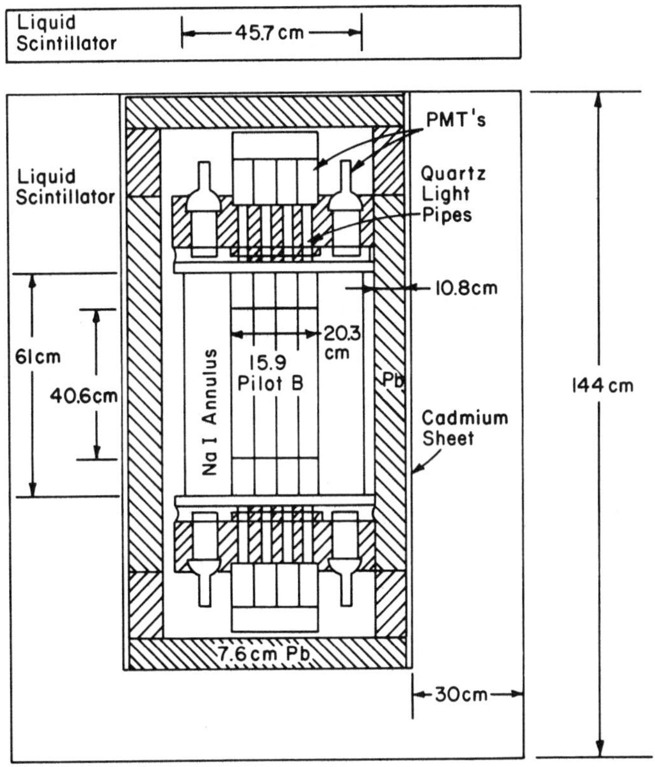

9 - Schematic of detector of the reactor experiment. The plastic target region is surrounded by veto counters and shielding.

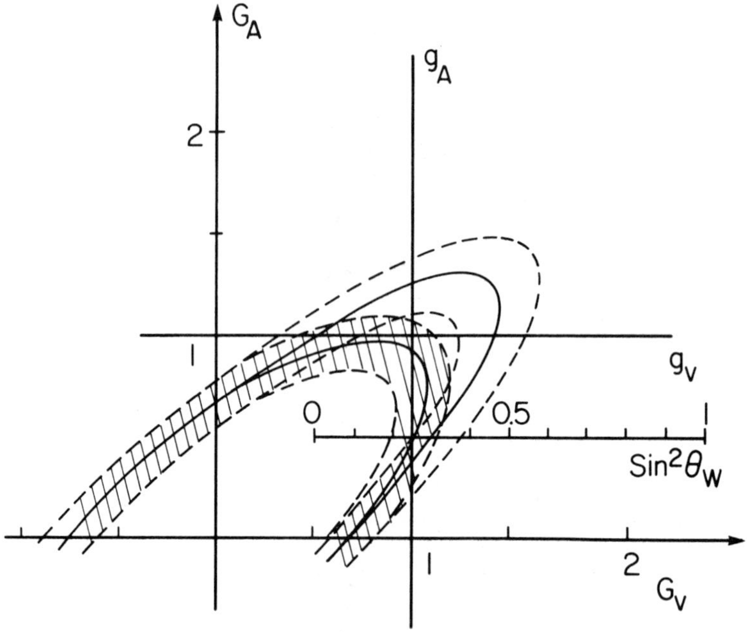

1.5 - 3. MeV $\quad Sin^2\theta_W = 0.26 \begin{smallmatrix} +0.05 \\ -0.06 \end{smallmatrix}$

3. - 4.5 MeV $\quad Sin^2\theta_W = 0.32 \pm 0.05$

10 - The $g_V g_A$ plot with the reactor experiment results on $\bar{\nu}ee^-$ cross-section in the two energy range 1.5 - 300 Mev and 3.0 - 4.5 Mev. These results are compatible with pure V-A charged current theory ($g_A = g_A = \nu$) and also with $Sin^2\theta = 0.29 \pm 0.06$.

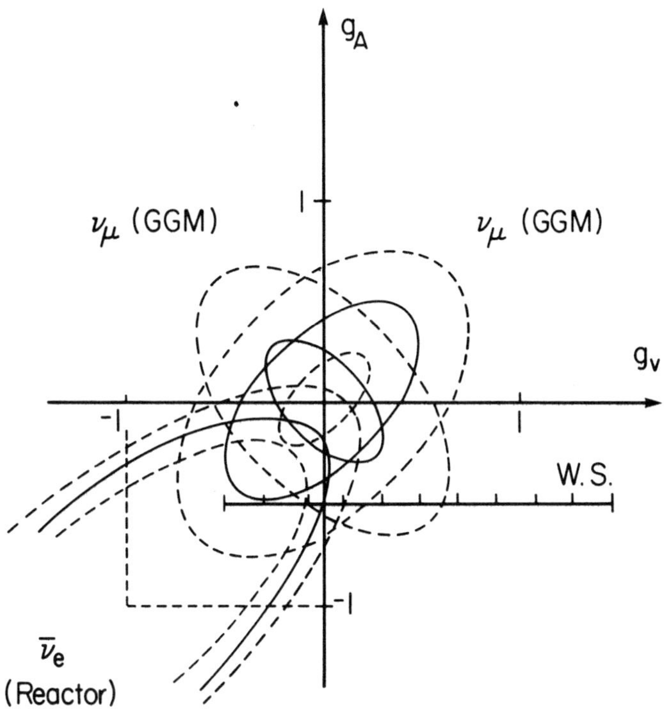

11 - $g_V g_A$ plot with the GGM and the reactor experiment results. These results are compatible with a value of $\mathrm{Sin}^2\theta w$ in the range 0.1 - 0.4.

ISOSPIN PROPERTIES OF THE WEAK NEUTRAL CURRENT

D. Haidt
CERN, Geneva, Switzerland

Introduction

At the Aachen Neutrino Conference 1976 musical terms were applied to describe neutral currents [1]. Undoubtedly, the discovery era 1973-1976 is an *allegro con brio*. The present era is devoted to study the detailed properties and thus likely to be a *lento moderato* with two main themes: the space-time theme and the isospin theme. Results from recent experiments start to reveal the structure of both themes.

The framework in which the experiments are interpreted, is set by writing down the Lagrangian for the weak neutral current process:

$$L^{NC} = -\frac{G}{\sqrt{2}}\bar{\psi}_\nu\gamma_\alpha(1+\gamma_5)\psi_\nu J^{NC}_\alpha$$

with:

$$J^{NC}_\alpha = g_V V^3_\alpha + g_A A^3_\alpha + s_V V^o_\alpha + s_A A^o_\alpha$$

$$= \bar{\psi}_u\gamma_\alpha\left[u_L(1+\gamma_5) + u_R(1-\gamma_5)\right]\psi_u +$$

$$\bar{\psi}_d\gamma_\alpha\left[d_L(1+\gamma_5) + d_R(1-\gamma_5)\right]\psi_d.$$

This assumes a SU_2 - symmetric world. Four coupling constants characterize the weak neutral current as an arbitrary superposition of a vector and an axial vector as well as an arbitrary superposition of an isovector and an isoscalar. The most popular theory, the Salam-Weinberg theory, predicts all four coupling constants in terms of one free parameter $\sin^2\theta_w$:

$$g_V = 1 - 2\sin^2\theta_w \qquad u_L = \frac{1}{2} - \frac{2}{3}\sin^2\theta_w$$

$$g_A = 1 \qquad u_R = -\frac{2}{3}\sin^2\theta_w$$

$$s_V = -\frac{2}{3}\sin^2\theta_w \qquad d_L = -\frac{1}{2} + \frac{1}{3}\sin^2\theta_w$$

$$s_A = 0 \qquad d_R = +\frac{1}{3}\sin^2\theta_w .$$

Inclusive scattering of ν and $\bar{\nu}$ off nuclei provides only information about the V, A structure. In order to reveal the isospin structure exclusive and inclusive pion production may be studied. Two recent experiments contribute essentially to this subject and will be subsequently described in some detail.

Exclusive Pion Production

The weak neutral current can induce four single pion reactions. They are listed in table 1 together with their amplitudes describing the isovector (A_1, A_2) and the isoscalar (S) transition.

Table 1: The amplitudes for weak single pion production

Channel	Amplitude
$\nu p \rightarrow \nu p \pi^o$	$2A_3 + A_1 - S$
$\nu n \rightarrow \nu n \pi^o$	$2A_3 + A_1 + S$
$\nu n \rightarrow \nu p \pi^-$	$\sqrt{2}(A_3 - A_1 - S)$
$\nu p \rightarrow \nu n \pi^+$	$\sqrt{2}(A_3 - A_1 + S)$

If the weak neutral current is either a pure isovector (S = 0) or a pure isoscalar ($A_3 = A_1 = 0$), then

$$\frac{\sigma(\nu p\pi^o)}{\sigma(\nu n\pi^o)} = 1 \quad \text{and} \quad \frac{\sigma(\nu p\pi^-)}{\sigma(\nu n\pi^+)} = 1.$$

Therefore, the observation of these ratios contains information about the isovector-isoscalar interference term.

The experiment [2] is carried out in the CERN PS neutrino beam using the bubble chamber Gargamelle filled with a 90% propane (C_3H_3)/10% freon (CF_3 Br) mixture. This liquid is dense enough to ensure a reasonable event rate. The good hadron identification and the high gamma detection efficiency allow for an analysis of all four single pion channels, including $\nu n \rightarrow \nu n\pi^o$ consisting of neutrals only. The liquid is on the other hand light enough to separate reliably all channels. The event selection requires one identified pion and-depending upon the channel - zero or one proton. Events with additional protons, visible down to about 8 MeV, are excluded. Table 2 shows the basic and final sample together with the relative cross sections.

Table 2: Results of the ν propane experiment

Channel	$\nu p\pi^o$	$\nu n\pi^o$	$\nu p\pi^-$	$\nu n\pi^+$
Basic sample	240	31	94	104
Final sample (before nucl. corr.)	248	99	134	141
Relative cross sections	1.00 ± 0.12	0.60 ± 0.14	0.80 ± 0.20	0.61 ± 0.10

Obviously, the extreme hypotheses of a pure isoscalar current resp. a pure isovector 3/2 requiring $\sigma(p\pi^o) : \sigma(n\pi^o) : \sigma(p\pi^-) : \sigma(n\pi^+)$ to be 1 : 1 : 2 : 2 resp. 1 : 1 : ½ : ½ are both unlikely. The two differences

$$\sigma(p\pi^o) - \sigma(n\pi^o) = - C \text{ Re } S^*(A_3 + \frac{1}{2} A_1) = 0.40 \pm 0.20$$

$$\sigma(p\pi^-) - \sigma(n\pi^+) = -\,C\ \mathrm{Re}\ S^*(A_3 - A_1) = 0.19 \pm 0.22.$$

differ from 0 by 2 resp. 1 standard deviation and measure the isovector-isoscalar interference term. This indicates that the weak neutral current has both an isovector and an isoscalar part. The strong excitation of the $\Delta(1236)$ - resonance (fig. 1) observed in the channel $\nu p \to \nu p \pi^o$ suggests a dominant A_3 amplitude.

Inclusive Pion Production

In this experiment [3] the ν and $\bar{\nu}$ interactions occur in Gargamelle filled with freon ($CF_3\,Br$) which is essentially an isoscalar target. The information about the isospin structure of the weak neutral current is contained in the pion charge multiplicities. The initial sample consists of 850 ν and 1091 $\bar{\nu}$ events. The π^+/p ambiguity is resolved by Monte-Carlo techniques. The corrected charge multiplicities disagree again with the hypothesis of a pure isoscalar current. Furthermore, in order to apply quark-parton model ideas the pions are selected in the current fragmentation region. This region is defined by requiring the hadron energy $E_{had} > 1$ GeV and the ratio of the pion energy to the hadron energy to be between 0.3 and 0.7. This definition turns out to be good in 80% of the cases as tested with a charged current sample. Then, the following ratios are obtained:

Table 3: Results

Beam	N^+/N^-	$(N^++N^-)/N^o$
ν	0.77 ± 0.14	1.90 ± 0.35
$\bar{\nu}$	1.64 ± 0.36	2.03 ± 0.45

These data are interpreted by L. Sehgal [4]:

$$\frac{N^+}{N^-} = \frac{\left(u_L^2 + \frac{1}{3}\,u_R^2\right) D_u^{\pi^+} + \left(d_L^2 + \frac{1}{3}\,d_R^2\right) D_u^{\pi^-}}{\left(u_L^2 + \frac{1}{3}\,u_R^2\right) D_u^{\pi^-} + \left(d_L^2 + \frac{1}{3}\,d_R^2\right) D_u^{\pi^+}}$$

for ν interactions. For $\bar{\nu}$ interactions the same expression holds with indices L and R exchanged. The theoretical expression depends upon the four coupling constants and the ratio of two pion fragmentation functions which are known from electro production and neutrino experiments. Together with the data from inclusive neutrino experiments Sehgal derived the squares [5] of the coupling constants (fig. 2) in good agreement with the predictions of the Salam-Weinberg theory for $\sin^2\theta_w \approx 0.3$.

Summary

Table 4 summarises evidence against the simple hypotheses of a pure isoscalar or pure isovector 3/2 weak neutral current.

Table 4: χ^2-probabilities for the hypothesis of a pure isoscalar (IS) and of a pure isovector 3/2 (IV 3/2) weak neutral current.

Lab	Target	Measured Quantities	χ^2-Probability IS	IV 3/2	Ref.
ANL	H_2, D_2	$\nu p\pi^o$, $\nu n\pi^+$, $\nu p\pi^-$	0.3	> 0.7	6
BNL	H_2, D_2	$\nu p\pi^+\pi^-$, $\nu n\pi^+\pi^-$	0.05		7
CERN	Freon	$\nu N\pi^o$, $\nu N\pi^-$	0.1		8
		$\bar{\nu} N\pi^o$, $\bar{\nu} N\pi^-$	0.01		8
CERN	Propane	$\nu p\pi^o$, $\nu n\pi^+$, $\nu n\pi^o$, $\nu p\pi^-$	< 0.0001	0.02	2
		$\bar{\nu} N\pi^o$, $\bar{\nu} N\pi^-$	0.08		9
CERN	Freon	$\nu\pi^+X$, $\nu\pi^-X$, $\nu\pi^oX$	0.01		3
		$\bar{\nu}\pi^+X$, $\bar{\nu}\pi^-X$, $\bar{\nu}\pi^oX$	0.0001		3

Table 5 summarises evidence for the isovector-isoscalar interference.

Lab	Target	Measured Quantities	Significance	Ref.
ANL	H_2, D_2	$\nu n\pi^+/\nu p\pi^- = 1.0 \pm 0.6$	none	6
BNL	H_2, D_2	$\nu n\pi^+\pi^-/\nu p\pi^+\pi^- = 0.53 \pm 0.21$	2.5 s.d.	7
CERN	Propane	$\nu p\pi^o - \nu n\pi^o = 0.40 \pm 0.20$	2 s.d.	2
		$\nu p\pi^- - \nu n\pi^+ = 0.19 \pm 0.22$	1 s.d.	2
CERN	Freon	$\nu\pi^+/\nu\pi^- = 0.77 \pm 0.14$	1.5 s.d.	3
		$\bar{\nu}\pi^+/\bar{\nu}\pi^- = 1.64 \pm 0.36$	1.5 s.d.	3

Finally, both in ν and $\bar{\nu}$ experiments [2, 9] the excitation of the $\Delta(1236)$ - resonance was observed.

References

[1] H. Faissner, Welcome to the 1976 International Neutrino Conference at Aachen.

[2] W. Krenz et al., (Neutrino Propane Gargamelle Collaboration) CERN/EP/PHYS 77-50.

[3] H. Kluttig et al., submitted to Physics Letters.

[4] L. Sehgal, Invited talk given at the 1977 International Symposium on Lepton and Photon Interactions at High Energies, Hamburg.

[5] Further analysis of the sign ambiguities see, G. Ecker Ref. TH 2406-CERN and P.Q. Hung and J.J. Sakurai, UCLA/77/TEP/17 University of California, Los Angeles.

[6] L.G. Hyman, Proceedings of the 1975 Paris Neutrino Conference.

P. Schreiner, Proceedings of the 1976 Aachen Neutrino Conference.

[7] A.G. Garfinkel, Invited talk at the 1977 International Symposium on Lepton and Photon Interactions at High Energies, Hamburg.

[8] G.H. Bertrand-Coremans et al., Phys. Letters 61B (1976) 207.

[9] D. Haidt, Invited talk at the 1977 International Symposium on Lepton and Photon Interactions at High Energies, Hamburg.

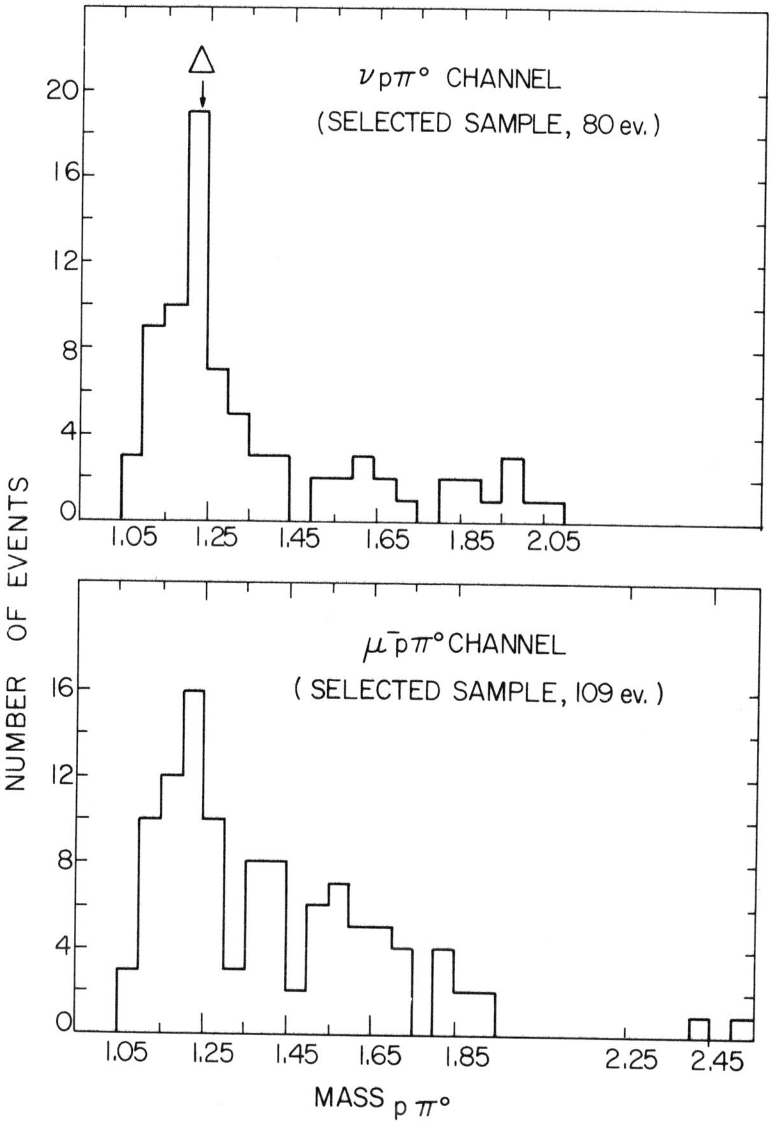

Fig. 1 Invariant mass distribution of the $p\pi^o$ system in $\nu p \to \nu p\pi^o$ and $\nu n \to \nu^- p\pi^o$ (from ref. [2]).

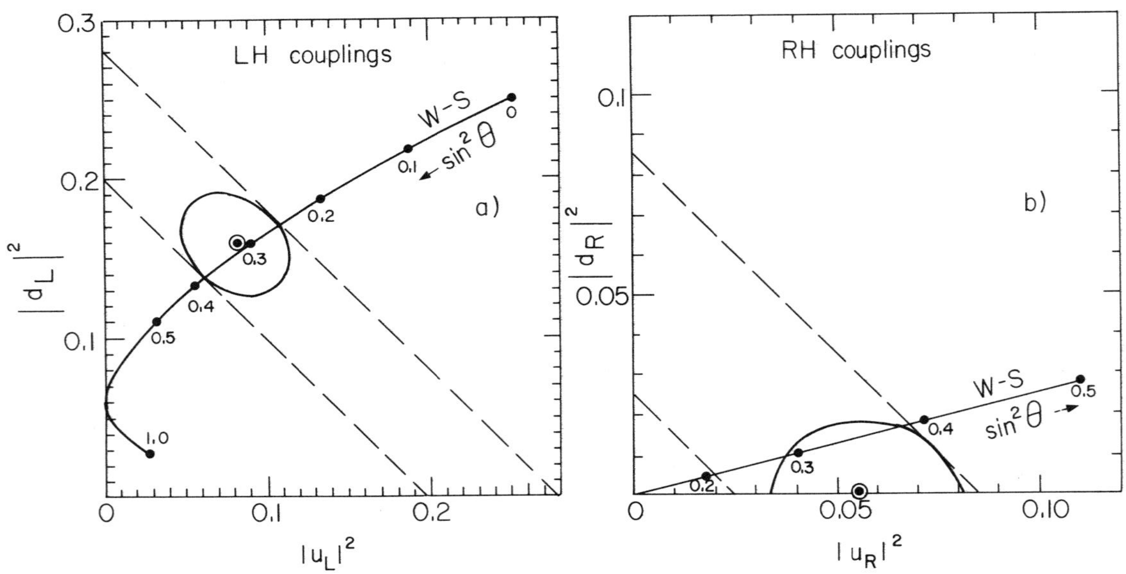

Fig. 2 Neutral coupling constants (from ref. [4]).

Analysis of Neutrino Electron Scattering Data

Helmut Faissner

III. Physikalisches Institut, Technische Hochschule Aachen,

Aachen, Germany

Abstract:

The Aachen-Padova Group measured, in a spark chamber exposed to the CERN PS neutrino beam ($<E_\nu> \simeq 2$ GeV), the total cross sections for both, $\nu_\mu e$ and $\bar{\nu}_\mu e$ scattering. The result is: $\sigma = (1.2 \pm 0.6)$ and $\bar{\sigma} = (2.4 \pm 1.1) \times 10^{-42}$ cm^2 $\times$ E_ν/GeV. It excludes a purely left-handed electron (V - A) by $\gtrsim 3$ standard deviations (s.d.). A completely right-handed electron is admitted; but the favored solution is a mixture of V and A, with a mixing parameter $x_W \equiv \sin^2\theta_W = 0.38 \pm 0.10$. A pure A (or V) current is disfavored by 1 s.d. on the basis of total cross sections alone. But the distributions of recoil electron energies are different for ν_μ and $\bar{\nu}_\mu$, at the 2 s.d. level. The respective average energies are 700 ± 170 and 1100 ± 190 MeV, and suggest a fraction of V - A of $\lesssim 20\%$ of the total interaction. If one generalizes the minimal $SU_2 \times U_1$ model by admitting a Higgs boson of arbitrary isospin I_Φ, the cross section ratio $\bar{\sigma}/\sigma = 2.0 \pm 1.0$ still yields a $x_W \simeq 0.35$. The absolute cross sections computed with this value, and an arbitrary neutral current coupling constant G^{NC}, have been compared to experiment. The result is $G^{NC}/G = 1.04 \pm 0.20$, excluding $I_\Phi > 1/2$. If one settles on $I_\Phi = 1/2$, the neutrino electron scattering data fix the weak isospin of the right-handed electron to zero.

The difficulties of doing neutrino electron scattering experiments have been expounded by André Rousset[1]. The present report deals with the implications of their results. Our faith in them is growing, since the three groups involved: Gargamelle[2], Aachen-Padova[3], and Irvine[4], have spent an unusual amount of time and effort to assess the background experimentally. Heavy liquid bubble chamber[2], multiplate spark chamber[3], and counter[4] groups agree that the background is mainly single gammas. The key point is to make it small a priori by suitable selection criteria, and to determine it a posteriori, with the aid of aptly chosen control samples.

The results on $\overset{(-)}{\nu}_\mu e$ scattering are given in Table I. The bubble chamber can discriminate (to some extent) between electrons and electron-pairs by direct inspection. The spark chamber group, instead, has to resort to tight kinematical acceptance criteria. The background remaining amongst their ten (eight) $\nu_\mu(\bar{\nu}_\mu)e$ scattering candidates was determined from seven (seven) control samples, which comprised several hundred of NC or CC produced 1γ or 2γ events. It was backed up by a Monte Carlo program, which computed rates and distributions of single gammas from π^o decay in any region of phase space and apparatus[5]. First it was used to extrapolate 1γ distributions into the kinematically accepted region from the outside. Recently it has given absolute numbers too[6]. They confirm the small background in $\bar{\nu}_\mu e$, and indicate that the $\nu_\mu e$ background in Table I might have been overestimated by not quite one event.

This good systematic knowledge of background is the basis for the high confidence with which both, $\bar{\nu}_\mu e$ and $\nu_\mu e$ scattering, have been established. The cross sections, instead, are known with an error determined by the statistics of a small number of events. In addition there are problems of acceptance, introduced by the kinematical cuts, combined with the limited accuracy in electron energy ($\simeq$ 20%) and angle ($\simeq$ 20 mrad). These resolution functions have been determined experimentally,

Table I :

Results of ν_μ and $\bar{\nu}_\mu$ electron scattering measurements.

	ν_μ	$\bar{\nu}_\mu$	
	AC - PD	AC - PD	GGM
No of candidates	10	8	3
Background:			
a) 1γ from π^0	3.7 ± 0.4	1.3 ± 0.3	0.26 ± 0.11
b) inverse β - decay	0.4	0.05	0.07 ± 0.04
c) $(\nu_e + \bar{\nu}_e)$ e scattering	0.06	0.03	0.11
No of background events	4.2 ± 0.4	1.4 ± 0.3	0.44 ± 0.13
percentage	42 ± 4	18 ± 3	15 ± 4
significance of effect	99.3 %	>99.99 %	99.9 %
Effect :	5.8 ± 3.2	6.6 ± 2.8	2.6 ± 1.7
Total cross-section 10^{-42} E_ν cm^2	1.2 ± 0.6	2.4 ± 1.1	$1.0^{+2.1}_{-0.9}$

by shooting electrons of known energy into the Al chambers[7)].* The acceptance can be computed, provided the energy and angle distributions of recoil electrons are known too. This introduces a certain model dependence, but fortunately it is not very pronounced. These problems, as well as the reconstruction problems emphasized by Rousset[1)], enter at the 10% level. The real problems lie somewhere else: in the small number of candidates, and in the systematics of background subtraction.

The $\nu_\mu e$ and $\bar{\nu}_\mu e$ scattering cross sections are visualized in Fig. 1. The $\nu_\mu e$ total cross section σ is plotted along the abscissa, the $\bar{\nu}_\mu e$ ($\bar{\sigma}$) along the ordinate. The Gargamelle $\bar{\nu}_\mu e$ result is effectively an upper limit, and defines together with a preliminary limit on $\nu_\mu e$[10)], roughly an allowed square of σ, $\bar{\sigma} \lesssim 3$ (10^{-42} cm^2 $\times$ E_ν / GeV). The Aachen-Padova values, with their errors, define an ellipse, situated inside that square. A change of acceptance criteria would move the AC-PD point somewhat, but in general not beyond the limits of the error ellipse given.

The straight lines in Fig. 1 correspond to a fixed cross section ratio $r_e = \bar{\sigma}/\sigma$. Since we analyse the data in terms of a mixture of vector and axialvector interaction, these lines imply a definite ratio of vector to axialvector electron current too. In fact,

$$r_e \equiv \frac{\bar{\sigma}}{\sigma} = \frac{A + 3B}{3A + B} , \qquad (1)$$

where the V - A coefficient A, and the V + A fraction B, are related to the usual V and A coupling constants by

---

* In fact, part of the larger cross section, reported last year[8,9)], was due to an overestimate of the energy of small showers. Other factors were: improved resolution and acceptance functions, sharper cuts, changed neutrino fluxes, and better statistics.

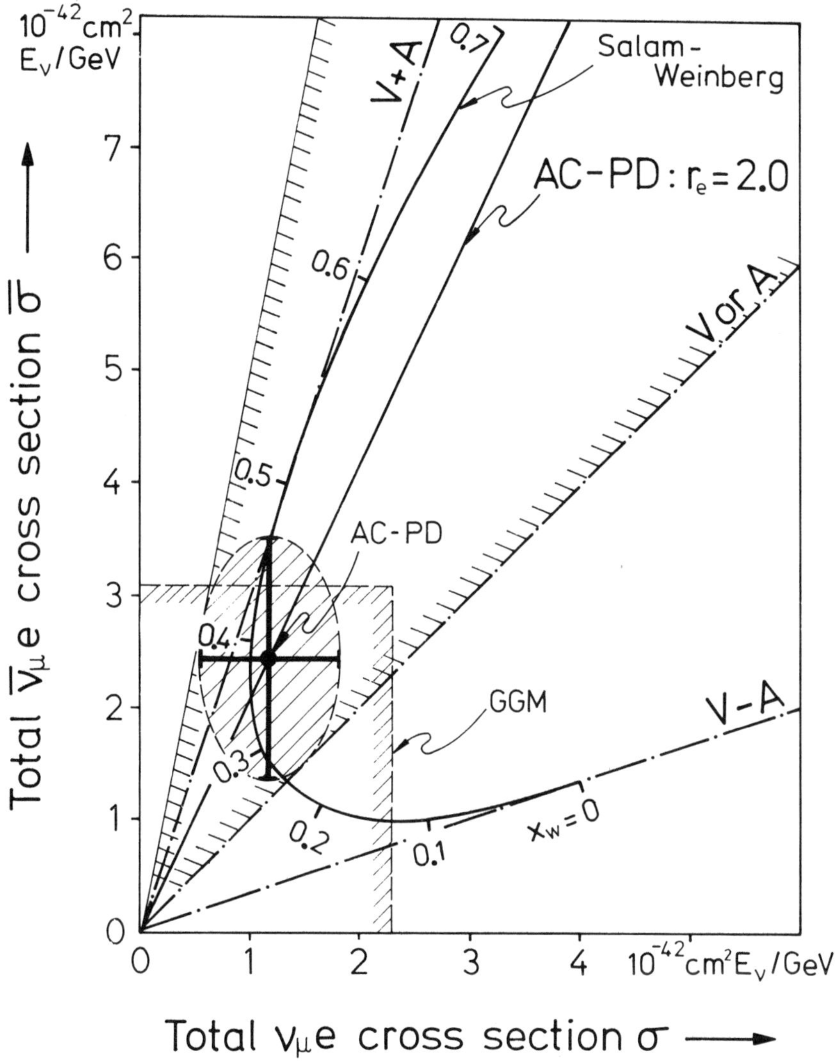

Fig. 1: ν_μe and $\overline{\nu}_\mu$e total cross sections, as measured by AC-PD and GARGAMELLE. Straight lines delineate simple V,A interactions, the curve is for the minimal gauge model.

$$A = \left(\frac{C_V + C_A}{2} \right)^2 \quad , \quad B = \left(\frac{C_V - C_A}{2} \right)^2 \quad ; \qquad (2)$$

and V - A means $C_V/C_A > 0$. The relative amount of V - A, or V + A, may be characterized by the coefficients

$$\alpha = \frac{A}{A + B} \quad , \quad \beta = \frac{B}{A + B} \quad . \qquad (3)$$

The sector allowed by V,A-theories is bounded by $\alpha = 1$ ($r_e = 1/3$) on the right-hand side, and by $\beta = 1$ ($r_e = 3$) on the left. The center line, pure V or A ($r_e = 1$), is parity conserving. In the minimal $SU_2 \times U_1$ gauge model of Salam and Weinberg[11], characterized by

$$C_V = -\frac{1}{2} + 2x_w \quad , \quad C_A = -\frac{1}{2} \quad , \qquad (4)$$

this is achieved by a vanishing vector part $C_V = 0$, and thus a mixing parameter $x_w \equiv \sin^2\theta_w = 1/4$.

Some <u>conclusions</u> from the Aachen-Padova (AC-PD) data can be read off Fig. 1 directly, as has been done already last year[8,9]:

- Pure V - A is excluded at the 3 standard deviation (s.d.) level.
- Pure V + A is admitted.
- The best solution is a mixture of V and A, with a small V - A fraction $\alpha = (17 \pm 33)\%$.

These statements follow from the cross section ratio $r_e = 2.0 \pm 1.0$ by itself. The most conspicuous observation to be made about the absolute cross sections is that

- they happen to fall right on the Salam-Weinberg curve.
- Less prominent, but still suggestive: they favor a mixing parameter x_w close to 0.4.

Table II :

Coupling constants and mixing parameters from measured cross section.

Quantity	Value	derived from
$\beta = \frac{B}{B+A}$	0.83 ± 0.33	r_e
A B	0.026 ± 0.046 0.13 ± 0.07	σ and $\bar{\sigma}$
C_V C_A	0.20 ± 0.15 -0.52 ± 0.15	B and A *
$\sin^2 \Theta_w$	$0.35^{+0.15}_{-0.10}$ 0.38 ± 0.10	r_e $\sigma, \bar{\sigma}$

*As evidenced by Fig. 3, there are three other solutions

Use of the absolute values strengthens the exclusion of pure V - A. All these statements are summarized in the coupling constants of Table II, and in their errors.

The most interesting question, if the neutral weak electron current could be pure A (or V) is not yet satisfactorily answered by these numbers. This parity conserving interaction would imply A = B, and thus $\bar{\sigma} = \sigma$, and this is disfavored only by 1 standard deviation (s.d.) by the cross sections. It seemed expedient, therefore, to look for other differences between neutrino and antineutrino data. Clearly, any difference between $\nu_\mu e$ and $\bar{\nu}_\mu e$ cross sections would manifest itself in the distributions of the recoil electron energies E_e.

The differential neutrino cross section is

$$\frac{d\sigma}{dy} = \sigma_e \left\{ A + B\,(1 - y)^2 \right\} , \quad y = \frac{E_e}{E_\nu} , \tag{5}$$

with $\sigma_e = \kappa^2 \frac{G^2}{\pi} 2 m_e E_\nu = \kappa^2\ 17.2 \times 10^{-42}\ \mathrm{cm}^2\ E_\nu/\mathrm{GeV}$.

For antineutrinos A and B interchange place. ($\kappa = G^{NC}/G$ allows for an overall normalization of the weak neutral current NC [12,13]). Since the neutrino spectrum is known, the ensuing recoil electron spectrum can be computed for any value of A and B. If they are equal, identical electron energy distributions are expected for neutrinos and antineutrinos, provided the two spectra are the same. Since the ν spectrum is somewhat harder ($\langle E_\nu \rangle$ = 2.34 GeV) than the $\bar{\nu}$ spectrum ($\langle E_{\bar{\nu}} \rangle$ = 2.05 GeV), one would have expected a corresponding difference in E_e. As a matter of fact, the experimental distributions (Fig. 2) do not at all bear out these expectations. They show a substantially harder electron spectrum for antineutrinos, and a soft tail, instead, for

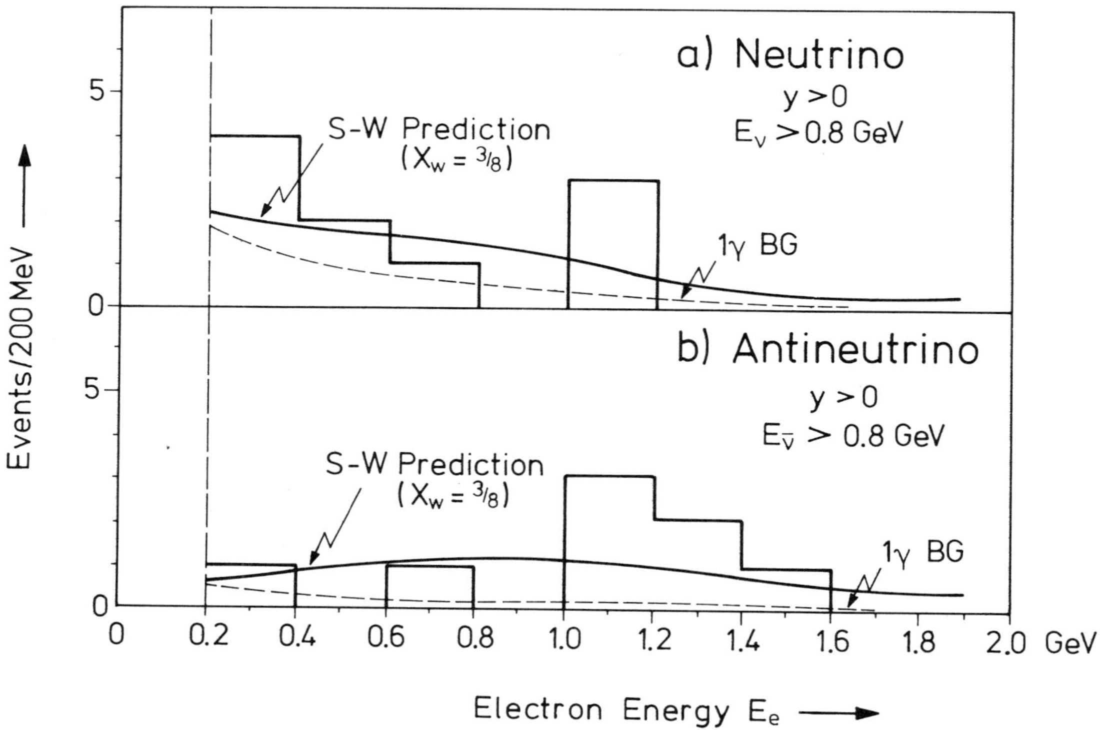

Fig. 2: Electron energy distributions for: a) $\nu_\mu e$, b) $\bar{\nu}_\mu e$ scattering. - Upper curve is the expectation from the minimal model (and x_W = 3/8); lower curve, the experimentally determined 1γ background.

neutrinos, and neither of these features can be explained by background. After background subtraction the two distributions do remain different (with $\geqq$ 90% confidence), if one applies statistical tests fit for small event numbers[3)].

Alternatively, the two distributions may be characterized by their first moments, the average electron energies $\langle E_e \rangle$ and $\langle \bar{E}_e \rangle$. As evidenced by Table III, they look different, if one considers the measurement errors only, and remain so, if one includes the statistical uncertainty. One can compute an average y from the average energies:

$$\langle y \rangle = \langle E_e \rangle / \langle E_\nu \rangle \quad , \quad \langle \bar{y} \rangle = \langle \bar{E}_e \rangle / \langle E_{\bar{\nu}} \rangle \quad . \tag{6}$$

As the energies are relatively better measured than the angles, these quantities are more reliable than an $\langle y \rangle$ derived from reconstructing y from event to event. But they are still biased by the experimental cuts. However, in order to decide the present issue there is no need to compute the biases - if the two differential cross sections were really equal, so would be the acceptances. Hence one may contrast $\langle y \rangle$ and $\langle \bar{y} \rangle$ directly, the $\langle E_e \rangle$'s can be compared after correction for the different $\langle E_\nu \rangle$. The difficulty lies in estimating the expected statistical uncertainty. This was done by Monte Carlo. A conservative estimate is

$$\Delta \langle E_e \rangle = 555 \pm 275 \text{ MeV} \quad .$$

This is a 2.0 s.d. effect, and hence the probability that the $\nu_\mu e$ and $\bar{\nu}_\mu e$ cross sections are equal is only about 2%.

Digressing briefly from our (V,A)-based analysis, we note that this result excludes also any other Dirac variant occuring by itself. The case of an arbitrary mixture of scalar (S) and pseudoscalar (P) is even less probable, since it would imply, in addition to the equality of $\nu_\mu e$ and $\bar{\nu}_\mu e$ cross sections, an absolute value of $\langle y \rangle = \langle \bar{y} \rangle = 3/4$, which is completely

Table III :

Average electron energies E_e, and related quantities

Quantity [E in GeV]	$\nu_\mu e$	$\bar{\nu}_\mu e$	Remarks
$\langle E_\nu \rangle$	2.34	2.05	weighted by $\sigma \sim E$ (official CERN spectrum)
$\langle E_e \rangle$	0.70 ± 0.07	1.10 ± 0.09	measurement error only
	± 0.17	± 0.19	estimated statistical error included.
$\langle y \rangle = \frac{\langle E_e \rangle}{\langle E_\nu \rangle}$	0.30 ± 0.07	0.54 ± 0.09	" " "
$\langle y \rangle = \langle f(E_e, \Theta_e) \rangle$	0.31	0.52	by $y = 1 - E_e \Theta_e^2 / 2m_e$ [errors!]

out of the experimental range (s. Tab. III). Thus, although the data cannot prove the absence of the variants other than currents, it does not supply any hint for their occurence either.

The corrected ratio of the average electron energies $\langle E_e \rangle$ (i.e. of average energy transfers $\langle y \rangle$):

$$r_y \equiv \frac{\langle \bar{y} \rangle}{\langle y \rangle} = \frac{\langle \bar{E}_e \rangle}{\langle E_e \rangle} \frac{\langle E_\nu \rangle}{\langle E_{\bar{\nu}} \rangle} = 1.80 \begin{matrix} + 0.85 \\ - 0.45 \end{matrix} \qquad (7)$$

is a particularly well measured and unbiased quantity. It can be used to test relations implied by the assumption of an interaction between vector and axialvector currents only.[15] Besides it serves to place a rather model-independent limit on the ratio of the respective coupling constants. This is clear, when one realizes that according to eq. (5) $\langle y \rangle$ depends only on the ratio $A/B \equiv a$, viz

$$4 \langle y \rangle = \frac{6a + 1}{3a + 1} \quad . \qquad (8)$$

$\langle \bar{y} \rangle$ is given by the same expression, except that $a = A/B$ is replaced by B/A. Thus the type of interaction can be read directly off $\langle y \rangle$, or $\langle \bar{y} \rangle$. Conversely, for a mixture of V and A relations must exist between r_y and r_e, and they provide the wanted tests.[15] As one easily checks, they are all fulfilled by the data of Table II and III.

Finally we use the energy ratio r_y to deduce an approximate value for a. It is based on (8), and it is valid for small a only, (and thus for $1 < r_y \leqq 2$):

$$a \simeq \frac{1}{6} \; \frac{2 - r_y}{r_y - 1} \qquad (9)$$

Insertion of (7) gives a left-to-right ratio $a = 0.04 \,{}^{+}_{(-)}\, 0.20$, a value very close to pure V + A. The 1 s.d. error may be transformed into the C_V, C_A - plane of Fig. 3. It implies that only a sector around the V + A line is allowed by the measured energy ratio. The sector boundaries are approximately given by

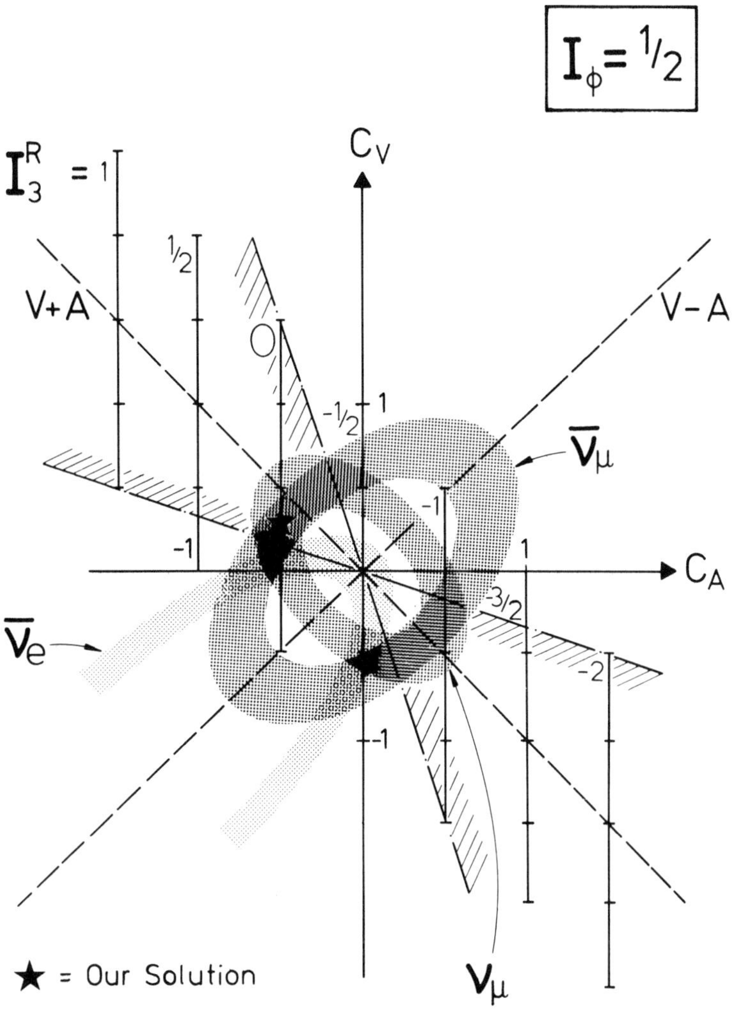

Fig. 3: Bounds on V and A coupling constants provided by ν_μe, $\bar{\nu}_\mu$e, and $\bar{\nu}_e$e total cross sections[3,4]. Vertical bars indicate allowed range of x_W for different I_3-assignments to right-handed electron, and Higgs isospin $I_\phi = 1/2$. Ratio of electron energies admits sectors around V + A axis. (All errors ± 1 s.d.)

$$C_V = - C_A/3 \, , \quad \text{and} \quad C_A = - C_V/3 \quad . \tag{10}$$

They illuminate once more that a pure V (or A) interaction is disfavored, at the 2 s.d. level, by the energy measurement, and that higher proportions of V - A are practically ruled out.

Having the space-time structure of the weak neutral electron current thus rather well defined, we should like to end with some statements about charge-like quantum numbers. Amongst the basic assumptions one has to make in gauge theories are the properties of the Higgs sector, in particular the isospin I_Φ of the Higgs boson Φ. As has been observed by Ross and Veltman[6], one can retain all the other properties of the minimal model[11], but regard I_Φ as a variable parameter. Thus one gets an infinity of $SU_2 \times U_1$ models, of the type originally introduced by Glashow[16]. A Higgs isospin $I_\Phi \neq 1/2$ causes an overall normalization $\kappa \neq 1$ of the neutral current, as has been introduced already in eq. (5). Ross and Veltman[6] gave an important relation, valid for arbitrary, but fixed, isospin I_Φ:

$$\kappa = (2 \, I_\Phi)^{-1} \tag{11}$$

They recommend to determine $x_W = \sin^2\theta_W$ from $r_e = \bar{\sigma}/\sigma$, which clearly is independent of κ. We have done this already in Table II. The result*) $x_W = 0.35^{+0.15}_{-0.10}$ is so close to the world average, Sakurai[17] and B.W. Lee[18] derived last year at the Aachen Neutrino Conference, that we decided to use it (without error) for predicting the absolute cross sections[19]. Confronting those with our experimental values (s. Table I) gives

$$\kappa = 1.04 \pm 0.20 \quad , \tag{12}$$

and excludes, with a confidence > 99%, any $I_\Phi \geqq 1$.

---

*) There is a second solution $x_W' > 1/2$, which we disregard in this analysis, since it would correspond to an unphysical $\kappa >> 1$.

Finally, we tested for the weak isospin I^R of the right-handed electron. In the standard model $I^R = 0$; if one chooses $I^R \neq 0$ instead, its third component I_3^R will enter the V + A coefficient B [21]. As a result, the allowed regions (= vertical lines) in the C_V, C_A - plane are shifted in steps of $^1/_2$, with respect to the minimal model ($I_3^R = 0$). This is illustrated in Fig. 3. A measured total cross section and its 1 s.d. error, defines an elliptic ring in this plane. The Aachen-Padova $\nu_\mu e$ and $\bar{\nu}_\mu e$ data display considerable overlap to both sides of the V - A axis, admitting $I_3^R = 0, -{}^1/_2$ and - 1. The situation is improved by adjoining the $\bar{\nu}_e e$ measurement of Fred Reines and his group[4]: It excludes the isovector solution ($I_3^R = -1$) by many standard deviations. The remaining ambiguity is removed by the Aachen-Padova energy ratio (7), which renders $I_3^R = -{}^1/_2$, a pure vector solution, improbably at a level of some percent. What remains, is the minimal model - and as far as we are concerned, with a mixing parameter around $^3/_8$ [19]. Of course, this solution is not unique in a mathematical sense. There are combinations of high I_Φ with high I^R still compatible with the data. But they are incidental, and devoid of any deeper reason.

On the other hand, the simple structure revealed by neutrinos may be a partial aspect of a larger embedding group. In fact, if atomic physics persists in evincing no trace of parity violation[22], then the minimal model has to be extended. This can be done, in a basically parity-conserving way, without losing the Salam-Weinberg predictions for our neutrinos[23]. In these models the parity violation, apparently observed in NC neutrino interactions, reflects just the fact that the primary particles are screwy. And the simplicity we see may come from our eyes not being open enough ...

Let us get consolation from Ben Lee's closing sentence at the Aachen Conference[18]: "It is not as important that we do not fully comprehend, and do not always agree on, what we see; it is very important that our vistas are constantly expanding by this endeavor."

References:

1) A. Rousset, this Conference.

2) J. Blietschau et al., Nuclear Phys. B 114, 189 (1976).

3) H. Faissner, E. Frenzel, T. Hansl, D. Hoffmann, E. Radermacher, H. Reithler, H. de Witt, M. Baldo-Ceolin, F. Bobisut, H. Huzita, M. Loreti, G. Puglierin, Proc. Internat. Neutrino Conf. Elbrus (1977), in press; also Aachen Report PITHA-91 (1977)
H. Reithler, Proc. Internat. Symposium on Lepton and Photon Interactions at High Energies, Hamburg (1977), in press; also Aachen Report PITHA-95 (1977).

4) F. Reines, H.S. Gurr, and H.W. Sobel, Proc. International Neutrino Conf. Aachen (1976) p. 217 and Phys. Rev. Letters 37, 315 (1976). The energy-averaged $\bar{\nu}_e e$ cross section, used in Fig. 3, was kindly communicated to us by Fred Reines.

5) H. de Witt, Diplomarbeit TH Aachen (1977), and Aachen Report, in preparation.

6) H. Reithler, W. Thomé, and H. de Witt, private communication.

7) H. Faissner et al., Nuclear Instr. and Meth., to be published

8) H. Faissner et al., pres. by F. Bobisut, Proc. Internat. Neutrino Conf. Aachen (1976) p. 223

9) H. Faissner, Proc. 18. Internat. Conf. High Energy Phys. Tbilisi (1976) Vol. II, p. B 114

10) P. Musset, Proc. Internat. Symposium on Lepton and Photon Interactions at High Energies, Hamburg (1977), in press

11) J.C. Ward and A. Salam, Phys. Letters 13, 168 (1964)
A. Salam, in Elementary Particle Theory, Ed. N. Swartholm (Stockholm 1969) p. 367
S. Weinberg, Phys. Rev. Letters 19, 1264 (1967); 27, 1688 (1971); Phys. Rev. D5, 1412 and 1962 (1972)

12) J. Bernabeu and C. Jarlskog, Phys. Letters 69 B, 71 (1977)

13) D.A. Ross and M. Veltman, Nuclear Phys. B 95, 135 (1975)

14) H. Faissner et al., Phys. Letters 68 B, 377 (1977) applied this argument for demonstrating $\bar{\sigma} \neq \sigma$ for NC produced π^0's.

15) L.M. Sehgal, ANL Report ANL-HEP-PR-75-45, also Aachen Report PITHA-87 (1975)
M. Gourdin, Proc. International Neutrino Conference Aachen (1976), p. 234
P.R. Rosen, Proc. International Neutrino Conference Downington (1974), p. 5

16) S.L. Glashow, Nuclear Phys. 22, 579 (1961)

17) J.J. Sakurai, Proc. International Neutrino Conference Aachen (1976), p. 519

18) B.W. Lee, Proc. International Neutrino Conference Aachen (1976), p. 704

19) It would not matter too much for the present purposes, if we took the currently fashionable values of x_W instead. But a value of 0.24 ± 0.02 as advertised by Holder et al.[20], is definitely at variance with our electron energy measurements.

20) M. Holder et al., Phys. Letters 71 B, 222 (1977)

21) A. de Rújula, H. Georgi, S.L. Glashow, and H.R. Quinn, Rev. Mod. Phys. 46, 391 (1974)
H. Fritzsch, Proc. International Neutrino Conference Aachen (1976), p. 571 and Preprint Ref. TH 2297-CERN

22) P.G.H. Sandars, this Conference

23) H. Fritzsch and P. Minkowski, Nucl. Phys. B 103 (1976) 61
R.N. Mohapatra and D.P. Sidhu, Phys. Rev. Lett. 38, 667 (1977)
B.R. Kim, Phys. Letters 70 B, 425 (1977)
summarized these attempts, and discussed possibilities to retain $SU_2 \times U_1$.

SEARCH FOR BARYON DECAY AND NEUTRINO OSCILLATIONS*

Frederick Reines
Department of Physics
University of California
Irvine, California 92717

Baryon Conservation

The title "Search for baryon decay and neutrino oscillations" is correctly taken to suggest that no violation of baryon number conservation has been found. The subject of baryon conservation had its start in 1938 with the suggestion by Stückelberg.(1) It was discussed by Wigner in 1949(2) by Feinberg and Goldhaber(3) (1959) and then (1973, 1974) in the context of gauge theories which unified the strong-electromagnetic and weak interaction by Pati and Salam(4) and others.(5) In this talk we summarize the status of experimental tests which have been made and give estimates of limits which might be achieved by such tests. We note at the outset that if baryon number is not conserved then at least one other conservation law must also be violated. Further we assume that the postulated violation occurs independently for each baryon whether it is bound in a nucleus or free.

As in tests of any conservation law it is first necessary to postulate the mode of violation and then to search out that mode. So, for example, we can retain all conservation laws except for baryon number, B, and lepton number, L, and then

*Supported by the United States Department of Energy

deduce the observable consequences. Under these assumptions a decaying proton would with high probability produce at least one energetic charged lepton which could be sought in a well shielded detector.

The Table summarizes the present experimental limits for lifetimes against B number violating nucleon decay under various assumptions regarding the concurrent validity of other conserved quantities e.g. lepton number, L, electric charge, Q, energy-momentum. Also listed are limits which we estimate to be experimentally achievable under these assumptions.

Baryon Decay Modes and Limits*

Non-conserved		n	p	τ(yrs.) Present Limit	Achievable
B, L	(a)	$\pi^+ + \pi^- + \nu$	$\pi^+ + \nu$	3×10^{30}	10^{32}
	(b)	$\mu^+ + \mu^- + \nu$	$\mu^+ + \nu$	3×10^{30}	10^{32}
	(c)	$e^+ + e^- + \nu$	$e^+ + \nu$	10^{29}	10^{31}
	(d)	——	$e^+ + \gamma$	10^{29}	10^{31}
B, L, Q	(e)	$\gamma + \nu$	$\gamma + \nu$	10^{29}	10^{31}
	(f)	ν's (odd no.)	ν's (odd no.)	2×10^{26}	10^{29}(?)
Everything!	(g)	?	?	3×10^{23} (for protons)	10^{29}(?)

*The inventive reader can suggest many more variants of each mode e.g. B, L $n \rightarrow 2\pi^+ + 2\pi^- + 3\nu$.

The experimental methods employed range from the search in a well-shielded underground detector for either single pulses or pulses in delayed coincidence to search for nuclei which would not be present except for baryon non-conservation.

Present Experimental Limits for Modes a,b[6]

The most restrictive limits are derived from the Case-Wits-Irvine South African deep mine atmospheric neutrino experiments. In these experiments a total of 67 ton years of liquid scintillation detector operation at a depth of 2 miles yielded 6 pairs of pulses identified as muons which stopped and decayed in the detector. Because of the detector geometry it was not possible to determine the muon multiplicity in a given nucleon decay. Atmospheric muons were responsible for at most 0.3 decays and the remainder of the delayed coincidences appear to be consistent with muons produced by atmospheric ν_μ, $\bar{\nu}_\mu$ which interacted in the rock surrounding the detector. Since the detection efficiency for such events is $\sim$ 50% for a μ stop energy > 15-20 MeV and a μ decay electron energy > 10 MeV, we can deduce that the mean life $\tau_{a\ \underline{or}\ b}$ for neutron or proton decay in mode (a) or (b) is

$$\tau_{a\ \underline{or}\ b} > \frac{N}{\frac{dN}{dt}} > \frac{67 \times 10^6 \times 6 \times 10^{23}}{12 \times 2} > 1.5 \times 10^{30}\ \text{yrs.}$$

However since the event rate is consistent with ν produced muons we conclude

$$\tau_{a\ \underline{or}\ b} \gtrsim 3 \times 10^{30}\ \text{yrs.}$$

Achievable Limits for (a) and (b)

The CWI detector geometry (60 liquid scintillator slabs each 12.5 x 60 x 550 cm^3) was chosen with atmospheric ν detection in mind and hence was not well suited for baryon conservation studies. The design could be significantly improved by increasing the detector mass, adding a cosmic ray anticoincidence which would also assure that the baryon decay sought originated in the detector and then making use of the difference in the μ spectrum produced by baryon decay and ν_μ, $\bar{\nu}_\mu$. In this way it seems reasonable to improve discrimination against single muons from ν_μ, $\bar{\nu}_\mu$ by a factor of ~ 100, so achieving a signal of 10/yr (neutron and proton decay rates assumed equal) with a 2000 ton detector and a signal/background ratio of 3/1 at a mean life $\tau_{a,b} \sim 10^{32}$ yrs.*

Present and Achievable Limits for Modes (c,d)

The best limits for this class of modes are again derived from the CWI detector in which we recorded 600 single counts exceeding 15-20 MeV with a mass-time product of 67 ton years. Cosmic ray muons are estimated to give $\sim 2/3$ of the observed counts and ν_μ, $\bar{\nu}_\mu$ to give the remaining $\sim 1/3$. If we assign 1/4 of the total as an upper limit on single events attributable to baryon decay, then

*This limit is for decay events producing single muons. Since low energy (< 1 GeV) neutrino interactions do not generally give rise to multiple muons (via π decay) a greater limit is achievable for higher multiplicity nucleon decay modes.

$$\tau_{c,d} \geq \frac{67 \times 10^6}{150} \times \frac{6 \times 10^{23}}{2} > 10^{29} \text{ yrs.}$$

The improved detector mentioned under modes a, b would also increase limits for c,d by $\sim 10^2$, or to 10^{31} years.

Limits for (e)

The limits for this γ-associated mode also result from the CWI neutrino experiment. The detector elements had an average thickness for isotropic radiation of 25 cm, or ~ 1 γ mean free path. Accordingly, present and achievable limits for modes c, d also apply to 3.

Limits for (f,g)

If we assume a more generalized breakdown of all the conservation laws the limits are less well known. The approach adopted is to assume that the given nucleon may vanish leaving the remaining nucleons to obey all the conservation laws, as usual. The case involving fewest assumptions regarding nuclear states following nucleon decay is(7)

$$D \rightarrow n + ?$$

Here the disappearance of a proton is signaled by capture of the neutron to which it was initially bound. Using a heavy water cerenkov detector in which neutron capture γ rays were observed a limit of $\tau > 3 \times 10^{23}$ years was derived. Our experience with He^3 neutron detectors immersed in D_2O suggests that this limit on proton decay can be increased by at least two orders of magnitude.

Other Limits

Lifetime limits set by the directly emitted charged particles or gammas resulting from nucleon rearrangement subsequent to nucleon decay are much shorter because of the low nuclear binding energies and high resultant backgrounds.

Other ways to search for nucleon decay involve isotope separation and analysis either of stable or radioactive products as suggested by Rosen,[8] Evans and Steinberg,[9] and Fireman.[10]

Stable Products

e.g. One searches[9] for Xe^{129} in a sample ($\sim 10^9$ yrs old) of Te^{130}

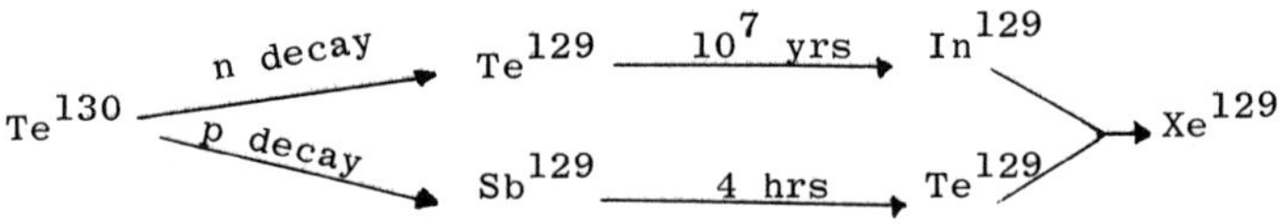

The observed abundance of Xe^{129} gives a measure of nucleon stability

$$\tau > 10^{25} \text{ yrs. (for n } \underline{\text{or}} \text{ p decay).}$$

Radioactive Products

Here, for example,[9, 10] one searches for A^{37} in a sample of K^{39}.

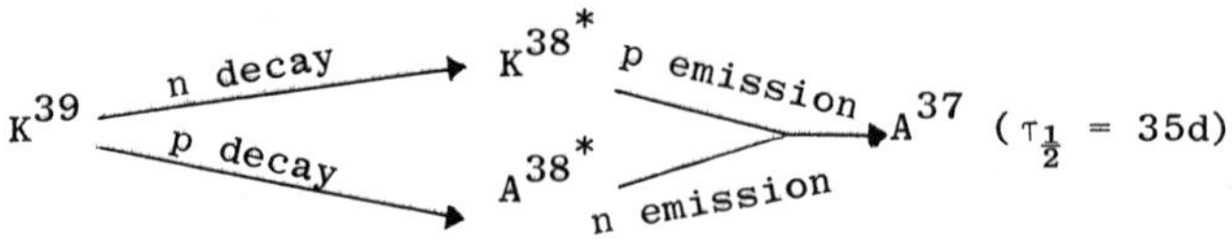

The lifetime derived for p or n decay using 1.7 tons of potassium acetate is

$$\tau > 2 \times 10^{26} \text{ yrs.}$$

According to Evans and Steinberg this method should be sensitive to $\tau \sim 10^{29}$ yrs.

Both methods are ultimately limited by backgrounds from atmospheric ν_μ, $\bar{\nu}_\mu$ reactions.

In the event of a positive indication of baryon decay it would then be necessary to prove that the signal in fact did not arise from any other source. Such an identification would require the use of direct counting techniques.

In summary, present experimental limits against baryon decay range from 3×10^{23} to 3×10^{30} years, depending on decay modes assumed. It appears unlikely that experiments will be improved in sensitivity beyond $\sim 10^{32}$ years even for the most readily studied decay modes.*

Neutrino Oscillations

The idea that neutrinos might oscillate from one observable state to another was suggested in analogy with K^o decay by B. Pontecorvo[(11)] (1968) and considered by many other authors since.[(12)]

---

*It is clear that the effort required to reach these "ultimate" limits is considerable. In the absence of any experimental hint or firm theoretical guide the continuation of the search will depend very much on the predilection of the experimentalist.

If by way of an illustrative example we consider $\bar{\nu}_e$ and $\bar{\nu}_\mu$ as the result of oscillations between two states $\bar{\nu}_1$, $\bar{\nu}_2$ characterized by rest masses m_1, m_2 then

$$\begin{pmatrix} \bar{\nu}_e \\ \bar{\nu}_\mu \end{pmatrix} = \begin{pmatrix} \cos\xi & -\sin\xi \\ \sin\xi & \cos\xi \end{pmatrix} \begin{pmatrix} \bar{\nu}_1 \\ \bar{\nu}_2 \end{pmatrix}$$

where $\bar{\nu}_1 = e^{-iE_1 t}$, $\bar{\nu}_2 = e^{-iE_2 t}$

$E = (p^2 + m^2)^{\frac{1}{2}}$, ξ = mixing angle

$(\hbar = c = 1)$

the probability of observing $\bar{\nu}_e$ at time t if we start with $\bar{\nu}_e$ is

$$|\bar{\nu}_e(t)|^2 = 1 - \frac{\sin^2 2\xi}{2} \{2 - \cos(E_2-E_1)t\} \quad .$$

If $p >> m$ then $(E_2-E_1)t = \frac{2\pi r}{\lambda}$ where

$\lambda \cong \frac{p}{40} \times 10$, $\Delta = m_2^2 - m_1^2$ (ev^2), p = MeV/c

λ = oscillation length (meters).

Choosing a reference energy E_o, $\lambda = \frac{\lambda_o}{E_o} E$ and

$$[\nu_e(r)]^2 = 1 - \frac{\sin^2 2\xi}{2} \left\{1 - \cos\left(\frac{2\pi E_o}{\lambda_o} \frac{r}{E}\right)\right\}$$

Experiment at Fission Reactor (H. W. Sobel, H. S. Gurr, F. Reines)

Fission reactors are attractive $\bar{\nu}_e$ sources with which to search for ν oscillations* because of the low energy of fission $\bar{\nu}_e$ (few MeV). The ideal experiment at a fission reactor would be to study the inverse beta reaction

$$\bar{\nu}_e + p \rightarrow n + e^+ \tag{1}$$

as a function of distance from the reactor. We report here on an experiment at only one distance (11.2 meters from the center of an 1800 MW Savannah River Plant reactor**). The detector we used was the 15.7 Kg plastic scintillator inside of a segmented 300 Kg NaI scintillator employed in the study of $\bar{\nu}_e + e^- \rightarrow \bar{\nu}_e + e^-$.*** The figure summarizes the experimental results.

---

*Accelerators[13] and cosmic rays have also been proposed to study ν oscillations. In general the higher energies require larger source-detector distances. This is somewhat balanced by the positive nature of the search e.g. can look <u>for</u> ν_e reactions instead of measuring the <u>absence</u> of ν_μ signals.

**The reactor may be treated as a point source (good to a few %) for our reactor (a cylinder 5 meters in diameter and 5 meters high) if $\lambda > 10$ meters and $r \geq 11.2$ meters.

***For details of the detector see F. Reines, H. S. Gurr, H. W. Sobel, Phys. Rev. Lett. <u>37</u>, 315 (1976).

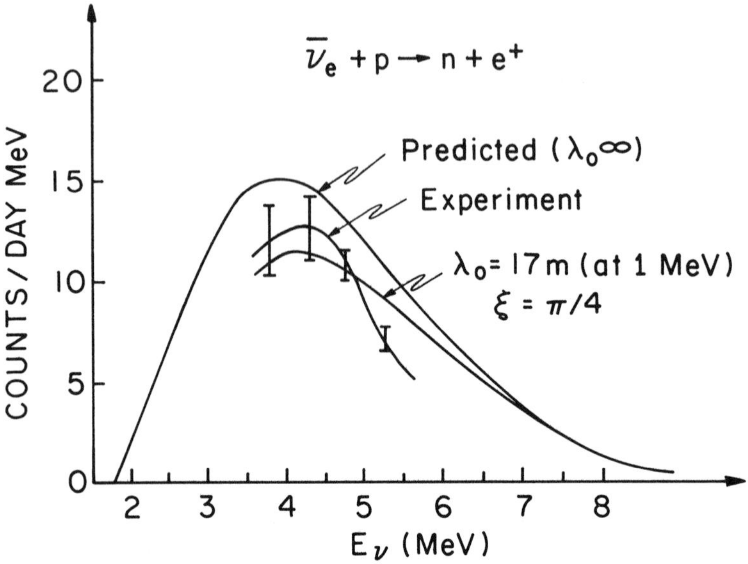

Curve 1 shows the rate predicted for our detector and source with λ_o ∞. The fission $\bar{\nu}_e$ spectrum used is that of Avignone(14).

Curve 2 shows our experimentally observed e^+ spectrum from reaction (1).

Curve 3 indicates the e^+ spectrum expected for λ_o = 17 meters (E_o = 1 MeV), two states ν_1 and ν_2 and a maximum mixing angle ($\xi = \pi/4$).

The observed and predicted spectral shapes differ, but not as predicted by ν oscillation theory. The agreement is fair in the range 3.75 → 4.75 MeV and if we disregard the observed spectrum > 4.75 MeV the results for maximum mixing are not inconsistent with $\lambda_o \geq 17$ meters i.e. $\Delta \leq \frac{p_o}{\lambda_o} \frac{10}{4} = \frac{1}{17} \times 2.5 = 0.15$ ev^2.

Analysis of additional data continues in the hope of understanding the difference between curves 1 and 3 in the range $E_\nu > 4.75$ MeV and a much improved experiment is in preparation to measure both the spectrum and the absolute rate as a function of the reactor-detector distance.

Limitations

The sensitivity of the method depends on mixing angle ξ, dropping by a factor of 10 as $\xi \to 10^o$. The limit on Δ increases from 0.15 to 0.30 ev^2 in the range $45 > \xi > 18^o$. Below $\xi = 10^o$ accelerators must be used e.g. search for ν_e in ν_μ beam.

References

(1) E. C. G. Stückelberg, Helv. Phys. Acta 11, 317 (1938).

(2) E. P. Wigner, Proc. Am. Philos. Soc. 93, 521 (1949); Proc. Natl. Acad. Sci. U.S.A. 38, 449 (1952).

(3) G. Feinberg and M. Goldhaber, Proc. Natl. Acad. Sci. U.S.A., 45, 1301 (1959).

(4) J. C. Pati and A. Salam. Phys. Rev. Lett. 31, 661 (1973); Phys. Rev. D8, 1240 (1973); Phys. Rev. D10 , 275, (1974).

(5) H. Georgi and S. L. Glashow, Phys. Rev. Lett. 32, 438 (1974); H. Georgi, H. R. Quinn, S. Weinberg, Phys. Rev. Lett. 33, 451 (1974).

(6) F. Reines and M. F. Crouch, Phys. Rev. Lett. 32, 493 (1974) and subsequent analysis (1977).

(7) F. E. Dix, Dissertation, Case Western Reserve University (1970).

(8) S. P. Rosen, Phys. Rev. Lett. 34, 774 (1975).

(9) J. C. Evans, Jr. and R. I. Steinberg, Science 197, 989 (1977); R. I. Steinberg and J. C. Evans, Jr. Proc. Int. Conf. on Neutrino Physics and Neutrino Astrophysics, Elbrus, U.S.S.R. (June 1977).

(10) E. L. Fireman, Proc. Int. Conf. On Neutrino Physics and Neutrino Astrophysics, Elbrus, U.S.S.R. (June 1977).

(11) B. Pontecorvo, Soviet Physics JETP, 26, 984 (1968); S. M. Bilenky and B. Pontecorvo, Int. Conf. on High Energy Physics, Tbilisi, U.S.S.R., (1976).

(12) H. Fritzsch and P. Minkowski, Phys. Lett. 62B, 72 (1976).

(13) A. K. Mann and H. Primakoff, Neutrino Oscillations and Neutrino Types, University of Pennsylvania preprint September (1976).

(14) F. T. Avignone III, Phys. Rev D2, 2609 (1970). An updated spectrum is in preparation by Dr. Avignone (private communication 1977). The $\bar{\nu}_e$ spectrum is inferred from measured β decay of fission fragments plus assumptions as to decay chains.

Parity Nonconservation in Atomic Processes

PARITY NON-CONSERVATION IN ATOMS*

G. Feinberg
Department of Physics, Columbia University

1. Introduction

The proposal of unified gauge theories of weak and electromagnetic interactions[1], followed by the recognition that these theories in some cases imply the existence of parity non-conserving interactions between leptons and hadrons[2] has stimulated both theoretical analyses and experimental searches for handedness in atoms.

By handedness is meant the non-invariance of a system under space reflection or mirror reflection, as typified by our left or right hands. Such handedness is well known for macroscopic systems. It was also discovered among molecules by Pasteur, in the 19th century. Handedness of molecules, such as tartaric acid, typically occurs because the molecule contains a tetrahedral arrangement in which a carbon atom is bound to four different complexes. (Fig.1)

This type of handedness is unrelated to parity non-conservation. Instead, it is a consequence of near degeneracy among the molecular energy levels, such that the parity eigenstates

$$\psi_1 \sim L+R$$

$$\psi_2 \sim L-R$$

where L and R are the handed configurations, are almost degenerate. As a result the states L and R are both

*This research was supported in part by the U.S. Dept. of Energy.

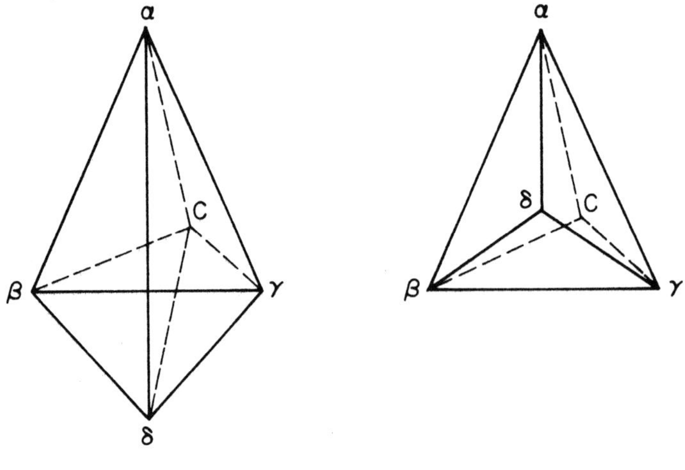

Figure 1.

Two forms of handedness of an organic molecule. C indicates carbon, α, β, γ, δ indicate four molecular groups.

quasistable states, with a very long tunneling time between them.

In atoms, where such degeneracies do not occur, we are instead interested in a case in which the Hamiltonian eigenstates themselves have some preferential handedness, and the reflected states are not Hamiltonian eigenstates at all. This can only occur if there are terms in the atomic Hamiltonian that do not conserve parity.

Such states can be characterized by a non-zero expectation value for some pseudoscalar quantity such as $h = \langle s.\hat{p} \rangle$ .The amount of handedness can be measured by this expectation value. We shall see that in typical cases it is expected to be 10^{-10} or less, indicating how small the effects that must be searched for are.

Even before the advent of unified gauge theories, there were a number of theoretical discussions of[3-8] and experimental searches for[9,10] handedness of atoms. These resulted in an upper limit of about 10^{-8} for the value of h.

2. Possible Experiments to Detect Atomic Handedness

Because the expected handedness of atoms is so small, it is essential to observe an effect linear in the handedness, which involves the measurement of a pseudoscalar quantity. Most proposed experiments, and the ones that have actualy been carried out involve the interaction of photons with the atom.

In this case, the pseudoscalar can be the helicity of the photon.

One possible experiment, proposed by Bouchiat and Bouchiat[11] involves the observation of resonance absorption of circularly polarized photons by a heavy atom such as cesium.The cross section is measured for left-handed photons ($s\cdot\hat{p}=-1$) and right-handed photons ($s\cdot\hat{p}=+1$) and compared. Any difference between the cross sections must be due either to a handedness of the atom, or to non-conservation of parity in the electromagnetic interaction, which is expected to be less important. Experiments along these lines are being carried out in cesium atoms by M.A. Bouchiat and coworkers[12] and by E. Commins and coworkers[13] in Thallium atoms.

An alternative way to get the same information, which is apparently more convenient experimentally, was proposed by several physicists independently[14,15]. It uses the relation between index of refraction and photon scattering amplitude

$$n^{\pm} = 1 + \frac{2\pi}{\omega^2} N f_{\pm}(\omega) \tag{1}$$

Here $N_{\pm}$ are the complex refractive indices for right-handed and left-handed photons in a medium of some type of atom. N is the atom number density, and $f_{\pm}(\omega)$ are the photon scattering amplitudes at energy ω. Near resonances we can write

$$f_{\pm} \simeq - \frac{\omega |V_{\pm}|^2}{\omega-\omega_0+\frac{i\Gamma}{2}} \tag{2}$$

where $V_{\pm}$ are the photon absorption matrix clements, ω_0 is the resonance energy, and Γ is the width of the line.

If n_+ and n_- are different, then linearly polarized light, which is a superposition of the two circular polarizations, will have its plane of polarization rotated when traversing the medium of these atoms. The angle of rotation is given by

$$\Delta\theta = \frac{k\ell \, \mathrm{Re}(n_+ - n_-)}{2} \qquad (3)$$

$$\sim \frac{N\ell(\omega-\omega_0)}{(\omega-\omega_0)^2+\frac{1}{4}\Gamma^2} \; \frac{2V\,\mathrm{Re}(V_+ - V_-)}{2} \qquad (4)$$

assuming that V_+ and V_- are nearly equal. In this formula ℓ is the path length traversed. Because of absorption, this path length is limited to a few times the absorption mean free path α, where

$$1/\alpha \sim 2k\,\mathrm{Im}\,n \sim \frac{2N|V|^2\Gamma/2}{(\omega-\omega_0)^2+\frac{1}{4}\Gamma^2} \qquad (5)$$

Therefore, the expected amount of rotation will be

$$\Delta\theta \sim \frac{\omega-\omega_0}{\Gamma/2} \; \frac{\mathrm{Re}(V_+ - V_-)}{2V} \qquad (6)$$

The "detuning factor" $\frac{\omega-\omega_0}{\Gamma/2}$ cannot be very large for experimental reasons. Therefore, the expected magnitude for $\Delta\theta$ will be

$$\Delta\theta \sim \frac{\mathrm{Re}(V_+ - V_-)}{2V}$$

and $\Delta\theta$ measures the same thing as the absorption experiments. However, $\Delta\theta$ seems somewhat easier to measure than does the absorption, and so the experiments to measure it in atoms have already given some results[16,17].

Other experiments to search for handedness of heavy atoms involve additional external electric fields. I shall not describe these experiments here as they are reported in other contributions to this conference [12,13].

3. Parity Non-Conservation in Unified Gauge Theories

An expected cause of handedness in atoms is the parity non-conserving interaction of neutral intermediate vector bosons (Z^o particles) with electrons and quarks that occurs in (some) unified gauge theories. Such interactions generate several types of terms in the effective Hamiltonian of atoms, as indicated in the diagrams (Figs.2,3). In addition to these effects, there may be other kinds of terms which modify the interaction of electrons with external photons. These terms, which are expected to be smaller,will be discussed later.

I will concentrate first on the parity violating electron-nucleus interaction, generated by Fig.1. The electron-electron parity violating interaction has been found to be less important in heavy atoms[18]. It is clearly irrelevant in hydrogen atoms. I return to this interaction later.

The electron-nucleon interaction can, under reasonable assumptions, be described with four parameters. The effective Hamiltonian can be written

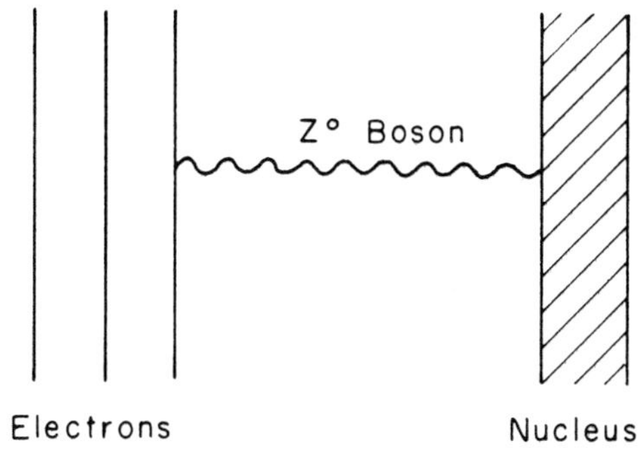

Figure 2.

A diagram that generates a parity violating electron-nucleus interaction.

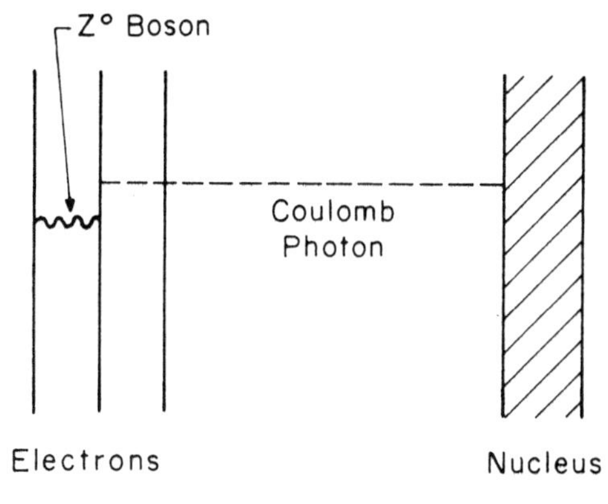

Figure 3.

A diagram that generates a parity violating electron-electron interaction.

$$H_{PV} = \frac{G_F}{\sqrt{2}} \sum_i \sum_J \Big[C_{1p}\, \delta^3(r_i - r_J)(\gamma_\alpha\gamma_5)_i^e (\gamma^\alpha)_J^p$$

$$+ C_{1N}\delta^3(r_i - r_J)(\gamma_\alpha\gamma_5)_i^e(\gamma^\alpha)_J^N$$

$$+ C_{2p}\, \delta^3(r_i - r_J)(\gamma_\alpha)_i^e(\gamma^\alpha\gamma_5)_J^p + C_{2N}\delta^3(r_i - r_J)(\gamma_\alpha)_i^e(\gamma^\alpha\gamma_5)_J^p \Big]$$

The sum i is carried out over the electrons in the atom, while the sum J is over the protons or neutrons in the nucleus. The constants C_{1p}, C_{1N}, C_{2p}, C_{2N}, are numbers that vary from one to another unified gauge theory[(19)]. In order to obtain H_{PV} it is necessary to neglect terms of relative order p/m_c and p/m_N, where p is a typical electron momentum in the atom. Such terms could arise through various types of radiative corrections to the diagrams of Figs. 2,3, involving weak, electromagnetic or strong interactions. There is reason to expect that the contribution of these terms is always of relative order α to the terms in H_{PV} They should therefore be treated along with other terms of this magnitude, and we defer their discussion until later.

When the nuclei are taken to be at rest, the first two terms, involving C_{1p}, C_{1N} are independent of the nuclear spin whereas the terms involving C_{2p}, C_{2N} are proportional to the nuclear spin. In this limit, and neglecting the vari- tion of the electron wave function over the nucleus, which is also expected to be a good approximation, the nuclear factors in the spin-independent terms reduce to just Z and N, the proton and neutron numbers. The spin dependent terms, on the other hand, reduce to the average spins of protons and neutrons in the nucleus. These latter quantities are much smaller than the neutron and proton numbers. The electron factors which multiply the two kinds of terms, $(\gamma_5)^e$ and $(\gamma_5\sigma_i)^e$ respectively are similar in magnitude. Therefore, if C_1 and C_2 are similar in magnitude, the C_1 terms should give much larger contributions, by a factor of 50 or so, to the handedness of heavy atoms than the C_2 terms, even if the nuclear spins have definite values, as in some hyperfine transitions. On the other hand, in light atoms such as H or D, the magnitudes of the two terms are comparable, and the observation of specific hyperfine transitions should allow for the observation of both terms[20].

In the standard model[1], in which the gauged symmetry is SU(2) x U(1), Fermions are in left-handed doublets and right-handed singlets, and a single Higgs doublet assumed, the parameters C_1, C_2 are

$$\begin{aligned}
C_{1p} &= \tfrac{1}{2}(1-4\sin^2\theta_w) + O(\alpha) \\
C_{1N} &= -\tfrac{1}{2} + O(\alpha) \\
C_{2p} &= \tfrac{1}{2}(1-4\sin^2\theta_w)\times 1.2 + O(\alpha) \\
C_{2N} &= -\tfrac{1}{2}(1-4\sin^2\theta_w)\times 1.2 + O(\alpha)
\end{aligned} \tag{8}$$

The expressions for C_{2p}, C_{2N} can be obtained by relating the neutral weak hadron axial current to the charged weak hadron axial current by isospin rotation, and then using the observed value of $\frac{G_A}{G_V}$. The terms of order α come from weak and electromagnetic radiative corrections, and have not yet been calculated explicitly, to my knowledge.

If the value of $\sin^2\theta_W \simeq .25$, suggested by recent neutrino experiments[21] is used, then we obtain

$$C_{1p} \simeq 0 \qquad C_{1N} \simeq -.5 \qquad (\sin^2\theta_W \simeq .25)$$
$$C_{2p} \simeq 0 \qquad C_{2N} \simeq 0 \tag{9}$$

which will obviously have important implications for experiments in H.

Values of the C_i for other gauge theories have been extensively discussed in the literature[19]. The result for some theories is given in Table I.

4. Calculations of Parity Mixing in Heavy Atoms.

Because experiments to look for handedness of heavy atoms are apparently easier to do, these are the experiments that have been done first[16,17]. However, the interpretation of the results are substantially more difficult. A heavy atom consists of Z electrons and a nucleus, all interacting through electrostatic forces. Only approximate results are available for the electron wave functions. One who is not an atomic theorist certainly may doubt whether the approximate methods available to do atomic calculations are reliable enough that one can make an accurate prediction of the expected results of an experiment, even assuming a specific set of values for the C_i. I describe here some of the calculations that have been done, and will indicate what I can about their expected level of accuracy.

The effective Hamiltonian H_{PV} reduces to:

$$H_{PV} \rightarrow \frac{G_F}{\sqrt{2}} \sum_i \sum_J \left\{ \left[C_{1p} \delta^3(r_i - r_J) + C_{1N} \delta^3(r_i - r_J) \right] (\gamma_5)_i^e - \left[C_{2p} \delta^3(r_i - r_J) \vec{\sigma}_J^{\,p} + C_{2N} \delta^3(r_i - r_J) \sigma_J^N \right] \cdot \vec{\alpha}_i^{\,e} \right\} \tag{10}$$

This is obviously a sum of single electron operators. This means that if we can write the atomic many-electron wave function as a sum of products of single electron wave functions, only a single electron at a time will be linked by H_{PV}. This approach is used in all of the calculations done so far.

The single electron matrix element can only link states whose wave function at the nucleus is non-zero, because of the presence of the delta functions. There must be $S_{1/2}$ and $P_{1/2}$ states. We use the notation of Bethe and Salpeter[22] in which

$$|S_{1/2}\rangle \sim \begin{pmatrix} g_s(r)u \\ -if_s(r)\sigma\cdot\hat{r}u \end{pmatrix}$$

$$|P_{1/2}\rangle \sim \begin{pmatrix} g_p(r)\sigma\cdot\hat{r}u \\ -if_p(r)u \end{pmatrix} \tag{11}$$

in which u is a normalized two-component spinor.

Thus we get $\langle S_{1/2}|H_{PV}|P_{1/2}\rangle =$

$$\frac{-G_F}{\sqrt{2}}\frac{1}{4\pi}\left(g_sf_p-g_pf_s\right)_{nuc}\left[C_{1p}Z + C_{1N}(A-Z) - \sigma^{(e)}\cdot I(C_{2p}K^P + C_{2N}K^N)\right] \tag{12}$$

Here the subscript nuc indicates that the electron wave functions are to be evaluated at the nucleus. The precise method of this evaluation appears to be unimportant, in that one can do the evaluation at the nuclear surface, or average of the nucleus in some way.

The quantities K^P, K^N measure the average spin of the proton and neutron in the nucleus, i.e.[23]

$$K^p \equiv \frac{\langle \Sigma_J \sigma^p_{3,J}\rangle}{I_3}$$

$$K^N \equiv \frac{\langle \Sigma_J \sigma^N_{3,J}\rangle}{I_3} \tag{13}$$

where $\vec{I}$ is the nuclear spin vector.

It can be seen that if C_1 and C_2 are of similar magnitude, than in a heavy atom, the terms with C_2 are much smaller than those with C_1, and we neglect them for the moment. Also, to a good approximation $(g_p)_{Nuc} = 0$, and so we neglect the terms with g_p as well. It is not a good approximation in heavy atoms to replace f_p, g_S by their non-relativistic limits, as the Dirac wave functions are much greater near the origin than their non-relativistic limits, and an enhancement factor of 5-10 arises from this source in heavy atoms.

Bouchiat and Bouchiat[11] have calculated these matrix elements using a modification of the Fermi-Segre theory of hyperfine structure. They directly calculate the electron wave function near the origin, by taking the nuclear potential as dominant there. This wave function is then joined to an expression for the wave function at large r, given in terms of the quantum defect ν_n of the state, defined by the equation

$$E_n = \frac{-Ry}{(n-\nu_n)^2} \tag{14}$$

They obtain for the matrix element (12)

$$\langle nS|H_{PV}|n'p\rangle \simeq \frac{iGm_e^2}{4\pi\sqrt{2}} \frac{\alpha^2(Z\alpha)^2 m_e\, Q_W(Z,A)k_r(Z)}{(\nu_n\nu_n')^{3/2}} \tag{15}$$

Here $Q_W(Z,A) = 2[C_{1p}Z + C_{1N}(A-Z)]$.

$k_r(Z)$ is a relativistic correction factor which comes from using the Dirac wave functions in (12). It is approximately 3 for $Z \sim 50$, and 9 for $Z \sim 80$. In (15) I have neglected the terms in C_2 .

Providing that C_{1p} and C_{1N} do not vanish, this matrix element can be seen to go as $Z^3 k_r(Z)$. This is one indication of the advantage of experiments in heavy atoms. If the nucleus were unshielded by the electrons, the matrix element would instead go as $Z^5 k_r(Z)$.

The actual mixing parameter is

$$\delta_{ns,n'p} \equiv \frac{\langle nS|H_{PV}|n'p\rangle}{E_{ns}-E_{n'p}} \tag{16}$$

For low lying state of valence electrons $E_{ns}-E_{n'p} \sim \frac{\alpha^2 m_e}{2} \sim Ry$

Therefore

$$\delta_{nS,n'p} \sim \frac{iGm_e^2}{\pi 4} (Z\alpha)^2 Q_w k_r(Z) \tag{17}$$

Since $\nu_n, \nu_{n'}$ are typically numbers of order 1. This gives $|\delta_{s,p}| \approx 10^{-10}$ for $Z = 80$ and the standard model for the C_i. This value is about 10 times bigger than the $2S_{1/2}-2P_{1/2}$ mixing in neutral hydrogen, to be discussed below. The large increase in the parity violating matrix element is almost entirely compensated by the much smaller energy difference in hydrogen ($\sim \alpha^5 m_e$).

In both of the types of experiment done in heavy atoms it is necessary to calculate an induced electric dipole matrix element $\mathcal{M}$(E1) given by

$$\mathcal{M}(E1) = \sum_n \frac{\langle i|H_{PV}|n\rangle\langle n|E1|f\rangle}{E_i-E_n} + \sum_n \frac{\langle i|E1|n\rangle\langle n|H_{PV}|f\rangle}{E_i-E_n} \tag{18}$$

If the states $|i\rangle, |f\rangle$ are states of the same parity and angular momentum, a photon transition between them can go as E1 only through a matrix element such as (18) in which parity mixing occurs explicitly. However, the transition can go as an M1 transition without parity mixing. It is the interference between these two channels that is sensitive to photon handedness. By comparing with our previous expression from δ, Eq.(16), it can be seen that the quantity $\frac{V_+ - V_-}{V}$ occurring in Equation (6) will be given by

$$\frac{\Delta V}{V} \sim \sum_n \delta_n x \frac{(E1)_n}{(M1)} \qquad (19)$$

where $(E1)_n$ is the allowed electric dipole matrix element from initial state to the opposite parity state $|n\rangle$, and (M1) is the allowed (M1) matrix element between initial and final states.

$$(E1) \sim \left\langle er_J \right\rangle$$

$$(M1) \sim \left\langle \frac{e}{m_e}(\sigma_J + \frac{1}{2} L_J) \right\rangle$$

The ratio (E1)/(M1) is typically quite large.

(E1)/(M1) ~ 100 if there is no change in principal quantum number

(E1)/(M1) ~ 10^6 if there is a change in principal

quantum number.

The reason for the latter result is that if we neglect relativity, and retardation, the magnetic dipole operator commutes with the atomic Hamiltonian, and so cannot induce transitions between orthogonal eigenstates.

These factors from (E1)/(M1) serve to enhance δ significantly, and to turn an impossible search into a difficult, but feasible one, in which the term depending on photon handedness is

$\sim 10^{-8}$ for an "Allowed M1 transition"

$\sim 10^{-4}$ for a "Forbidden M1 transition"

The experiments of Fortson, et al[16], and of Sandars[17] are of the first type while those of Bouchiat et al[12], and of Cummins et al[13] are of the second type.

It can be seen from Eq.(18) that to calculate $\mathfrak{M}$(E1) an infinite set of terms must (in principle) be summed. Two distinct approaches have been used to do this calculation.

1. This method has been used by Bouchiat and Bouchiat[11,24] and by Novikov, Sushkov and Kriplovich[25].

The matrix elements of H_{PV} between individual $S_{\frac{1}{2}}$ and $P_{\frac{1}{2}}$ states are calculated by the "quantum defect" method described above. The analysis is complicated to describe, especially in the relativistic case. The authors maintain that the matrix elements of H_{PV}, calculated in this way for valence electrons, in atoms such as Cs or Tl agree to about 10% accuracy with the results of Hartree-Fock calculations of the same quantity.

Assuming this to be true, it is then necessary to know corresponding matrix elements of the El operator r_i between similar states. These matrix elements can either be calculated [24], using numerical wave functions, or, in many cases taken from experiment [25]. Using these inputs, a (small) number of the intermediate states contributing to M(El), can be summed, mostly the discrete excited states of the valence electrons. This has been done by both groups for thallium atoms, and by N-S-K for bismuth and lead. The latter cases are more complex because there are several equivalent electrons, whose effects must be disentangled. The results of the calculations, where they overlap, agree to about 10-20%, for the terms considered thus far.

It is necessary to estimate corrections to these values. Such corrections can come from higher excited states of the valence electrons, from excited states of the core electrons, from corrections to the approximate wave functions, etc. It appears difficult to estimate such corrections systematically in this approach, but both groups have attempted to do so. In Cs, Bouchiat and Bouchiat[24] find a correction of only a few percent from excitation of higher states and of core states, and estimate a correction of perhaps 20% from these sources in Tl. On the other hand N-S-K[25] find quite a large correction, of order 70% from excitation of core states in Tl, and their final estimates (Table II) include this effect. It is hard for me to decide how accurate the estimates made by this method really are.

2. The second approach, also mentioned by B. and B. [24], has been applied in detail to Cs, Tl and Bi by Henley and coworkers[26,27] and by Sandars and coworkers[28,29]. In this method, the electrostatic interaction between electrons is replaced by an approximate one body central field, and the difference between this field and the true interaction is treated as a perturbation. This can be done either by the Hartree-Fock method[26,27,30], or by directly choosing the parameters of the central potential to fit the energy levels of the atom[28,29]. In either approach, the main advantage is to reduce the calculation of $\mathcal{M}(E1)$ to a sum of single electron excitations, which can be calculated using the solutions to the one body Dirac equation.

Furthermore, in order to avoid calculating the infinite sum over states in M(E1), a trick originally due to Sternheimer and others is used. The quantity $\mathcal{M}(E1)$ is re-written

$$\mathcal{M}(E1) = \langle i|H_{PV}|\psi_f\rangle + \langle\psi_i|H_{PV}|f\rangle \tag{20}$$

where $$|\psi_i\rangle = \sum_n \frac{|n\rangle \langle n|E1|i\rangle}{E_n - E_i} \tag{21}$$

and similarly for $|\psi_f\rangle$. Here E_i, E_f, E_n are the eigenvalues of the central field Dirac Hamiltonian. These states $|\psi_i\rangle$, $|\psi_f\rangle$ satisfy

$$(H_0 - E_i)\,|\psi_i\rangle = (E1)\,|i\rangle \tag{22}$$

etc.

This is a single particle, inhomogeneous equation, which can be solved numerically. The results can then be substituted into (20), and $\mathcal{M}$(E1) calculated by quadratures. In their approach, the contributions of all excited and continuum states are automatically included, within the central field approximation. Again, in the case of atoms with several valence electrons, such as Bi the procedure is actually somewhat more complicated.

The results of different calculations, indicated in Table **II**, agree fairly well with each other, and with those of the other method. However, one cannot accept them on those grounds alone. Especially in the atoms with several valence electrons, it is necessary to investigate systematically the sources of correction to these answers, arising from the neglected residual interaction ΔV

$$\Delta V = V_{central} - \frac{1}{2} \sum_{i,J} \frac{e^2}{r_{iJ}} \qquad (23)$$

There are at least two important potential sources of correction to the results obtained by this method.

1. Configuration mixing among states of the valence electron, either ground or excited, due to the residual interaction[31]

2. Higher order effects of the residual interaction involving excitations of the core states[29].

I consider these effects within the framework of method 2, although they also play some role in method 1.

The effects of configuration mixing in Bi which were suggested as possibly being large[31] have been treated in a recent paper by Henley et al[27]. The ground state of Bi, is usually described as $(6s)^2\ (6p)^3$ and the parity mixed states as $(6s)^2\ (6p)^2\ 7s$, etc. The configuration mixing involves replacing one or more of these electrons by another orbital, with the same parity, such that the overall angular momentum is the same. For example, the initial state might include the configuration $(6s\ 6d)(6p)^3$, etc. H-K-W find that the contribution of such mixed configurations to $\mathcal{M}(E1)$ is B_i are quite small, ~4%. The correction due to configuration mixing of this type can be written as a sum of terms of the form

$$\frac{V_{nn'}}{E_i-E_n}\ \frac{\langle f|E1|n\rangle}{\langle f|E1|n'\rangle} \tag{24}$$

where $V_{nn'}$ is the matrix element of the configuration mixing interaction between the states n,n' that are being configuration mixed, while E_i-E_n is the energy difference between initial state and intermediate state. (This is the term that occurs from configuration mixing in the intermediate state.) In Bi the ratio $V_{nn'}/E_i-E_n$ is quite small, accounting for the small effect.

The residual interaction also has matrix elements that are not small compared to energy splittings between some states of the central potential. As a result, there are possibly significant higher order contributions to $\mathcal{M}(E1)$ proportional to one or more powers of ΔV. For example, there are 3rd order perturbation terms such as:

$$\sum_{n,m} \frac{\langle i|H_{PV}|n\rangle \langle n|\Delta V|m\rangle \langle m|E1|f\rangle}{(E_i - E_n)(E_i - E_m)} \qquad (25)$$

as well as terms in which the order of the operators H_{PV}, ΔV, E1 are permuted. The states $|n\rangle$, $|m\rangle$ can involve excitation and deexcitation of the core electrons. There are also 4th and higher order terms involving 2 or more factors of ΔV.

Calculations of terms such as (25) have recently been carried out by Loving and Sandars[29] and are reported by Sandars at this conference. They find that the terms of the form (25) give a large contribution to $\mathcal{M}$(E1) of opposite sign to those of (18), which have the effect of more than cancelling (18). However, based on a physical identification of these new terms as due to shielding of the external photon by the atomic electrons, they go on to calculate a selected set of terms of order $(\Delta V)^n$, and find that these terms restore part of the value of (18), leading to the value quoted in Table II, under L-S.

The fact that a subset of the higher order graphs gives an important contribution to $\mathcal{M}$(E1) inevitably raises the question of whether other terms, not yet calculated, may also make large contributions to this quantity. I believe that a <u>necessary</u> condition for believing the accuracy of a calculation is that there is some indication that the terms omitted from the calculation should be small compared to the terms included. This does not seem to have been shown for the published calculations, although the new results of Ref. 29 improve this situation somewhat.

To my mind, the discovery of these new large effects makes the interpretation of the Bi optical rotation experiments somewhat less clear. It will take some sorting out of the calculations to be sure whether there is a discrepancy between the experiments and the standard model. Both better (i.e., further) calculations, as well as the detection of an effect, or much lower experimental limits would help considerably here. Also, experiments in Tl and in Cs, where there are fewer "important" electrons may be easier to interpret theoretically when they give results, because the higher order effects may be smaller.

5. Experiments in H and D.

In light atoms such as H or D, the atomic physics calculations are simple so that if the handedness can be detected, its interpretation is straightforward. The parity mixing is greatest between $nS_{1/2}$ and $nP_{1/2}$ states, because of the small energy difference bewteen these states. For example, between $2S_{1/2}$ and $2P_{1/2}$, one gets[20,23,32,33]

$$\delta_{2S_{1/2},2P_{1/2}} = \frac{-iGm_e^2}{\sqrt{2}} \frac{1}{4\pi\alpha}(Z\alpha)^5\frac{\sqrt{3}}{8}m_e \\ \cdot\frac{\left[C_{1p}Z + C_{1N}(A-Z) - \sigma_e\cdot I(C_{2p}K_p + C_{2N}K_N)\right]}{E_{2S_{1/2}} - E_{2P_{1/2}} + \frac{i}{2}\Gamma_{2P_{1/2}}} \tag{26}$$

In this formula $K_P = \frac{<\sigma_3{}^P>}{<I_3>}$ in the nuclear state, and similarly for $K_N, K_P = 2$, $K_P = 0$ for H, while $K_P = K_1 = 1$ for D. In (26) I have neglected the relativistic correction factor, which is unimportant in H and D. However, I have included a term in the energy dominator involving the width of the $2P_{\frac{1}{2}}$ state, which is essential in discussing some of the experiments.

When there is no external magentic field, $E_{2S_{\frac{1}{2}}} - E_{2P_{\frac{1}{2}}}$ is the Lambshift

$$\simeq \frac{\alpha}{6\pi} (Z\alpha)^4 m_e F(z)$$

where F(z) is around 7.7, and decreases slowly with Z.

Furthermore $\Gamma_{2P_{\frac{1}{2}}} \simeq \frac{1}{10} (E_{2S_{\frac{1}{2}}} - E_{2P_{\frac{1}{2}}})$.
The corresponding zero field value of $\delta_{2S,2P}$ is in H

$$\delta_{2S_{\frac{1}{2}},2P_{\frac{1}{2}}} \simeq \frac{i\, 10^{-11} \{C_{1p} - 2C_{2p}(F(F+1) - 3/2)\}}{1 + \frac{i}{20}} \tag{27}$$

This is obviously quite small, and some methods of enhancing it is called for. In order to do this, one possibility is to use external magnetic fields to decrease the energy difference. The Breit-Rabi diagram for hydrogen (Fig.4) indicates a variety of level crossings between $2S_{\frac{1}{2}}$ and $2P_{\frac{1}{2}}$ in various hyperfine states, for external magnetic fields of around 10^3 gauss. At such level crossings, the quantity δ increases in magnitude by a factor of 20 or so, and also changes in phase from imaginary to real.

A further interesting feature of working at level

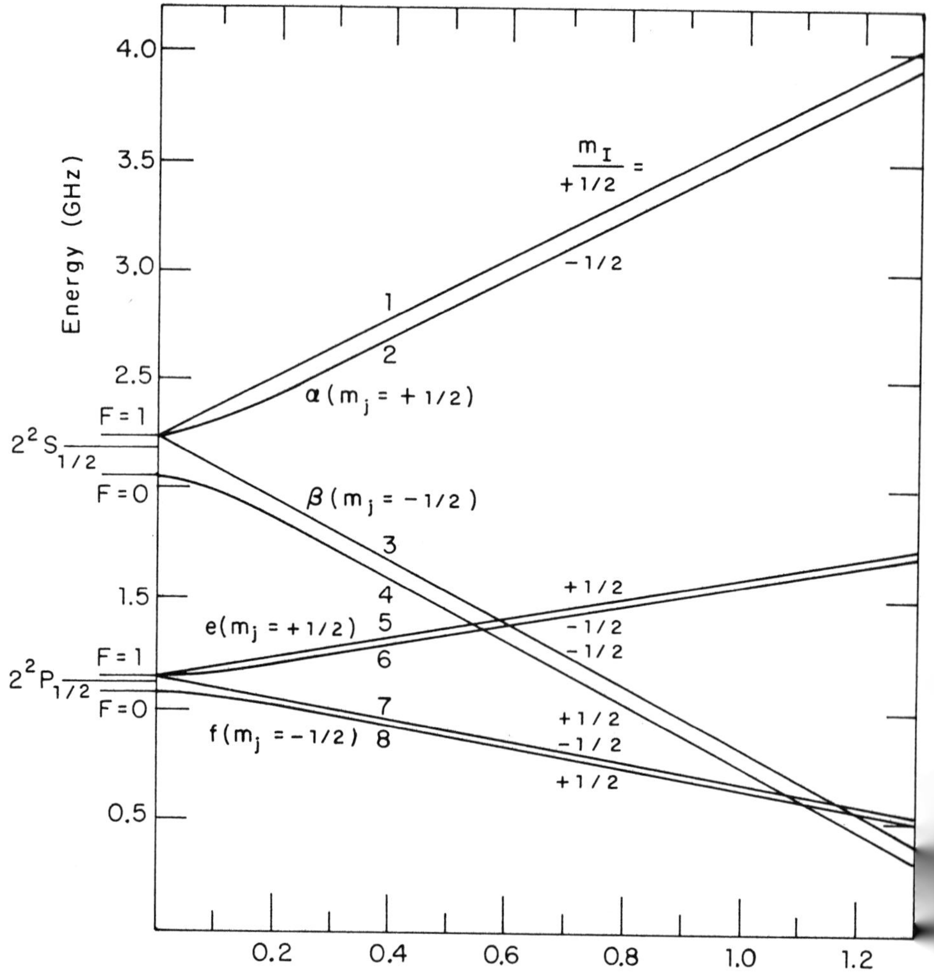

Figure 4 Breit-Rabi Diagram for Hydrogen

The Breit-Rabi diagram for hydrogen, with n = 2.

crossings is that the mixing at different level crossings involves different linear combinations of C_1 and C_2. For example, at an external magnetic field of about 550 gauss, there is a crossing in hydrogen between the $2S_{\frac{1}{2}}$ F=0 state and the $2P_{\frac{1}{2}}$, F=1, M_F=0 state[33]. At this crossing, only the term with C_2 contributes, giving a mixing of

$$\delta \sim 5\ C_{2p} \times 10^{-10} \qquad (28)$$

Similarly, at 1175 gauss, there is another crossing of the $2S_{\frac{1}{2}}$, F=1, M_F=-1 with the $2P_{\frac{1}{2}}$,F=1, M_F=-1 at which

$$\delta \simeq 2(C_{1p}-C_{2p})\times 10^{-10} \qquad (29)$$

Similar effects occur, at different external fields, in deuterium.

If both of these mixing parameters can be observed, it would be possible to infer both C_{1p}, C_{2p}. Similar experiments in deuterium would give information about $C_{1p}+C_{1N}$, and modulo some nuclear physics ambiguities of the spin matrix elements, also give information about $C_{2p}+C_{2N}$.

Experiments along these lines are in progress at Michigan[20,33], at Yale[34], and at University of Washington[35]. The experiments are fairly similar, and in each case use a number of external electromagnetic fields to define a handedness that can interact with the handedness of the hydrogen atom.

In the Michigan experiment, hydrogen atoms in the $2S_{\frac{1}{2}}$ state are passed through external static electric and magnetic fields, as well as an external electric microwave frequency field. The static magnetic field is chosen to give the 550 gauss level crossing mentioned above. The microwave electric field can induce transitions between the $2S_{\frac{1}{2}}$ atoms and that $2P_{\frac{1}{2}}$ hyperfine component which H_{PV} mixes with another $2S_{\frac{1}{2}}$ state. Alternatively, a $2S_{\frac{1}{2}} \rightarrow 2S_{\frac{1}{2}}$ transition can be induced through the microwave field in combination parity mixing by the external static electric field of a $2P_{\frac{1}{2}}$ component with a different m_F. The latter mixing does not of course involve any parity violation. The amplitudes for these transitions can be written symbolically as

$$M_{E.F.} \sim E_s \cdot \varepsilon_m$$

$$M_{P.V.} \sim \delta \, B_s \cdot \varepsilon_m \qquad (30)$$

Note that $M_{P.V.}$ appears odd under T as well as P, which is necessary because of the extra factor of i in δ at the level crossing.

The expected ratio of $M_{P.V.}$ to $M_{E.F.}$ is

$$\sim 2.5 \, C_{2p} \times 10^{-6} \, \frac{E_s}{1v/cm} \qquad (31)$$

The static field cannot be made too large because of quenching effects. It is possible to enchance the signal to noise ratio by varying the signs of $E_s \cdot \varepsilon_m$ or $B_s \cdot \varepsilon_m$ through field reversals, each of which changes the sign of $M_{E.F.}$ relative to $M_{P.V.}$ It can be seen from (31) that a measurement of the transition rate to 1 part in 10^6 is necessary to detect the P violating effect. The experimenters believe that such a sensitivity is feasible. Further details are given in Ref.(33).

The experiment at the University of Washington is similar to the one just described, while that at Yale, which is done at the 1099 gauss level crossing, is somewhat different in detail[36] but similar in sensitivity to $C_{1p} + 1.1\ C_{2p}$, instead of C_{2p}. In each case, similar experiments can be done in deuterium.

These light atom experiments appear to be more difficult than the heavy atom experiments, but have the advantage of much simpler atomic physics, as well as greater short term capability for extracting them for constants C_i. We eagerly await the results!

C. Other forms of parity violation in atoms.

a. Effect of the C_2 terms in heavy atoms.

The nuclear spin dependent terms in H_{PV} have been analyzed for some heavy atoms by Novikov, et al.[37], who find surprisingly large effects. For the $6P_{3/2} \rightarrow 6P'_{3/2}$ transition in Bi, between the F=6 and F=7 hyperfine levels, they find a contribution from the C_2 term of order $+3 \times 10^{-9}$ for $\Delta\theta$, compared to -2×10^{-7} for the C_1 terms. Since the present experimental limits in B_i are already $\sim 3 \times 10^{-8}$, an order of magnitude improvement might be enough to show the effects of the C_2 terms. In particular, a model in which $C_1=0$ because of the existence of right handed currents involving electrons and heavy leptons would give $C_2 \neq 0$, and so might show up in this way. In order for this measurement to be feasible, it is necessary to isolate an individual hyperfine component of the transition, which is already the case for the Oxford experiment, and will soon be the case for the University of Washington experiment.

b. Parity non-conserving electron-electron interaction.

This has the form for the interaction between two electrons,

$$H_{PV}^{ee} = \frac{G_F}{\sqrt{2}} C_{ee}^{VA} \left[(\gamma_\alpha)_1^e (\gamma^\alpha \gamma^5)_2 + (\gamma^\alpha \gamma^5)_1^e (\gamma_\alpha)_2 \right] \delta^3 (r_{12}) \qquad (30)$$

where $C_{ee}^{VA} = -\frac{1}{2} + 2\sin^2\theta_W$ in the standard model, and $C_{ee}^{VA} \simeq 0$ for $\sin^2\theta_W = \frac{1}{4}$. The effect of this interaction in Cs has been considered in Ref.(11), where it is found that the contribution to $\delta_{s,p}$ is at most a few percent of the contribution of the electron-nucleus interaction. It would be of interest to make a more precise estimate of this effect, also in Bi, especially in view of the indication that the e-nucleus interaction may be very much smaller than expected in Bi.

If there are measurable parity non-conserving effects in helium, or heliumlike ions, they might be a good place to look for H_{PV}^{ee} .

c. Vertex and self energy corrections

There are graphs (Fig.5) giving P violating effects

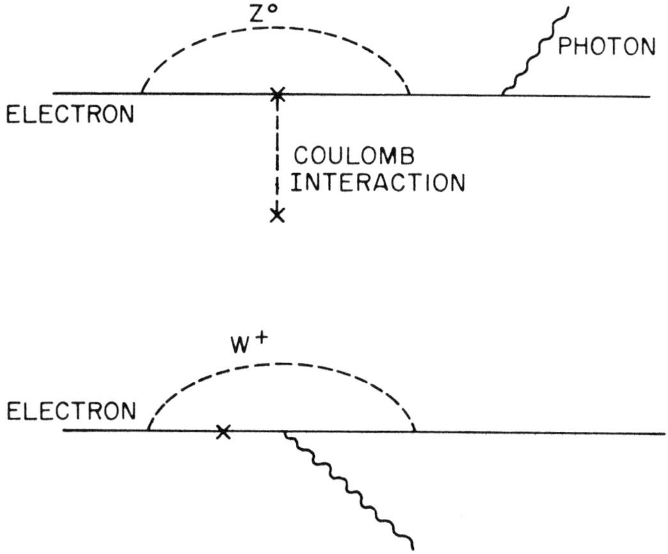

Figure 5.

Some diagrams that modify the electromagnetic interactions of electrons.

in which the virtual Z boson or W boson is reabsorbed by the object emitting it, nucleus or electron, rather than by another particle. These graphs have not been systematically calculated. They involve things such as P violating terms in the effective electron-photon interaction, and in the nucleus-photon interaction. The ones that I have looked at appear to give effects of relative order α compared to the spin-independent terms in H_{PV}. However, if these leading terms are about, it is possible that the effects mentioned here may become dominant, and they should be calculated.

d. Higher order corrections.

If the leading terms are not present, or if experiments eventually can be done on the one present level of accuracy, it will be necessary to calculate various radiative corrections that can give P violation. These include a number of graphs of the type in Fig.6, including exchange of two intermediate vector bosons, of an intermediate vector boson and a photon, etc. This is especially important in theories for which there is a symmetry that insures parity conservation in leading order. It would be very helpful to know in what circumstances the radiative corrections are of order $G_F \alpha \ln \frac{m_Z}{m_e}$, of order $G_F^2 \Delta_m^2$, etc. To my knowldege, there are as of yet no systematic answers to these questions.

If the experiments in Bi, and in other elements persist in giving their "low" values, then all concerned,

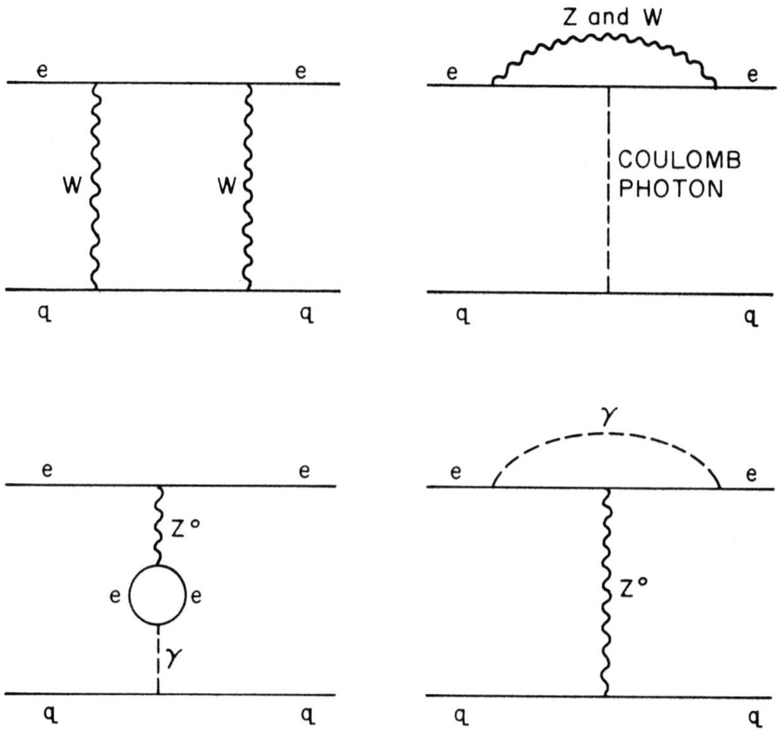

Figure 6.

Some diagrams that give higher order corrections to H_{PV}.

gauge theorists, atomic theorists, and experimenters will have our work cut out for us in the coming years. It will be exciting to see what happens.

Acknowledgement

The writing of this summary has been greatly assisted by conversations that I had with Dr. H.M. Foley. I would also like to thank Drs. E.M. Henley and P. Sandars for helpful discussions.

References

1. S. Weinberg, Phys. Rev. Lett.19, 1264, 1967
2. C.C. and M.A. Bouchiat, Phys. Lett. 48B, 111, 1974
3. G. Feinberg, Phys. Rev. 112, 1637, 1958
4. E.E. Salpeter, Phys. Rev. 112, 1642, 1958
5. Ya. B. Zeldovich, JETP 9, 682 (1959)
6. F.C. Michel, Phys. Rev. 138B, 408, 1965
7. R.A. Carhart, Phys. Rev. 132, 2337, 1953
8. B. Sakitt and G. Feinberg, Phys. Rev. 151, 1341, 1966
9. M. Lipeles, et al., Phys. Rev. Lett.15, 690, 1965
10. L.C. Bradley and N.S. Wall, Nuovo Cimento 25, 48,1962
11. M.A. and C. Bouchiat, Jour. de Phys. 35, 899, 1974
12. M.A. Bouchiat, report at this conference
13. E. Commins, report at this conference
14. I.B. Khriplovich, JETP Lett.20, 315. 1974
15. P.G.H. Sandars in Atomic Physics IV and, ed. Zu Putlitz, et al.Plenum Press, N.Y. 1975,also E.N. Fortson, Bull. Am.Phys.Soc.20,491, 1975
16. L.L. Lewis, et al., Phys. Rev.Lett.39, 795,1977
17. P.E.G. Baird, et al., Phys. Rev. Lett.39, 798, 1977
18. See Ref.11, Appendix III.
19. See Table I, taken from a paper by R.N. Cahn and G.L. Kane, Phys.Lett., to be published
20. R.R. Lewis and W.L. Williams, Phys. Lett.59B, 70,1975
21. M. Holder, et al., CERN preprint (1977)

22. H.A. Bethe and E.E. Salpeter, Handbook de Physik, ed. S. Flugge, Vol.35, P.88, Springer, Berlin,1957.

23. See G. Feinberg and M.Y. Chen, Phys. Rev.D10,3789,1974.

24. M.A. Bouchiat and C. Bouchiat, Jour.de Physique,36, 493, 1975. This paper is denoted B. & B .

25. V. Novikov, et al., Soviet Physics, JETP 44, 872, 1976, This paper is denoted N-S-K.

26. E.M. Henley and L. Wilets, Physics. Rev.A.14,1411,1976.

27. E.M. Henley, M. Klapisch and L. Wilets, Phys.Rev.Lett. 39, 994, 1977. This paper is denoted H-K-W.

28. M.W.S.M. Brimicombe, et al., J. Phys. B.9, L237, 1976. This paper is denoted B-L-S.

29. C.E. Loving and P.G.H. Sandars, preprint. This preprint is denoted L-S.

30. See E.g., C.F. Fischer, the Hartree-Fock Method for Atoms, J. Wiley and Sons, New York, 1977.

31. R.D. Cowan, S. Meshkov, and S.P. Rosen, preprint.

32. Ya Azimov, et al.JETP 40, 8, 1975.

33. R.W. Dunford, et al., Parity Non-conservation in the Hydrogen Atom, II, III, preprints.

34. E.A. Hinds, and V.W. Hughes, Phys. Lett.67B, 487,1977.

35. E.N. Fortson, private communication.

36. V.W. Hughes, private communication.

37. V.N. Novikov, et al., Novosibirsk, preprint, 1977.

Table I.

	C_{1p}	C_{1n}	C_{2p}	C_{2n}
WSGIM	-0.17	-0.5	-0.21	+0.21
b	+0.83	0	-0.07	+0.14
t	-0.67	-1.5	-0.14	+0.07
b,t	+0.33	-1	0	0
E^o	0	0	+0.42	-0.42
E^o,b	0	0	0.13	-0.28
E^o,t	0	0	+0.28	-0.13
E^o,b,t	0	0	0	0
Mohapatra-Sidhu SU(2)xSU(2)xU(1)	0	0	0	0
DGG SU(2)xSU(2)xU(1)	-0.08	-0.25	-0.10	+0.10
Lee-Weinberg SU(3)xU(1)	0	0	0.22	-0.10

The values of C_i in various gauge theories. The notation is defined by Kahn & Cane, Ref.19.

Table II

Comparison of Calculations and Experiments in Heavy Atoms

Element (transition)	Experiment[16,17]	B-L-S.[28]	L-S[29]	H-K-W[27]	N-S-K[25]
Bi(8757)	-.07±.32	-2.3	-1.2	-2.3	-1.6
Bi(6477)	+.27±47	-3	-1.6		-1.7
Tl					

Numbers are 10^7 times Im $\frac{\mathcal{M}(E1)}{\mathcal{M}(M1)}$

Most of these calculations are done for $\sin^2 \theta_W \approx .35$. Use of the value $\sin^2 \theta_W \approx .25$ decreases the numbers by a factor of about .8.

THE SEARCH FOR PARITY NON-CONSERVING OPTICAL ROTATION IN ATOMIC BISMUTH

P. G. H. Sandars
Clarendon Laboratory, Oxford University

We report here on the present status of the "bismuth problem": the apparent disagreement between the results of the bismuth optical rotation experiments at Oxford[1] and Washington[2] and the predictions of the Weinberg-Salam model plus relativistic central field atomic theory[3,4,5].

1. PNC optical rotation

According to straightforward theory one expects the PNC optical rotation close to an allowed M1 transition to be given by

$$\phi^{PNC} = -4\pi \frac{(n-1)}{\lambda} L\ R \qquad (1)$$

where n is the refracture index, L is the path length in the optical medium and λ is the wavelength of the light. R = Imag. $E1^{PNC}/_{M1}$ is the ratio of E1 the PNC transition matrix element caused by the weak electron-nucleon interaction and M1 the usual magnetic dipole element. ϕ^{PNC} is therefore proportional to (n-1) and has a rapid dispersion-like wave-length variation close to resonance. From the relation between refractive index and absorption one can show that at one atomic absorption length, the optimum from a signal to noise point of view, the rotation at the peak of the dispersion curve is R radians. Since the absorption falls off more rapidly than the refractive index the rotation angle for one absorption length increases as one moves away from resonance, but this is at the expense of the rapid wavelength variation which is essential to discriminate $\emptyset^{PNC}$ from background.

The Oxford and Washington experiments have both been carried out close to allowed M1 transitions within the ground $6p^3$ configuration of atomic bismuth. The Oxford group chose the $J=3/2 \rightarrow 5/2$ line at $\lambda=648$nm in the visible, while the Washington group preferred the $J=3/2 \rightarrow 3/2'$ line at $\lambda=876$nm in the near infra-red. The advantage of the 648nm line is that it is in a region which is covered by narrow band tunable dye lasers whereas at the time the experiments were begun only a pulsed parametric laser was available at the 876nm wavelength. This has a much broader spectral output than the dye laser and as a result the Washington experiment is unable to resolve the bismuth hyperfine structure in contrast to the Oxford work which is on a single hyperfine component. In compensation, the 648nm line is overlayed by molecular absorption which reduces by a factor 20 or so the intensity transmitted through the bismuth. The molecular absorption is absent at 876nm which allows the Washington group to use longer optical path lengths and correspondingly greater angles than are possible in the Oxford experiment.

It is convenient to take as a 'standard' estimate of the value of R for the two transitions a calculation based on a relativistic central field model of the atom combined with the Weinberg-Salam model of the weak interactions. Computations using a self-consistent potential[4] and using a semi empirical parametric potential[3] agree closely and yield the values listed in Table 1.

2. The experiments

Since reports of both experiments have recently been published[1,2] I shall be rather brief here, concentrating on the principles employed and the main difficulties encountered to date.

The two experiments employ the same basic method, though the detailed execution differs markedly. Both make use of the fact that the transmission of polarized light through a pair of crossed polarizers is proportional to the square of the angle through which the plane of polarization of the light is rotated in the region between the polarizers. An intensity term, $2\phi^F\phi^{PNC}$, linear in ϕ^{PNC} is obtained by adding to the bismuth optical rotation an additional controlled rotation, ϕ^F, produced via the Faraday effect in water. Discrimination of the interference term $2\phi^F\phi^{PNC}$ from a background term $2\phi^F\phi^R$ due, for example, to residual misalignment of the polarizers, optical rotation in optical components etc., is achieved by requiring that the signal change sign with the very small ($\delta\lambda/\lambda \approx 10^{-6}$) change in wave-length required to move from one side to the other of the ϕ^{PNC} dispersion curve. Remaining background effects can be removed by subtracting that part of the signal which remains in the absence of the bismuth.

A very important feature of sensitive experiments of this type is the method of calibration. In both cases this is achieved by applying a small magnetic field to the bismuth and measuring the resultant Faraday rotation angle ϕ^{FBi}, detected in just the same way as the optical rotation ϕ^{PNC}. By comparing one angle with another we remove uncertainty both in our instrumental calibration and in the optical depth of the bismuth vapour, provided, of course, we thoroughly understand the bismuth Faraday effect. In order to check this we have done a subsidiary experiment[6] in which we have investigated the detailed Faraday spectrum (see Fig.2) for the 648nm line. Agreement between theory and experiment is excellent.

The two laser optical rotation experiments have enormous potential sensitivity because of the large photon fluxes available - greater than 10^{15}/sec in the Oxford experiment - but both suffer from 'optical noise' which at present considerably reduces the attainable sensitivity. In the Oxford experiment this additional noise takes the form of slow (~ 10 minutes) variations in the wavelength dependent residual angle ϕ^R caused, we believe, by changes in the spatial distribution of the laser light leading to slight differences in the effective crossing angle of the polarizers, which are of course never perfect. A related though different problem has been found in the Washington experiment. As mentioned earlier, the absence of molecular absorption at 876nm allows the use of a long optical path length involving considerable absorption near the resonance centre. The Washington group have observed anomalously high 'angle noise' in this region which they have avoided by working on the wings of the dispersion curve.

The first stages of the two experiments have now been completed and neither group has seen optical rotation of the expected magnitude. The published results are listed in Table 1 and it will be seen that the experimental limits are now appreciable lower than the predicted values. In both cases the errors quoted are found by straightforward analysis of the data obtained in a large number of separate runs.

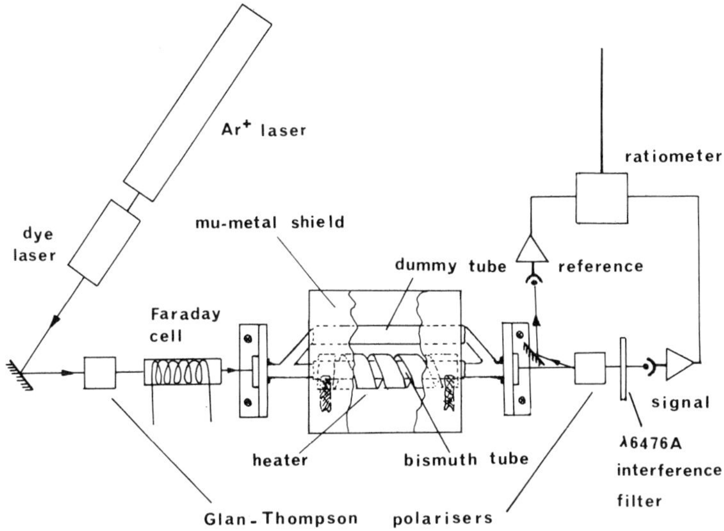

Fig.1. Schematic of the Oxford optical rotation apparatus.

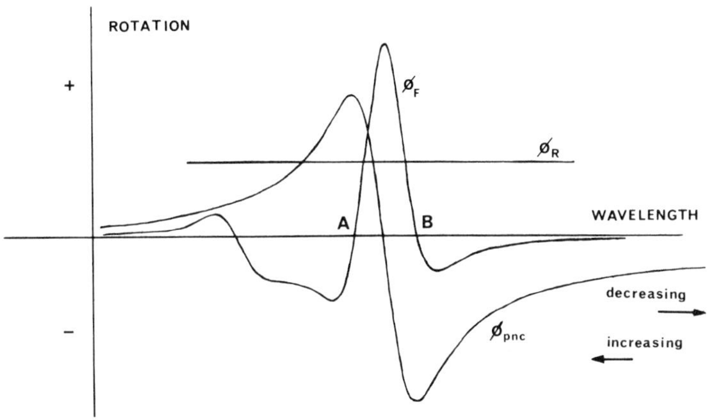

Fig.2. Wavelength dependence of the optical rotation angle $\emptyset_{PNC}$, the bismuth Faraday angle $\emptyset_F$, and the residual angle $\emptyset_R$ for the F=6 → 7 hyperfine component of the 648nm transition.

	$(R = \text{Imag } E1^{PNC}/_{M1}) \times 10^8$	
	Theory	Experiment
Washington		
$3/2 \rightarrow 3/2'$; 876nm	-23	-0.7 ± 3.2
Oxford		
$3/2 \rightarrow 5/2$; 648nm	-30	$+2.7 \pm 4.7$

Table 1 Theory and experiment for $R = \text{Imag.}E1^{PNC}/_{M1}$ in atomic bismuth. The theoretical values are obtained using the Weinberg-Salam model ($\text{Sin}^2\emptyset_W=0.35$) combined with a relativistic central field approximation for the atom. The experimental results from the optical rotation experiments, refs. 1 and 2, have error limits quoted at the 90% confidence level.

3. Atomic Many Body Effects

The diagreement between theory and experiment contained in Table 1 is both startling and surprising in view of the excellent agreement between the results of neutrino experiments and the Weinberg-Salam predictions[7]. Clearly, before any very strong implications for particle physics can be drawn from the bismuth results, the reliability of the intevening atomic theory must be subject to scrutiny.

The relativistic central field calculations are based on an independent particle model which neglects

$$H_1 = \sum_{i>j} V_{ij} - \sum_i V(\mathbf{r}_i)$$

the part of the two particle electron-electron interaction which has not already been included in the central field. To date there have been four different approaches to the problem.

(i) Inclusion of potentially dangerous near lying configurations

Henley, Klapisch and Wilets[5] have diagonalized the full atomic Hamiltonian including H_1 within a subspace which includes not just the central field configurations of interest but also a number of potentially dangerous nearlying ones. These configuration interaction states are then used in a perturbation calculation of $E1^{PNC}$ for the $J=3/2 \rightarrow 3/2'$ transition. Rather surprisingly, Henley et al find that the inclusion of this configuration interaction makes very little difference to $E1^{PNC}$; the correction factors are of order 1%.

(ii) Polarization

An indication that inclusion of near lying configurations may not represent the whole story comes from a static field polarization argument. One knows that when an atom is placed in a static uniform external field it becomes polarized and the polarization produces an internal induced field which tends to cancel the applied field. Furthermore this cancellation must be quite large since the induced and applied fields must exactly cancel at the nucleus which, being a charged particle in equilibrium, must experience zero net electric field. The implication of this is that we may expect some reduction in $E1^{PNC}$ since the transition involves the 6p electrons whose main density is in a region where the polarization shielding will have made a noticeable modification to the field.

At Oxford we have looked into this polarization[8] using the statistical polarization and time-dependent self-consistent Hartree techniques which are familiar in atomic physics. Both methods indicate that $E1^{PNC}$, and hence R, should be reduced by a factor of order 0.5 from the unpolarized central field calculation.

(iii) Many body perturbation theory

An alternative and relatively well-defined way of investigating the effect of the residual perturbation H_1 is to use diagrammatic many-body perturbation theory. In order to make the problem more practicable we have adopted a variant in which the PNC interaction is included in the zeroth order central field Hamiltonian. This has the important effect of reducing the perturbation order by one, enormously reducing the complexity. We have nearly completed a detailed calculation[9] of all diagrams of lowest order in H_1. We find that the only resultant contributions which are seriously large are those that correspond to polarization and which are already included in (ii) above. The remaining first order terms yield a correction to R of +5% for the 876nm line and of +15% for the 648nm line.

(iv) Semiempirical approach

An alternative approach which has been adopted by Khriplovich and his-coworkers[10] is to use empirical data as far as is possible. Unfortunately, some of the relevant data is not available for bismuth and has to be taken from other heavy elements. On the basis of their analysis, Khriplovich et al quote values for $E1^{PNC}$ which are lower than the central field model value by a factor $\approx$ 0.7. Comparison between the E1 matrix elements used by Khriplovich et al with the results to be expected in the shielding model suggest that part at least of the reduction found by Khriplovich et al[10] has its origin in the shielding effect discussed in (ii).

4. Discussion

While there is now reason to believe that the standard relativistic central field model overestimates the magnitude of R, it is not at all clear that many body effects can completely resolve the discrepancy between theory and experiment. A great deal more work both theoretical and experimental is urgently required to resolve the issue.

1. P.E.G. Baird, M.W.S.M. Brimicombe, R.G. Hunt, G.J. Roberts, P.G.H. Sandars, D.N. Stacey, Phys.Rev.Lett. 39, 798 (1977).

2. L.L. Lewis, J.H. Hollister, D.C. Soreide, E.G. Lindahl, E.N. Fortson, Phys.Rev.Lett. 39, 795 (1977).

3. M.W.S.M. Brimicombe, C.E. Loving and P.G.H. Sandars, J.Phys. B.9 L1 (1976) and to be published.

4. E.M. Henly and L. Wilets, Phys.Rev. A14; 1411 (1976).

5. E.M. Henley, M. Klapisch, L. Wilets, Phys.Rev.Letters 39, 994 (1977).

6. P.E.G. Baird, R.G. Hunt, G.J. Roberts, P.G.H. Sandars and D.N. Stacey to be published.

7. P. Musset, Invited talk at 1977 International Sympossium on Lepton and Photon Interactions at High Energies 25-31 August 1977.

8. C.E. Loving and P.G.H. Sandars to be published.

9. C.E. Loving and P.G.H. Sandars to be published.

10. I.B. Khriplovich, V.N. Novikov, O.P. Sushkov J.E.T.P. 71, 1665 (1976)

SEARCH FOR PARITY NON-CONSERVATION IN THE $6^2P_{1/2}$ - $7^2P_{1/2}$ FORBIDDEN M1 TRANSITION IN THALLIUM

E. D. Commins
Univ. of California, Berkeley, CA 94720

In this lecture, I will give a progress report on an experiment being carried out at Berkeley by P. Bucksbaum, S. Chu, R. Conti, and myself to search for PNC in the highly forbidden $6^2P_{1/2}$ - $7^2P_{1/2}$ M1 transition (292.7 nm.) in atomic Tℓ vapor. The experiment is of a type originally suggested by C. Bouchiat and M. Bouchiat.[1] I will first explain why the transition in question is chosen for study, and then describe the experimental method and results to date.

1. Properties of Thallium

Thallium (Z = 81) has a ground configuration $(1s)^2....(5d)^{10}$ $(6s)^2 6p$, and the low lying energy levels are as shown in fig. 1. To a reasonably good approximation one can regard Tℓ as a single electron outside of spherically symmetric closed shells. Thus one can hope to calculate the properties of Tℓ (in particular the PNC effects) fairly accurately by the one electron central field approximation (OECFA).

David Neuffer of UCB has carried out numerical construction of OECFA Dirac wave-functions and energy levels using a modified "Tietz" potential chosen to fit the $6^2P_{1/2}$, $7^2P_{1/2}$ energy levels.[2] He has then used these wave-functions to calculate hfs splittings(see Table I) allowed E1 matrix elements(see Table 2), Stark transition amplitudes for the $6^2P_{1/2}$ - $7^2P_{1/2}$ transition in an external electric field, the $6^2P_{1/2}$-$7^2P_{1/2}$ M1 transition amplitude M, and the g_J anomaly in the $6^2P_{1/2}$ state. In the latter two quantities, relativistic effects, configuration interaction, Breit interaction and hyperfine interaction effects are taken into account. The agreement between our calculations and the observed quantities is in all cases very good, and provides confidence that an estimate of the PNC E1

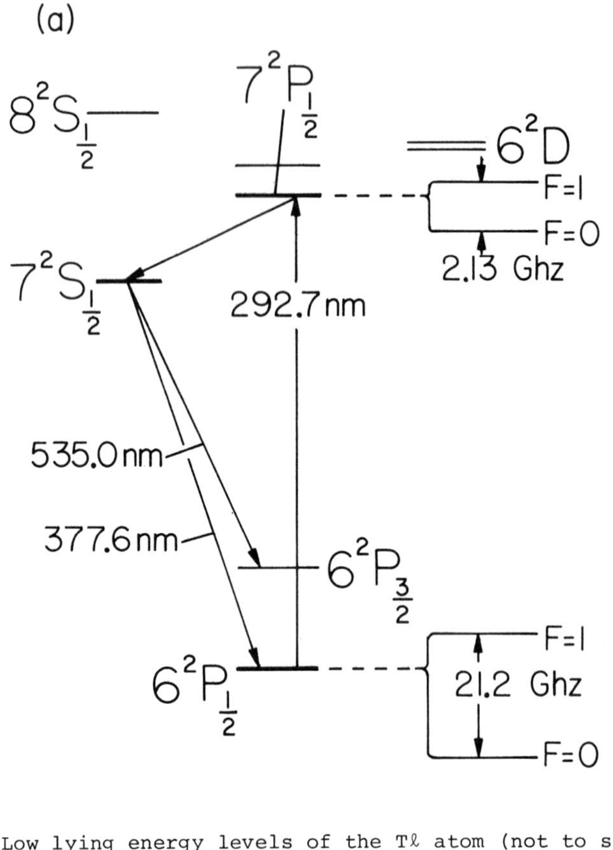

Figure 1. Low lying energy levels of the Tℓ atom (not to scale). The hyperfine structure splittings of $6^2P_{1/2}$, $7^2P_{1/2}$ states are shown. Absorption of the $6^2P_{1/2}$ - $7^2P_{1/2}$. M1 photon (292.7 nm) is detected in an earlier version of our experiment by observing fluorescence at 535 nm accompanying decay of the $7^2P_{1/2}$ state. In the present version atoms in the $7^2P_{1/2}$ state are optically pumped to the $8^2S_{1/2}$ level and decay fluorescence ($8^2S_{1/2} \rightarrow 6^2P_{3/2}$) at 323 nm is observed.

Table I. Calculated energy levels and hyperfine structure splittings (from ref. 2). The large discrepancy between calculated and observed values of the $6^2P_{3/2}$ hfs splitting is removed by taking into account configuration interaction.

Spectroscopic Level Designation	Fitted Energy Level (Ionization Energy, $m_ec^2 = 1$)	Spectroscopic Energy Level	Valence Electron Hyperfine Splitting (GH_z)	Observed Hyperfine Splitting (GH_z)
$6p\,^2P_{1/2}$	1.1939×10^{-5}	1.1953×10^{-5}	21.8	21.3
$6p\,^2P_{3/2}$	9.8745×10^{-6}	1.0062×10^{-5}	3.27	.528
$7p\,^2P_{1/2}$	3.6756×10^{-6}	3.6648×10^{-6}	2.71	2.13
$7p\,^2P_{3/2}$	3.3933×10^{-6}	3.4219×10^{-6}	.494	.62
$8p\,^2P_{1/2}$	1.9199×10^{-6}	1.9158×10^{-6}	.989	
$8p\,^2P_{3/2}$	1.8155×10^{-6}	1.8254×10^{-6}	.187	
$7s\,^2S_{1/2}$	5.4164×10^{-6}	5.5289×10^{-6}	14.3	12.4
$8s\,^2S_{1/2}$	2.5169×10^{-6}	2.5521×10^{-6}	4.32	
$9s\,^2S_{1/2}$	1.4650×10^{-6}	1.4796×10^{-6}	1.90	
$10s\,^2S_{1/2}$	9.594×10^{-7}	9.626×10^{-7}	1.01	
$11s\,^2S_{1/2}$	6.772×10^{-7}	6.811×10^{-7}	0.59	

Table II. Allowed E1 transition rates (from ref. 2)

Transition	A-Coefficient (Gallagher & Lurio) 10^7 sec^{-1}	A-Coefficient (this work) 10^7 sec^{-1}
$7^2S_{1/2} - 6^2P_{1/2}$	6.25 ± .31	5.78
$8^2S_{1/2} - 6^2P_{1/2}$	1.78 ± .16	1.75
$9^2S_{1/2} - 6^2P_{1/2}$	.78 ± 0.10	0.777
$10^2S_{1/2} - 6^2P_{1/2}$	-----	0.412
$11^2S_{1/2} - 6^2P_{1/2}$	.31 ± 0.06	0.244
$7^2S_{1/2} - 6^2P_{3/2}$	7.05 ± .32	8.30
$8^2S_{1/2} - 6^2P_{3/2}$	1.73 ± 0.18	2.30
$9^2S_{1/2} - 6^2P_{3/2}$	0.80 ± 0.08	1.01
$10^2S_{1/2} - 6^2P_{3/2}$	0.57 ± 0.06	.534
$6^2D_{3/2} - 6^2P_{1/2}$	12.6 ± 1	16.04
$7^2D_{3/2} - 6^2P_{1/2}$	4.4 ± 0.5	6.39
$8^2D_{3/2} - 6^2P_{1/2}$	1.89 ± 0.3	3.19
$9^2D_{3/2} - 6^2P_{1/2}$	.98 ± 0.22	1.82
$10^2D_{3/2} - 6^2P_{1/2}$	.58 ± 0.15	1.14
$6^2D_{3/2} - 6^2P_{3/2}$	2.20 ± .23	2.88
$7^2D_{3/2} - 6^2P_{3/2}$	0.76 ± .08	1.01
$8^2D_{3/2} - 6^2P_{3/2}$	0.37 ± .04	0.498
$9^2D_{3/2} - 6^2P_{3/2}$	0.19 ± .02	0.279
$6^2D_{5/2} - 6^2P_{3/2}$	12.4 ± 1.5	16.3
$7^2P_{5/2} - 6^2P_{3/2}$	4.2 ± 0.5	6.06
$8^2P_{5/2} - 6^2P_{3/2}$	1.7 ± 0.2	2.96

amplitude ϵ_{PNC} for the $6^2P_{\frac{1}{2}}$ - $7^2P_{\frac{1}{2}}$ transition can be made with 15-20% accuracy at least. Using the Salam-Weinberg model and summing over discrete, continuum and auto-ionizing intermediate S states by means of a Green's function technique we find for the circular dichroism:

$$\delta = \frac{\mathrm{Im}(\epsilon_{PNC})}{M_{Expt}} = 2.6 \times 10^{-3} \qquad (1)$$

This result agrees with an independent calculation done by Sushkov, Flambaum, and Khriplovich[3] (δ=2.5x10^{-3}) and an earlier estimate by Bouchiat and Bouchiat ($\delta \cong 2 \times 10^{-3}$). It is obvious why the present transition is chosen for study: δ can be estimated reliably, and it is large.

2. Experimental Method

The simplest experimental approach to detect δ would be to illuminate Tℓ vapor with circularly polarized 292.7 nm laser light and look for a dependence of the fluorescence intensity in the 535 nm line ($7^2S_{\frac{1}{2}} \rightarrow 6^2P_{3/2}$) on the 292.7 nm circular polarization. (Note that the 377.6 nm line ($7^2S_{\frac{1}{2}} \rightarrow 6^2P_{\frac{1}{2}}$) cannot be used for this purpose; 377.6 nm photons are resonantly trapped at high Tℓ densities and converted to 535 nm photons). However, this method is not practical because the oscillator strength f of the 292.7 nm M1 transition is very small; (we have measured f = 1.8x10^{-15}!). Collisions between Tℓ atoms in the vapor give rise to momentary electric fields which mix the $6^2P_{\frac{1}{2}}$, $7^2P_{\frac{1}{2}}$ states with S and D states, thus allowing the transition to go by a collisional (E1) Stark amplitude considerably stronger than the M1 amplitude M. This provides a large background. Thus, at an early stage, we recognized the necessity of studying M and ϵ_{PNC} by observing interference effects when an external electric field is applied.

These effects may be understood as follows: First consider the case when incoming laser light is linearly polarized, with polarization vector $\hat{\varepsilon}$ parallel to static field E (see figure 2). With axis of quantization z as shown, only hfs transition F = 0 → F'=0, or F=1 → F' = 1, $\Delta M_F = 0$ can occur. The $7^2P_{\frac{1}{2}}$ state is polarized along the z axis because of $E1_{Stark}$, M1 interference. The relevant Stark matrix element in this case is

$$\alpha = \frac{e^2E}{q}\sum_{ns} R_{7P,\,ns}R_{6P,\,ns}\left(\frac{1}{E_{7P}-E_{ns}}+\frac{1}{E_{6P}-E_{ns}}\right) + \frac{2e^2E}{q}\sum_{nD_{3/2}} R_{7P,\,nD}R_{6P,\,nD}\left(\frac{1}{E_{7P}-E_{nD}}+\frac{1}{E_{6P}-E_{nD}}\right) \tag{2}$$

where the R's are dipole radial integrals. In atomic units, (but E in volts/cm), we calculate:

$$\alpha_{CALC.} = 7.4\times 10^{-8}E. \tag{3}$$

The polarization of the $7^2P_{\frac{1}{2}}$ state in the F=1 → F'=1 transition is $P_{1-1} \cong \frac{4M}{3\alpha}$, inversely proportional to E for E >> 1 volt/cm. Of course, the total intensity in the transition is porportional to E^2.

Next consider $\hat{\epsilon}\perp$ E (see figure 3). Then the hfs transitions F=0 → F'=1, $\Delta m_F = \pm 1$ and F=1 → F'=1 or F'=0, $\Delta m_F = \pm 1$ are possible. For the 0 → 1 transition, interference between M and the Stark matrix element

$$\beta = \frac{e^2E}{q}\sum_{nS} R_{7P,\,nS}R_{6P,\,nS}\left(\frac{1}{E_{6P}-E_{nS}}-\frac{1}{E_{7P}-E_{nS}}\right) + \frac{e^2E}{q}\sum_{nD_{3/2}} R_{7P,\,nD}R_{6P,\,nD}\left(\frac{1}{E_{7P}-E_{nD}}-\frac{1}{E_{6P}-E_{nD}}\right) \tag{4}$$

occurs. Again in atomic units (but E in volts/cm)

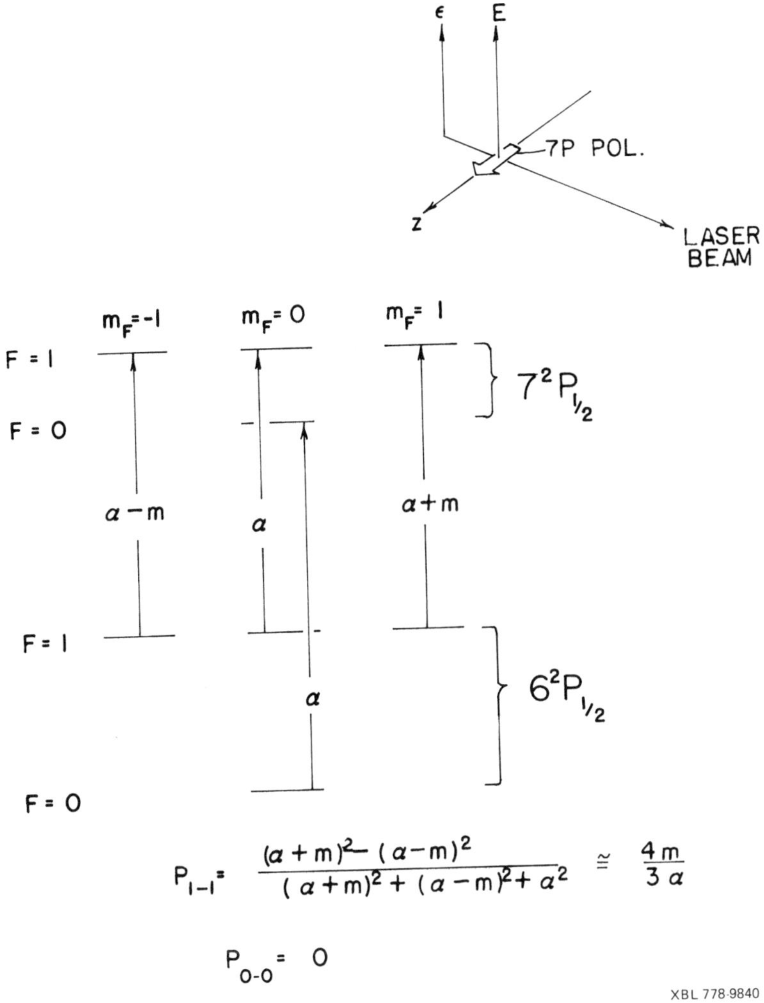

Figure 2. Transitions and ($7^2P_{1/2}$) polarization when 292.7 nm light is linearly polarized parallel to static electric field. α, M are defined in text.

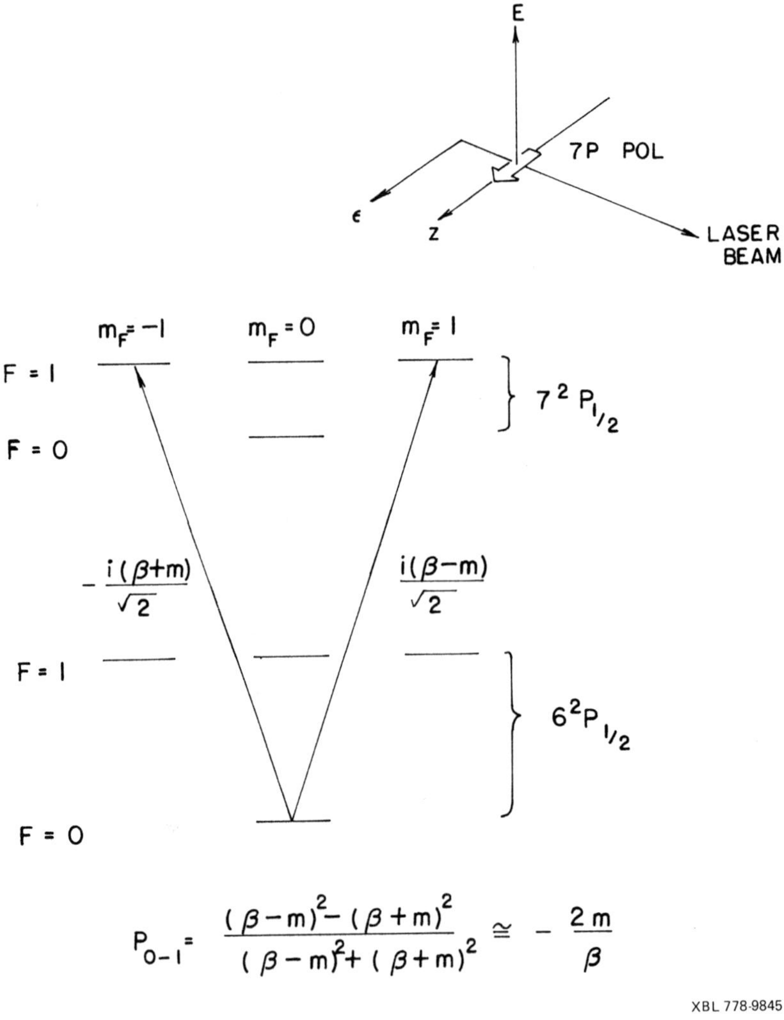

Figure 3. Transitions and ($7^2P_{\frac{1}{2}}$) polarization when 292.7 nm light is linearly polarized perpendicular to static electric field. β is defined in text.

$$\beta_{calc} = 6.0 \times 10^{-8} \ E \qquad (5)$$

Once more the $7^2P_{\frac{1}{2}}$ polarization is along z, with

$$P_{0-1} \cong \frac{-2M}{\beta}$$

for E>>1 volt/cm.

The $7^2P_{\frac{1}{2}}$ polarization can be detected by analyzing the ensuing decay fluorescence at 535 nm which is circularly polarized along the z axis.

It is important to be sure that the detection scheme functions as it should. To calibrate the experimental apparatus we utilize the Hanle effect, illustrated in figure 4. Here the incoming 292.7 nm light is circularly polarized. This produces a component of $7^2P_{\frac{1}{2}}$ polarization along the laser beam axis for the F=1 → F'=1 transition only (because of interference between the α and β amplitudes). In the presence of an external magnetic field B parallel to E, this $7^2P_{\frac{1}{2}}$ polarization precesses, giving rise to a circular polarization of the decay fluorescence which is large, dependent on B, but independent of E and M.

Next consider parity non-conservation. Fig. 5 shows the most important effect which occurs if the incoming 292.7 nm light is circularly polarized. As in the case of incoming linear polarization, the $7^2P_{\frac{1}{2}}$ state has polarization along the z axis, which changes sign with E. That part of the $7^2P_{\frac{1}{2}}$ polarization which also changes sign with the CP of 292.7 radiation is proportional to δ.

Before we proceed further, let's consider some experimental details. Figure 6 is a schematic of our earlier apparatus. It consists of a flash-lamp pumped pulsed dye laser which delivers ½μs pulses of light at 585.4 nm (tunable),

HANLE EFFECT

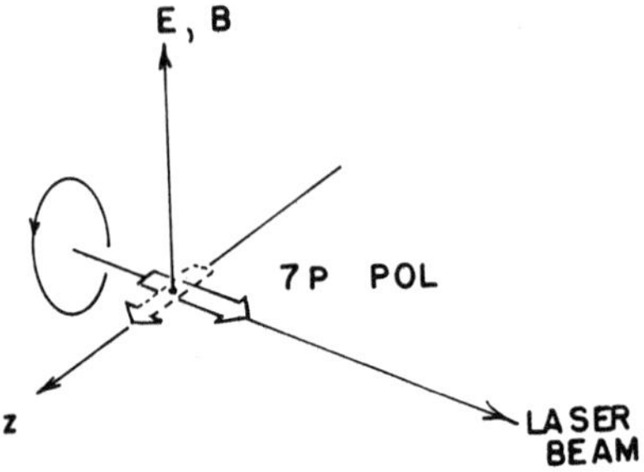

$$\left\langle P(z) \right\rangle = \frac{\int_0^{\infty} P\, e^{-\lambda t}\, dt}{\int_0^{\infty} e^{-\lambda t}\, dt}$$

$$P = \mp \frac{4\,\alpha\,\beta}{3\,\alpha^2 + 2\,\beta^2} \cdot \sin \omega t$$

FOR 1–1 TRANSITION

XBL 778-9841

Figure 4. Hanle effect

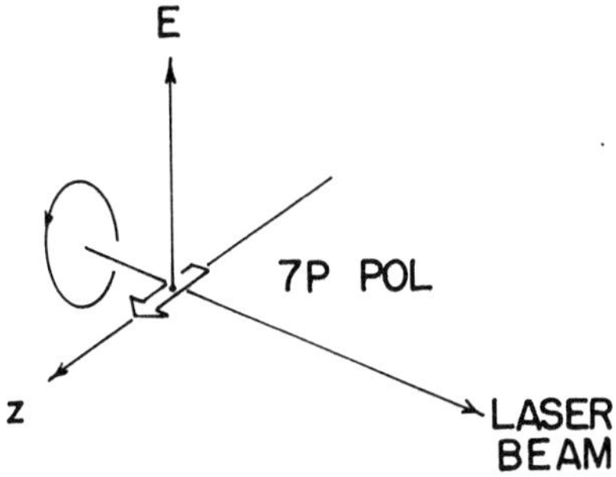

TEST FOR PARITY NON CONSERVATION

$$P_{1-1} = \frac{(4\alpha - 2\beta)m}{3\alpha^2 + 2\beta^2}(1 \mp \delta/2) = \frac{.56\,(1 \mp \delta/2)}{E}$$

$$P_{0-1} = -\frac{2m}{\beta}(1 \mp \delta/2) = -\frac{2.57\,(1 \mp \delta/2)}{E}$$

$$\text{W-S:} \quad \delta = 2.6 \times 10^{-3}$$

XBL 778-9843

Figure 5. The PNC experiment. When the incoming 292.7 nm light is circularly polarized the $7^2P_{1/2}$ state has polarization given by P_{1-1} (for F=1 → F'=1) or P_{0-1} (for F = 0 → F'=1).

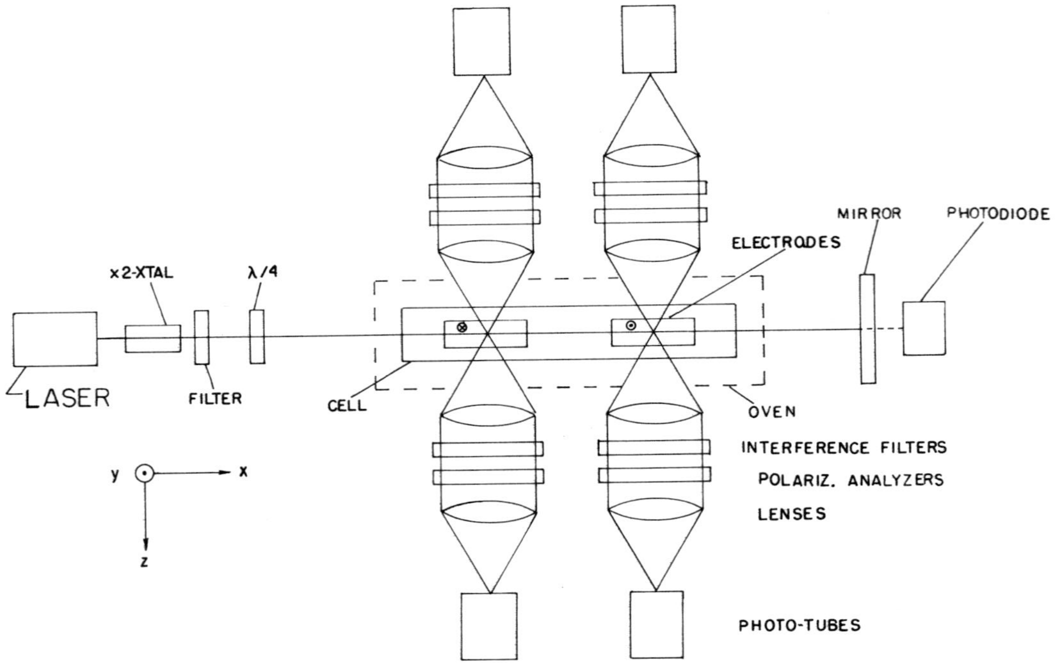

XBL 778-9846

Figure 6. Schematic of older version of experiment.

bandwidth ~ 1.5GhZ, rep. rate ~ 10/s. This visible light passes through an ADA doubling crystal which generates second harmonic (292.7 nm) pulses. The visible light is then filtered out, and the U.V. light is circularly polarized with a rotatable $\lambda/_4$ plate. The beam then passes through a quartz cell containing Tℓ vapor at T ~ 1000^oK ($2x10^{14}$atom/cm^3).

There are two interaction regions with electric fields in opposite directions, and each field is reversed after every second pulse, while the 292.7 nm circular polarization is reversed after every pulse. The (circularly polarized) fluorescence at 535 nm is analyzed and detected in a conventional way.

Figure 7 shows a more realistic sketch of the cell.

Figure 8 shows results obtained for the Hanle effect which are in good agreement with expectations.

Figure 9 shows typical Stark resonances obtained for the F=0 → F'=0 transition (∥ linear polarization) and F=0 → F'=1 transition (⊥ linear polarization).[4] The ratio α^2/β^2 is equal to the ratio of peaks of these resonance curves, which are separated by the hfs splitting of the $7^2P_{\frac{1}{2}}$ state (2.13GhZ). We obtain α/β expt = 1.19, compared to α/β calc = 1.23. Polarization of the $7^2P_{\frac{1}{2}}$ state for linearly polarized 292.7 nm light leads to an experimental asymmetry Δ in the counting rates for 535 nm detectors on opposite sides of a given electric field region (see figure 6), which is used to measure M. Experimental values of Δ are plotted in figure 10 versus E field for the indicated hfs transitions. These data yield a value of the M1 amplitude: $M = (-2.11 \pm 0.30)x10^{-5}\mu_B$ in accord with our theoretical estimate:

$$M = (-3\pm 1) \times 10^{-5}\mu_B; \text{ (here } \mu_B = \frac{|e|\hbar}{2mc} \text{).}$$

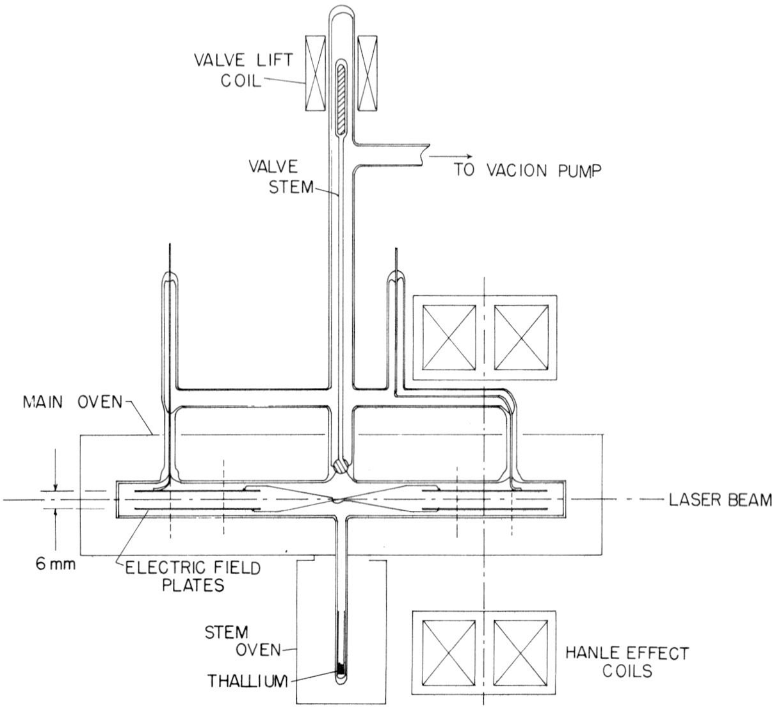

Figure 7. Sketch of original cell.

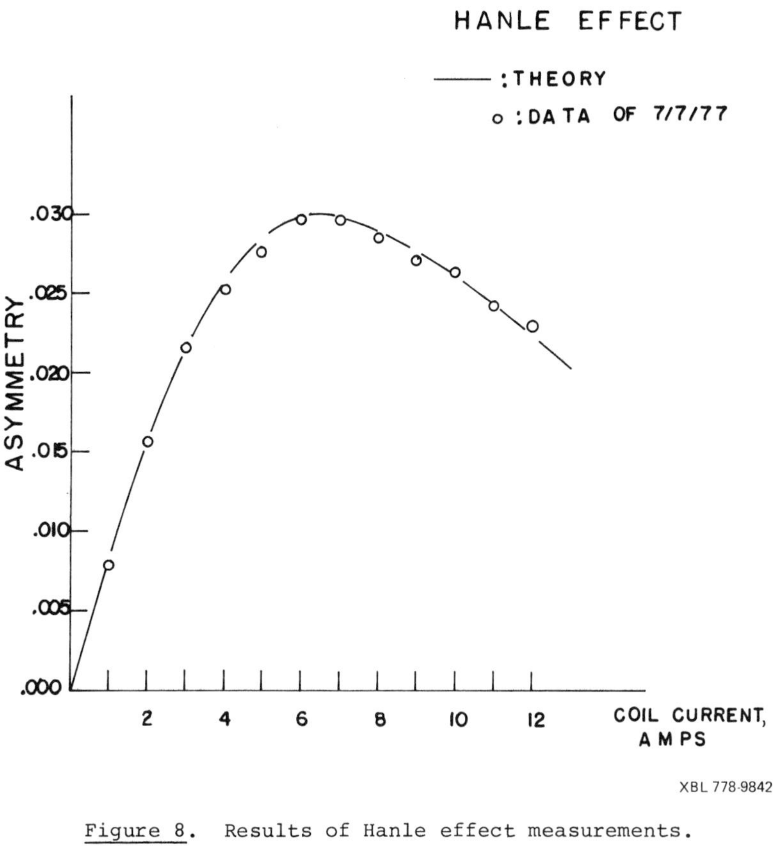

Figure 8. Results of Hanle effect measurements.

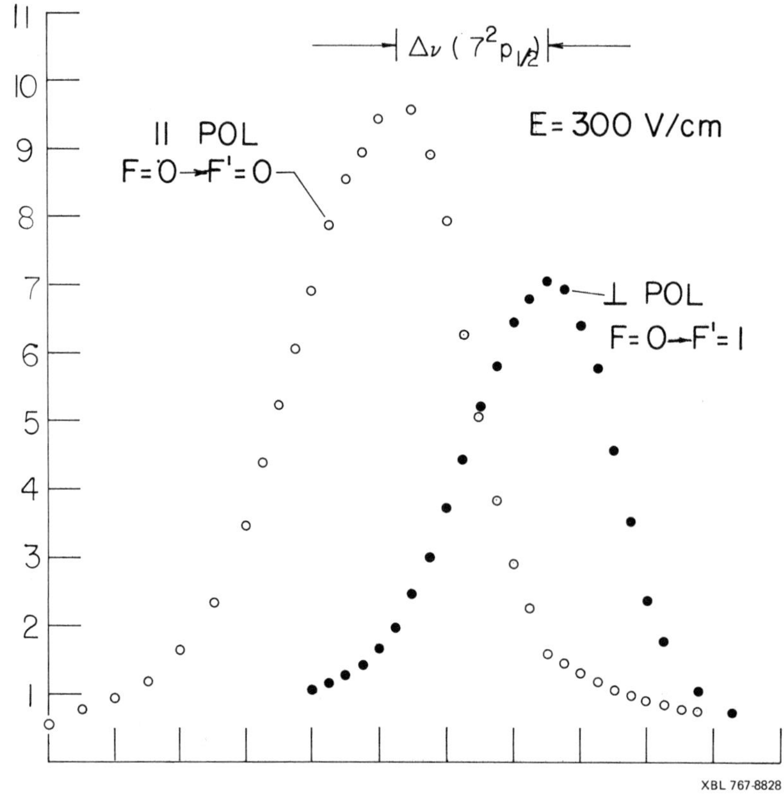

Figure 9. Typical Stark resonances obtained for F=0 → F'=0 and F=0 → F'=1 (from ref. 4)

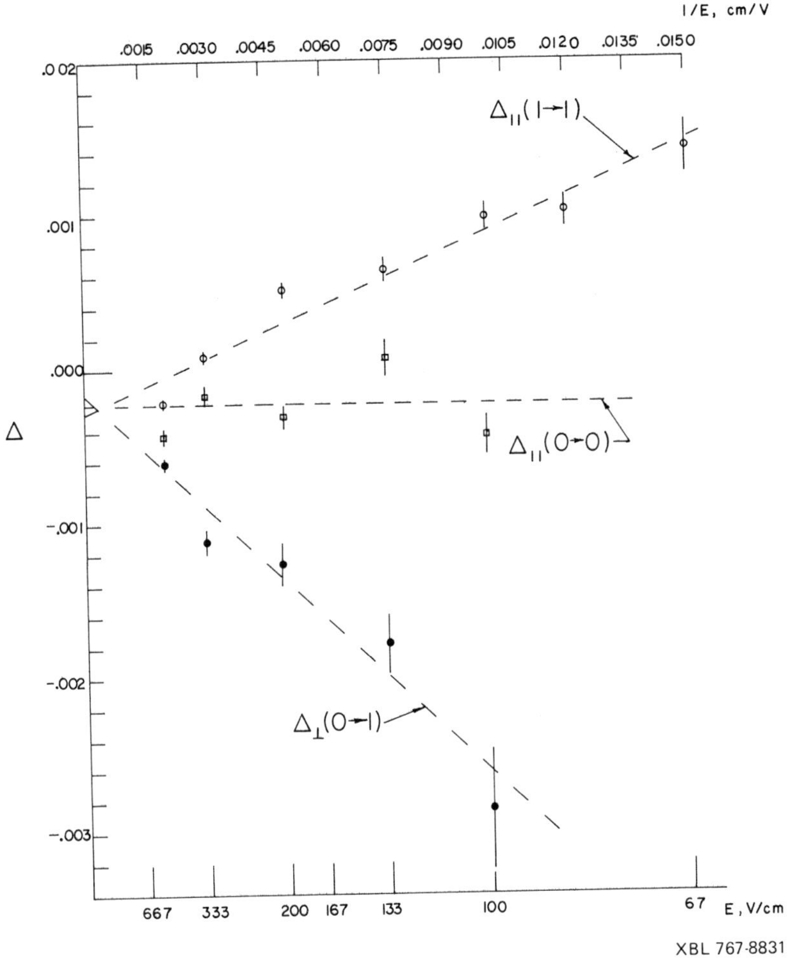

Figure 10. Experimental asymmetries obtained for linearly polarized 292.7 nm light. M_{expt} was obtained from this data (Ref. 4).

From 24 hours of data on the PNC experiment using CP 292.7 light, we have established an upper limit on δ one order of magnitude larger than the predicted value according to Salam and Weinberg. This preliminary value is not very useful; however, we have now installed a much improved version of the experiment which can be understood by consideration of figure 11. Let us concentrate on the energy level diagram in the upper left hand corner first. In the older version of the experiment, described above, polarized $7^2P_{\frac{1}{2}}$ atoms first decay to the $7^2S_{\frac{1}{2}}$ state (we thereby lose a factor of 3 in polarization). The polarization loss in the CP of the ensuing 535 nm fluorescence would be an additional factor of 2, except that the $7^2S_{\frac{1}{2}}$-$6^2P_{\frac{1}{2}}$ transition at 377.6 nm has a branching ratio of ~ 50%, and is resonantly trapped and depolarized. Thus an additional factor of 2 is lost, making a total polarization loss of a factor of 12. This loss is avoided in the new scheme, which makes use of optical pumping from the $7^2P_{\frac{1}{2}}$ to the $8^2S_{\frac{1}{2}}$ state by a crossed beam of circularly polarized laser light at 2.18μ (near-infrared). It can easily be seen that all of the polarization information is preserved if one observes the <u>intensity</u> of the decay fluorescence from the $8^2S_{\frac{1}{2}}$ state, at 323 nm.

We have tried this scheme and it works. Figure 12 shows a Hanle-like experiment without magnetic field in which UV and IR beams are collinear. The $7^2P_{\frac{1}{2}}$ polarization for the F=1 → F'=1 line is as shown in figure 4; inserting values of α and β we have P ~ 0.67. Our quickly measured polarization is in agreement with this prediction.

Finally, our newly installed and functioning cell is shown schematically in figure 13. With the gain of 12 in polarization (a factor of 144 in counting time) we are optimistic that an interesting result for δ can be obtained in the near future.

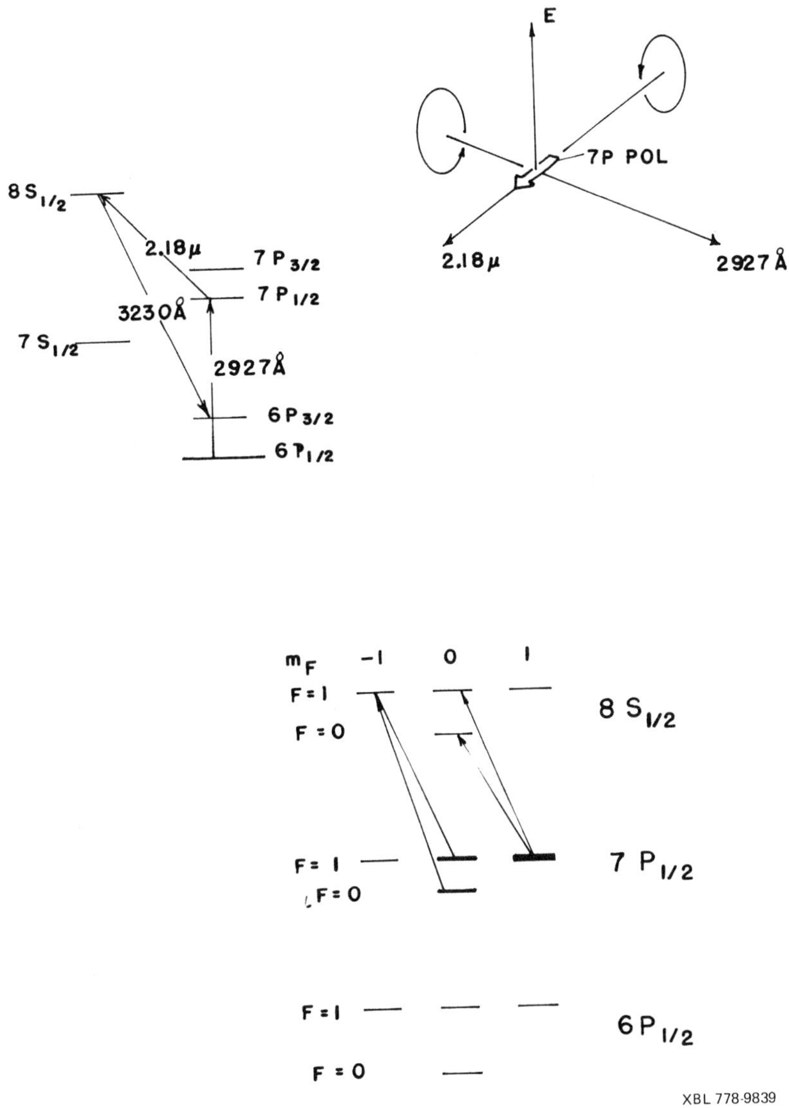

Figure 11. Schematic illustration of present version of our experiment.

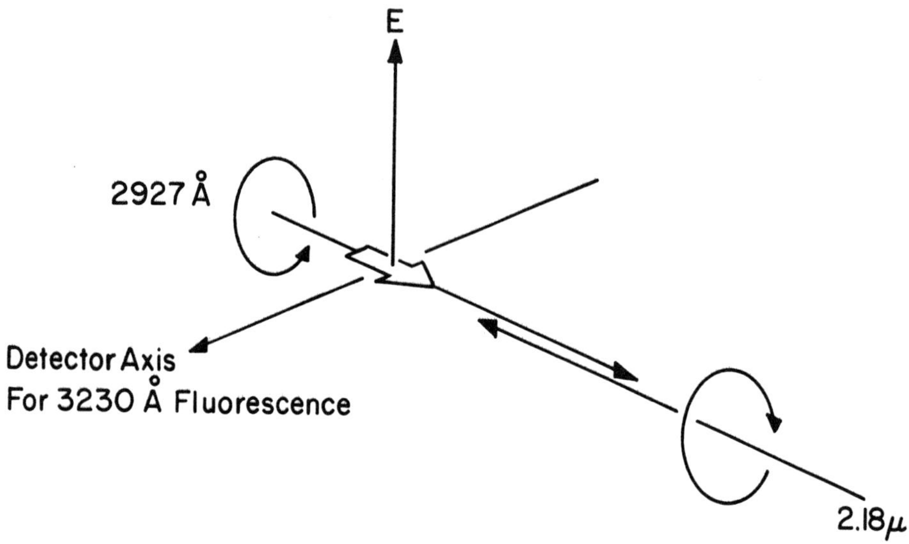

$$P(7^2P_{1/2}) = \frac{4\alpha\beta}{3\alpha^2 + 3\beta^2} \approx 67\%$$

$$P_{OBS} = 50 \pm 15\%$$

For $F=3 \rightarrow F=1$ Transition

Figure 12. An experimental test of the new method: a Hanle-like measurement without magnetic field.

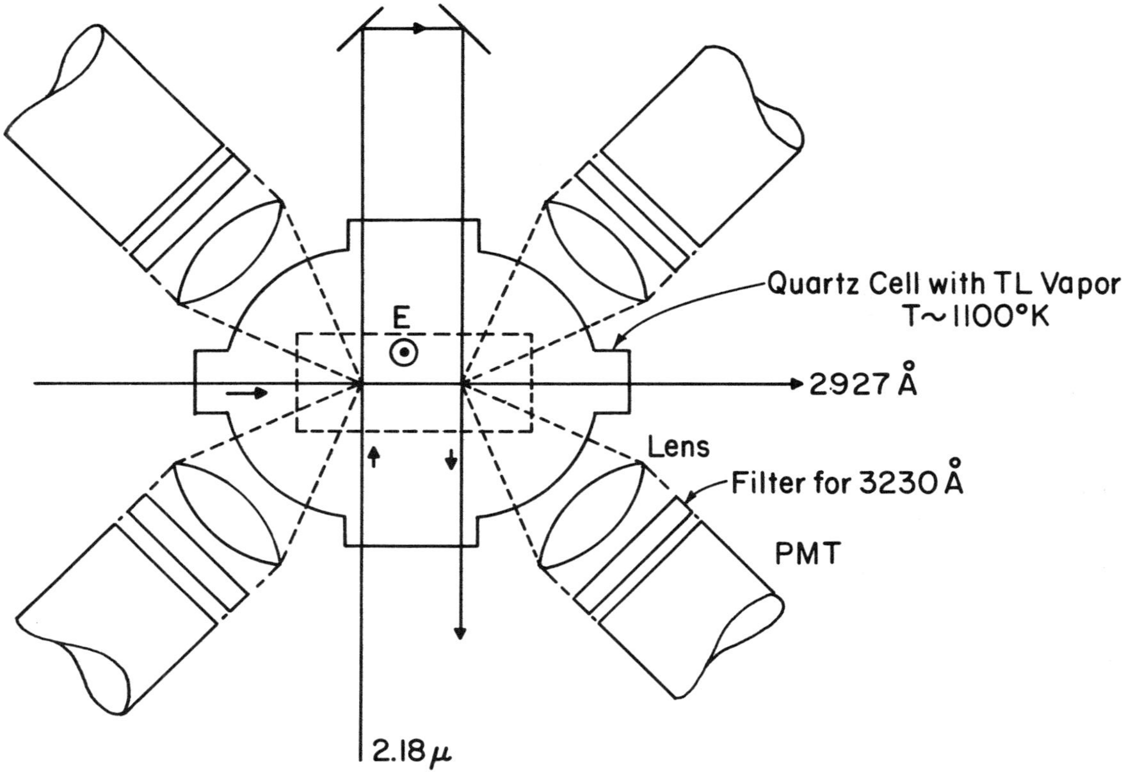

Figure 13. Schematic diagram of new cell.

References:

1. M.A. Bouchiat and C.C. Bouchiat
Phys. Lett. 48B, 111, (1974)
J. Phys. (Paris) 35, 899(1974)
36, 493(1975)

2. D.V. Neuffer and E.D. Commins
Phys. Rev. A 16, 844, (1977)

3. O.P. Sushkov, V. Flambaum, I.B. Khriplovich
JETP Lett. 24, 502, (1976)

4. S. Chu, E.D. Commins and R. Conti, Phys. Lett. 60A, 96, 1977

PROGRESS REPORT ON PARITY VIOLATION IN ATOMIC CESIUM

M.A. Bouchiat and L. Pottier
Laboratoire de Spectroscopie Hertzienne de l'Ecole Normale Supérieure
24, rue Lhomond, 75231 Paris Cedex 05 - France

Research on Parity Violation in atoms started in 1973, at E.N.S. in Paris, with a theoretical work showing that parity violating effects due to neutral currents are considerably enhanced in heavy atoms : they were shown to follow a Z^3 law in the frame of the Weinberg-Salam model ([1]). Among the possible ways of detecting the parity violating effects, the circular dichroïsm in twice forbidden magnetic transitions was analyzed in detail theoretically. Proposals for different experiments, including nS-n'S transitions in Cesium and nP-n'P transitions in Thallium were published ([1])-([2]).

The present paper is limited to a progress report on the experiment relative to the $6S_{1/2}$-$7S_{1/2}$ Cesium transition presently in progress at E.N.S.

Our choice for this particular transition was motivated by several reasons. Very important were the simple atomic structure of the Cesium atom and of this particular transition and also the existence of many well interpreted spectroscopic data. This offers favourable conditions for computing the parity violating effects in a reliable way. Also the atomic number Z = 55 is relatively high and we benefit by a factor 10^6 of enhancement on the right-left asymmetry to be searched for, because of the Z^3 law. Note that Cesium is the heavy atom having the simplest atomic structure : a single s electron outside a closed shell. Another reason for this choice was the large hindrance factor affecting the electromagnetic amplitude as a result of the orthogonality of the 6S and 7S wave functions The M_1 amplitude results from a combined effect of configuration mixing and spin-orbit interaction and is 10^4 times smaller than for normal magnetic dipole transitions. Owing to the Z^3 enhancement and M_1 hindrance factors, the expected circular dichroïsm reaches the level of 10^{-4}.

Other practical considerations were taken into account, such as for instance the availability of a c.w. tunable laser delivering a large flux of photons to excite the transition at 5400 Å. From an experimental point of view, it was also important to notice that the sensitive method of fluorescence detection was usable and could, at least partially, compensate for the small absorption cross-section due to the large hindrance factor. In our experiment we monitor the fluorescence light emitted in the $7S_{1/2}$ to $6P_{1/2}$ decay. This is an allowed transition with a characteristic wavelength of 1.36 μ (see fig. 1). Actually, the sensitivity can be further increased by monitoring the circular polarization which reflects a spin polarization in the 7S state. It will be shown below how such a spin polarization can be created when a d.c. electric field is applied to the Cesium vapour. Figure 2 represents a very schematic view of the experiment : the fluorescence is monitored in a direction perpendicular both to the exciting light beam and to the electric field. Figure 3 shows the fluorescence spectrum obtained; the fluorescence yield is plotted versus the laser frequency for two different orientations of the linear light-polarization $\vec{\varepsilon}$ with respect to the electric field $\vec{E}_0$: either $\vec{\varepsilon}\ //\ \vec{E}_0$ (fig. 3-a) or $\vec{\varepsilon} \perp \vec{E}_0$ (fig. 3-b). The ground state and excited state hyperfine structures are fully resolved. When $\vec{\varepsilon}$ and $\vec{E}_0$ are parallel, only transitions conserving the total angular momentum are allowed.

To explain the basic principle of the search for parity violation in the present experiment ([3]), let us first review the different dipole moments associated with the 6S-7S transition in a d.c. electric field :

- the magnetic dipole moment induced along the electron spin ($\vec{S} = (\hbar/2)\vec{\sigma}$) : $\vec{\mu} = M_1\vec{\sigma}$.
- the electric dipole moment induced by parity violation, $\vec{D}_1^{pv} = E_1^{pv}\,\vec{\sigma}$, also directed along the spin, but $\pi/2$ out of phase with respect to $\vec{\mu}$, as a consequence of T-reversal invariance : $\vec{D}_1^{pv} = i\ (\mathrm{Im}\ E_1^{pv})\ \vec{\sigma}$.
- the electric dipole moment induced by the d.c. electric field. $\vec{D}^{ind}$ splits into two parts. First the spin-independent induced dipole $E_1^{ind}\,\vec{E}_0\,/|\vec{E}_0|$ is directed along $\vec{E}_0$ and in phase with $\vec{\mu}$. It is associated

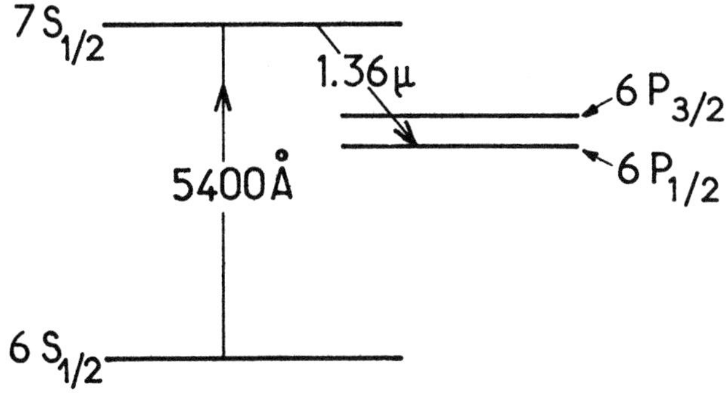

FIGURE 1 : ATOMIC ENERGY LEVELS INVOLVED IN THE CESIUM PARITY EXPERIMENT

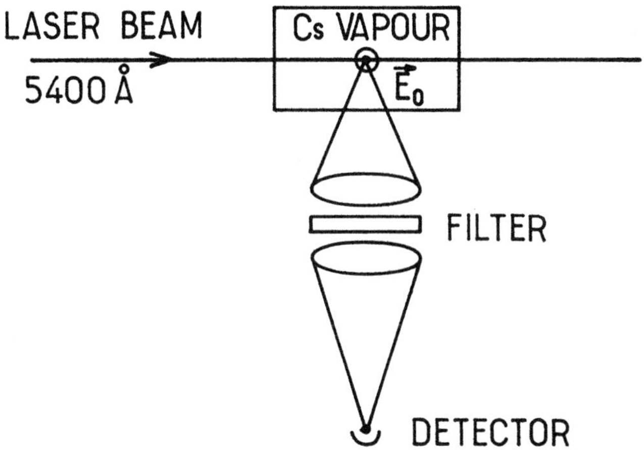

FIGURE 2 : SCHEME OF THE EXPERIMENT

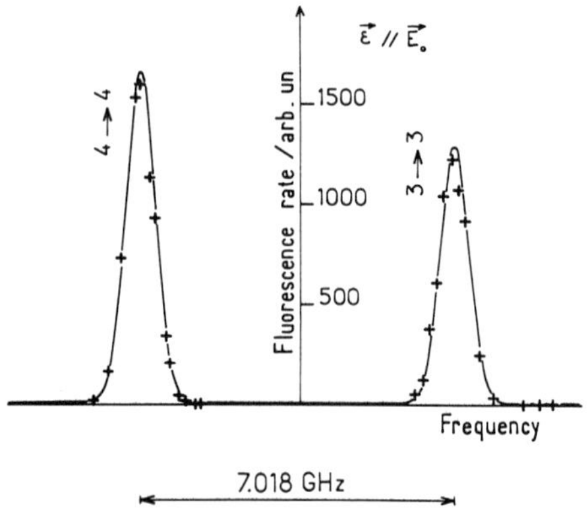

FIG. 3-A

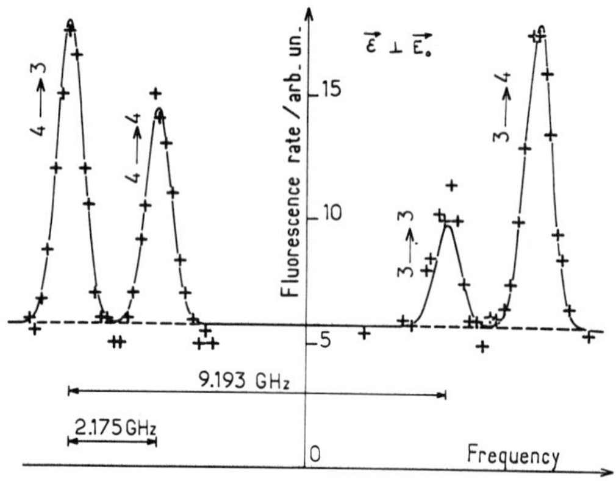

FIG. 3-B

FIGURE 3 : FLUORESCENCE RATE (ARBITRARY UNITS) VERSUS INCIDENT LASER FREQUENCY. ELECTRIC FIELD 1000 V/CM. SOLID LINE : THEORETICAL SPECTRUM; CROSSES : EXPERIMENTAL POINTS.
(A) $\vec{\mathcal{E}} \,//\, \vec{E}_0$, (B) $\vec{\mathcal{E}} \perp \vec{E}_0$.

with the scalar part of the polarizability. Whereas the second part, the spin-dependent induced dipole (associated with the vector polarizability) can be shown, from simple symmetry considerations, to behave differently : it is directed along the vector product of the spin and the field, $\pi/2$ out of phase; besides, it is 10 times smaller. We shall write this part : $i\beta.\vec{\sigma} \wedge \vec{E}_0$.

So we expect finally four different amplitudes of transition which can interfere one with each other. Note that each has specific symmetry properties, so that their interferences must give rise to very different, well distinguishable physical effects.

In the absence of a d.c. field, there is only one interference effect present, involving M_1 and $E_1{}^{pv}$; it gives rise to a circular dichroïsm typical of the parity violating $E_1{}^{pv}$ amplitude. This is the analog of the rotatory power but for the imaginary part of the index of refraction instead of the real part. This is not the kind of effect we are looking for in the present experiment.

Among the other interference effects, we shall concentrate our attention on all those which involve a spin dependent amplitude and the single spin independent one, $E_1{}^{ind}$:

$$M_1 \; E_1{}^{ind} \quad ; \quad E_1{}^{pv} \; E_1{}^{ind} \quad ; \quad \text{and} \quad i\beta E_0 \; E_1{}^{ind}.$$

(Note that $E_1{}^{ind}$ is also the largest amplitude, with a proper choice of the electric field).

It can easily be predicted that all these interference effects will manifest themselves on the spin orientation in the final state : when resonant light is absorbed they can all contribute to create a spin polarization in the 7S state.

We find indeed the three corresponding contributions in the expression giving the spin polarization of the upper state, following the absorption of a circularly polarized resonant photon :

$$\vec{P}_e = \underbrace{\left[\frac{M_1}{E_1{}^{ind}} + \xi_i \frac{\mathrm{Im}\, E_1{}^{pv}}{E_1{}^{ind}} \right] \hat{k}_i \wedge \hat{E}_o}_{P_e^{(1)}} + \underbrace{\xi_i\, P_e^{(2)}\, \hat{k}_i}_{P_e^{(2)}} \quad . \qquad (\star)$$

Here $\hat{k}_i$ and $\hat{E}_0$ denote unit vectors in the directions of the photon momentum and the electric field, ξ_i is the circular polarization of the photons; the above expression is valid when $E_1{}^{ind} \gg M_1$ and $E_1{}^{pv}$.

We expect each contribution to exhibit physical properties of its own and it is interesting to see how this can be used experimentally. There is one component along the vector product $\hat{k}_i \wedge \hat{E}_0$ and one along $\hat{k}_i$. The first one results from the interference between the mixed $M_1 - E_1{}^{pv}$ amplitude and the d.c. field induced amplitude. Since the vector product $\hat{k}_i \wedge \hat{E}_0$ has the transformation properties of an angular momentum under space and time reflection, the presence of a term involving the product of this quantity by ξ_i, which is a T-invariant pseudoscalar, is a clear indication that the atomic hamiltonian contains a P-odd T-even piece. It receives contribution from the axial vector electronic, vector nucleonic component of neutral currents. By observing a component of the spin polarization in a direction perpendicular to $\vec{k}_i$ and $\vec{E}_0$ while exciting with linearly polarized light (i.e. $\xi_i = 0$), we have actually been able to measure the ratio $M_1/E_1{}^{ind}$. Since $E_1{}^{ind}$ can be computed theoretically, the result thus yields the M_1 amplitude in magnitude and in sign. In this way we have obtained (4) :

$$M_1 = (-4.24 \pm 0.34) \times 10^{-5}\ |\mu_B|\ /\ c.$$

Using now circularly polarized light, from the observation of a ξ_i-dependence of the spin polarization in the same direction, we can, in principle, obtain the ratio $\mathrm{Im}\ (E_1{}^{pv} / M_1)$.

($\star$) *For simplicity of the presentation, we have omitted in this expression a numerical factor* $\frac{8}{3} \frac{F(F+1)}{(2I+1)^2}$ *, involving the nuclear spin I and depending on the hyperfine transition under consideration.*

Let us now look at the component $\vec{P}_e^{(2)}$ of the spin polarization pointing in the direction of the incident photon momentum. It results from the interference between the two components of the d.c.-field-induced electric dipole, i.e. the spin dependent and the spin independent ones. Note that it can be unambiguously distinguished from $\vec{P}_e^{(1)}$: first, it is created in a perpendicular direction (as a consequence, the behaviour in a d.c. magnetic field -the so-called "Hanle effect"- is different); then, owing to the fact that it is independent of $\vec{E}_0$, while $\vec{P}_e^{(1)}$ reverses with the electric field. Furthermore the magnitude of $\vec{P}_e^{(2)}$ is known from independent spectroscopic measurements, and the detection of $\vec{P}_e^{(2)}$ in our experiment offers a very convenient way of calibrating correctly optical signals in terms of spin polarization.

The method just described, based on the search for a parity violating d.c.-field-induced spin polarization has several advantages :

- It leads to a complete elimination of the background coming from the radiative transitions induced by collisions, i.e. :

$$\text{Photon} + \text{Cs (6S)} + \text{Cs (6S)} \rightarrow \text{Cs (6S)} + \text{Cs}^{*} \text{ (7S)}.$$

Indeed the collision-induced dipole moment varies randomly in orientation from one collision to another and cannot contribute, on the average, to a macroscopic spin polarization.

- The circularly polarized fluorescence intensity associated with the parity violating spin polarization is proportional to E_0. Thus it is possible to optimize the signal-to-noise ratio by an appropriate choice of the electric field. This means that the sensitivity of the method is increased, like in an heterodyne experiment.
- Another interesting feature of the present experiment is that the right-left asymmetry under search is the same for the incident light beam and for the same beam reflected back through the vapour. Indeed the parity violating signal involves the photon angular momentum $\xi_i \vec{k}_i$ which is conserved in light reflection at normal incidence. This offers the possibility of using multipasses of the beam to enhance the parity violating signal. On the contrary, it can be shown that certain possible systematic errors cancel out when the light passes back.

The experimental set-up is described in ref. ([5]). Let us simply recall here that we : excite with a single-mode c.w. laser; modulate the circular polarization of the incident beam; modulate at a different frequency the circular polarization of the observed fluorescence light; perform a synchronous detection at the sum and difference frequencies; and average digitally during runs of about 1 hour while periodically reversing the d.c. electric field.

Our newest results were deduced from 10 hours' data acquisition. The measured value

$$|E_1^{pv} / M_1| = (.56 \pm 1.8) \times 10^{-3}$$

is compatible with a true value equal to zero. The uncertainty is quoted at 90% confidence level. The result can be expressed in terms of an upper limit for E_1^{pv} :

$$|E_1^{pv}| < 2.8 \times 10^{-10} \; |e| \; a_0 \quad \text{at 90\% confidence level,}$$

or for the average weak vector charge of the nucleons :

$$\overline{C}_V < 6.2 \quad \text{at 90\% confidence level,}$$

$\overline{C}_V$ being defined as :

$$\overline{C}_V = \frac{Z}{A} \; C_{Vp} + \left(1 - \frac{Z}{A} \right) C_{Vn} \; .$$

Let us compare this result with theory. The theoretical value predicted, using the Weinberg-Salam model, is (2)(6) :

$$E_1^{pv} = -|e| < 6S|z|7S >$$

$$= -i \; 1.7 \times 10^{-11} \; |e| \; a_0 \quad \text{supposing } \sin^2\theta = .35$$

(or 1.3 instead of 1.7 with $\sin^2\theta = .25$).

The theoretical error on these numbers coming from the atomic wave function is believed to be less than 20% because of the simple atomic structure of Cesium.

We conclude that our experimental limit is presently a factor 15 to 20 higher than the value predicted in the frame of the Weinberg-Salam model.

It is of some interest to compare this limit with those derived from other experiments. For instance, the Bi experiment performed by the Washington group (7) yields :

$$|E_1^{pv}| < 10^{-10} \; |e| \; a_0 .$$

One can remark that the absolute sensitivity is not very different for the Cs and the Bi experiments, but the parity violating amplitude is expected to be larger for Bi than for Cs by one order of magnitude, because of Z enhancement effect. This is the price we have to pay for working with a simple atom where theoretical calculations are much more reliable.

Finally, we would like to discuss different improvements planned in our experiment. The limitation is presently not due to any spurious effect or systematic error but rather to a thermal noise coming from the detector, so that the signal-to-noise ratio is proportional to the fluorescence light intensity. We are now working to improve our signal-to-noise ratio in several directions.

First, we have built a multipass cell with internal mirrors, which results in a very substantial increase of the effective intensity of the exciting light. Second, we have found a way of reducing the intrinsic noise of our infrared detector. We can also lengthen the integration time by servo-controlling the frequency and intensity of the laser. So we expect to obtain an improved result as soon as the new version of our experiment is under operating conditions.

References

(1) M.A. Bouchiat and C.C. Bouchiat, Phys. Letters, 48B, 111 (1974); and J. de Phys. 35, 899 (1974)

(2) M.A. Bouchiat and C.C. Bouchiat, J. de Phys. 36, 493 (1975)

(3) M.A. Bouchiat and L. Pottier, Phys. Letters, 62B, 327 (1976)

(4) M.A. Bouchiat and L. Pottier, J. Phys. Lettres, 37, L-79 (1976)

(5) C. Bouchiat, M.A. Bouchiat and L. Pottier : in Atomic Physics 5, ed. by R. Marrus, M. Prior and H. Shugart (Plenum Press, 1977), p. 1

(6) C.E. Loving and P.G. Sandars, J. Phys. B, 8, L-336 (1975)

(7) L.L. Lewis, J.H. Hollister, D.C. Soreide, E.G. Lindahl and E.N. Fortson, Phys. Rev. Letters, 39, 795 (1977)

Parity Nonconservation in Nuclei

NUCLEAR PARITY MIXING - AN EXPERIMENTAL REVIEW

E. G. Adelberger
Physics Department, University of Washington
Seattle, Washington 98195

Abstract: Experiments which detect the isospin structure of parity nonconservation (PNC) in nuclear states may provide information on the space-time and isospin structure of the neutral weak current. Nuclear experiments which measure the $\Delta I=0$, $\Delta I=1$ and $\Delta I=2$ PNC interactions between two nucleons are reviewed.

In this talk I will begin by discussing what we hope to learn from nuclear parity mixing experiments and draw some comparisons and contrasts to the very interesting atomic parity violation studies we heard about yesterday. Next I will give a very biased review of experimental results and conclude by summarizing what has been learned, what is still needed and finally what new results we may expect in the next few years. A complete understanding of nuclear PNC requires first that we be able to obtain the PNC nucleon-nucleon (N-N) force from a weak interaction model and secondly to connect this force to the experimental PNC effects in nuclei. The first of these is a difficult problem in particle physics and will be mentioned only for heuristic purposes. My talk will be concerned primarily with the second point.

Why do we study nuclear parity nonconservation (PNC)? Because it probes the parity and isospin properties of the weak N-N force, and these properties reflect, for example, the nature of the neutral weak current. It is practical to study the parity and isospin properties of the PNC force because nuclear levels are essentially eigenstates of parity and isospin. That nuclear levels have parity as a good quantum number comes as no surprise, but since this audience may not be familiar with the properties of nuclear states I thought it might be interesting to show in fig. 1 one example of just how good the isospin symmetry is for a typical nuclear level. The upper part of the figure shows the cross sections as a function of bombarding energy for the elastic scattering of protons by the I=0 nucleus ^{12}C. Broad resonances are seen which have widths Γ on the order of MeV. These correspond to I=1/2 states in the compound system ^{13}N. The arrows point to very narrow glitches in the cross section. They are caused by isospin forbidden resonances corresponding to the formation of I=3/2 levels in ^{13}N. These I=3/2 levels have no I conserving decays and hence have only small widths. The bottom part of fig. 1 shows one of these I=3/2 resonances on a very expanded scale. Its width is only 0.86±0.12 keV[1] so that within a factor of two, the $^{12}C + p$ resonance is fractionally as narrow as the ψ resonance in e^+e^-.

Since weak interactions do not conserve isospin, the PNC interaction will in general be different in the nn, np and pp systems. To completely determine the interaction we must separately measure $\Delta I=0$, $\Delta I=1$ and $\Delta I=2$ components of the N-N interaction. What is the connection between the isospin properties of the PNC N-N force and interesting fundamental problems, such as the nature of the neutral currents, which are the main topic of this conference? Of course we must realize that hadronic weak processes, such as $NN \to NN$ or even $K \to 2\pi$, are not yet understood at the quantitative level. Therefore I shall give a qualitative discussion based upon a phenomenological weak Lagragian

$$L \propto (J_\mu^+ J^\mu + J_\mu J^{\mu +}) + (Z_\mu^+ Z^\mu + Z_\mu Z^{\mu +})$$

where J represents the familiar charged or Cabibbo weak current and Z the neutral weak current. The charged hadronic current, J, contains two components: $J=\cos\theta_C J^{(1)}+\sin\theta_C J^{(1/2)}$ where $J^{(1)}$ is the $\Delta S=0$, $\Delta I=1$ current active in nuclear

$^{12}C + p$ elastic scattering

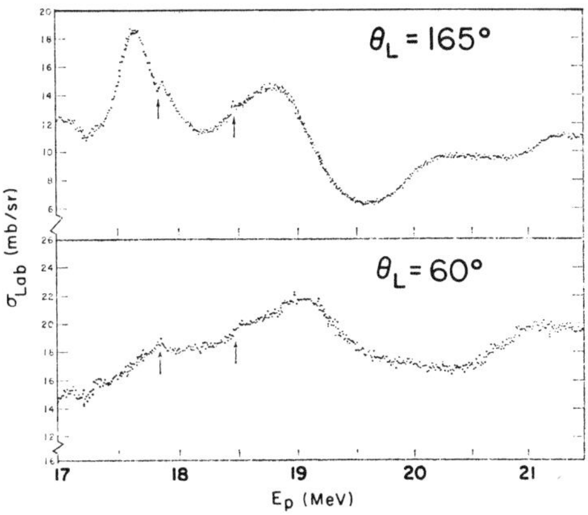

M.J. Levine & P.D. Parker
PR 186 (1969) 1021

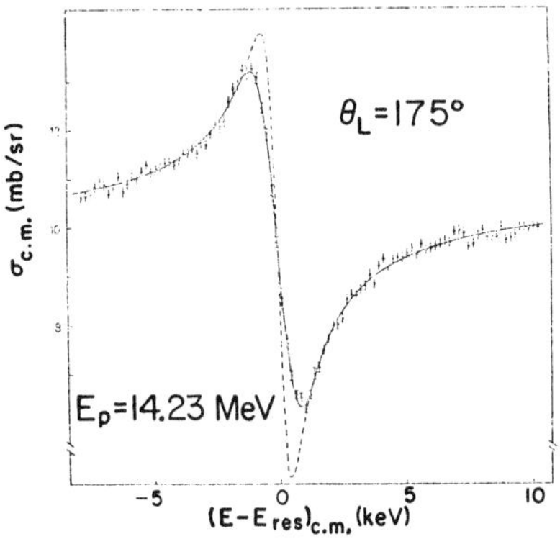

F. Hinterberger *et al.*
NP A 253 (1975) 125

Fig. 1. Cross sections for elastic scattering of protons by ^{12}C.
upper figure -- scattering cross section from E_p=17-21.5 MeV showing broad I-1/2 resonances. The narrow glitches indicated by the arrows are I=3/2 resonances.
lower figure -- an I=3/2 resonance plotted on a very expanded scale. This resonance has a width of 0.9 keV.

β decay, $J^{(1/2)}$ is the $\Delta S=1$, $\Delta I=1/2$ current active in Λ decay, and θ_c is the Cabibbo angle. For this discussion we assume that the hadronic neutral current, Z, also has two components: $Z=\alpha Z^{(0)}+\beta Z^{(1)}$ where $Z^{(0)}$ carries $\Delta S=0$, $\Delta I=0$ and $Z^{(1)}$ carries $\Delta S=0$, $\Delta I=1$. The weak N-N force arises from the products

$$\cos^2\theta_c J^{(1)}\otimes J^{(1)} + \sin^2\theta_c J^{(1/2)}\otimes J^{(1/2)} + \alpha^2 Z^{(0)}\otimes Z^{(0)} + \beta^2 Z^{(1)}\otimes Z^{(1)} + 2\alpha\beta Z^{(0)}\otimes Z^{(1)}$$

where $\otimes$ denotes the symmetric product. Notice the isospin properties of the different products. $Z^{(0)}\otimes Z^{(0)}$ is obviously $\Delta I=0$. $J^{(1)}\otimes J^{(1)}$ and $Z^{(1)}\otimes Z^{(1)}$ are symmetric products of two identical I=1 objects and thus carry $\Delta I=0$ and $\Delta I=2$ but not $\Delta I=1$ which is antisymmetric. Similarly $J^{(1/2)}\otimes J^{(1/2)}$ and $Z^{(0)}\otimes Z^{(1)}$ are $\Delta I=1$. Because the Cabibbo angle is so small ($\tan^2\theta_c \sim 1/20$) the charged currents lead to a $\Delta I=1$ PNC NN force, $\sin^2\theta_c J^{(1/2)}\otimes J^{(1/2)}$, which is quite suppressed(2). On the other hand, the neutral current contribution to the $\Delta I=1$ PNC force, $2\alpha\beta Z^{(0)}\otimes Z^{(1)}$, may be dominant(3) if the neutral currents violate parity and are isoscalar-isovector mixtures as predicted by the Weinberg-Salam model(4). If the neutral currents do not satisfy both of these conditions they will of course have no effect on the $\Delta I=1$ PNC interaction.

In contrast to atomic PNC, nuclear PNC is a considerably more complicated phenomenon. Not only does it involve hadronic, as compared to semileptonic, weak precesses, but also the charged currents as well as the neutral currents are involved.

The main goal of nuclear parity mixing experiments is to determine the isospin dependence of the PNC NN force. Although it is in principle possible to do this with a series of experiments in the NN system, the expected effects are so small that with one exception no one has yet observed positive results. Since experiments in the two body system are so difficult, the first measurements were in heavy nuclei where experiments to detect the circular polarization (CP) of γ-rays from randomly oriented radioactive sources were feasible. (See ref. 5 for a review.)

The early experiments revealed a first order weak interaction between nucleons as predicted by the current-current theory. However controversies arose over the quantitative interpretation of the results--for a time these seemed to be much larger than expected. The difficulties were due to the complex structure of nuclei which were selected for experimental convenience rather than theoretical tractibility. Consider, for example, the well known case of the 482 keV transition in ^{181}Ta which is shown in fig. 2. Experiments from a number of different laboratories[(5)] yield $P_\gamma = -(5\pm1)\times10^{-6}$. The CP arises from the interference of a PNC E1 transition with the allowed M1 transition. Some of the difficulties in calculating P_γ given the fundamental PNC NN interaction are obvious from simple perturbation theory arguments. Then the wavefunctions of the ground state and 482 keV level are

$$|g.s.\rangle + |7/2^+\rangle + \sum_i \alpha_i |7/2^-;i\rangle \qquad \alpha_i = \frac{\langle 7/2;-i|H_{PNC}|7/2^+\rangle}{E_{7/2;i}-E_{7/2+}}$$

$$|482\rangle + |5/2^+\rangle + \sum_j \beta_j |5/2^-;j\rangle \qquad \beta_j = \frac{\langle 5/2;-j|H_{PNC}|5/2^+\rangle}{E_{5/2;j}-E_{5/2+}}$$

and the PNC E1 matrix element is

$$\langle g.s.|E1|482\rangle = \sum_j \beta_j \langle 7/2^+|E1|5/2^-;j\rangle + \sum_i \alpha_i \langle 7/2^-;i|E1|5/2^+\rangle$$

Unfortunately no single level essentially exhausts the sums over i and j (although the giant dipole resonances play an important role in spite of their large energy denominators because of their very large E1 matrix elements) so that one is faced with computing a coherent sum of many contributions. It makes the situation in the Bi atom look simple!

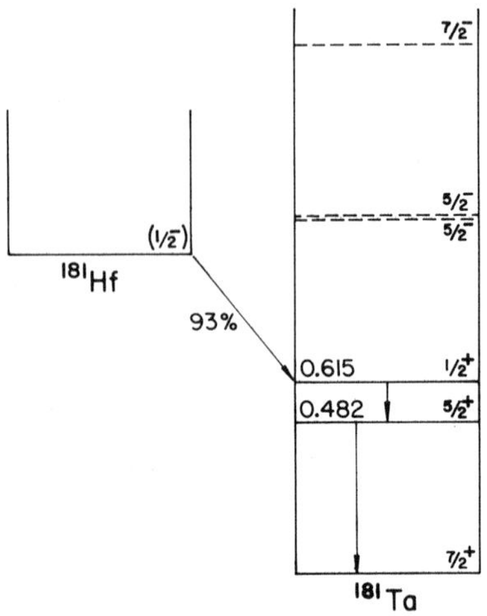

Fig. 2. Levels of ^{181}Ta involved in the measurement of the CP of the 482 keV γ-ray. Solid lines indicate levels established experimentally. No negative parity levels are known experimentally which could be admixed into the ground and 482 keV states. A few of the possible admixed states expected theoretically are shown as dashed lines.

A real advance in the study of nuclear PNC was made when techniques were developed for using activities induced by reactor and accelerator beams rather than β decay. This development considerably enlarged the number of nuclei accessible to experiments and for the first time made it possible to study systems which reveal the isospin properties of the PNC interaction. And we shall see that some of these nuclear systems turn out to be so remarkably simple that the uncertainties from the many body problem are quite small. This of course is quite similar to the pattern in atomic PNC, where the first experiments were done in heavy atoms which are most attractive from an experimental viewpoint. However, controversies arose over the interpretation of the results so new experiments are done to investigate theoretically attractive cases such as Ti and Cs. Finally efforts are begun to study the two body system which has a simple interpretation but requires the detection of tiny effects.

I will now briefly review some of the more interesting nuclear experiments. Although there are quite a number of PNC experiments that one can imagine doing, there are only a limited number of independent PNC quantities which one can expect to determine reliably from these experiments. This is easily seen by recalling that the nucleons inside complex nuclei are moving very slowly ($E_F \sim 30$ MeV) so that one is essentially at the low-energy limit in the NN interaction. Then the only important PNC transitions are the five involving s and p states

$$^3S_1 \rightarrow {}^3P_1 \qquad \Delta I = 1$$

$$^1S_0 \rightarrow {}^3P_0 \qquad \Delta I = 0,1,2$$

$$^3S_1 \rightarrow {}^1P_1 \qquad \Delta I = 0$$

where we have labelled the states by $^{2S+1}L_J$. Ideally one would like to find 5 experiments, each of which measured one of the transitions above. This is not possible, but nature has provided us with nuclear systems which involve linear combinations of the various transitions which are sufficiently difficult to allow in principle the unfolding of the different contributions. A particularly "clean" case occurs in the low energy np → dγ system; there are two pseudoscalar observables, P_γ the CP of γ-rays induced by unpolarized neutrons incident on unpolarized protons, and A_γ the asymmetry of the photons with respect to the spins of polarized neutrons incident on unpolarized protons. Danilov(6) has shown that at low energies P_γ measures the ΔI=0 and ΔI=2 PNC interactions while A_γ measures only the ΔI=1 interaction.

Measurements of P_γ and A_γ have been made using thermal neutrons from reactors. The P_γ experiment was done at a reactor in Leningrad by Lobashov *et al.*(7). Neutrons were captured in a H_2O target located in the center of the reactor, and the CP of the capture radiation was analyzed in a transmission Compton polarimeter. Such polarimeters employ the $\sigma_\gamma \cdot \sigma_e$ dependence of the cross section for Compton scattering of photons from polarized electrons in a ferromagnet. The analyzing power is necessarily small since only 2/26 of the electrons in Fe are polarized. In fact Lobashov's polarimeter had an analyzing power of ≈5%. The flux of gamma rays was measured using an integral technique developed by Lobashov and coworkers in previous experiments. To avoid the rate limitation inherent in pulse-by-pulse counting schemes, Lobashov measured the integrated detector current. A disadvantage of this technique is that one loses any pulse height discrimination so that one cannot select from a background of uninteresting counts only those events having a particular energy. In the np → dγ experiment a particularly troublesome background arises from bremsstrahlung of electrons from β^- decays of the fission products, since the β^-'s are left handed the bremsstrahlung has a large negative polarization. Lobashov *et al.* find that $P_\gamma = -(1.30 \pm 0.45) \times 10^{-6}$. This is much larger than expected--most calculations predict effects $\sim$100 times smaller[8].

An upper limit on A_γ has been obtained using the polarized cold neutron beam at the Grenoble reactor[9]. Neutrons with a polarization of $\sim$70% were produced by conducting a beam of neutrons thermalized in liquid deuterium through a magnetized Fe-Co mirror guide tube. The neutron spins were reversed every 0.8 second by non-adiabatic transitions in external magnetic guide fields[10]. Neutrons were detected in plastic scintillators viewed by phototubes and integral detection methods were employed. The present measured value $A_\gamma = (0.6 \pm 2.1) \times 10^{-7}$ is unfortunately not sufficiently precise to test the theory. For example Lassey and McKellar[8] predict $A_\gamma \sim (4-7) \times 10^{-9}$ from the charged current contribution and $A_\gamma \sim (6-11) \times 10^{-8}$ when the Weinberg-Salam model of the fundamental weak interaction is employed.

The last experiment in the two nucleon system is a very sophisticated measurement at Los Alamos which has placed an impressively small upper limit on a pseudoscalar observable in the $\vec{p}$+p system at 15 MeV[11]. The experiment searches for longitudinal analyzing power $A_L = \frac{\sigma_+ - \sigma_-}{\sigma_+ + \sigma_-}$ in the total cross section $\sigma_\pm$ for the scattering by hydrogen of longitudinally polarized protons with + or - helicity. Since this experiment will be discussed later in this session I will only note that the p+p system is sensitive to the $\Delta I=0$ and $\Delta I=2$ amplitudes and also the short range part of the $\Delta I=1$ interaction. The present experimental upper limit[12] on A_L ($\sim 10^{-7}$) is comparable to the theoretical predictions[13] and already places strong constraints on the $\Delta I=0$, 1 and 2 amplitudes.

Our knowledge of the $\Delta I=0$ interaction is largely based on the rate of the PNC decay (see fig. 3) of the 8.87 MeV 2^- I=0 level of ^{16}O to the ground state of $^{12}C(0^+I=0) + \alpha(0^+I=0)$. Since this is the only known pure $\Delta I=0$ effect

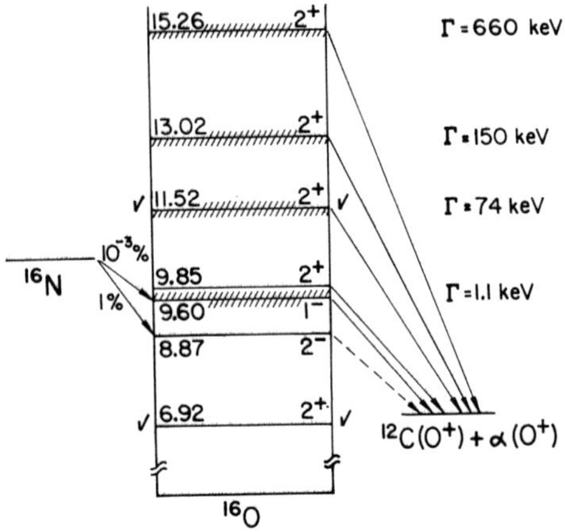

Fig. 3. Levels of ^{16}O involved in the PNC α decay of the 8.87 MeV 2^- level. The two 2^+ levels believed to play the dominant role in the parity impurity are denoted by check marks.

let me spend a moment on this case. Wäffler and collaborators at Mainz[14] selectively populated the 8.87 MeV level of ^{16}O by the β^- decay of the short lived isotope ^{16}N which they produced continuously on a small accelerator. The spectrum of α particles emitted following the β^- decay of ^{16}N showed a very small peak which has precisely the energy and width expected for the PNC α decay of the 2^- level. The PNC α width is $\Gamma_\alpha^{PNC} = (1.03\pm0.28)\times10^{-10}$ eV. This width is interpreted in perturbation theory as arising from the parity conserving decays of nearby 2^+ states (see fig. 3) mixed into the 2^- level by the $\Delta I=0$ PNC interaction. (Parity impurities in the ground states of ^{12}C and ^4He are neglected since neither of these nuclei has 0^- levels below 18 MeV excitation energy.) The amplitudes for α decay of all but one of these 2^+ levels is known (except for a sign) from the decay widths of the levels. The decay amplitude for one level cannot be measured directly since it lies below the $^{12}C + \alpha$ threshold and it must be taken from theory or inferred with large errors from nuclear reaction data. According to theory (which is not contradicted by the data) the subthreshold level plays the dominant role in the PNC α decay with the next most important 2^+ level being the 11.5 MeV state which interferes destructively. The predicted α width is $\sim1\times10^{-10}$ eV which is in good agreement with experiment[5]. However because of the destructive interference, the prediction is sensitive to the decay amplitude of the subthreshold state, which is not well tied down by experiment.

Now let us examine the systems which tell us about the $\Delta I=1$ PNC interaction. First consider the case of ^{19}F which we are currently studying in Seattle[15]. A previous version of this experiment has already been published[16]. Because of the favorable energy splittings (see fig. 4) the parity impurities in the ground ($1/2^+$) and 110 keV ($1/2^-$) states are well

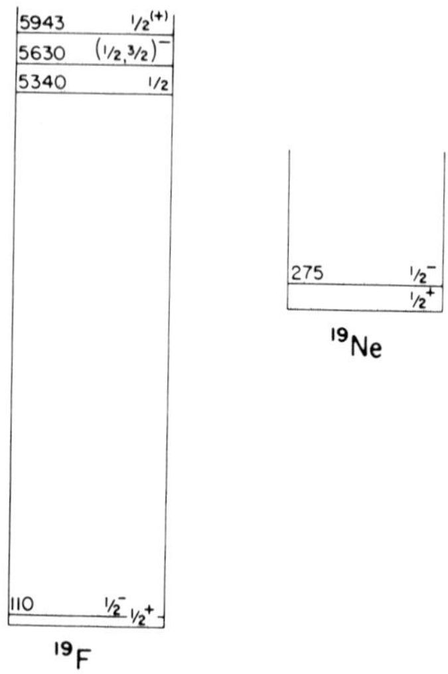

Fig. 4. Low lying J=1/2 levels of ^{19}F. The parity admixtures in the ground state doublet are well approximated by simple two state mixing.

approximated by simple two state mixing

$$|110\rangle = |-\rangle + \varepsilon|+\rangle \qquad \varepsilon = \frac{\langle -|H_{PV}|+\rangle}{110\ \text{keV}}$$
$$|g.s.\rangle = |+\rangle - \varepsilon|-\rangle$$

so that a measure of the parity impurity in this case directly yields the PV matrix element. In this case both the $\Delta I = 0$ and $\Delta I = 1$ PV interactions are involved since the levels have $I = 1/2$. To measure ε one needs to observe a pseudoscalar quantity. We chose to polarize the ^{19}F nuclei in the 110 keV state and measure the PNC anisotropy of the 110 keV γ rays: $W(\theta) = 1 + \delta P_F\ \vec{\sigma}_F\cdot\vec{k}_\gamma$, where P_F is the net polarization of the $^{19}F^*$'s. The relation between δ and ε is found by considering the mixed M1/E1 γ-transition between the 110 keV and ground states

$$\langle g.s.|E1 + M1|110\rangle = \langle +|E1|-\rangle +$$
$$\varepsilon\{\langle +|M1|+\rangle - \langle -|M1|-\rangle\} + \theta(\varepsilon^2).$$

We find δ from the M1/E1 interference

$$\delta = 2\varepsilon\frac{\{\langle +|M1|+\rangle - \langle -|M1|-\rangle\}}{\langle +|E1|\ \rangle}$$

Since the E1 matrix element is determined by the radiative lifetime of the 110 keV state and the M1 matrix elements are determined by the magnetic moments of the ground and 110 keV levels, a measurement of δ directly measures $\langle -|H_{PV}|+\rangle$ (Ref. 16). The experimental effect is magnified because the M1 matrix element is very strong (1.6 W.u.) while the E1 is retarded ($\sim 10^{-3}$ W.u.).

There have been several calculations of the asymmetry coefficient δ. A complete account of the nuclear physics aspects and of other calculations is given in ref. 18. In the Cabibbo model with factorization (no neutral current contribution to the PNC) the $\Delta I=0$ PV amplitude (P exchange) dominates and ref. 18 obtains $\delta \sim + 7\times 10^{-5}$, which is in $\sim\pm 30\%$ agreement with other calculations. It is thought that in the Weinberg-Salam (W-S) model the $\Delta I=1$ PV amplitude dominates because it is mediated by π exchange and not significantly suppressed by short range repulsion of the nucleons. In the W-S framework ref. 18 (the sign of f_π has been taken from ref. 3) obtains $\delta = \delta_{(0)} + \delta_{(1)} = (+9-40)\times10^{-5} = -31\times10^{-5}$. Our 1975 experiment, corrected for an improved value of the $^{19}F^*$ polarization, yielded $\delta = -(24\pm12)\times10^{-5}$. Within large errors this is consistent with the W-S model predictions but not with the Cabibbo model predictions of ref. 18. Clearly a measurement with smaller errors would be desirable. We are currently running a new experiment in ^{19}F and expect to reduce the errors in δ by a factor of 3-4.

The experimental geometry is shown in fig. 5. Polarized $^{19}F^*$'s are produced in the $^{22}Ne(p,\alpha)$ reaction induced by polarized protons. This reaction transfers 0.55±0.07 of the proton polarization to the $^{19}F^*$'s. Gamma rays are detected in two 10 $cm^2\times7$ mm thin-window planar Ge(Li)'s and digitized in two fast (7 μsec/event) pulse height analyzers. The outputs of the ADC's are routed into different regions of memory according to spin state. A gamma ray spectrum is shown in fig. 6. The main changes from our previous experiment[16] are in the polarized ion source and in the on-line polarization monitor. The proton spin is switched rapidly every millisecond using a very elegant scheme developed at Los Alamos[19] for the $\vec{p}$+p experiment mentioned above. This permits the proton spin to be reversed by applying, to neutral H(2s) atoms, a magnetic field having $\int B\cdot bd \approx 10$ gauss-cm. As a result, compared to ref. 16, we now have vastly smaller spin correlated modulations of the beam emittance and intensity and the proton polarization has increased from 42% to 78%. The proton spin magnitude and direction are continuously monitored by 4 surface barrier detectors which count protons

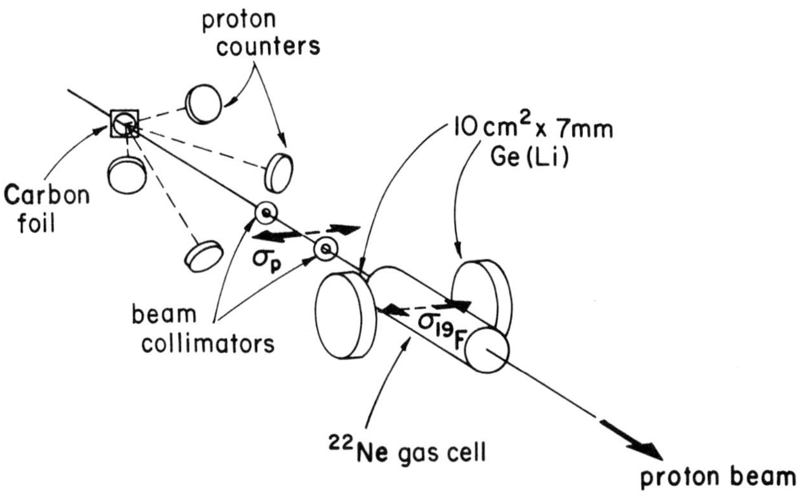

Fig. 5. Schematic layout of the new experiment to measure the parity mixing in ^{19}F.

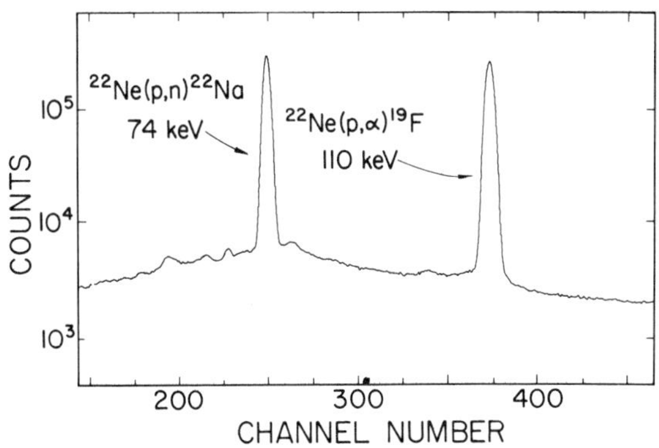

Fig. 6. Spectrum of γ-rays observed in the Ge(Li) detectors. We study the PNC asymmetry in the 110 keV γ-ray from ^{19}F. The 74 keV γ-ray provides null calibration as described in the text.

scattered at $\theta=52°$, $\phi=0°$, $90°$, $180°$ and $270°$ from a thin Au plus C target. Under these conditions the scattering from Au is Rutherford, while the scattering from C has a very large analyzing power $A = -0.82$. We use the polarimeter combined with an on-line computer as a digital feedback device which locks the spin direction to an accuracy of $0.1°$.

The sensitivity to many spurious modulations is greatly reduced by design of the experiment. We have chosen a bombarding energy so that in addition to the 110 keV radiation from $^{22}Ne(p,\alpha)$ we also produce 74 keV radiation from $^{22}Ne(p,n)$. The 74 keV γ-rays come from the decay of a $J = 0$ level and hence are rigorously isotropic, which permits us to measure the anisotropy of the 110 keV photons and at the same time check our techniques for a null effect in the 74 keV γ-rays. Preliminary results of our new experiment are very encouraging. The apparent anisotropy of the 74 keV photons is now smaller than the anisotropy of the 110 keV γ's. Our preliminary interim value from the new experiment is $\delta = -(6.0\pm3.3)\times10^{-5}$, which agrees in sign, but has a smaller magnitude than the Weinberg-Salam model calculations mentioned above.

We must emphasize that our value comes from the on-line analysis and is subject to revision from more careful analysis. We present it here only to demonstrate that the effect in ^{19}F is considerably smaller than predicted.

A particularly important pseudoscalar observable is the CP of the 1081 keV γ-ray in ^{18}F (see fig. 7). This transition probes only the $\Delta I=1$ PNC interaction, as did A_γ in $\vec{n}+p\rightarrow d+\gamma$, but in ^{18}F the expected effects are larger by a factor of $\sim10^5$ due to an "accidental" near degeneracy of the 1081 keV 0^+ $I=1$ and 1042 keV 0^- $I=0$ levels[20]. The two levels are separated by only 39 keV! Furthermore the M1 decay of the 0^+ $I=1$ level is isospin favored

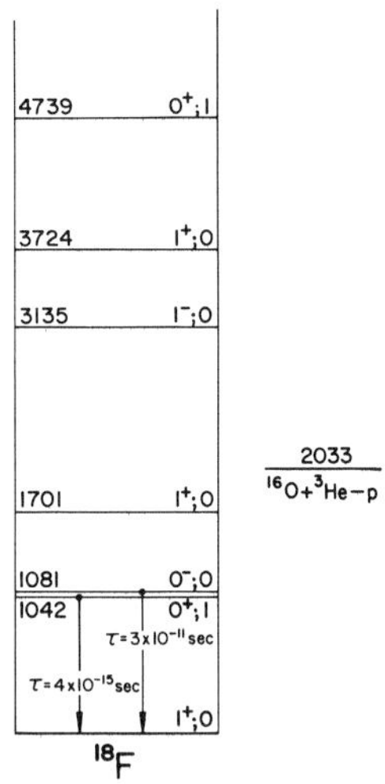

Fig. 7. Spectrum of low-lying J=0 and J=1 levels of ^{18}F. The CP of the 1081 keV γ-ray is sensitive only to the $\Delta I=1$ interaction.

while the E1 decay of the 0^- I=0 level is isospin forbidden. This causes the measured E1 lifetime to be $\sim 10^5$ times longer than the measured M1 lifetime, and means that a very small admixture of the 0^+ state into the 0^- level will cause a relatively large CP. In fact, the CP of the 1081 keV γ-rays is well described by two state mixing

$$CP_{1081} = 2\frac{\langle -|H_{PNC}|+\rangle}{39\ \text{keV}} \quad \frac{\langle E1\rangle}{\langle E1\rangle}$$

where $|+\rangle$ and $|-\rangle$ denote the 0^+ and 0^- levels and $\langle M1\rangle$ and $\langle E1\rangle$ the reduced matrix elements for γ-decay of the 1042 and 1081 keV levels respectively. Hence a measurement of CP_{1081} yields a value of $\langle -|H_{PNC}|+\rangle$ with essentially no uncertainties from the nuclear physics. The PNC in ^{19}F and ^{18}F should be contrasted to the PNC in the Bi atom where the largest uncertainty lies in the calculated irregular E1 transition speed. Because of the validity of the two level mixing approximation in ^{19}F and ^{18}F one can take the irregular M1 matrix elements directly from experiment.

Details of an experiment to measure CP_{1081} will be reported later in the session. Here I will merely note that the measurement gives an upper limit of $CP_{1081} = -(0.6\pm2.0)\times10^{-3}$. The most recent calculation of the nuclear physics aspects of this is by Gari *et al.*[20]. They predict $|CP_{1081}| = 3.6\times10^{-4}$ from the charged currents alone (Cabibbo model) and $|CP_{1081}| = 5.6\times10^{-3}$ if a neutral current contribution based on the Weinberg-Salam model is included. Further theoretical and experimental work on this case would be very valuable. In particular it would be valuable to understand just how sensitive the Cabibbo model predictions are to details of the nuclear wavefunctions.

What have we learned so far from the nuclear PNC experiments? First of all, with the exception of the Lobashov experiment, the orders of magnitudes of the observed effects make sense--they are neither 100 times larger nor smaller than one expects. Second, the observed effects in a number of different nuclei are apparently consistent with one another. In fact it is possible to account for all of the experimental results in those cases with tractable nuclear physics in terms of a consistent set of N-N PNC transitions[21]. This shows that the strictly nuclear aspects of the problem are reasonably well understood. The signs of the pseudoscalar observable in ^{19}F, ^{41}K, ^{175}Lu and ^{181}Ta, nuclei with an odd number of protons and an even number of neutrons, are all consistent with a PNC interaction having a large $\Delta I=1$ component. It is interesting that this is not expected from the charged weak currents alone but in qualitative accord with estimates[3,21] employing the Weinberg-Salam model of the neutral currents. Third, there is evidence from ^{18}F and ^{19}F that the enhancement of the $\Delta I=1$ component of the PNC interaction is not as large as predicted by some calculations based on the Weinberg-Salam model of the fundamental weak interaction.

What new information would we like to have? First of all we need considerably more experiments. So far most of the interesting cases have been done only once and with results which are either only a couple of error bars away from zero or else merely upper limits. It is very important to have these few interesting cases confirmed by other and hopefully more precise measurements.

In particular it is essential to repeat the Lobashov experiment. Porrman and Gari[22], and Zeigler have suggested that it may be easier to do this experiment backwards, to photodisintegrate the deuteron using circularly polarized bremsstrahlung from longitudinally polarized electrons. This very clever proposal gets around two of the biggest difficulties in the Lobashov experiment, the background in the reactor of circularly polarized bremstrahlung from β^- decays and the small analyzing power of Compton polarimeters. It is also important to have more precise results in the p+p system since the present datum already provides a strong constraint on the $\Delta I=0$, $\Delta I=1$ and $\Delta I=2$ amplitudes. It would be especially valuable to have a more precise measurement of a pure $\Delta I=1$ transition in a theoretically tractable system such as $\vec{n}$+p or ^{18}F.

To attain much higher precision in nuclear PNC experiments requires 1) either a clever new idea, such as that by Porrmann and Gari(22), or 2) further developments in the technology such as polarized ion sources and polarized neutron beams, or 3) sufficient beam time on high current accelera-tion and high flux reactors and in all cases 4) much hard work. It is particularly unfortunate that so many small high current accelerators are being shut down that it is now very hard to find sufficient beam time on machines suitable for many of the interesting experiments.

However it is very encouraging that a number of new experiments are under discussion or actually being performed. Among these are

1) Improved measurements in $\vec{p}$+p at LASL at 15 MeV as well as a new experiment at 800 MeV(12).
2) New measurements of $\vec{p}$+p and $\vec{p}$+d at ~50 MeV at Texas A&M(23) and SIN(24). The predicted effect has a maximum in the neighborhood of 50 MeV.
3) A pure $\Delta I=1$ transition in $^6\vec{Li}(\alpha,\gamma)^{10}B$ is being investigated at Argonne by a clever technique involving a polarized target(25).
4) New measurements of the pure $\Delta I=1$ transition in ^{18}F are under way at Mainz(26) and Dubna(27).

5) A pure $\Delta I=1$ transition in $^{16}O(\alpha,\gamma)^{20}Ne$ is being studied at MSU[28].

6) In ^{21}Ne the parity mixing of $(1/2^-, I=1/2)$ and $1/2^+, I=1/2)$ levels is being investigated at several laboratories. This case is similar to ^{19}F with the very important difference that it will be the first odd neutron nucleus to be studied. If the PNC interaction is predominantly isovector one should be able to determine whether the PNC interaction is predominantly isoscalar or isovector by comparing the signs of the effects in $I_3=+1/2$ and $I=-1/2$ nuclei.

It is a pleasure to report that we can expect so many interesting new experimental results in the next few years.

References

1. R.E. Marrs, E.G. Adelberger, and K.A. Snover, Phys. Rev. C 16 (1977) 61.
2. E.M. Henley, Phys. Lett. 28B (1968) 1.
3. As mentioned, the value of the $\Delta I=1$ PNC NN interaction is difficult to calculate in the W-S model. The following papers give different methods for doing this. All nuclear calculations described in this talk use the W-S value of f_π as given by Gari and Reid. M. Gari and J. Reid, Phys. Lett. 53B (1974) 237; J. Donoghue, Phys. Rev. D 13 (1976) 2064 and D 15 (1977) 184; B. Desplanques and J. Micheli, Phys. Lett. 68B (1977) 339.
4. S. Weinberg, Phys. Rev. Letts. 19 (1967) 1264; 27 (1971) 1688. A. Salam, Elementary Part. Theory (Nobel Symposium No. 8), ed. N. Svartholm (John Wiley, New York, 1969).
5. M. Gari, Phys. Rep. 6C (1973) 319.
6. G.S. Danilov, Phys. Lett. 18 (1965) 40.
7. V.M. Lobashov, D.M. Kaminker, G.I. Kharkevich, V.A. Kniazkov, N.A. Lozovoy, V.A. Nazarenko, L.F. Sayenko, L.M. Smotritsky, and A.I. Yegorov, Nucl. Phys. A197 (1972) 241.

8. K.R. Lassey and B.H.J. McKellar, Nucl. Phys. A260 (1976) 413.

9. J.F. Cavaignac, B. Vignon, and R. Wilson, Phys. Lett. 67B (1977) 148.

10. P. Liand, R.I. Steinberg, and B. Vignon, Nucl. Inst. Meth. 125 (1975) 7.

11. J.M. Potter, J.D. Bowman, C.F. Hwang, J.L. McKibben, R.E. Mischke, D.E. Nagle, P.G. Debrunner, H. Frauenfelder, and L.B. Sorensen, Phys. Rev. Lett. 33 (1974) 1307.

12. J.D. Bowman, private communication (1977).

13. V.R. Brown, E.M. Henley, and F.R. Krejs, Phys. Rev. Lett. 30 (1973) 770; M. Simonius, Nucl. Phys. A220 (1974) 209.

14. K. Neubeck, H. Schober, and H. Wäffler, Phys. Rev. C 10 (1974) 320.

15. E.G. Adelberger, H.E. Swanson, T.A. Trainor, Z. Iqbal, and R. Von Lintig, private communication (1977).

16. E.G. Adelberger, H.E. Swanson, M.D. Cooper, J.W. Tape, and T.A. Trainor, Phys. Rev. Lett. 34 (1975) 402.

17. E. Maqueda, Phys. Lett. 23 (1960) 571.

18. M.A. Box, A. J. Gabric, and B.H.J. McKellar, Nucl. Phys. A271 (1976) 412.

19. J.L. McKibben and J.M. Potter, Proc. Fourth Int. Symp. on Polarization Phenomena in Nucl. Reactions, ed. W. Grüebler and V. König, Birkhäuser Verlag Basel (1976) 863; T.A. Trainor, W.B. Ingalls, H.E. Swanson, and E.G. Adelberger, IEEE Trans. Nucl. Sci. NS-24 (1977) 1527.

20. M. Gari, J.B. McGrory, and R. Offerman, Phys. Lett. 55B (1975) 277.

21. M.A. Box, A.J. Gabric, K.R. Lassey, and B.H.J. McKellar, Journ. Phys. G: Nucl. Phys. 2 (1976) L107; B. Desplanques and J. Missimer, submitted to Nucl. Phys. (1977).

22. M. Porrmann and M. Gari, Phys. Rev. Lett. 38 (1977) 947.

23. R. E. Tribble, private communication (1977).

24. M. Simonius, private communication (1977).

25. S. J. Freedman, private communication (1977).

26. H. Wäffler, private communication (1976).

27. D. Chultem, private communication (1977).

28. R.G.H. Robertson, private communication (1977).

NUCLEAR PARITY VIOLATION IN NEUTRON CAPTURE

Richard Wilson*
Department of Physics
Harvard University
Cambridge, MA 02138

ABSTRACT

Measurements of asymmetries in neutron capture γ rays on nuclei are discussed. Also, an experiment giving an upper limit in np capture is described.

As I give this report I will also, at the request of my friend, Professor Yuri Abov, bring to your attention the work of his group at the Institute of Theoretical and Experimental Physics (ITEP) at Moscow.

The first experiment which showed parity violation in nuclei was, in fact, done by Professor Abov and collaborators. He studied the capture of polarized neutrons by Cd^{113} and measured the asymmetry of the ground state gamma ray, the 9.1 Mev γ ray in Cd^{114} as shown in the level scheme of figure 1.[1] The 9.1 Mev state has $J^p = 1^+$ and the ground state 0^+, so this is a magnetic dipole transition which we might expect to be hindered with respect to electric dipole transitions by $(c/v)^2$ where v is the velocity of nucleons in the nucleus (= c/10). We expect several 1^- states above 9.1 Mev as part of the giant dipole state, so that we expect a mixing of these states due to parity violation.

Abov used the thermal neutron beam from the ITEP reactor, and polarized it by reflection from Co mirrors as shown in figure 2.

Abov did indeed find an asymmetry of 3×10^{-4} of the γ ray with respect to the direction of the polarized neutron and therefore with respect to the

*Aided by a contract from the National Science Foundation.

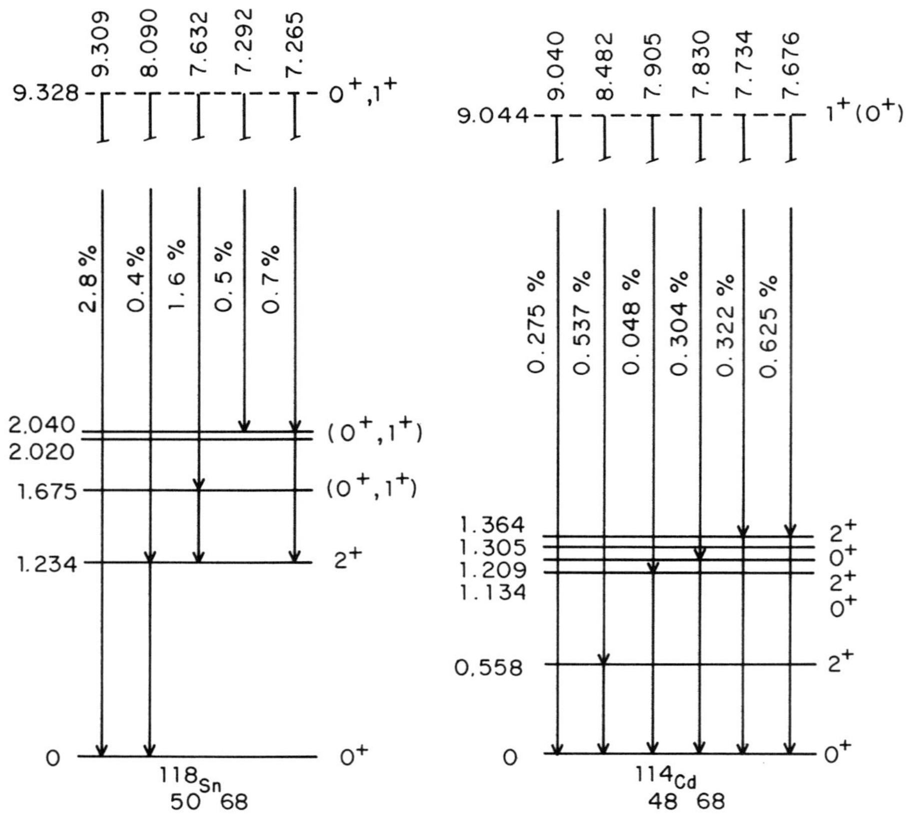

FIG. 1 COMPARISON OF LEVEL AND DECAY SCHEMES

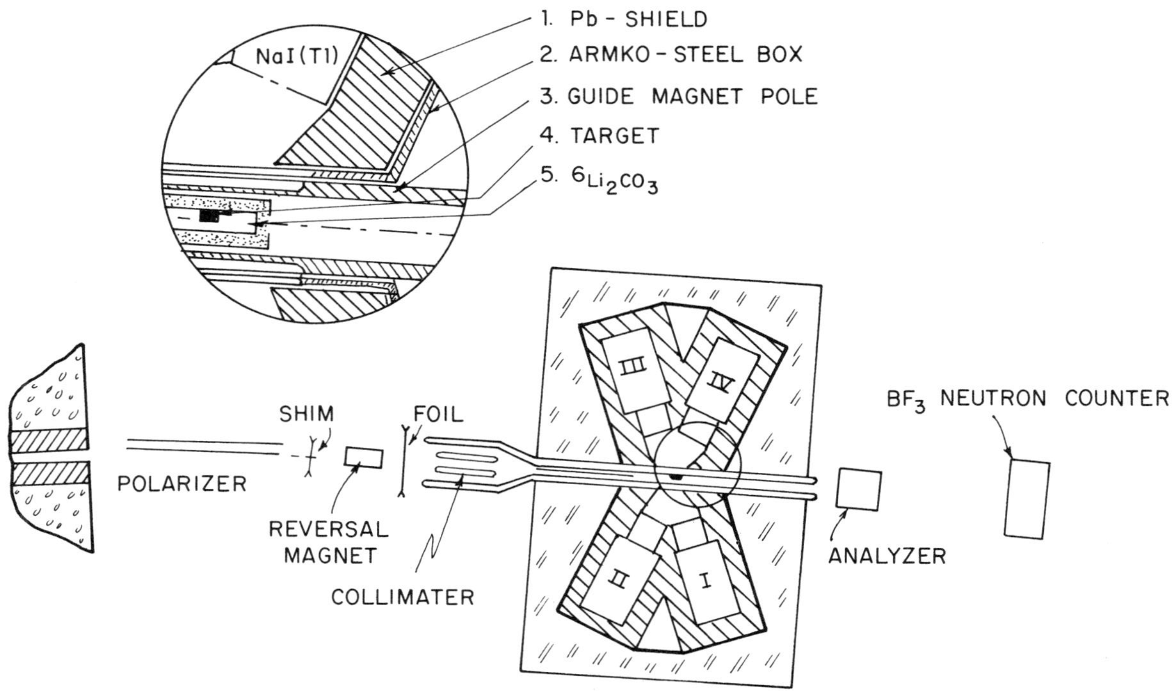

FIG. 2 EXPERIMENTAL ARRANGEMENT

direction of the polarized state of Cd^{114} at 9.1 Mev. Other workers failed to find an asymmetry. I will present a possible explanation shortly.

We can also produce the Cd^{114} in an unpolarized state; there will then be no asymmetry of the γ ray but the parity violation can manifest itself as a circular polarization of the γ ray. Alberi, et. al.[2] found this circular polarization in an experiment at the National Bureau of Standards. If we look at the theory of the subject[3] we find that the circular polarization is related to the asymmetry by P = A. As we look again at the level scheme of figure 1, we notice that there is also a transition 8.5 Mev to the 1st excited state of Cd^{114} with $J^P = 2^+$. The parity violating asymmetry of the transition to this state is expected to be of opposite sign to the asymmetry of the ground state transition. Experimentally, the detector resolution cannot always separate the γ rays and an asymmetry experiment can mix the asymmetries and get a zero result. I believe that this is what happened in the attempts to repeat Abov's experiment.[4] Fortunately in the circular polarization experiment, the polarizations are expected to be of the same sign. For the 8.5 Mev transition we expect P = -2A (assuming no electric quadrupole admixture).

In table 1 I list all known nuclear parity violating transitions and a few interesting null results. I take $-(4 \pm 1) \times 10^{-4}$ for the asymmetry in the reaction Cd^{113} (nγ) Cd^{114} being an average of several results. I call attention to the recent results on Sn^{117} (nγ) Sn^{118} both at ITEP[5] and Grenoble.[6] Sn^{118} is a nucleus similar to Cd^{114}, but the first excited state is well separated from the ground state so that the experimental problems are less once a target of tin enriched in the isotope Sn^{117} is available.

The magnitude of the asymmetry in Sn^{117} (nγ) Sn^{118} is similar to that in Cd^{113} (nγ) Cd^{114}, but is of opposite sign. If we assume that the corresponding

Table 1

Laboratory			$\times 10^{-6}$	W-S Theory
LASL	$p + p \quad \sigma_{tot}$	a	0.0 ± 0.15	$\simeq 0.1$
LASL	$p + d \quad \sigma_{tot}$	a	0.0 ± 0.15	$\simeq 0.1$
GRENOBLE	$N + p \rightarrow D + \gamma$	a	0.06 ± 0.2	$\simeq 0.2$
LENINGRAD	$N + p \rightarrow D + \gamma$	P_γ	-1.3 ± 0.5	
MAINZ	$O^{16}(2^-) \rightarrow \alpha C^{12}$	$\sqrt{\Gamma_\alpha}$	$(10 \pm 1)\ \sqrt{ev}$	$\simeq 10$
CAL TECH	$F^{18}\ (0^- \rightarrow 1^+)$	P_γ	-500 ± 2000	$\simeq 3500$
U WASH	$F^{19}\ \frac{1}{2}^- \rightarrow \frac{1}{2}^+$	a	-60 ± 33	$\simeq -170$
NBS	$Ne^{21}\ (\frac{1}{2}^+ \rightarrow \frac{3}{2}^+)$	P_γ	in progress	
LENINGRAD	$K^{41}\ (\frac{7}{2} \rightarrow \frac{3}{2}^+)$	P_γ	20 ± 4	
NBS	$Cd^{114}\ (1^+ \rightarrow 0^+)$	P_γ	-600 ± 150	
ITEP	$Cd^{114}\ (1^+ \rightarrow 0^+)$	a	-400 ± 100	
GRENOBLE	$Sn^{118}\ (1^+ \rightarrow 0^+)$	a	440 ± 60	
ITEP	$Sn^{118}\ (1^+ \rightarrow 0^+)$	a	850 ± 150	
LENINGRAD	$Lu^{175}\ (\frac{9}{2}^- \rightarrow \frac{7}{2}^+)$	P_γ	55 ± 7	
DARMSTADT	$Hf^{180}\ (8 \rightarrow 6)$	P_γ	-2400 ± 300	
LASL	$Hf^{180}\ (8 \rightarrow 6)$	a	-1400 ± 700	
LENINGRAD	$Ta^{181}\ (\frac{5}{2}^+ \rightarrow \frac{7}{2}^+)$	P_γ	-5.2 ± 0.5	
ITEP	Fission (U_{235})	a	170 ± 40	
ITEP	Fission (P_{239})	a	-480 ± 80	

states are similar, this would at first sight appear to present a problem. I think it can be resolved by assuming that the closest 1^- state which is mixed with the 1^+ state is at an energy above the 9.1 Mev state in one case, and below it in another, so that the admixture

$$\frac{\langle 1^+ \mid V_{pv} \mid 1^- \rangle}{E_{1^+} - E_{1^-}}$$

changes sign.

I am unable to account for the difference in magnitude measured in the two experiments. In our case, we are sure that the spurious asymmetries are low, both because of the internal tests in the experiment, and because the same apparatus has been used to measure a null effect on hydrogen 1000 times smaller.

Abov's group also have measured the asymmetry in uranium and plutonium fission. Again the states are uncertain, and so is the detailed interpretation.

This now brings me to the measurements of neutron capture on hydrogen. It is obvious that we know the wave functions in this case somewhat better than for the heavy nuclei. Following an original calculation of Danilov,[7] Wolfenstein[8] shows what states can be involved in the capture process.

Allowed by parity are

NP $^1S \rightarrow D + \gamma$ (M1) (1)

$^2S \rightarrow D + \gamma$ (M1) (2)

$^3S \rightarrow D + \gamma$ (E2) (3)

Forbidden by parity

$^1S \rightarrow D + \gamma$ (E1) (4)

$^2S \rightarrow D + \gamma$ (E1) (5)

$^3S \rightarrow D + \gamma$ (M2) (6)

The first of these is the dominant allowed transition.

To find the circular polarization we need to mix a parity violation transition starting with the same initial state--transition 4. This is clearly mixing by an isotopic spin conserving potential $\Delta I = 0$ or $\Delta I = 2$.

To find an asymmetry from a polarized initial state we need to have a state which can remember the direction of the initial polarization. Thus we must mix 5 or 6, and 5 is more probable. This mixing occurs with the isotopic spin changing potential $\Delta I = 1$.

The measurement of parity violation by observation of the transition[9] O^{16} $(1^-) \rightarrow \alpha + C^{12}$ shows agreement with theory suggesting that in this case (where the mixing has $\Delta I = 0$) the parity mixing potential is correctly calculated.

However, this is at once at odds with the circular polarization measured by Lobashov[10] (table 1) which gives a result 100 times larger than theory. Although this _might_ be $\Delta I = 2$ mixing, there is no theoretical reason for it.

It is my personal belief that the Lobashov experiment is wrong, and that he was measuring the (negative) circular polarization of a few bremsstrahlung gamma rays. If a few of these gamma rays produced by bremsstrahlung from the beta rays emitted by uranium fission products can scatter into the apparatus, a spurious effect can occur. Lobashov depended on a difficult Monte Carlo calculation to exclude this and he may be wrong.

For the last three years we, at Grenoble, have been measuring the asymmetry in $np \rightarrow d\gamma$.[11] We use a beam of cold (~5 Å) neutrons from the cold source in the high flux reactor at Institut Laue-Langevin (ILL). These are transported by reflection on the surfaces of a guide tube to the experimental hall. There they are reflected from an assembly of magnetized cobalt mirrors. Neutrons with one direction of spin are reflected and pass onwards to the apparatus, those with the other direction are transmitted and ultimately lost.

The beam polarization was reversed every second.

The polarized beam is guided onto a target of liquid parahydrogen. Cold parahydrogen has an angular momentum of 0, and cold neutrons will not therefore change their direction of spin in scattering from it. Ultimately the neutrons will either be captured in the hydrogen or in the walls of the target--lined with lithium to prevent gamma ray emission on capture (figure 3).

The light from the two liquid scintillation counters was integrated and digitized; the outputs were recorded on magnetic tape and the asymmetries corrected for gain fluctuations and beam fluctuations (figure 4).

The only possible large spurious asymmetry would be caused by the Stern Gerlach deflection of the neutrons in passing through the inhomogeneous fields of the spin flipper. This will be unequal for the two spin directions and could cause an asymmetry of up to 10^{-8}.

The particular interest in mixing with $\Delta I = 1$ is that these are expected to be dominated by the one pion exchange transition. Since the pion has $I = 1$, these are multiplied for charged currents by the square of the sine of the Cabibbo. angle $\sin^2 \theta_c$ and for neutral current transitions by $\cos^2 \theta_c$--an increase of a factor of 20 or so. Weinberg will at this meeting throw doubt on this factor, but not on the general motivation. For the asymmetry in $np \to d\gamma$, calculations agree on a value of 10^{-8} using the Cabibbo theory and on 5 to 20 times this using the Weinberg-Salam model. Our error ($\pm 2 \times 10^{-7}$) is therefore still too big.

In the next two years we expect to improve this by:

- Installing a new polarizer built by Gavin Williams of Rutherford Laboratory with higher efficiency and larger solid angle. This can increase polarization from 70% to 95% and counts by a factor of 6.

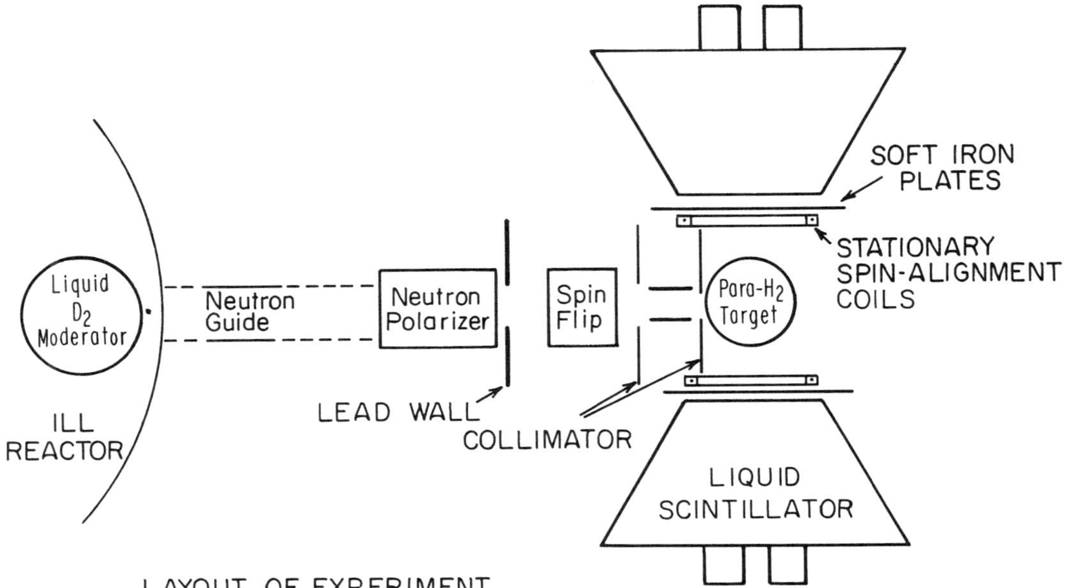

LAYOUT OF EXPERIMENT

Figure 3

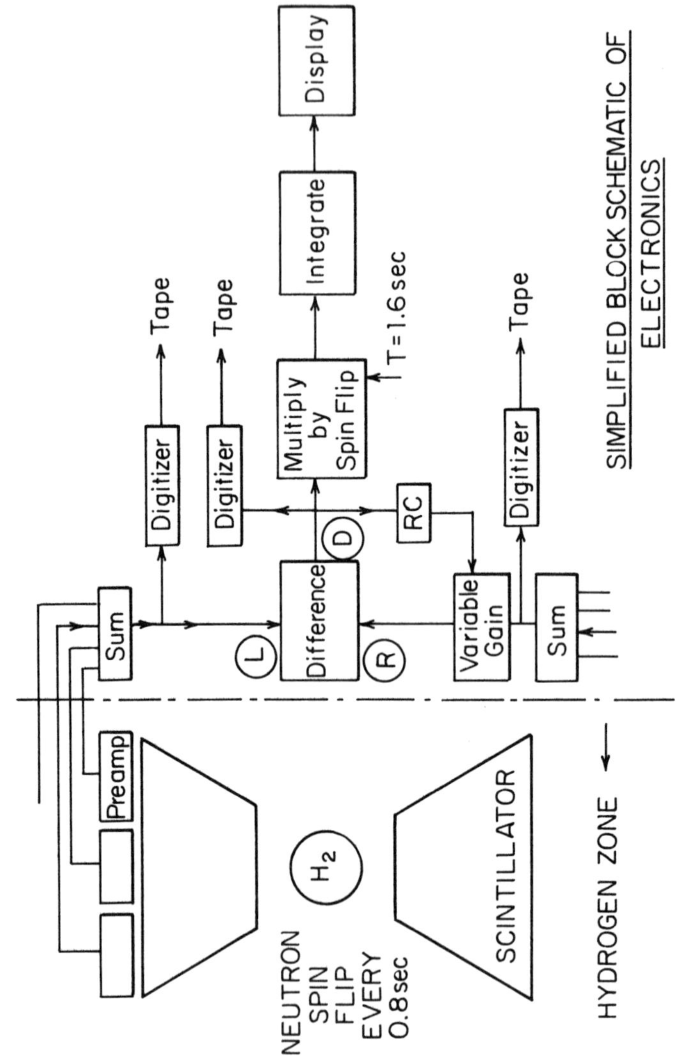
Preamp
Sum
Digitizer
Tape
Digitizer
Tape
Difference
L
R
D
Multiply
by
Spin Flip
T=1.6 sec
Integrate
Display
RC
Variable
Gain
Digitizer
Tape
Sum
H_2
NEUTRON
SPIN
FLIP
EVERY
0.8 sec
SCINTILLATOR
HYDROGEN ZONE
SIMPLIFIED BLOCK SCHEMATIC OF
ELECTRONICS

Figure 4

- Counting longer (x 3)

- Improving the light collection uniformity of the scintillator so that each gamma ray has the same weight and statistics are not degraded.

Overall, we might expect a factor of 5 improvement to $\pm 4 \times 10^{-8}$.

In a chance discussion with Dr. Pilt of Oxford University in the summer of 1976, another possibility presented itself. In figure 5 I show the decay scheme[12] for the odd neutron nucleus Ne^{21}. There are at 2.789 and 2.796 two states with the same spin and opposite parity. Because they are close together they will mix and because they have different transition probabilities to the ground state there will be a large circular polarization ($\simeq$ 1%) of the weaker transition.

A search is underway for this at the National Bureau of Standards[13] (NBS) using the same apparatus as used for the Cd^{113} $(n\gamma)$ Cd^{114} experiment but producing Ne^{21} by Ne^{20} (d, p) Ne^{21}. Bruce McKellar tells me that the wave functions for these $\frac{1}{2}^+$ and $\frac{1}{2}^-$ states are similar to the wave functions of the $\frac{1}{2}^+$ and $\frac{1}{2}^-$ states in F^{19} which Adelberger has discussed. Ne^{21} is Ne^{20} plus a neutron; F^{19} is Ne^{20} minus a proton. In one case the mixing potential has

$$\Delta V = \Delta V\ (I = 0) + \Delta V\ (I = 1)$$

and in the other

$$\Delta V = \Delta V\ (I = 0) - \Delta V\ (I = 1)$$

Therefore we may be able to isolate ΔV (I = 1) from this pair of measurements.

Finally a word for the future. So far, and in keeping with the main thrust of this conference, I have been talking about making measurements in the <u>simple</u> nuclear cases where we know the wave functions, in order to throw light upon weak interaction processes. We may soon know these processes.

But we must also know the nuclear physics. One big uncertainty--which

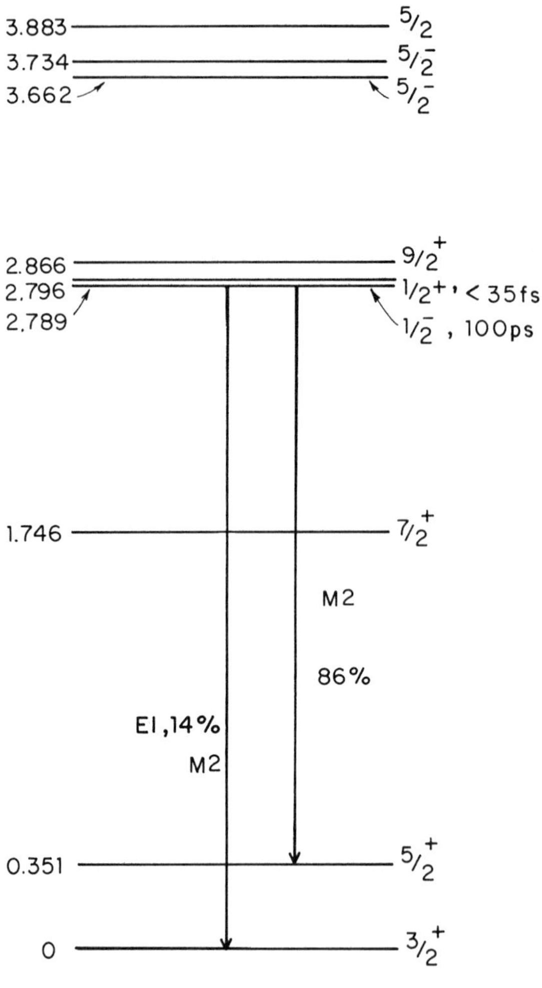

FIG. 5 DECAY SCHEME OF Ne^{21}

curve in particular when the calculations were made for the nucleus Ta^{181}--is the short range character of the forces between nucleons in nuclei. Because of the repulsive core, most nuclear phenomena are insensitive to the short range behavior. But the weak interaction is a weak, zero range force and is a probe of this short range behavior. Once we know the weak interaction physics we have an exciting nuclear physics tool.

In this connection I mention a proposal[14] in which I am honored to participate to put a polarized electron source in the 400 Mev Bates Electron Accelerator at MIT. The first experiments proposed are to study elastic scattering from carbon and hydrogen and understand the weak interaction. But the source can also be used to excite any of the ground state transitions in table 1! This is a more sensitive experimental scheme than looking for circular polarization because (1) the source has 50% polarization whereas the analysing power of a polarization analyser is typically 2%; and (2) the polarization analyses involve transmission through iron and intensity losses of 10 or 20. These give an overall intensity of electron scattering of about 5000.

I would like to thank by friends and collaborators listed in the various references for making this work possible.

References

1. Y.G. Abov, P.A. Krupitschky, and Y.A. Oratovskii, Sov. J. Nucl. Phys. 1, 341 (1965).

2. J.L. Alberi, I. Schroder, and Richard Wilson, Phys. Rev. Lett. 29, 518 (1972). (Dr. Abov pointed out to us an error in sign of this result.)

3. R.J. Blin-Stoyle, Phys. Rev. 120, 181 (1960).

4. See also, J. Alberi, I. Schroder, and Richard Wilson, in press.

5. G.V. Danilyan, V.V. Novitsky, V.S. Pavlov, S.P. Borovlyev, B.D. Vodennikov, and V.P. Dronyaev, JETP 24, 34 (1976), English translation JETP Letters (1977).

6. H. Benkoula, J.F. Cavaignac, J.L. Charvet, D.H. Koang, B. Vignon and Richard Wilson, Phys. Letts. in press.

7. G.S. Danilov, Phys. Letts. 35B, 579 (1971).

8. P. Herczeg and L. Wolfenstein, Phys. Rev. D 4051 (1973).

9. K. Neubeck, H. Schober, H. Waffler, Phys. Rev. C10, 320 (1974).

10. V.M. Lobashov, et. al., Nucl. Phys. A197, 241 (1972).

11. E. Jeenicke, Ph.D. Thesis, Harvard University, Cambridge, MA 02138 (1976). J.F. Cavaignac, B. Vignon and Richard Wilson, Phys. Letts. 67B, 148 (1977).

12. C. Rolfs, H.P. Trantretter, E. Kuhlmann and F. Ricis, Nucl. Phys. A189, 641 (1977).

13. By LeMays, O'Brien and I. Schroder.

14. Proposal, October 1977 to Bates Linear Accelerator by V. Hughes and collaborators.

A SEARCH FOR NEUTRAL WEAK CURRENT EFFECTS IN THE NUCLEUS ^{18}F

C. A. Barnes, M. M. Lowry, J. M. Davidson and R. E. Marrs*
California Institute of Technology, Pasadena, California 91125 U.S.A.

F. B. Morinigo
California State University, Los Angeles, Los Angeles, California 90032 U.S.A.

E. G. Adelberger and H. E. Swanson
University of Washington, Seattle, Washington 98195 U.S.A.

Abstract: The circular polarization of the γ-rays from the 1.08-0.0 MeV transition of ^{18}F has been measured to be $(-0.6 \pm 2.0) \times 10^{-3}$, a value significantly smaller than predicted by calculations which assume a Weinberg-Salam form for the neutral weak current.

The discovery of effects attributable to neutral weak currents in reactions induced by high energy neutrinos[1] has sharply increased interest in theories of neutral weak currents. One feature that may help in choosing between alternative theories[2] is the magnitude of the parity mixing arising from the $\Delta I = 1$ component of the nucleon-nucleon weak force. If neutral weak currents are parity-violating, and include both isoscalar and isovector components, the $\Delta S = 0$, $\Delta I = 1$ parity mixing in nuclei should be considerably enhanced over that predicted on the basis of charged weak currents only. For brevity, we refer to charged currents alone as the Cabibbo model.[3] In light nuclei, where isospin is expected to be nearly conserved, all reported cases of nuclear parity mixing are either insensitive to the $\Delta I = 1$ component of the weak force[4,5] or do not differentiate between $\Delta I = 1$ and $\Delta I = 0$ components.[6]

A promising situation for the study of $\Delta S = 0$, $\Delta I = 1$ parity mixing in nuclear states is provided by the 0^-, $I = 0$, 1.08 MeV and 0^+, $I = 1$, 1.04 MeV

*Present address: Department of Physics, Indiana University, Bloomington, Indiana 47401 U.S.A.

states of ^{18}F (see Fig. 1). Their energy separation ΔE is only 0.039 MeV, and they are nearly 4 MeV from the next known state of spin 0, which implies that any parity admixture in these states arises mainly from mixing between these two states. The long lifetime of the isospin-forbidden 1.08 to 0.0 MeV electric dipole transition suggests also that these states are relatively pure in isospin. The $\Delta I = 1$, parity-nonconserving component V^{PNC} in the nucleon-nucleon interaction thus results in a circular polarization (CP) of the 1.08 MeV γ-ray,

$$(CP) = -2\langle 0^+, I=1|V^{PNC}|0^-, I=0\rangle(||M1||/||E1||)\,\Delta E^{-1},$$

where $||M1||$ and $||E1||$ denote reduced matrix elements for the (very strong) magnetic dipole and (highly retarded) electric dipole transitions that de-excite the 1.04 and 1.08 MeV states, respectively.

During the planning and development phases of this experiment, the most detailed calculation of the matrix element of V^{PNC} available was that of Gari and coauthors,[7] who calculated a CP for the 1.08 MeV gamma transition of $\pm 3.6 \times 10^{-4}$ on the basis of the Cabibbo model. They also calculated that neutral currents would enhance the CP by a factor of 16, on the basis of the Weinberg-Salam model. Earlier estimates of the CP of this gamma transition were given by Henley, and by Wambach, Gari, and Kümmel,[8] and a semiempirical value has been inferred recently by Box et al.[9] from the results of other parity-violation experiments.

The layout of our experiment to measure the CP of the 1.08 MeV γ-ray is shown in Fig. 2. Excited ^{18}F nuclei were produced by the $^{16}O(^3He,p)^{18}F^*$ reaction, by bombarding a flowing water target with a 10-15 μA beam of 4 MeV $^3He^+$ ions from the Van de Graaff accelerator at the California State University, Los Angeles. The CP of the γ-rays was measured by two transmission-type

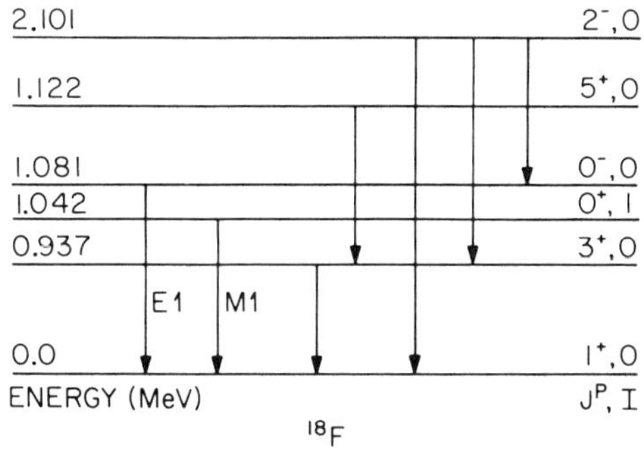

Fig. 1. Energy levels in ^{18}F. (not to scale.) Only those levels relevant to this experiment are shown.

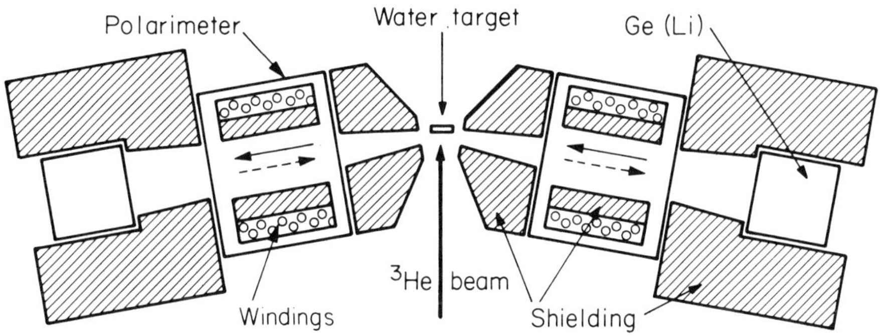

Fig. 2. Schematic layout of the experiment, shown in horizontal section.

Compton polarimeters, 8.28 cm long, made from an alloy of 65% iron and 35% cobalt. The polarimeters were positioned with their axes in the horizontal plane, at angles of 100° with respect to the beam.

The target was part of a closed circulating water system of total volume 20 liters. The flow removed residual ^{18}F activity from the target region and cooled the 7500 Å nickel foil that isolated the water from the vacuum system. The frequency of foil ruptures was reduced to a tolerable level (about 3 per day) when the beam was diffused over a spot of 3.2 mm diameter by passing it through 3 m of low pressure argon. Fast acting safety valves protected the accelerator vacuum system in the event of a foil rupture.

Pulse height spectra from the two Ge(Li) detectors were recorded in separate multichannel analyzers. The spectra for the two possible senses of the polarimeter magnetization, which was reversed every two seconds, were stored in different regions of memory. In addition to the peak corresponding to 1.081 MeV γ-rays, the recorded spectra included peaks arising from γ-rays of 0.937, 1.020, 1.042 and 1.163 MeV, as shown in Fig. 3a. The accumulated spectra were written on magnetic tape, and the memories cleared, at intervals of about two hours, for a total of about 1500 hours of counting.

The analyzing power (AP) of the polarimeters was measured in a separate experiment, in which they were placed in series between an intense ^{60}Co source and one Ge(Li) detector, one polarimeter serving as polarizer and the other as analyzer. The AP was determined to be $(2.31 \pm 0.11) \times 10^{-2}$ and $(2.19 \pm 0.12) \times 10^{-2}$ for γ-rays of energy 1.33 and 1.17 MeV respectively. An extrapolation with the slope given by Chessler[11] yields an AP of $(1.84 \pm 0.06) \times 10^{-2}$ for 1.08 MeV gamma rays.

For each of the spectral regions shown in Fig. 3a, an asymmetry

$$A = \frac{1}{4}\left[1 - (R_- L_+ / R_+ L_-)\right]$$

was calculated. In this expression, R_-, R_+, L_-, L_+ denote the numbers of counts recorded in a given spectral region by the Right or Left counter for the (+) or (-) sense of the polarimeter magnetization. To lowest order, A is independent of differences in the efficiencies of the γ-ray detectors and in the γ-yield produced by the beam during the counting times corresponding to the two senses of the magnetization. Since the magnetizations of the polarimeters were parallel, the asymmetry is related to the circular polarization and the analyzing power by A = (CP)(AP).

Since the CP expected for the 1.08 MeV γ-ray corresponds to an asymmetry less than 10^{-4}, it is necessary to demonstrate that instrumental effects are either insignificantly small or correctable. We have shown that the electronic system does not introduce detectable asymmetries at a level of 10^{-5}, in extensive runs using pulsers and a ^{60}Co source placed at the target site. We have identified a number of mechanisms that may result in asymmetries. Three of these arise from stray magnetic fields in the target region and along the beam path, that reverse direction as the sense of the polarimeter magnetization is changed; they are

i) displacements of the centroid of the beam spot on the target,

ii) rotations of the beam about a vertical axis, and

iii) precession of excited ^{18}F nuclei.

To investigate these three effects we have taken the following measures:
(1) The reversing component of the external polarimeter fields was carefully measured at the target site and along the beam path. The magnitude of the beam displacements, rotations, and the effects of precession in the excited ^{18}F were calculated and found to be negligible, with the possible exception of the 0.937 MeV γ-ray.
(2) As an extra precaution, field compensating coils were mounted on the out-

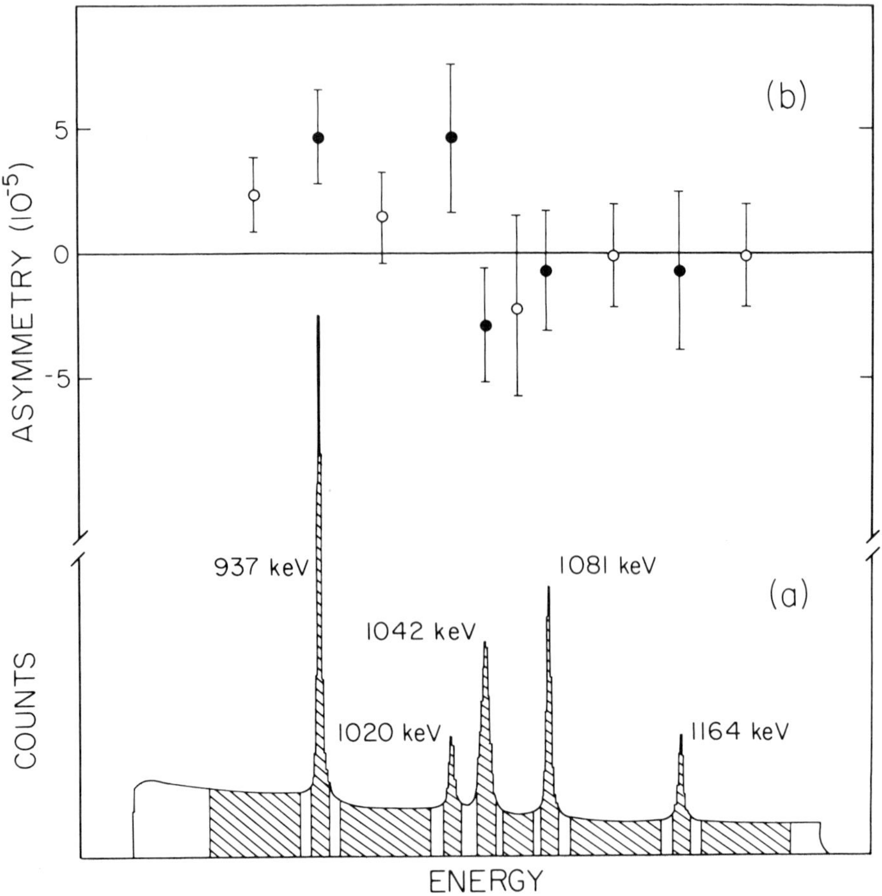

Fig. 3. (a) Typical pulse height spectrum. The ten spectral regions for which asymmetries were determined are shaded.

(b) The measured asymmetries with their statistical standard deviations. The open and closed circles are for the background and peak regions, respectively, shown in part (a).

side of the polarimeters, and powered in series with them, to reduce the reversing fields along the beam path and at the target by substantial factors. (3) Data were collected with the reversing fields artificially enhanced by a factor of about 100. Because the experimental station was constructed from nonferromagnetic materials, the enhanced field data provide a scale factor to correct for the effects of the reversing magnetic fields. These measurements show that the external polarimeter fields cannot produce significant asymmetries, except possibly for the 0.937 MeV γ-ray, for which the combination of a highly anisotropic angular distribution, and the large g-factor and long lifetime of the 5^{+} ^{18}F state which feeds into it, may result in a detectable asymmetry.

Energy-dependent background asymmetries may also result from circularly polarized γ-rays arising from in-flight annihilation of positrons, or from positron-induced bramsstrahlung. Because of the low end point energy of the positrons from ^{18}F, which is the dominant radioactivity produced, this source of asymmetry can affect only the lower part of the region shown in Fig. 3.

Independently of all the preceding considerations, the recorded spectra included γ-rays of 1.02, 1.04, and 1.16 MeV, and five background regions which serve as asymmetry monitors. The asymmetries observed for the ten regions of the spectrum shown in Fig. 3a are given in Fig. 3b, with their statistical standard deviations.

We have extracted the asymmetry A_p for the 1.08 MeV peak from the measured asymmetries by means of the expression

$$(1 - f)A_p = A_t - f A_b ,$$

where A_t is the asymmetry observed for the narrow region of the spectrum that includes both the peak and its underlying background, f is the fraction of the total counts in that region that are background counts ($f = 0.35$), and A_b is the asymmetry assumed for the background, as determined from a fit to the measured asymmetries for the 1.02, 1.04, and 1.16 MeV peaks, and the five background regions. The values obtained for A_b at the position of the 1.08 MeV peak for assumed constant, linear, and quadratic energy dependences are shown in Table I, together with the reduced χ^2 for the three fits. All are consistent with $A_b = 0$, and with one another.

The experimental values for the CP of the 1.08 MeV γ-ray deduced for the three values of A_b are listed with their statistical standard deviations in Table I. Our preferred value, CP = $(-0.6 \pm 2.0) \times 10^{-3}$, is based on the

Table I. The circular polarization deduced for the 1.08 MeV γ-ray for alternative fits to the background asymmetry.

$A_b(10^{-6})$	Reduced χ^2, A_b fit	$A_p(10^{-6})$	CP(10^{-3})
7.4 ± 7.7[a]	0.96	-14.7 ± 37.0	-0.8 ± 2.0
1.9 ± 8.5[b]	0.78	-11.7 ± 37.0	-0.6 ± 2.0
-3.5 ± 11.6[c]	0.84	-8.9 ± 37.3	-0.5 ± 2.0

[a]Background asymmetry assumed constant.

[b]Background asymmetry assumed to have linear energy dependence.

[c]Background asymmetry assumed to have linear plus quadratic energy dependence.

linear fit to the background asymmetries; the other fits give values for CP which are not significantly different. This CP is 2.5 standard deviations smaller than the theoretical value, CP = $\pm 5.6 \times 10^{-3}$, obtained by multiplying the Cabibbo model prediction of Gari et al.[7] by their neutral weak current enhancement factor of 16. During the later stages of our experiment, other more complete evaluations of the enhancement factor have become available. Although it is clear that the predicted enhancement depends on the form assumed for the neutral weak currents, and is in any case difficult to evaluate precisely, we adopt an enhancement factor of 10 as representative of the theoretical predictions for the Weinberg-Salam form of the theory.[11] When taken together with the Cabibbo model prediction of Gari et al.,[7] this yields a predicted CP of $\pm 3.6 \times 10^{-3}$. The present experimental result is consistent with CP = 0, and is smaller in magnitude than this theoretical prediction by 1.5 or 2.1 standard deviations, depending on the sign assumed for the theoretical prediction.

The Cal State, L.A., Van de Graaff accelerator laboratory is operated with State of California funds. This work was supported in part by the National Science Foundation (grant no. PHY76-83685) at Caltech, and by the Energy Research and Development Agency (contract no. EY-77-C-06-1388) at the University of Washington. During the three years that this work has been in progress, many persons have contributed including D. Brown, B. Chang, W. Davis, L. Gin, P. Ingalls, C. Malone, D. Scott, and H. Winkler. We acknowledge discussions with R. Bond, M. Gell-Mann, E. Henley, P. Herczeg, and S. Weinberg.

References

1. F. J. Hasert et al., Phys. Lett. 46B, 138 (1973); A. Benvenuti et al., Phys. Rev. Lett. 32, 800 (1974).
2. See, for example, S. Weinberg, Phys. Rev. Lett. 19, 1264 (1967); A. Salam, in Elementary Particle Theory, ed. by N. Svartholm (Almqvist and Wiksell,

Stockholm, 1968) p. 367; H. Fritzsch, M. Gell-Mann, and P. Minkowski, Phys. Lett. 59B, 256 (1975).

3. N. Cabibbo, Phys. Rev. Lett. 10, 531 (1963).

4. V. M. Lobashov et al., Nucl. Phys. A197, 241 (1972).

5. K. Neubeck, H. Schober, and H. Wäffler, Phys. Rev. C10, 320 (1974).

6. E. G. Adelberger et al., Phys. Rev. Lett. 34, 402 (1975).

7. M. Gari, J. B. McGrory, and R. Offermann, Phys. Lett. 55B, 277 (1975).

8. E. M. Henley, Phys. Lett. 28B, 1 (1968); V. M. Wambach, M. Gari, and H. Kümmel, Phys. Lett. 33B, 253 (1970).

9. M. A. Box et al., J. Phys. G: Nucl. Phys. 2, L107 (1976).

10. R. B. Chessler, Nucl. Instr. Meth. 37, 185 (1965).

11. B. Desplanques and J. Micheli, Phys. Lett. 68B, 339 (1977), and references cited there; E. M. Henley, Proc. Intl. Conf. on Nuclear Structure, Tokyo, September 1977, to be published; S. Weinberg, private communication, and this conference.

ANALYSIS OF NUCLEAR PARITY NONCONSERVATION

Dubravko Tadić
University of Zagreb, Zagreb, Croatia, Yugoslavia
and
"Rudjer Bošković" Institute, Zagreb, Croatia, Yugoslavia

1. Introduction

Soon after weak-interaction physics came of age through the discovery of parity nonconservation (PNC)[1,2] the existence of parity-violating (PV) nuclear forces was considered.[3-5] Such forces are a necessary consequence of any universal models of weak interactions.[6] A general weak-interaction Hamiltonian can be divided into leptonic, semileptonic, strangeness-violating nonleptonic, and strangeness-conserving nonleptonic sectors

$$H_W = H_L + H_{SL} + H_{NL}(\Delta S=1) + H_{NL}(\Delta S=0) \ . \tag{1.1}$$

The existence of the last piece, which is responsible for PV nuclear forces, seems to be well established experimentally, as described in numerous general reviews of the subject.[7-15,135] Each term in (1) corresponds to a particular type of weak process which can be tested through either decay or scattering experiments, as indicated in Table 1.

TABLE 1

	Experiment performed: Yes, No	
Hamiltonian	Decay	Scattering
H_L	Yes	Yes
H_{SL}	Yes	Yes
$H_{NL}(\Delta S=1)$	Yes	No
$H_{NL}(\Delta S=0)$	Yes	(Yes)

As all weak-interaction models lead to the same general structure (1), (see, for example, Refs. 16-25), one has to

* Greater part of this talk was prepared while visiting the School of Mathematical and Physical Sciences, the University of Sussex, Falmer, Brighton.

test a particular theory in full in order to establish its validity. Taking a historical point of view, we might say that some of the models, such as intermediate-scalar-boson theory including heavy leptons, seemed to encounter considerable difficulties in accommodating all of the experimental facts.[131] A peculiar parity-violating effect may appear in a model where colored strong gluons are mixed with weak intermediate vector bosons. Although this effect is semiweak, it is expected to vanish if hadrons are color (i.e., SU"(3)) singlets and it may appear only through symmetry breaking.[16] The presence of hadronic neutral currents, which characterize the gauge models of weak interactions[16-25] in nonleptonic processes, becomes apparent only through the $H_{NL}(\Delta S=0)$ piece. The study of decay processes led to the perception of the existence of weak interactions and revealed their basic properties. Much more has been learned from recent neutrino scattering experiments; these led indirectly to the discovery of charm and helped to reveal the exciting new structure in the strong dynamics. However, to accommodate the properties learned from the first two rows of Table 1, one has to construct models which actually contain more details than necessary for that purpose. All the models can account, at least qualitatively, for strengeness-changing nonleptonic decays. For further testing of a model and for some insight into its underlying dynamical structure, the processes indicated in the last row of Table 1 seem to be indispensable*.[8-15,26-28] Further, they might, in due course, become useful tools in studying nuclear physics proper, in the same way as happened with nuclear decay. Parity-violating weak interactions seem to be sensitive to the short-range behaviour of nuclear wave functions and also dependent on mesonic effects and corrections in nuclei.

---

*Strangeness-changing hadron scattering could also give more information than the nonleptonic decay amplitude, which is only the value at one energy point. For a very preliminary discussion see Ref. 29.

Extraction of all this valuable information is at present by no means a simple, easy, or uniquely well-defined task. The chain of reasoning, as sketched in Fig. 1, is lengthy and involved. Its main link - the weak PV nonrelativistic two-body nucleon-nucleon potential V_{12} - connects particle physics proper with nuclear physics. Uncertainties, questions, inputs, and calculational techniques are more numerous than can be indicated in one figure, where only some are selected for illustrative purposes. All this material will be elaborated in the following sections.

2. Parity-Violating Potentials

PV potentials are useful in coding elementary-particle-physics results in a form that can be conveniently used in nuclear physics and whicn is mainly nonrelativistic. The step leading from the simplified Feynman diagram to the corresponding PV potential usually does not introduce any serious uncertainty. Uncertainty is associated with the deduction of the effective Feynman diagram itself. The general PV nucleon-nucleon scattering amplitude should be approximated in the usual way by one-particle, two-particle, etc., exchanges. As indicated in Fig. 2, one of the one-boson-exchange (OBE) vertices is weak and should be either calculated or deduced semiempirically from other experiments.[8-12]

CP invariance and isospin (I) impose serious restrictions as long as nucleons are on the mass shell.[8-12,30] Vector-meson exchanges satisfy the selection rule $\Delta I=0,1,2$ and thus receive contributions from the appropriate parts of $H_{NL}(\Delta S=0)$. Pion exchanges satisfy the $\Delta I=1$ rule.[30] In the off-mass-shell case, which is actually common for nuclei, the situation is changed in such a way that, for example, the $\Delta I=0$ part of $H_{NL}(\Delta S=0)$ can contribute to pion exchange.[31-33] As the contribution is proportional to the proton-neutron mass difference, which is an isovector, the CP restriction is not broken. Although the off-mass-shell effects are rather large relative to the strength of PV potentials, it is dif-

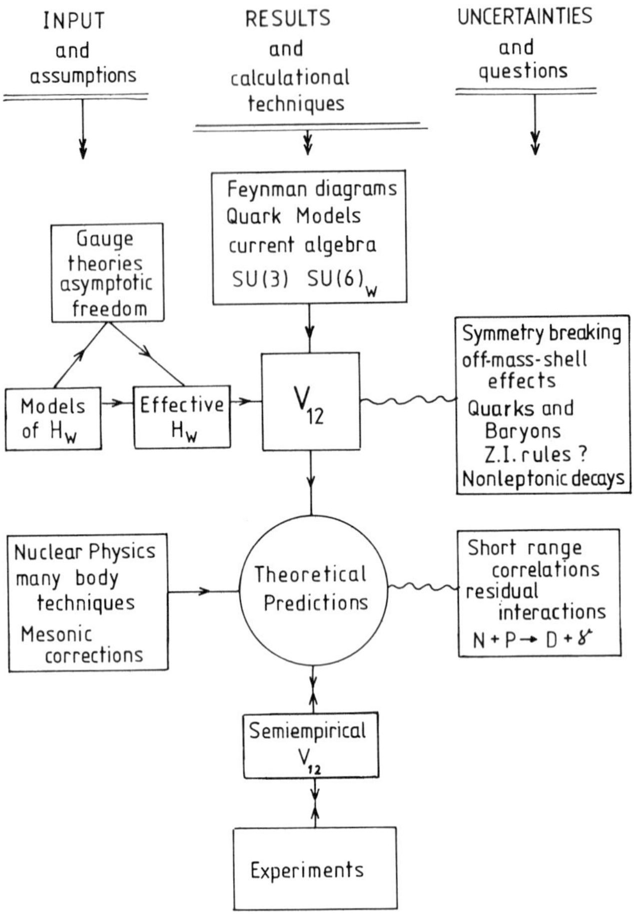

FIG. 1. Outline of the present scheme for extracting information on weak interactions from parity-violating effects in nuclei. Wavy lines indicate theoretical uncertainties and/or open questions.

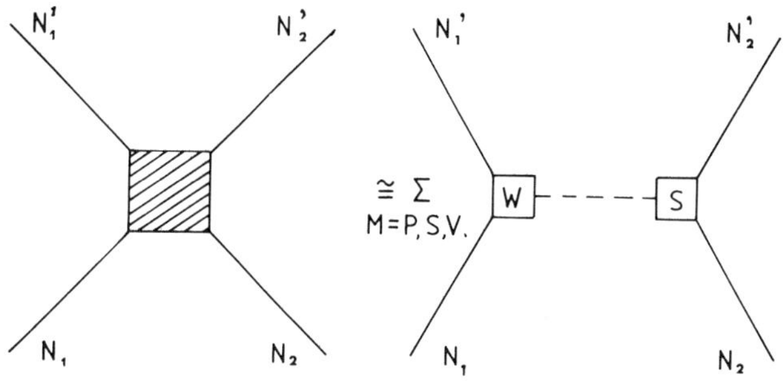

FIG. 2. One-meson-exchange contributions to parity-violating internucleon potentials (V_{12}). Weak and strong vertices are indicated. The dashed line represents the pseudoscalar or the vector meson, and the full lines represent nucleons.

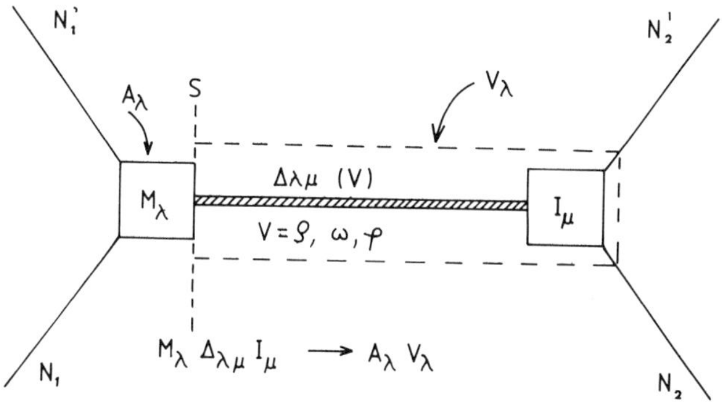

FIG. 3. The one-vector-meson-exchange contribution to V_{12}. The thick line corresponds to the vector-meson propagator. The weak and strong vertices are M_λ and I_λ, respectively. In the factorization approximation, the full-lined box corresponds to the axial-vector-current matrix element, while the dashed box symbolizes the vector-current matrix element.

ficult to say how important they are[31,32] in the present situation in which potential strengths are rather uncertain.

2.1. Vector-Meson Exchange

Theoretical uncertainty is best demonstrated by vector-meson-exchange contributions (i.e., ρ, ω, ϕ, etc.) whose PV vertices have to be calculated completely theoretically since they could be connected only to the strangeness-violating scattering experiments, which, however, have not been performed as yet. The corresponding potential was first obtained by the simple factorization approximation.[28] As indicated by dashed lines in Fig. 3, two nucleon lines emerge from either the matrix element of the vector current V or the matrix element of the axial vector current A. The PV weak Hamiltonian contains the usual current-current product $A_\lambda \times V_\lambda$.

$$\langle N_1' N_2' | H_W | N_1 N_2 \rangle \cong \frac{G}{\sqrt{2}} \langle N_1' N_2' | A_\lambda V_\lambda | N_1 N_2 \rangle$$
$$\cong \frac{G}{\sqrt{2}} \langle N_1' | A_\lambda | N_1 \rangle \langle N_2' | V_\lambda | N_2 \rangle \ . \qquad (2.1)$$

Such a product can be lifted "raw" from the given weak Hamiltonian[8-12] or the asymptotically free gauge invariant theories of Quantum Chromo Dynamics (QCD)[34,35] can be used to find an effective weak Hamiltonian.[37-40]

The factorization approximation was supported by the analyses based on current algebra (CA) and current field identity (CFI).[77,10] This support, however, was disputed,[41] but there are rebuttals of these disputations![10,39,126] Convergency problems, associated with the old nongauge type weak theories[42] are now laid to rest. An attempt to relate the NNπ amplitude to the observable hyperon decay using SU(6)$_W$ and pole dominance[43] suffers from ambiguities since some of the relevant reduced matrix elements do not occur in the matrix elements for processes involving pions. However, this attempt showed the inadequacy of the factorization approximation by indicating that the neutral vector meson should be emitted from the products of charged currents. This is impossible if the factorization approximation is used.

These problems have been elucidated in the quark-model approach, employed recently by several authors.[38,39,44] Diagram a) of Fig. 4 corresponds to the factorization approximation and all relevant constants are most naturally adjusted in such a way as to produce its results. The intermediate vector boson (IVB) propagator is, for heavy IVB, practically constant. Diagram b) corresponds to the nonfactorizable or nonseparable (NS) part and can also be estimated relatively easily by doubling the strength of the PV potential. Neutral vector bosons can be emitted from the products of charged currents owing to the diagram b) of Fig. 4. If neutral currents are present, as it may happen in some models[8-12] and in unified-field-theory models in particular[16-24], neutral-vector-boson vertices follow already from the factorization approximation.

Sophisticated unifled-field-theory models generally strengthen vector-meson-exchange contributions,[38,39] but this strengthening is not enough to account for the γ circular polarization measured in the $n+p \rightarrow d+\gamma$ experiment.[45] The same of course holds for the vector-like quark models which have been investigated recently.[87] The analysis conducted in the first paper of Ref. 87 was actually based on the factorization approximation, so the influence of strange (s) or unusually flavored quarks (i.e., b, h, etc., see the Appendix for some details) was not felt. The contributions from neutral currents were roughly equal to those for the Salam-Weinberg model. Some technical details and an example of the potential based on the Salam-Weinberg model are given in the Appendix. An obvious task for the future theoretical efforts is to include in the calculation the contributions from the quark-antiquark sea through the formalism of Ref. 83, which.will be briefly discussed in Sec. 2.3. An interesting result of asymptotic-freedom analysis is the fact that in such a picture the $\Delta I=0,1$ parts of the effective H_W are always enhanced compared with the $\Delta I=2$ part.[39] The $\Delta I=2$ part is always contained in the sym-

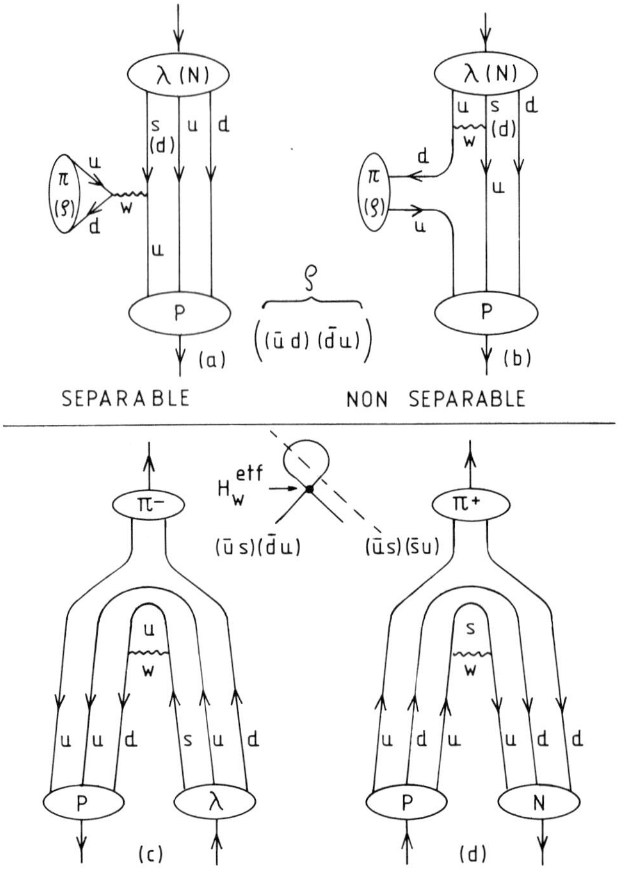

FIG. 4. Nonleptonic weak amplitudes in the quark model. The full lines symbolize quarks, while the wavy line shows where the weak interaction acts. The ellipses symbolize hadrons. When the symbols in the brackets are read, diagrams 4a,b depict the weak nucleon-nucleon vector-meson amplitude. When H_W^{eff} is used, the IVB wavy line colapses into a point, as indicated.

metric representation of the group underlying a particular weak-interaction model and thus always hindered by QCD effects. Thus the concept of octet dominance is expressed in a different language. It should also be mentioned that the appearance of the short-range (i.e., due to vector-meson exchange) $\Delta I=1$ PV nuclear force is a definite sign of the presence of neutral currents in the weak nonleptonic Hamiltonian.

Earlier ideas, taking into account actual nucleon form-factors instead of vector-meson propagators,[12] can be easily incorporated into the quark-model approach. However, such refinements are not relevant in the present rather uncertain situation. This is rather unfortunate, as the $\Delta I=0,2$ potentials are crucial in explaining either very accurate α-decay[46] and ^{181}Ta [47] measurements or the very important $n+p\rightarrow d+\gamma$ experiment. A selection of experiments on parity violation is presented in Table 2. For a more complete presentation, we refer the reader to Refs. 13, 14 and 48. The column of Table 2, with the heading Theoretical predictions (NPC approx.), is explained in Sec. 3. This column gives values obtained using the Cabibbo model.[71] While the α-decay experiment together with nuclear-scattering experiments are roughly in agreement with the simple theoretical $\Delta I=0,2$ potentials, the second and third experiments do not seem to be in agreement with such potentials. This conclusion is qualitatively unchanged even when using vector-meson-exchange potentials based on unified-field-theory and QCD corrections.[38-40] Such a situation naturally led to efforts to improve either nuclear-physics calculations or the form of the PV potentials; it also suggested the introduction of semiempirical PV potentials, which will be discussed in Sec. 2.5.

2.2. Two-Pion Exchange

Two-pion exchange contributions contain $\Delta I=1$ (Fig. 5a-d) and $\Delta I=0,2$ (Fig. 5e,f) parts. This follows if one accepts the dynamical prediction that PV $NN^*\pi$ coupling is almost entirely an isovector.[10] Historically, diagram e) of Fig. 5 was used in

TABLE 2

Experiment	Experimental result	ΔI	Theoretical prediction (NPC approx.)	Refs. Exp: Theory
$^{16}O(2^-, 8.88\ \mathrm{MeV})$ $\to {}^{12}C+\alpha$	$\Gamma_\alpha=(1.01\pm0.13)\times10^{-10}$ (in eV)	0	$0.5\times10^{-10}<\Gamma_\alpha$ $\Gamma_\alpha \geq 7\times10^{-10}$ (in eV)	46 52
n+p→d+γ				
Circular polarization	$P_\gamma=-(1.3\pm0.45)\times10^{-6}$	0,2	$P_\gamma \sim +2\times10^{-8}$	45
Asymmetry	$A_\gamma=(6\pm21)\times10^{-8}$	1	$A_\gamma \sim 0.5\times10^{-8}$	53-62
$^{19}F^*\to{}^{19}F+\gamma$ 110 keV $\frac{1}{2}^- \to \frac{1}{2}^+$	$A_\gamma=-(114\pm70)\ 10^{-6}$	0,1	$A_\gamma=-(110\pm70)$ $\times10^{-6}$	50,8 67,75, 103
$^{41}K^*\to{}^{41}K+\gamma$ 1.29 MeV $\frac{7}{2}^- \to \frac{3}{2}^+$	$P_\gamma=(1.9\pm0.3)\times10^{-5}$	0,1,2	$P_\gamma=-4.4\times10^{-6}$	47,127 64,128, 94
$^{175}Lu^*\to{}^{175}Lu+\gamma$ 396 keV $\frac{9}{2}^- \to \frac{7}{2}^+$	$P_\gamma=(5.5\pm0.5)\times10^{-5}$	0,1,2	$P_\gamma=-1.5\times10^{-5}$	47,129 63,64, 130,94
$^{181}Ta^*\ {}^{181}Ta+$ 482 keV $\frac{5}{2}^+ \to \frac{7}{2}^+$	$P_\gamma=-(5.2\pm0.5)\times10^{-6}$	0,1,2	$P_\gamma=+(0.5\sim1.0)$ $\times10^{-6}$	47 11,63- 65
pp scattering 15 MeV	$a=(0.05\pm1.4)\ 10^{-7}$	0,1,2	a=0 $a=0.3\ 10^{-7}$ octet dominance	51,8 68-70
$^{180}Hf^*\to{}^{180}Hf+\gamma$ 501 keV $8^-(K=8)\to6^+(K=0)$	$P_\gamma=-(2.3\pm0.6)\times10^{-3}$ $A_\gamma=-(1.66\pm0.18)\%$	0,1,2	$\lvert P_\gamma\rvert \leq 2\times10^{-3}$	49,10- 12 66
$^{18}F^*\to{}^{18}F+\gamma$ 1.08 MeV $0^- \to 1^+$	$P_\gamma \lesssim 36\times10^{-4}$	1	$\lvert P_\gamma\rvert \sim 3.6\times10^{-4}$	8,48 103,107

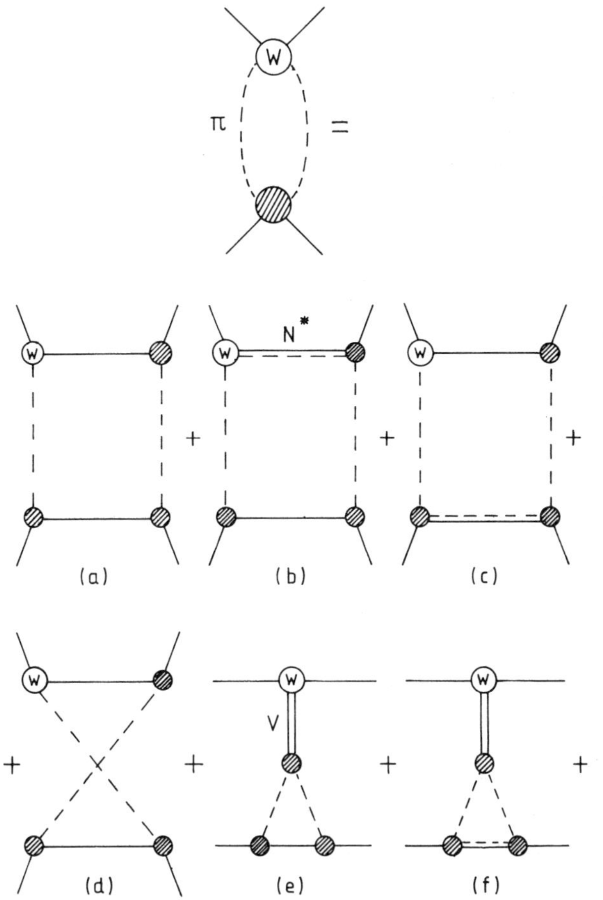

FIG. 5. Two-pion-exchange contributions to V_{12}. The dashed lines symbolize pions, the full and dashed lines are baryonic resonances and the double lines are vector mesons. The weak vertex is always indicated by W.

the first explicit calculation of the PV potential.[72] Several calculations including other diagrams have been carried out,[73] the most extensive analysis being made in Ref. 74. In this reference, standard PV vertices are used, as deduced by the factorization approximation (ρ vertex, preceding section) or by SU(3) analysis (ω vertex, Sec. 2.3), and the conclusion is drawn that two-pion exchange contributions satisfying either $\Delta I=1$ or $\Delta I=0,2$ are small, being about 30% of a one-pion exchange contribution. There are still some doubts about the adequacy of two-pion-exchange calculations[75] concerned mainly with the double-counting problem. This is due to the fact that, for example, diagram a) in Fig. 4 would also be generated from the one-pion-exchange contribution via the action of the ordinary, strong, nuclear potential.

2.3. One-pion Exchange

It was assumed that this particular contribution could be estimated with highest accuracy[27,8-12] but that has become less certain in view of some more recent investigations. The contribution was first estimated by relating the PV NN amplitude $A(n^o_-)$ to the observable hyperon decays using current algebra.[76] This led eventually to the sum rule

$$-A(n^o_-) = A(p^+_+) = \frac{\alpha}{\sqrt{3}} \, |4A(\Lambda^o_-) - 2A(\Xi^-_-)| \, . \tag{2.2}$$

In the above expression, all notation for parity-violating amplitudes has the usual meaning. Octet dominance has been assumed, but $\Delta I = \frac{3}{2}$ corrections can be easily added to the sum rule.[12,55] The parameter α measures the strength of the octet part, which is a characteristic of the particular model of weak interactions. In order to define α, the effective nonleptonic weak Hamiltonian should be normalized as follows

$$H_W^{eff} = T^8(I = \tfrac{1}{2}, \Delta S=1) + \alpha T^8(I = 1) + \beta T^8(I = 0,2) \, .$$

The hope that α can be larger for some models than the Cabibbo-model value (θ_C is the Cabibbo angle)

$$\alpha = \frac{\sin^2 \theta_C}{\sqrt{2}}$$

first prompted suggestions about the importance of one-pion exchange.[27] This might well be so in the case of unified-field-theory models, which contain neutral currents. Unfortunately, as will be explained later, the situation is too complicated for any unique conclusion to be drawn. The same result would follow by using SU(3) symmetry and CP invariance.[10,12,55] At first sight, this result appears to be very powerful, but one should be wary of it knowing that CA does not lead to satisfactory simultaneous fits of s-wave and p-wave hyperon decay amplitudes.

The physical reason for this discrepancy can be traced to SU(3)-symmetry breaking, which is responsible for nonleptonic decays in the first place. Approaches based on current algebra and/or SU(3) symmetry assume that all octet baryon masses are equal and work actually for off-mass-shell pions, i.e., the pion momentum $q \to 0$ or $q^2=0$ instead of $q^2=m_\pi^2$. Difficulties can be compensated by either introducing vector-meson or baryon-pole terms, which vanish in the $q \to 0$ limit,[78] or by considering final-state interactions.[79] The final result, which contains the standard current-algebra term (CAT), is modified either by addition of pole terms (PT), as indicated in Fig. 6, or by appropriate final-interaction factors. This results in a modified sum rule,[80] which always leads to a smaller $A(n^0_-)$ and thus to a PV pion vertex up to three times smaller than that obtained by the CAT alone. It should be mentioned that the PT (vector) term was introduced a long time ago when discussing hyperon decays[81] and that it corresponded to the factorization approximation which was put forward as a convenient estimate of $A(n^0_-)$.[32]

Analysis by using quark models, illustrated in Fig. 4, has led to further insight. Earlier results[82] were already in disagreement with the sum rule (2.2). Inadequacies of the sum rule stemming from SU(3)-symmetry breaking were forcefully stressed in Ref. 38. As indicated in Fig. 4, diagrams a), b), and c) can

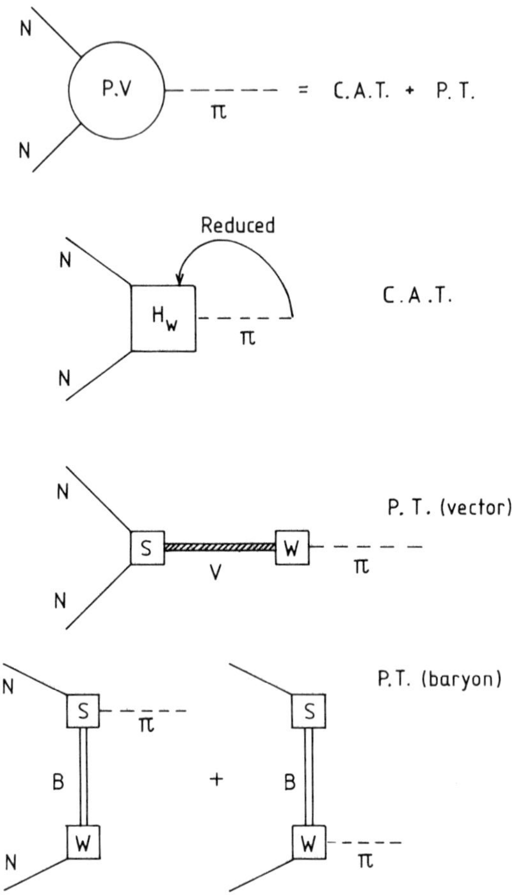

FIG. 6. Diagrams showing current-algebra (C.A.T.) and pole-term (P.T.) contributions to the PV NNπ amplitude. In the current-algebra limit ($q\to0$) and/or strict SU(3) symmetry limit (all baryon masses belonging to the same multiplet are equal). P.T. contributions vanish.

contribute to $A(\Lambda^{o}_{-})$ with a nonlocal Hamiltonian containing currents $(\bar{u}s)$ and $(\bar{d}u)$. When the VB propagator collapses, as it happens when a local effective Hamiltonian stemming from gauge theories is used, the contribution from diagram c) is zero. However, the amplitude $A(p^{+}_{+})$ receives the contribution only from diagram d), which vanishes for a local effective Hamiltonian. In unified $SU(2)\otimes U(1)$ field theory, neutral currents lead to the nonvanishing $A(p^{+}_{+})$ [38]

$$A(p^{+}_{+}) = -1.9 \sin^2\theta_W \, A(\lambda^{o}_{-}) = -2.6\times10^{-7} \qquad (2.3)$$

in appropriate units ($\hbar=1$, $c=1$). This value is actually larger by an order ot magnitude than the straightforward estimate based on the Cabibbo model

$$A(p^{+}_{+})\,(\text{Cabibbo; CAT}) = 1.8\times10^{-8} \ . \qquad (2.4)$$

It is hoped that recent work on the quark sea and related problems[83] might help to clarify some points raised by the discussion of the diagrams in Fig. 4. In these diagrams, no gluon exchanges were allowed for. This avoidance of the closed loops involving hadrons only is similar to the well-known Zweig-Iizuki (ZI) rule.[84] The quark model also allows, at least in principle, the determination of the sign of the $A(p^{+}_{+})$ amplitude,[85] which was not possible in CA or symmetry-based approaches. By enriching the usual valence baryon configuration $3q$ with quark-antiquark pairs in the form

$$|B\rangle = |3q\rangle + \sum_{\alpha} C_{\alpha}|3q,\bar{q}q\alpha\rangle \quad , \qquad (2.5)$$

it is possible to account for the f/d ratio and the $\Delta I=1/2$ rule for hyperon nonleptonic decays.[83] The estimate of Ref. 85 was essentially in the same spirit[83] and besides the sign of $A(p^{+}_{+})$ it also gave an enhancement over the NPC approximation between 4.7 and 8 times. In all these approaches the pion was contracted first, so the rotated weak Hamiltonian acted between baryon states, as shown in Fig. 7. No closed hadron loops were considered.

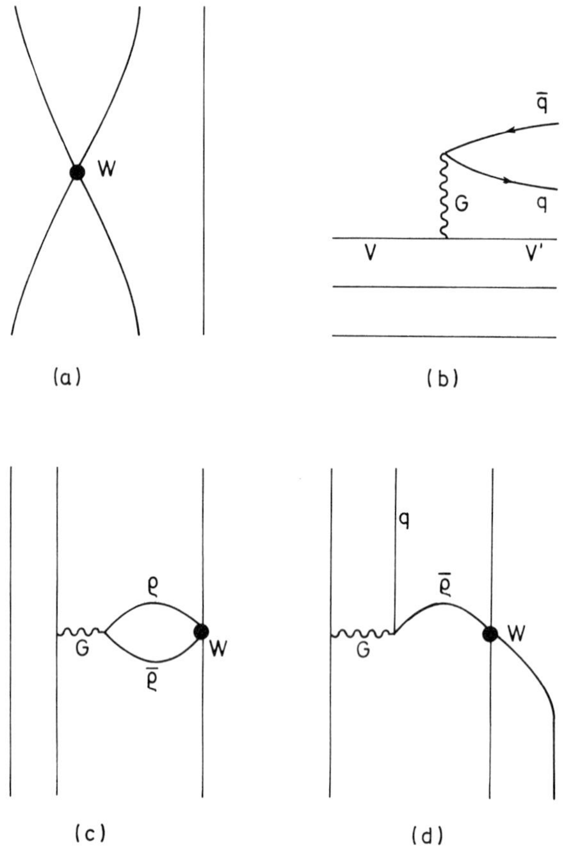

FIG. 7. The weak Hamiltonian (W) matrix element between baryon states after pion reduction. Diagram (a) is the contribution of valence quarks. Diagram (b) depicts valence quarks plus quark-antiquark pair. Diagrams (c) and (d) show the quark-antiquark pair (i.e., "sea") contributions to the process. G symbolizes the exchanged gluons.

There have also been attempts to predict the $A(p^+_+)$ amplitude for unified-field-theory models without referring to the quark structure but relying on symmetries alone.[86,142] All these attempts are beset by ambiguities, because neutral currents produce reduced matrix elements containing unusually flavored quarks (charm, etc.) which conserve strangeness and thus do not contribute to hyperon nonleptonic decays. In the context of Fig. 4, the contribution of such matrix elements would vanish. Even standard quarks (u,d,s) may appear in combinations contributing uniquely to the $A(p^+_+)$ amplitude only. If quarks are, for example, classified by SU(3) (SU(4)) symmetry and if octet (15-plet) dominance is assumed, an octet tensor (15-dimensional tensor) can be constructed from the product of the two octet (15-plet) currents or from the product between an octet (15-plet) and a singlet current. Naturally, these two nonequivalent tensors have unequal reduced matrix elements. In this case the $A(p^+_+)$ amplitude cannot be connected with the experimental amplitudes by any symmetry arguments alone and therefore additional dynamical arguments are needed.[40,86]

Such difficulties make the results[87] deduced from the vector-like quark models[22-25] somewhat uncertain. The negative outcome of the atomic PV experiment[88] together with the results of neutrino scattering experiments[87,89] are at present better support in favor of such models. As already stated in Sec. 2.1, vector-like quark models do not help with the $n+p \to d+\gamma$ experiment. In Ref. 87, better agreement with experimental γ-circular polarization measured for medium and heavy nuclei (^{19}F, ^{41}K, ^{175}Lu, and ^{181}Ta) has been achieved owing to the large pion-exchange contribution. This is mostly due to the presence of the charged-current term $(\bar{u}b)_R(\bar{b}u)_R$, which has no Cabibbo suppression. Relegating technical details to the Appendix, we just mention that the first paper of Ref. 87 assumes that the PV matrix element for $NN\pi$ states of the term $(\bar{u}b)_R(\bar{b}u)_R$ is equal to the PV matrix elements of the standard term $(\bar{u}s)_L(\bar{s}u)_L$, which is of importance in the NPC estimate. All uncertainties and difficulties elaborated in this section apply in addition to that assumption.

The asymptotic-freedom analysis may lead to the appearance of two-quark operators.[37-40,90,91] While such an operator is suppressed by the GIM mechanism for $\Delta S \neq 0$ processes, it is not necessarily so for $\Delta S=0$ processes.

Problems with two-quark operators lead us to the question, raised recently,[91] whether the Wilson expansion should be performed before or after the pion has been contracted by PCAC. If the contraction is performed first and if the charmed quark is extremely heavy, two-quark operators appear to be dominant for both $\Delta S \neq 0$ and $\Delta S=0$ processes.[91] In such a case, however, the effective strength of the isovector part of the Hamiltonian, i.e., of $A(p_+^+)$ is the same as for the Cabibbo model. Some additional details on weak PV potentials are given in the Appendix.

In summarizing this subsection, we may say that predictions based on the Salam-Weinberg model[38-40,86,92] span too large a range of values:

$$10^{-9} < |A(p_+^+)| < 10^{-6} \ . \qquad (2.6)$$

This reflects the complex situation already outlined, which is illustrated graphically in Fig. 8. We believe that only future experiments, especially the ones involving $\Delta I=1$ isospin changes, can definitely clarify the situation. It seems at present that the overall fits of existing experiments,[93,94] performed by using semiempirical potentials, produce a PV pion coupling which is larger by about an order of magnitude than the one that would be obtained from the sum rule (2.2) for the Cabibbo model.

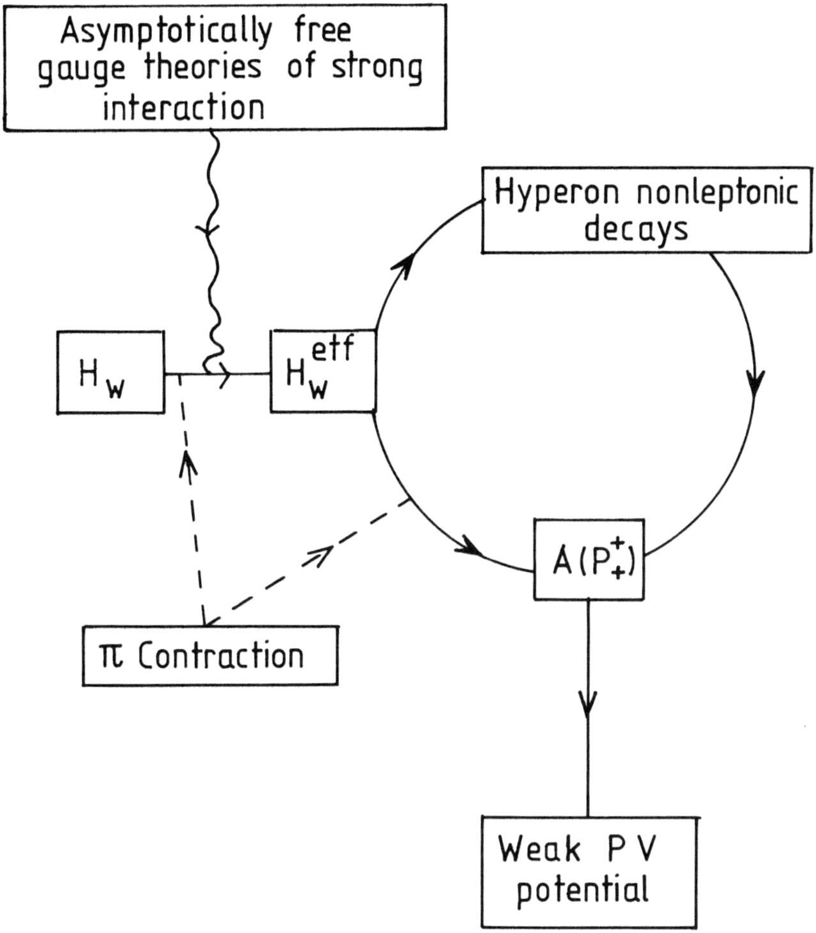

FIG. 8. Outline of the theoretical scheme(s) for the prediction of the PV NNπ amplitude. The circle involves the diagrams of Fig. 6.

Theoretical prediction of $A(p_+^+)$ requires detailed handling of the elementary-particle-physics dynamics and thus can seriously test theoretical models used for describing hyperon non-leptonic decays.

2.4. Gauge Invariance and PV Photon Coupling

Parity-violating photon coupling appears in two ways. The one is through the influence of weak interactions, as depicted in diagrams of Fig. 8, while the other would follow from the genuine parity-violating "intrinsic" electromagnetic vertex. At present there is no particular reason to suppose the existence of such a vertex. Some indication in its favor comes from the fact that considerable theoretical difficulties are encountered in explaining PV experiments with gamma emission, such as ^{181}Ta or $n+p \rightarrow d+\gamma$. The PV electromagnetic vertex might lead to photon emission only for nucleons that are off the mass shell and it might also contribute to virtual photon exchanges.[95] Contributions represented in Fig. 9 are the ones which are expected from standard weak and electromagnetic (or unified) theories; these have been widely used to explain circular polarization experiments on parity violation. Any OBE PV potential leads automatically to all possible diagrams of the type a) but including only the positive energy part of the nucleon propagator. The negative energy parts, or "pair contributions" in the language of mesonic nuclear physics, have to be included togeth-

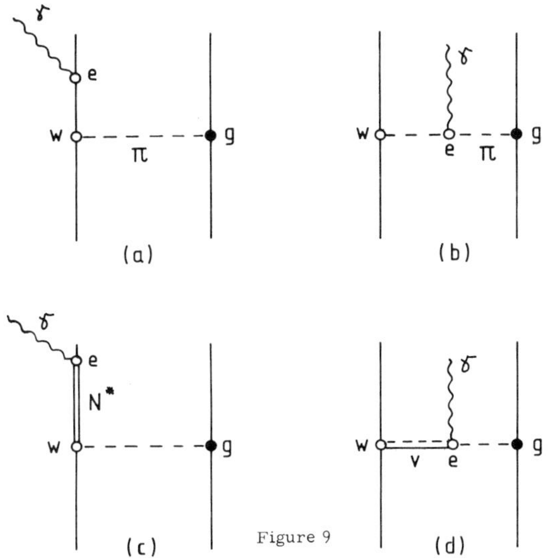

FIG. 9. Diagrams of PV effects in photon emission. Diagram (c) shows a baryonic resonance contribution, while diagram (d) is due to the vector-meson-pion-photon coupling ($V \rightarrow \pi + \gamma$).

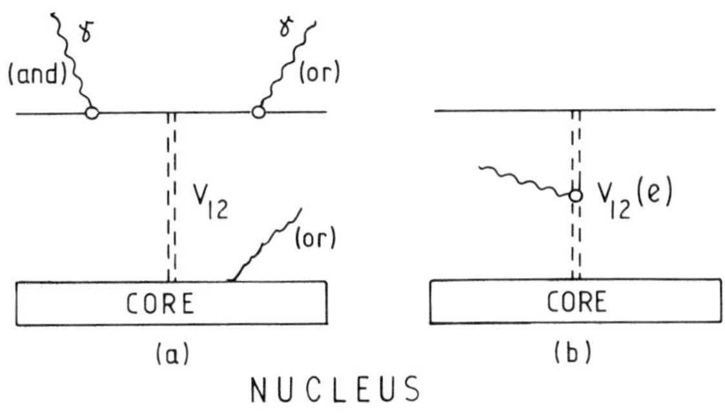

FIG. 10. Diagrams symbolizing PV effects in nuclear emission. The result depends on the PV interaction (V_{12}) of the valence nucleon with the nucleons in the core. The photon can be emitted from the valence nucleon, from the exchanged mesons or from the core. Diagram (b) corresponds to the "pair contribution" and to diagram 9b.

er with diagram b) in order to achieve gauge invariance. Translated into the weak PV potential language, this corresponds to diagrams a) and b) of Fig. 10 and will be discussed in Sec. 3. Diagrams of the type c) and d) are separately gauge invariant. It has been argued that the PV $NN^*\pi$ vertex is very small.[96] Diagram d) has been evaluated several times.[97,133] Even when a very large semiempirical value was used for the PV $NN\rho$ vertex, its contribution to circular polarization in the $n+p\rightarrow d+\gamma$ process turned out to be very small.

2.5. Semiempirical PV Potentials

It was shown in the preceding section that the theoretical derivation of a PV weak potential is a tedious process full of difficulties and ambiguities. Nevertheless, attempts at the derivation ought to be made, as through them we learn useful and relevant data for elementary particle physics. Since it has proved very difficult to obtain consistent fits to available experimental data using theoretical potentials, it is reasonable to raise the question whether it is possible to construct an empirical potential capable of producing such consistency. As general symmetry considerations[12,72,98] still offer too wide a range of potential forms, some selection should be made that is based on the theoretically deduced potentials. In Ref. 93, the potentials used are of the form as given by OBE and the overall strengths of the particular pieces are treated as adjustable parameters. There are three parameters corresponding to $\Delta I=0,2$, ρ-meson exchange, and pion exchange, respectively. They are sufficient to fit data provided that $\Delta I=1$ and especially $\Delta I=2$ contributions are appreciably stronger than the old naïve prediction based on the Cabibbo model using the factorization (2.1) and the sum rule (2.2) (NPC). The enhanced isotensor part ($\Delta I=2$) is of fundamental importance, as it seems to be at variance with the picture projected from a unified field theory with the Wilson expansion based on QCD. The approach of Ref. 94 is based on a suitable parametrization of the PV nucleon-nucleon scattering amplitude[99] for the short-range part of the interaction (vector-meson exchange). Hard-core corrections are included in the effective coupling constant, while the spatial dependence has the form of a δ function. A larger variety of parameters has introduced more variety in spin and isospin dependence. The long-range part of the potential has the standard one-pion-exchange form. The same type of parametrization but with all constants fixed would follow by the methods of Refs.[38,39,44] starting from the theoretical potentials. Thus again direct comparison with a theoretical potential is possible. According to Ref. 94, data can also be fitted by strengthening both short-range and long-range $\Delta I=1$ pieces instead of the isotensor.

It is interesting to note[94,134] that the short-range amplitudes decrease to the magnitude consistent with theoretical predictions[38] if P_γ in the $n+p \rightarrow d+\gamma$ process is assumed to be smaller by an order of magnitude. While with the direct use of the PV nucleon-nucleon potential, such as (A 20) or (A 18) one has to include short-range correlations in nuclear-physics calculations,[9-12] the situation is simplified by parametrizing nucleon-nucleon scattering amplitudes.[94] Details of semiempirical potentials, together with one version of the theoretical PV potential, are shown in the Appendix.

The approach using semiempirical potentials is a very useful one in systematizing experimental data and in discovering the general nature, such as spin, isospin dependence, etc., of PV nuclear forces. Additional data, especially those coming from $\Delta I=1$ experiments and from scattering experiments, will fix free parameters in a more unique way and hopefully resolve the isotensor vs isovector question. Preliminary results on P_γ measurements in ^{18}F and on the $\Delta I=1$ transition in ^{16}O seem to rule against too strong $\Delta I=1$ components in the PV potentials.[125] This means that large pion-exchange enhancements[38,40,86] as well as the third fit of Ref. 94 (i.e., fit 5 in Table 3) might already be ruled out.

2.6. Second-Class-Current Effects

Although there is growing interest in second-class current (SCC) effects in β decay,[100] their role in PV nucleon-nucleon forces has recently received little attention. Undoubtedly, this is due to the already complex enough situation in the field. Investigation using semiempirical potentials would, of course, automatically include SCC contributions in its results. As has been stated in the literature, the presence of charged SCC in a nonleptonic weak Hamiltonian would give rise to an isovector potential of the same general form as the pion exchange potential, but presumably of much shorter range.[10,12] The theoretical prediction of the $A(p_+^+)$ amplitude would also be affected.[143] Better $\Delta I=1$ type experiments and much more reliable theoretical approaches are needed to drawn any conclusion on SCC from parity violating in nuclei.

3. Nuclear-Physics Calculations

Weak PV potentials, be they theoretical or semiempirical, are treated as a perturbation for the nuclear-physics problem determining the opposite-parity admixture to the particular nuclear state, say ψ_i. Then the new state, with undetermined parity, is

$$\hat{\psi}_i = \psi_i + \sum_{k \neq i} \frac{\langle k | V_{12} | i \rangle}{E_i - E_k} \psi_k = \psi_i + \phi_i . \tag{3.1}$$

In the case of complex nuclei, an accurate expression for the ψ_i state will, in general, not be available. Further, the summation should be performed over all k states that can contribute. Even using a simple, single-particle model to describe nuclei such as ^{181}Ta might mean summation over more than 50 states when neutral currents are involved.[10-12,63] More sophisticated models, involving the effects of the giant dipole resonance, increase complexity even further and it is by no means certain that the calculations using only a few k states close to the unperturbed i state are really reliable.[64,75] Residual interactions such as pairing should also be considered[11,64,65]. As part of the PV potential has extremely short range, it actually tests the short-range behaviour of the nuclear wave function, which is usually not crucial for other applications. This point is best illustrated by the two-body case discussed below. Nuclear complexities prevent theoretical exploitation of the strikingly large PV effect measured in $8^- \rightarrow 6^+$ transition in ^{180}Hf (see Table 2). Eightfold K-forbiddeness is too difficult for any ambitious theoretical analysis and an estimate of one order of magnitude only has been provided so far.[66] However, it is important to emphasize that the theoretical knowledge of nuclear structure was used to predict a large effect in these perticular nuclei. This also supports the assumption that PV effects in nuclei are due to the opposite-parity admixtures in nuclear wave functions rather than to an intrinsic PV NNγ coupling.

Complete gauge invariance requires contributions from both diagram a) and diagram b) of Fig. 10 to be considered when dealing with electromagnetic transitions. When normal transitions are magnetic multipole (ML), the contribution of the admixed irregular electric multipole (EL) can be calculated in a straightforward way. The Siegert theorem takes care of both the "pair" contribution coming from diagram a) and of the exchange contribution symbolized by diagram b). However, when the opposite is the case, i.e., with EL, (ML) combinations, these pieces lead to an effective two-body PV electromagnetic Hamiltonian analogous to the ordinary magnetic exchange effects.[10,63,101] In calculating the asymmetry in $\frac{1^-}{2} \rightarrow \frac{1^+}{2}$ 110-keV transition in ^{19}F, Ref. 75, for example, neglects these effects estimating that they will make a contribution of a few per cent in this case. A general opinion is that such effects are negligible for ^{19}F.[132] However, the transition in ^{175}Lu ($\frac{9^-}{2} \rightarrow \frac{7^+}{2}$, 396 keV) is susceptible to the "pair" term contribution (see Figs. 9a and/or 10a), which can reduce γ circular polarization by about 20%.[132] All other exchange contributions are much smaller and it is generally safe to neglect them.[92,132,133]

As systematized a long time ago,[5] nuclear experiments can be divided into those where

i) there is violation of an absolute parity selection rule, such as parity-forbidden α decay and

ii) a pseudoscalar quantity is observed.

This might involve

iiA) photon emission, as in the case of γ-circular polarization measurements

iiB) or might not as in the case with scattering experiments using polarized particles.

Considerable complexity of theoretical nuclear-physics calculations gives special importance to the two-body problem, to which we turn next. This complexity is not alleviated by introduction of semi-empirical potentials. However, such potentials do help to systematize results of the interaction between experimental data and theoretical calculations in a way that is very instructive for the theory of weak interactions.

3.1. The $n+p\to d+\gamma$ Reaction

In this reaction, the emitted photon shows circular polarization.[45] (See also the Appendix.) If one of the initial nucleons is polarized, γ emission will be asymmetric with respect to its spin.[45] An important theorem shows that the first process depends only on the $\Delta I=0,2$ short-range part of the PV potential, while the second process depends on the $\Delta I=1$ pieces in the weak potential.[54] This reaction is thus of crucial importance for the test of strangeness conserving nonleptonic weak interactions. The measurement of γ-circular polarization reported in Table 2 has already had a very strong influence on theoretical efforts.[38-44] As already mentioned in Sec. 2.5, this experiment can be explained only by assuming that the PV vector-meson-exchange potential ($\Delta I=0,2$) is very strong. However, if this particular result is excluded, much weaker $\Delta I=0,2$ potentials with the strength predicted theoretically could explain other experiments.[94,132,134] The asymmetry measurement might have a similar impact if and when its accuracy is sufficiently improved.

Although this is a two-body problem, and although the summation indicated in Eq. (3.1) can be avoided, theoretical analysis is by no means simple.[54-62] For a while, existing theoretical calculations were at variance with each other but they have now settled, with all corrections including relavistic ones taken into account, at about the value

$$P \cong 2\times10^{-8} \; . \tag{3.2}$$

This value for circular polarization, obtained using NPC potentials, is in complete disagreement with experimental data ($P_\gamma = -(1.3\pm0.45)\times10^{-6}$, see Table 2). The main difficulty is associated with the nature of the strong potential used to solve the Schrödinger equation for NN scattering and the deuteron. This potential has to reproduce N-N scattering phase shifts, which are essentially long-range features, while PV $\Delta I=0,2$ potentials are of the very short-range nature. This extreme short-range sensitivity is nicely illustrated in Ref. 56. It is shown in this reference that unitary phase equivalent transformation (PET) can change the theoretical value of P_γ by an order of magnitude without disturbing N-N phase shifts. Recent investigations[144] have, however, shown that such PET's give exchange-current corrections to the total cross section for $n+p\rightarrow d+\gamma$ in serious disagreement with the existing theory[145] which seems to have explained the experimental value for the cross section.

The discrepancy with experimental data has also prompted some speculations about nonorthogonality of the bound and continuous two-body states due to the possible energy dependence of the strong potential.[102,103] However, strong arguments have been put forward claiming that a properly treated system of quantized pion and nucleon fields should always give proper orthogonal wave functions in the nonrelativistic approximation.[104]

The experiment with polarized neutrons[45] will yield information that is hopefully not too dependent on the vagaries of theoretical calculations since the result depends mostly on the long-range pion exchange. Even when the short-range $\Delta I=1$ piece becomes strong, as in the case with some semiempirical potentials, it does not influence the result by more than about 20%. The contribution from pion exchange seems to be rather independent of the calculational method. Values for γ asymmetry are in the range

$$A_\gamma \approx k_\pi (0.5\text{-}0.6)\times 10^{-8} , \quad (3.3)$$

where k_π measures the strength of the pion potential, being unity for NPC potentials. According to Refs. 38-40, 86 and 103, larger $|k_\pi|$ favors unified-field-theory models with neutral currents. The existing experimental value ($A_\gamma = (0.6 \pm 20)\times 10^{-8}$, see Table 2) carries too large an error to draw any definite conclusion.

The inverse of this experiment, i.e., the e+d→e+p+n reaction using longitudinally polarized electrons, seems to be measurable.[139] The asymmetry for backscattering at low momentum transfer is estimated as $A \leq 4\times 10^{-8}$ for the W-S model.[16,17*]

3.2. Parity-Violating α Decay

The α decay of the 2^- state in ^{16}O has been measured several times, yielding finally satisfactorily precise data.[46] This decay is important because it involves only the $\Delta I=0$ part of the weak interaction. α-decay theory is, unfortunately, not understood enough to allow precise calculation of α-decay widths. Estimates are made by calculating the ratio of irregular (i.e., parity-forbidden) and regular decay widths, and results depend on accounting for short-range hard-core effects. In spite of considerable calculational complications involved we may take some comfort from the fact that existing theoretical calculations approximately agree.[52] However, errors of factors of 4 for the rate are possible in such calculations.[93]

Thus far it has not yet been agreed upon whether experimental data indicate an $\Delta I=0$ PV potential strength that is of the order of magnitude predicted by the NPC choice[93] or whether differences exist.[94] In the latter case, differences are due to the richer isospin structure of the potential. As can be seen in the

---

*Scattering on ^{18}F or ^{19}F or some other nuclei could also give profitable results.

Appendix (A 8), the potential of Ref. 94 has two isoscalar terms. (For the NPC choice, $V_1 \equiv 0$.) There might be cancellations between parts of the potential corresponding to either ρ exchange (V_2) or ω exchange (V_1), with each being individually large.*

There are additional interesting possibilities involving α particles. The upper limit on the PV parameter $\;^2 < 4\times 10^{-10}$ has been found for the $\Delta I=1$ reaction[106]

$$\alpha + d \rightarrow {}^6\mathrm{Li} + 3.562\ \gamma\ . \tag{3.4}$$

The particular level in ^{6}Li could have been excited only through the violation of parity. This corresponds to the irregular width

$$\Gamma \leq 8\times 10^{-4}\ \mathrm{eV}\ , \tag{3.5}$$

while the rough theoretical estimate[86,103] for the W-S model[16,17] with the strong $\Delta I=1$ potential gives

$$\Gamma \lesssim 10^{-8}\ \mathrm{eV}\ . \tag{3.6}$$

Another reaction that is being studied[136] is the following:

$$p + {}^{19}\mathrm{F} \rightarrow {}^{20}\mathrm{Ne}^*(13.65\ \mathrm{MeV}\ J^P=0^+\ I=1(?);13.48\ \mathrm{MeV}(1^+,0)) \rightarrow {}^{16}\mathrm{O} + \alpha(0^+,0)\ , \tag{3.7}$$

which involves both $\Delta I=0$ and $\Delta I=1$ forces. For the moment, experiments seem to be yielding a negative result. The theoretical estimate, obtained by using the potential[86,103] with strong $\Delta I=1$ components is $\Gamma \sim 10^{-3}$ eV for the W-S model.[16,17] It might be expected that this experiment will set a limit on the strength of the $\Delta I=1$ component of the weak PV potential. Some other interesting possibilities are reviewed in Ref. 103.

There is also a possibility of doing the reverse of the ^{16}O decay experiment and determining PV by detecting a γ ray as in the experiment (3.4).[137]

---

*Some doubts have been raised in Ref. 135 that there might exist a 0^+ level lying very close to the 2^- level. Recent experiments do not support this conjecture.[146]

3.3. Parity Violation in Electromagnetic Transitions

In this section we are concerned with the effects described in 3.1 but involving large nuclei. There are three kinds of effects:[8-15]

i) Circular polarization of γ radiation from unpolarized nuclei (see the Appendix).

ii) Angular distribution of γ radiation from polarized nuclei.

iii) Up-down asymmetry of γ radiation following the capture of polarized thermal neutrons.

The effect iii) has been measured and theoretically analyzed many times, but so far complexity of the necessary nuclear calculations has not allowed the extraction of significant information about the weak interaction.[12] New interesting results involving capture of thermal neutrons by nuclei ^{118}Sn, ^{114}Cd, ^{236}U, and ^{240}U were presented at the Zürich Conference.[48] (In the last two cases, asymmetry for the light fragments has been measured.) A theoretical study of such cases can, in principle, give information different from the one available from γ decay in complex nuclei.[132]

Theoretical investigation of processes i) and ii) runs through all the difficulties already mentioned at the beginning of this section. The analysis is rather more complicated than in the two-body case owing to the fact that the result is affected in an important way by the PV interaction of the valence nucleon with the nucleons in the core (as symbolized in Fig. 10). The effect is proportional to the product of matrix elements such as

$$\sum_{k\neq i} \frac{\langle f|H_{el.mg.}|k\rangle\langle k|V_{12}|i\rangle}{E_i-E_k} , \tag{3.8}$$

with the summation runing over all nucleons in the nuclei. As every contribution is important, the wave function of the nuclei not only must include core effects but must be known more accurately than if calculating processes in which valence nucleons are of primary importance.[10-12,75] Core excitations also influence the result,[64] as indicated in Fig. 10a.

(Some additional technical details are to be found in the Appendix.) It appears[132] that the correlations responsible for the existence of the giant dipole resonance introduce an additional factor

$$\left(\frac{\omega_o}{\omega_D}\right)^2 \cong 0.3 \tag{3.9}$$

for complex nuclei. Here ω_D is the resonance energy, while ω_o is the characteristic excitation energy of the opposite parity admixtures in the nuclear wave function. The factor (3.9) is to be considered as an average expectation.

Besides short-range correlations, which have to be estimated by solving approximately the Bethe-Goldstone equation, corrections due to pairing[9-12,64,65,103] also have to be introduced.

Calculations have been performed for many nuclei including ^{19}F, ^{41}K, ^{175}Lu, and ^{181}Ta.[105] The case of ^{181}Ta, which has been investigated extensively both experimentally and theoretically, seems to indicate the need for stronger $\Delta I=0,2$ pieces in weak interactions that predicted by either the NPC approximation or deduced from unified field theories. Conclusions are somewhat obscured by the fact that the $\Delta I=1$ piece also contributes, and there have been claims stating that if the $\Delta I=1$ piece is sufficiently strong, the experimental data are explainable.[87] The ^{181}Ta experiment played an important role in a semiempirical determination of the strength of the pion-exchange PV force.[94] Investigations using semiempirical potentials seem to indicate that all pieces ($\Delta I=0,1,2$) could be appreciably larger than the first theoretical estimates.[93,94] Their sign is also puzzling.

Useful information will follow from experiments on light nuclei where the isospin content can be such as to lead to a strict isospin selection rule.[8,12,106] Besides the $n+p\rightarrow d+\gamma$ and ^{16}O experiments a good example for that is the γ decay of the 1.08 state in ^{18}F, which allows a change of the $\Delta I=1$ part only.[107] The present experimental upper limit on P_γ ($P_\gamma \lesssim 36\ 10^{-4}$, see Table 2) is already larger than the estimate based on the W-S model,[103,138] which is

$$|P_\gamma| \sim 57\times10^{-4} \quad . \tag{3.10}$$

In this estimate, the strong pion-exchange force has been used.[86] Owing to the long range of such a force the result is relatively insensitive to the short-range nucleon-nucleon correlations, being without correlations $|P_\gamma| \sim 61\times 10^{-4}$. If the potential (Ref. 94, see also (A21) in the Appendix) had been used, the situation might have been different. Finding relative isotopic strengths of the weak interaction through such and/or related scattering experiments will "show" neutral currents in nuclei.[94,103] Once the weak PV potential is fixed by cross-checking empirical data, the task of the theoretician is to reproduce these semiempirical findings using the whole apparatus of elementary particle physics. "It is obvious that reliable conclusions about the nature of weak Lagrangian can be reached only if the results of several experiments are evaluated and compared."[55]

3.4. Parity-Violating Effects in Nuclear Scattering

Nuclear scattering at thermal or low energies may be an invaluable tool for finding and cross-checking semiempirical weak potentials.[8,68-70,94,28,108-114] Scattering of nucleons can provide us with spin-isospin states which do not occur in decay experiments. PV observables for proton (p-p) and neutron-proton (n-p) scattering at low energies emerge as a useful by-product of the calculations for the $n+p\to d+\gamma$ reaction.[8,99,108] Semiempirical analysis shows[8,94,132] that the experiments depend on six parameters. One is connected with long-range pion exchange, while the others are matrix elements of short-range potentials. The situation is summarized in Table 3. (The notation used is ${}^{2s+1}L_J$.) At large energies and large momentum transfers the PV effect might cease to be weak and the study of such experiments can then even provide new insights into the proton structure of elementary particles.[115-123]

Coherent scattering of polarized thermal neutrons by heavy targets was first proposed and investigated in Refs. 28 and 109. Effects associated with the molecular structure of the target material were discussed in Ref. 110. Possible experiments are under serious investigation,[111,113] while theoretical re-

TABLE 3

Range	Exchanged particle	Matrix element	Particle content	ΔI
long	π		pn	1
short	ρ,ω	$\langle {}^1S_0 \vert V(\Delta I=0) \vert {}^3P_0 \rangle$	pp,nn,pn	0
short	ρ	$\langle {}^1S_0 \vert V(\Delta I=1) \vert {}^3P_0 \rangle$	pp,nn,pn	1
short	ρ	$\langle {}^1S_0 \vert V(\Delta I=2) \vert {}^3P_0 \rangle$	pp,nn,pn	2
short	ρ,ω	$\langle {}^3S_1 \vert V(\Delta I=0) \vert {}^1P_1 \rangle$	pn	0
short	ρ	$\langle {}^3S_1 \vert V(\Delta I=1) \vert {}^3P_1 \rangle$	pn	1

search[112] indicates that they might help to fix some, as yet arbitrary, parameters of semiempirical potentials. Thermal PV scattering also offers a possibility of observing neutron-nucleus resonances from a new angle.[111,114]

The nonrelativistic PV potential can describe scattering experiments up to 300 MeV.[68-70] Few experiments in this range have been performed[8] (see Table 2), measuring the dependence of the total cross section on the helicity of the incident beam. If total cross sections positive (negative) helicity are denoted by σ_+ (σ_-), then the asymmetry caused by parity-violating forces is

$$a = \frac{\sigma_+ - \sigma_-}{\sigma_+ + \sigma_-} \quad .$$

Such experiments were suggested in Ref. 68 and analyzed theoretically for N-N scattering in Refs. 68-70. Recent calculations have been extended to pd, p^3He, and p^4He scattering in DWBA.[8] Existing experimental values carry large errors and seem to be consistent even with the null effect. These values might be in agreement with "strong" short-range semiempirical $\Delta I=0,1$, or 2 PV potentials.[93,94]

At very high energies, in the GeV region, the effects coming from short-range weak interactions might become more pronounced if investigated at large momentum transfers.[115] As energy and momentum transfers increase beyond the presently accessible range, weak interactions may eventually become dominant.[115,119,121] At already available energies an asymmetry as large as 10^{-3}-10^{-4} may be expected.[122] Moreover, extensive theoretical investigations indicate that for a given weak-interaction model the size of the PV effect may strongly depend on the way strong interactions are described and thus it may be used to discriminate among various parton models of strong interactions.[123] Latest theoretical investigations are thus leading us into the region in which the PV nucleon-nucleon potential is no longer a significant theoretical tool.

3.5. Other Parity-Violating Nuclear Processes

Other experiments capable of detecting parity violation in nuclei have been suggested. In a β-γ correlation measurement a forward-backward asymmetry is to be expected if there is mixing in nuclear states.[12] Parity violation in nuclei leads to effects closely analogous to parity violation in β decay.

Parity impurities in nuclear states will lead to longitudinal polarization of internal conversion electrons[12,28] or to up-down asymmetry if the initial spin is polarized.[124] Expected effects are of the same order as circular polarization in γ transitions.

Mössbauer effect can be used to compare intensities of the components of the hyperfine spectrum, the PV effect being directly related to the magnitude of the circular polarization of the emitted radiation.[12]

Although all these effects are relatively small and/or not very suitable for measurements at present, their time might come And there is no doubt that human ingenuity will in due course think of other possible PV effects.

4. Conclusion and Outlook

Although considerable work has been done, many tasks are still in front of theorists studying nuclear parity violation. If some skeptical voices among experimentalists[137] are neglected, it may be said that the very fact that nuclear parity violation has been measured and that the effects have strength characteristic of weak interactions (see Table 2) already restricts our freedom in the weak-interaction model building. This is especially so for the models in which semiweak effects are possible[16,131] and which were mentioned in the Introduction.

Although the well-known case of ^{180}Hf (Ref. 49) is still beyond the possibility of a serious theoretical analysis, it carries great qualitative significance. It has been found exactly where it should have been found if the effect had been due to the opposite parity admixtures in the nuclear wave function. This mixing is caused by a weak nuclear force, so the whole effect could not be produced by any other cause, as for example, intrinsic PV nucleon-nucleon-photon coupling. Speaking with some optimism, we might say that at the moment we do have a qualitative understanding of experimental facts, although a definite quantitative account is still eluding us.

Future theoretical efforts have to start on the one hand from the field theory of weak interactions and on the other hand from nuclear theory in describing experiments. If nucleon-nucleon scattering at high energies[140] is further investigated, the theoretical methods of high-energy particle physics will be called for[8,115-123] thus taking us outside of the realms of nuclear parity violation in the narrow sense.

The step leading from the weak-interaction field theory to the PV nuclear potential has not yet been clarified satisfactorily. Even if the latest advances in QCD are employed to produce an effective weak Hamiltonian, much has to be understood about the strong dynamics influencing weak-interaction effects.[30-44,76-87,90-92] Such a situation does not allow for any definite conclusions regarding weak neutral currents. Nonleptonic effects of such currents are felt, according to gauge models,[16-25] only through nuclear parity violation and related effects. The existence of the short-range (i.e., vector-meson-exchange) $\Delta I=1$ potential is a definite indication for neutral weak hadronic currents.* Unfortunately, their contribution is mixed up with the pion exchange $\Delta I=1$ potential which appears even without any neutral currents. The strength

*This potential, however, could also be produced by the effects of charged second-class currents.

and sign of weak pion exchange are not yet fixed definitely theoretically, although many promising efforts have been made.[76-87,90-91] If we are allowed some guessing, we might say that the relatively strong pion exchange can be readily explained only as an effect of weak neutral currents. Our present theoretical knowledge does not exclude even rather weak pion exchange with neutral currents present. However, such a situation seems rather unlikely unless a model is constructed in which only vector (or axial-vector) neutral currents appear. This is puzzling, as it seems that some experiments in progress ($^{18}F^* \rightarrow {}^{18}F+\gamma$, $^{20}Ne^* \rightarrow {}^{16}O+\alpha$) tend to indicate that the weak pion coupling cannot be very strong. Such a situation lays a strong emphasis on the theoretical efforts at a better description of experiments, at suggesting new experiments, and at parametrizing and systematizing experimental results by introducing semi-empirical phenomenological nucleon-nucleon potentials. This study has been done on the basis of the first seven experiments listed in Table 2 (excluding A_γ measured for the $n+p \rightarrow d+\gamma$). There is a remarkable consistency among the results for the odd proton nuclei such as ^{19}F and ^{175}Lu (E1(M1))*, ^{181}Ta (M1(E1), and ^{41}Ka (M2(E2)), which indicates that a certain degree of control has gained over the pure nucleon part of the calculation,[132] i.e., within a factor of 2-3. All these transitions seem to be dominated by a strong PV pn force, as the strength of the PV pp force derived from the measurement of a in p-p scattering is probably too small to explain the above experiments. The required strength of the PV p-n force might be due to the relatively strong PV pion exchange. A further answer to that conjecture might come by improving some recent experiments measuring the $\Delta I=1$ component of the weak force (i.e., A_γ in $n+p \rightarrow d+\gamma$, P_γ in $^{18}F^*$ decay, and $^{20}Ne^* \rightarrow {}^{16}O+\alpha$).[125,132] A further important remark is that the semiempirical strength of the PV pion coupling is very dependent on the theoretical understanding of the ^{181}Ta experiment.[94] Similarly, the P_γ measurement for the $n+p \rightarrow d+\gamma$ process has a very particular influence on the strength of the $\Delta I=0,2$

*Parity-allowed and parity-forbidden multipoles are indicated.

parts of the PV nuclear force.[93,94] Should one, for its understanding, need a very strong isotensor component $\Delta I=2$ (Ref. 93) in the PV force, this result would be quite puzzling from the point of view of unified field theory. The α-decay experiment ($^{16}O \to ^{12}C+\alpha$) exerts certain but not absolute control over the $\Delta I=0$ pieces of PV nuclear interactions. Some logical possibilities are listed in Table 4. Further details on these novel and interesting

TABLE 4

Experiment	If taken into account	If disregarded
^{19}F, ^{41}K, ^{175}Lu, ^{181}Ta	"Strong" PV pn force might but need not contain "strong" $\Delta I=1$ piece.	One loses constraint on PV pn forces.
$n+p \to d+\gamma$ P_γ	Strong $\Delta I=0$ and/or $\Delta I=2$ force, signs of the phenomenological potentials opposite to theoretical expectations.	"The magnitude of all short-range isospin amplitudes and the relative signs of isoscalar amplitudes will be closer to theoretical expectations[94]
$^{16}O^* \to ^{12}C+\alpha$	"Weak" $\Delta I=0$ PV force. If the preceding experiment is to be fitted without the strong $\Delta I=2$ piece, there should be cancellation among two large $\Delta I=0$ contributions.	One loses constraint on PV isoscalar forces.
pp a	"Weak" PV pp force	One loses constraint on PV pp forces.

approaches[93,94] are in the preceding text. Not all of technical details seem to be settled at the moment.[125]

A recent analysis, based on the suitably weighted effective one-body potential,[132] (see the Appendix) indicates that there is a possibility for a one-parameter fit to the experiments in ^{19}F, ^{41}K, ^{175}Lu, and ^{181}Ta. This actually means more freedom in semiempirical analysis and, in particular, makes possible a reasonable fit with a relatively weak $\Delta I=1$ component in the PV nuclear force.[94]

Inspection of Table 4 in juxtaposition with Table 2 shows the direction in which the future experimental and theoretical efforts should go. Constraints on the $\Delta I=1$ pieces of the PV force and on the PV nn force should be found. Existing experimental values should, however, be rechecked and improved. Experiments in this direction involving either new decays or nuclear scatterings are already in progress or under preparation, and they were described to some extent on the preceding pages. Theoretical efforts continue, with new vistas being opened in the region of very high energies. It seems that we can look to the future with certain, but skeptical, optimism.

The author would like to thank Professor R.J. Blin-Stoyle for hospitality and for reading the manuscript. Professor E. Fischbach has also read the manuscript and has made some useful suggestions. The author also had useful communications with Professors B. Desplanques, E. Fiorini, M. Gari, E.M. Henley, B.H.J. McKellar, and J. Missimer.

Appendix

This short appendix is not meant to serve as an introduction to the formalism used in the study of weak interactions in nuclei. For that purpose we refer the reader to existing reviews and books.[8-12] In this Appendix we want to provide some formal illustrations of the points which were discussed in the main text.

Asymptotic Freedom and the Effective Weak Hamiltonian

H_W^{eff} is found from the second-order contribution in weak interactions[37-40]

$$H^2 = \int d^4x D_W(x) \,|T(J_\mu^+(x) J_-^\mu(0)) + T(J_\mu^-(x) J_+^\mu(0))| + \int d^4x D_Z(x) T(J_\mu^Z(x) J_Z^\mu(0)) \tag{A1}$$

Here the propagators correspond to intermediate vector bosons, i.e., W or Z and Higgs scalar exchanges have been neglected.

In the Weinberg-Salam (W-S) model, hadronic currents are made of left-handed (L) doublets

$$\begin{pmatrix} u \\ \cos\theta_C d + \sin\theta_C s \end{pmatrix}_L \qquad \begin{pmatrix} c \\ \cos\theta_C s - \sin\theta_C d \end{pmatrix}_L \quad , \tag{A2}$$

while vector-like quark models with a "bottom" quark b (i.e., U-B models) contain in addition right-handed (R) doublets and other exotic quarks, i.e.,

$$\binom{u'}{d'}_L \quad \binom{u}{b}_R \quad \binom{c}{h}_R \quad \binom{u'}{b'}_R \quad \ldots\ldots \tag{A3}$$

The hadronic currents in the W-S model are the following:

$$\begin{aligned} J^+ &= \cos\theta_C(\bar{d}u)_L + \sin\theta_C(\bar{s}u)_L + \cos\theta_C(\bar{s}c)_L - \sin\theta_C(\bar{d}c)_L \ , \\ J^Z &= \tfrac{1}{2} J^0 - 2\sin^2\theta_W J^\gamma \ , \\ J^0 &= (\bar{u}u - \bar{d}d + \bar{c}c - \bar{s}s)_L , \\ J^\gamma &= |\tfrac{1}{2}(\bar{u}u - \bar{d}d) + \tfrac{1}{6}(\bar{u}u + \bar{d}d) - \tfrac{1}{3}\bar{s}s + \tfrac{2}{3}\bar{c}c|_{L+R} \ . \end{aligned} \tag{A4}$$

The U-B models have additional terms corresponding to various doublets from (A3). For the u,b combination, the charged current is changed as follows

$$J^{+}(U\text{-}B) = J^{+}(W\text{-}S) + (\bar{b}u)_R \quad . \tag{A5}$$

The first two terms in the charged current (A4) correspond to the Cabibbo-model current. The $\Delta S=1$ hyperon nonleptonic decays are thus due to the combination

$$H_S \sim \sin\theta_C \cos\theta_C (\bar{s}u)_L(\bar{u}d)_L \ , \tag{A6}$$

while the $\Delta S=0$ $NN\pi$ amplitude obtains the contribution from

$$H_1 \sim \sin^2\theta_C (\bar{s}u)_L(\bar{u}s)_L \tag{A7}$$

and from many terms appearing in the product of neutral currents. In the case of the U-B models, (A5) leads to (A7) plus a term which is not suppressed by the Cabibbo factor, i.e.,

$$(\bar{b}u)_R(\bar{u}b)_R \quad . \tag{A8}$$

The enhancement, predicted on the basis of (2.2) is (-24). (See also Sec. 2.3.)

The vector-meson- exchange potential has so far been calculated using either the factorization approximation or diagrams a and b in Fig. 4. In either case, baryonic and mesonic states which were used, contained only valence quarks, so only the term

$$H_{0,2} \sim \cos^2\theta_C (du)_L(ud)_L \tag{A9}$$

could contribute from charge currents (A4) or (A5). Similar considerations apply also to neutral currents. In the case of neutral currents the product of isovector pieces like

$$\bar{u}u - \bar{d}d \tag{A10}$$

with the isoscalar pieces like

$$\bar{u}u + \bar{d}d \tag{A11}$$

leads to the isovector $\Delta I=1$ vector-meson-exchange potential in addition to the $\Delta I=0,2$ contributions.

It is generally assumed that the dominant contribution comes from the leading terms in the Wilson expansion

$$T(J(x)J(0) = \sum_n C_2^n(x)O_2^{(n)}(0) + \sum_n C_4^n(x)O_4^n(0) + \dots \quad . \qquad \text{(A12)}$$

The four-quark operators, which are usually considered, are the ones multiplied by $C_4^n(x)$. They contain the local current-current product which is very much of the same form as the old current-current theory Hamiltonian.[6] The factors are of the form

$$C(x) = C_o(x)(1 + b \ln(\mu^2 x^2)) \quad . \qquad \text{(A13)}$$

Here b has to be found by solving the Callan-Symanzik equation, μ^2 depends on the subtraction point, and $C_o(x)$ is the value of the factor for free-quark fields, i.e., when gluon exchanges are switched off. A net result is the current-current Hamiltonian with an effective (renormalized) coupling constant replacing the Fermi coupling constant. Quark combinations are given, for example, in (A6-A9).

The analysis of Ref. 91 starts with the pion contraction, as symbolized in Figs. 6 and 8, and assumes that the dominant contribution comes from the isospin rotated product of currents

$$\langle\pi(q)N|H^2|N\rangle_{q^2\to 0} \int d^4x D(x) e^{iqx} \, C_{ABC}\langle N|T(J_\mu^B(x)J_A^\mu(0))|N\rangle + \dots \; . \qquad \text{(A14)}$$

A further assumption is that the main term in the Wilson expansion (A12, A13) of this equal-time-commutator contribution is the two-quark operator proportional to the very heavy charmed quark mass. Loosely speaking, such a term emerges from the product of currents in which each current contains a charmed-quark field. These fields are "contracted"

$$\sin\theta_C \cos\theta_C (\bar{d}c)(\bar{c}s) \to \sin\theta_C \cos\,\theta_C (\bar{d}s) \; . \qquad \text{(A15)}$$

In the $\Delta S=0$ sector, no such term appears in the product of neutral currents, as their charmed quarks are always in pair $\bar{c}c$ (see

(A4)). The only possibility is the last term in the charged current

$$\sin^2\theta_C(\bar{d}c)(\bar{c}d) \to \sin^2\theta_C(\bar{d}d) \quad . \tag{A16}$$

As this term suffers a typical suppression, the hypothesis of Ref. 91 predicts no enhancement for either W-S or U-B models.

PV Vector-Exchange Potential

Starting from the current-current Hamiltonian, the elementary or the effective one, the vector-meson-exchange potential can be calculated from the diagrams in Fig. 4. The example[39] quoted below is for illustrative purposes and it does not mean that it is preferable to the others existing in the literature.[38,44,86] It has the following isospin dependence:

$$\begin{aligned} V(\text{vector}) &= V_1(\Delta I=0) + V_2(\Delta I=0)\vec{\tau}_1\cdot\vec{\tau}_2 \\ &+ V_3(\Delta I=2)(3\tau_{1z}\tau_{2z} - \vec{\tau}_1\cdot\vec{\tau}_2) + V_4(\Delta I=1)(\tau_{1z}+\tau_{2z}) \\ &+ V_5(\Delta I=1)(\tau_{1z}-\tau_{2z}) \quad . \end{aligned} \tag{A17}$$

Its spin-spatial dependence is given by

$$\begin{aligned} V_1 &= (2.41\ v_- + 2.53\ w)\cdot 10^{-8} \ , \\ V_2 &= -2.54|v_- + (1+\mu_v)w|\cdot 10^{-8} \ , \\ V_3 &= 0.42|v_- + (1+\mu_v)w|\cdot 10^{-8} \ , \\ V_4 &= (0.89\ v_- + 1.17\ w)\cdot 10^{-8} \ , \\ V_5 &= 0.93\ v_+\cdot 10^{-8} \ , \end{aligned} \tag{A18}$$

where the following notation has been used

$$\begin{aligned} v_i &= \vec{\sigma}_i \cdot \{\vec{p}_{12}, \frac{e^{-m_\rho r}}{r}\}_+ \ , \\ v_\pm &= v_1 \pm v_2 \ , \\ w &= i(\vec{\sigma}_1\times\vec{\sigma}_2)\cdot |\vec{p}_{12}, \frac{e^{-m_\rho r}}{r}|_- \quad . \end{aligned} \tag{A19}$$

Here $\vec{p}_{12} = \vec{p}_1 - \vec{p}_2$; $r = |\vec{r}_1 - \vec{r}_2|\mu_\nu = 3.7$ and m_ρ is the ρ-meson mass. The units $\hbar = c = 1$ have been used.

The semiempirical potential of Ref. 93 follows from (A17) by the following substitution

$$V_1 = V_4 = V_5 = 0 \, , \qquad \text{(A20)}$$
$$V_{2,3} = C_{2,3}|v_- + (1+\mu_\nu)w| \, .$$

The values of the constants $C_{2,3}$ are displayed in Table 5 (in fm).

The spin-spatial dependence of the semiempirical potential of Ref. 94 is given by

$$V_i = C_i(\vec{\sigma}_1 - \vec{\sigma}_2) \cdot \{\vec{p}_{12}, \delta(\vec{r})\}_+ \, , \qquad i = 1,2,3,4 \qquad \text{(A21)}$$
$$V_5 = C_5(\vec{\sigma}_1 + \vec{\sigma}_2) \cdot \{\vec{p}_{12}, \delta(\vec{r})\}_+$$

As before, the coupling constants are given in Table 5 (in fm^3).

TABLE 5

	$C_i \times 10^8$					
Potentials	C_1	C_2	C_3	C_4	C_5	C_π
Fit 1 (a)	0	1.0485	-12.0734	0	0	1.1429
Fit 2 (a)	0	-1.0405	-11.9145	0	0	1.9540
Fit 3 (b)	5.3592	24.8800	-6.4583	-3.3599	0	2.7940
Fit 4 (b)	5.1915	23.3678	-7.4861	0	-3.4328	2.7940
Fit 5 (b)	5.2732	35.0458	0	-24.2194	20.9091	2.7940

(a) Ref. 93

(b) Ref. 94

PV Pion Exchange Potential

The theoretical expression for the pion exchange potential is of the form

$$V_\pi = C_\pi(\vec{\sigma}_1 + \vec{\sigma}_2) \cdot |p_{12}, \frac{e^{-m_\pi r}}{r}| \, i(\vec{\tau}_1 \times \vec{\tau}_2)_z^{\;*} \, . \qquad \text{(A22)}$$

The constant C_π depends on the amplitude $A(p_+^+)$ (or $A(n_-^o)$) whose

* $-i/2 \; (\tau_1 \times \tau_2)_z = T_{12}^{(-)} = \tau_{1+}\tau_{2-} - \tau_{1-}\tau_{2+}$.

value can be determined in numerous ways discussed in Sec. 2.3. Its standard NPC value is

$$C_\pi \cong 0.15\times 10^{-8}\ \text{fm} . \tag{A23}$$

This corresponds to the $A(p_+^+)$ value (2.5).

Semiempirical potentials use the form (A22) with the respective constants C_π listed in Table 5 (in fm).

The theoretical calculation based on the quark model[85] predicts C_π positive and falling in the range

$$1.4\times 10^{-8} < C_\pi < 2.4\times 10^{-8} , \tag{A24}$$

thus giving the enhancement (E) of 9.4 to 16 times over the NPC value (A23).

The most pesimistic range of values for $A(p_+^+)$ given by (2.6) corresponds roughly to (A23) being multiplied by

$$10^{-1} \le |E| \le 10^2 . \tag{A25}$$

Circular Polarization of γ Radiation

As an example for parity-violating effects in nuclei we select circular polarization of the emitted photons.

The matrix element responsible for nuclear γ decay is

$$\langle f|e\, A_\mu^s j_\mu|i\rangle \Longrightarrow$$

$$\Longrightarrow e\sum_L i^L\,(2(2L+1))^{1/2}\left(\frac{L}{L+1}\right)^{1/2}\frac{k^L}{(2L+1)!!} \tag{A26}$$

$$\times\ \langle f|s(EL)^s + (ML)^s|i\rangle .$$

It is obtained by the nonrelativistic approximation for the electromagnetic current j_μ and by the multipole expansion of the electromagnetic potential A_μ. Here $s = \pm 1$ denotes the circular polarization of the photon with energy k, while (EL) and (ML) symbolize the electric and magnetic multipole operator, respectively. The emitted radiation shows circular polarization if there

there is an excess of photons with the particular polarization, i.e., if

$$P_\gamma = \frac{\tilde{P}(s=1) - \tilde{P}(s=-1)}{\tilde{P}(s=1) + \tilde{P}(s=-1)} \quad ,$$

$$\tilde{P}(s) = |\langle eA^s_\mu \, j_\mu \rangle|^2 \tag{A27}$$

is different from zero. In the numerator, only the cross terms between electric and magnetic multipoles can survive. For example, the 482-keV transition in ^{181}Ta is a mixture of M1 and E2, to which V_{12} admixes E1, thus leading to

$$P_\gamma = -2 \, \frac{\langle\langle E1 \rangle\rangle}{\langle M1 \rangle (1 + \delta^2)} \quad . \tag{A28}$$

Here δ measures M -E2 mixing, while <<E1>> has to be found by the standard perturbation theory, as discussed in Sec. 3.3. It is given by

$$\langle\langle E1 \rangle\rangle = \sum_n \frac{\langle f|V_{12}|n\rangle\langle n|E1|i\rangle}{E_f - E_n}$$

$$+ \sum_k \frac{\langle f|E1|k\rangle\langle k|V_{12}|i\rangle}{E_i - E_k} \quad . \tag{A29}$$

If the main contribution to (A29) comes from the states n and k with an excitation energy $\approx \omega_o$, (A29) can be approximated by

$$\langle\langle E1 \rangle\rangle \approx -\frac{1}{\omega_o} \langle f|\{V_{12}, E1\}_+|i\rangle \quad . \tag{A30}$$

In the giant dipole resonance approach, (A29) can be approximated by

$$\langle\langle E1 \rangle\rangle \approx -\frac{1}{\omega_D} \langle f|\{V_{12}, E1\}_+|i\rangle \quad . \tag{A31}$$

Comparison of the two approximations shows a factor ω_o/ω_D. More careful consideration[64,132] leads to this factor squared, as mentioned in Sec. 3.3.

Gauge-invariance requirements can be illustrated for the case of the pion-exchange potential as follows: "Pair contribution" in diagram a) of Fig. 10 and in diagram b) of the same figure replace the pion-exchange potential V_π (i.e., plain diagram in Fig. 2) by

$$V_\pi \to V_\pi + i \sum_\ell \frac{1}{2} |E1(\ell), V_\pi| \quad . \tag{A32}$$

Here we have introduced simplified notation. For complete definition we refer the reader to Refs. 10, 11, and 63. In the commutator (A32), $E1(\ell)$ means a dipole operator depending on the coordinate of the ℓth particle. One finds

$$\begin{aligned}(i\omega)<<E1>> = {} & <f|i\,\frac{1}{2}\sum_\ell |E1(\ell), V_\pi||i> \\ & -i\sum_n \frac{(E_n-E_i)<f|V_\pi|n><n|\sum_\ell E1(\ell)|i>}{E_f-E_n} \\ & -i\sum_k \frac{(E_f-E_k)<f|\sum_\ell E1(\ell)|k><k|V_\pi|i>}{E_i-E_k} \quad .\end{aligned} \tag{A33}$$

As

$$\begin{aligned} & E_i-E_f-\omega = 0 \ , \\ & E_n-E_i = E_n-E_f - \omega \ , \\ & E_f-E_k = E_i-E_k - \omega \ , \end{aligned} \tag{A34}$$

there is cancellation leading to the expression proportional to (A29). For an irregular magnetic multipole transition the second term has a more complicated structure than the one sketched in (A32). There are no cancellations analogous to (A33) and (A34). At the end, one is left with an expression analogous to (A29) (replace E1 by M1!) plus a matrix element of an effective parity-violating two-body electromagnetic operator.

Effective Single-Particle PV Potential

In the fundamental paper[28] the matrix elements of weak PV two-body potentials were simplified by averaging over the nuclear core, i.e.,

$$\langle f|V_{12}|i\rangle = \int d\tau \sum_s u_f^*(1)u_s(2)V_{12}(u_i(1)u_s(2)$$

$$- U_i(2)u_s(1)) \rightarrow \int U_f^* V_{eff} u_i d\tau \quad . \tag{A35}$$

Such a procedure, analogous to the construction of the nuclear optical potential, leads to the form

$$V_{eff} \sim \text{const } \{\rho(r), \vec{\sigma}\cdot\vec{p}\} \ . \tag{A36}$$

Here $\rho(r)$ is the nuclear density, while the operators refer to valence nucleons. It is possible to find general rules which determine the constant in front of (A36) for any particular piece in V_{12}. One of the advantages of using (A36) is that the corresponding Schrödinger equation for the nuclear system can be solved exactly numerically[132] and in some cases even analytically,[114] thus avoiding the perturbational approach with its extensive summations. The results obtained by such an effective single-particle potential are usually in good qualitative agreement with more sophisticated calculations.[63] * If the single-particle potential (A36) is properly weighted by a selection of the "realistic" nuclear density $\rho(r)$, results can be improved even further.[132] In addition, the correction factor $(\omega_o/\omega_D)^2$ for the effective charge multiplying E1 and the pairing factor $u_i u_f \pm v_i v_f$ should also be introduced.[64]

---

*It should be noted that the results of Ref. 141 are too large because of the erroneous selection of the particle hole phases.

References

[1]D.T. Lee and C.N. Yang, Phys. Rev. 104, 254 (1956).

[2]C.S. Wu et al., Phys. Rev. 105, 1413 (1957);
R.L. Garwin et al., Phys. Rev. 105, 1415 (1957);
J.I. Friedman and V.I. Telegdi, Phys. Rev. 105, 1681 (1956).

[3]Quotations from Ref. 1: "If parity is not strictly conserved all atomic and nuclear states become mixtures consisting mainly of the state they are usually assigned together with small percentages of states possessing the opposite parity .... Parity nonconservation implies the existence of interactions which mix parities....."

[4]N. Tanner, Phys. Rev. 107, 1203 (1957).

[5]D.H. Wilkinson, Phys. Rev. 109, 1603, 1610, 1614 (1958).

[6]R.P. Feynman and M. Gell-Mann, Phys. Rev. 109, 193 (1958).

[7]W.D. Hamilton, Progr. Nucl. Phys. 10, 1 (1968);
High Energy Physics and Nuclear Structure, Eed. S. Devons (Plenum Press, New York - London) 696 (1970).

[8]E.M. Henley, Ann. Rev. Nucl. Sci. 19, 367 (1969);

[9]B.H.J. McKellar, High Energy Physics and Nuclear Structure, Ed. S. Devons (Plenum Press, New York - London) 682 (1970);
Invited talk, Vancouver Conference 1977.

[10]E. Fischbach and D. Tadić, Phys. Rep. 6C, 123 (1973).

[11]M. Gari, Phys. Rep. 6C, 317 (1973).

[12]R.J. Blin-Stoyle, Fundamental Interactions and the Nucleus (North-Holland/American Elsevier, 1973).

[13]E. Fiorini, Rivista del Nuovo Cim. 4, 401 (1974).

[14]E. Kuphal, Interaction Studies in Nuclei, Eds. H. Jochim and B. Ziegler (North Holland) 357 (1975).

[15]M. Konuma, K. Ohya, T. Oka, and Y. Yamamoto, Suppl. Progr. Theor. Phys. 60, 203 (1976).

[16]A. Salam and J.C. Ward, Phys. Lett. 13, 168 (1964);
A. Salam, Proceedings 8th Nobel Symposium, Ed. N. Svartholm, (Almquist Forlag, Stockholm) (1968);
J.C. Pati and A. Salam, Phys. Rev. D8, 1240 (1973); D10, 275 (1974); Phys. Rev. Lett. 34, 613 (1975).

[17]S. Weinberg, Phys. Rev. Lett. 19, 1264 (1967); 27, 1688 (1971).
[18]B.W. Lee, Phys. Rev. D6, 1188 (1972).
[19]J. Prentki and B. Zumino, Nucl. Phys. B47, 99 (1972).
[20]H. Georgi and S.L. Glashow, Phys. Rev. Lett. 28, 1494 (1972); B.W. Lee, J.R. Primack,and S.B. Treiman, Phys. Rev. D7, 510 (1973); Fayyazuddin and Riazuddin, Nucl. Phys. B76, 125 (1974).
[21]A. De Rújula, H. Georgi, and S.L. Glashow, Phys. Rev. Lett. 35, 69 (1975); Phys. Rev. D12, 3589 (1975).
[22]R.L. Kingsley et al., Phys. Rev. D12, 2768 (1975).
[23]H. Fritsch et al., Phys. Lett. 59B, 256 (1975).
[24]S. Pakvasa et al., Phys. Rev. Lett. 35, 703 (1975).
[25]R.M. Barnett, SLAC-Publ. 1821 (1976).
[26]R.J. Blin-Stoyle, Phys. Rev. 118, 1605 (1960).
[27]R. Dashen, S. Frautschi, M. Gell-Mann, and Y. Hara, in "The Eightfold Way" (W.A. Benjamin Inc., New York 1964), p. 254.
[28]F.C. Michel, Phys. Rev. 133, B329 (1964).
[29]D.S. Beder, Nucl. Phys. B49, 45 (1972).
[30]G. Barton, Nuovo Cimento 19, 512 (1961).
[31]E.M. Henley, Phys. Rev. Lett. 27, 542 (1971); E.M. Henley, T.E. Keliher, W.J. Pardee, and D.U.L. Yu, Phys. Rev. D8, 1503 (1973) and D9, 755 (1974).
[32]L. Schülke, Nucl. Phys. B40, 386 (1972).
[33]A. Andraši, B. Eman, J. Missimer, and D. Tadić, Phys. Rev. D11, 2484 (1975).
[34]See review paper, R.L. Jaffe, Nature 268, 201 (1977).
[35]D.J. Gross and F. Wilczek, Phys. Rev. Lett. 30, 1343 (1973); H.D. Politzer, Phys. Rev. Lett. 30, 1346 (1973); S. Weinberg, Phys. Rev. Lett. 31, 494 (1973) and Phys. Rev. D8, 4482 (1973).
[36]D.J. Gross and F. Wilczek, Phys. Rev. D8, 3633 (1973) and D9, 980 (1974); H. Georgi and D. Politzer, Phys. Rev. D9, 416 (1974).
[37]M.K. Gaillard and B.W. Lee, Phys. Rev. Lett. 33, 108 (1974); G. Altarelli and L. Maiani, Phys. Lett. 52B, 351 (1974);

G. Altarelli, R.K. Ellis, L. Maiani, and R. Petronzio, Nucl. Phys. B88, 215 (1975).
[38]J.F. Donoghue, Phys. Rev. D13, 2064 (1976); D15, 184 (1977).
[39]H. Galić, B. Guberina, and D. Tadić, Z. Phys. A276, 65 (1976);

Phys. Rev. D14, 2327 (1976).
[40]H. Galić and D. Tadić, Fizika 8, 99 (1976).
[41]P. Olesen and J.A. Rao, Phys. Lett. 29B, 233 (1969); M.A. Perez, Lett. Nuovo Cimento 12, 692 (1975).
[42]G.S. Danilov, Nucl. Phys. B60, 221 (1973); B. Guberina, J. Missimer, and D. Tadić, Phys. Rev. D9, 2456 (1974).
[43]B.H.J. McKellar and P. Pick, Phys. Rev. D7, 260 (1973); D6, 2184 (1972).
[44]H.J. Pirner and M.L. Rustgi, Nucl. Phys. A239, 427 (1975).
[45]V.M. Lobashov et al., Nucl. Phys. A197, 241 (1972); J.F. Cavaignac et al., Phys. Lett. 67B, 148 (1977).
[46]N. Neubeck, H. Shober, and H. Wäffler, Phys. Rev. C10, 320 (1974).
[47]V.M. Lobashov et al., J.E.T.P. Letters 3, 173 (1966); J.E.T.P. Letters 5, 59 (1967); Sov. Journ. Nucl. Phys. 13, 313 (1971); J.C. Vanderleeden and F. Boehm, Phys. Lett. 30B, 467 (1969); Phys. Rev. C2, 748 (1970); P. Bock and B. Jenschke, Nucl. Phys. A160, 550 (1971); E.P. Lipson et al., Phys. Rev. C5, 932 (1972).
[48]E.G. Adelberger, this Conference; R. Wilson, this Conference; C. Barnes, this Conference.
[49]E. Lipson, F. Boehm, and J.C. Vanderleeden, Phys. Rev. C5, 932 (1972); K.S. Krane, C.E. Olsen, J.R. Sites, and W.A. Steyert, Phys. Rev. C4, 1907 (1972); there are additional references in Refs. 10-12.
[50]E.G. Adelberger et al., Phys. Rev.Lett. 34, 402 (1975).
[51]J.M. Potter et al., Phys. Lett. 33, 1307 (1974).
[52]E.M. Henley et al., Phys. Rev. Lett. 23, 941 (1969); M. Gari and H.Kummel, Phys. Rev. Lett. 23, 26 (1969); M. Gari et al., Nucl. Phys. A161, 625 (1971); B. Apagyi, G. Fái, and J. Németh, Nucl. Phys. A272, 301 (1976).
[53]R.J. Blin-Stoyle and H. Feshbach, Nucl. Phys. 27, 395 (1961); F. Partovi, Ann. Phys. (N.Y.) 27, 114 (1964).
[54]G.S. Danilov, Phys. Lett. 18, 40 (1965); 35B, 579 (1971); Sov. J. Nucl. Phys. 14, 443 (1972).
[55]D. Tadić, Phys. Rev. 174, 1694 (1968); corrected values are in Ref. 59.

[56]E. Hadjimichael and E.Fischbach, Phys. Rev. D3, 755 (1971); B.A. Craver, E. Fischbach, Y.E. Kim, and A. Tubis, Phys. Rev. D13, 1376 (1976); E. Fischbach, private communication.

[57]K.R. Lassey and B.H.J. McKellar, Lett. Nuovo Cim. 11, 161 (1974); Aust. J. Phys. 27, 637 (1974); Phys. Rev. C11, 349 (1975); Phys. Rev. C12, 721(E)(1975); Nucl. Phys. A260, 413 (1976).

[58]M.L. Rustgi and H.J. Pirner, Phys. Rev. C10, 2099 (1974).

[59]B. Desplanques, Nucl. Phys. A242, 423 (1975). (His sign has been generally accepted so far.)

[60]G.N. Epstein, Univ. of Oxford report, 1975 (unpublished).

[61]M.A. Pérez, Ann. Phys. (N.Y.) 94, 374 (1975).

[62]J.P. Leroy, J. Micheli, and D. Pignon, Nucl. Phys. A28o, 337 (1977).

[63]B. Eman and D. Tadić, Phys. Rev. C4, 661 (1977).

[64]B. Desplanques and N. Vinh Mau, Phys. Lett. 35B, 28 (1971); B. Desplanques, Phys. Lett. 47B, 212 (1973) and Thesis, Université de Paris-Sud, Centre d'Orsay, 1975 (unpublished).

[65]J. Missimer, A.D. Jackson, and E. Fischbach, Nucl. Phys. A209, 221 (1973).

[66]P. Vogel, Caltech report CALT-63-155 (1971).

[67]M. Gari, A.H. Huffman, J.B. McCrory, and R. Offerman, Phys. Rev. C11, 1485 (1975);
M.A. Box and B.H.J. McKellar, Phys. Rev. C11, 1859 (1975).

[68]M. Simonius, Phys. Lett. 41B, 415 (1975); Nucl. Phys. A220, 269 (1974); Interaction Studies in Nuclei, Eds. H. Jochim and B. Ziegler (North Holland) 3 (1975).

[69]V.R.Brown, E.M. Henley, and F.R. Krejs,Phys. Rev. Lett. 30, 770 (1973); Phys. Rev. C9, 935 (1974).

[70]G.N. Epstein, Phys. Lett. 55B, 249 (1975).

[71]N. Cabibbo, Phys. Rev. Lett. 10, 531 (1963).

[72]R.J. Blin-Stoyle, Phys. Rev. 118, 1605 (1960).

[73]R. Lacaze, Nucl. Phys. B4, 657 (1968); B. Desšlanques, Phys. Lett. 41B, 461 (1972); H.J. Pirner and D.O. Riska, Phys. Lett. B44, 451 (1973).

[74]H. Chemtob and B.Desplanques, Nucl. Phys. B78, 139 (1974).

[75]M.A. Box, B.H.J. McKellar, P. Pick, and K.R. Lassey, J. Phys. G:

Nucl. Phys. 1, 493 (1975).

[76] B.H.J. McKellar, Phys. Lett. 26B, 107 (1967); E. Fischbach, Phys. Rev. 170, 1398 (1968); E. Fischbach and K. Trabert, Phys. Rev., 174, 1843 (1968).

[77] E. Fischbach, D. Tadić, and K. Trabert, Phys. Rev. 186, 1688 (1969); High Energy Physics and Nuclear Structure, Ed. S. Devons (Plenum Press, New York - London) 742 (1970).

[78] M. Gronau, Phys. Rev. D5, 118 (1972); M.D. Scadron and L.C. Thebaud, Phys. Rev. D8, 2190, 2329 (1973).

[79] J.H. Reid and N.N. Trofimenkoff, Nucl. Phys. B40, 255 (1972).

[80] J. Reid, Lett. Nuovo Cim. 9, 21 (1974); A. Andraši, B. Eman, J. Missimer, and D. Tadić, Phys. Rev. D11, 2484 (1975).

[81] B.W,. Lee and A.R. Swift, Phys. Rev. 136, B228 (1964); J. Schwinger, Phys. Rev. Lett. 12, 630 (1964); 13. 355 (1964); J.J. Sakurai, Phys. Rev. 156, 1504 (1967).

[82] J.C. Pati and C.H. Woo, Phys. Rev. D3, 2920 (1971); J.G. Körner, Nuovo Cim. 11A, 597 (1972) and Phys. Lett. B44, 361 (1973).

[83] J.F. Donoghue, E. Golowich, and B. Holstein, Phys. Rev. D15, 1341 (1977); J.F. Donoghue and E. Golowich, Phys. Rev. D15, 3421 (1977) and University of Massachusetts preprint COO-3066-94 (1977); C. Schmid, Phys. Lett. 66B, 353 (1977); J.F. Donoghue, private communication.

[84] G. Zweig, CERN report (1964); J. Iizuki, Supp. Progr. Theor. Phys. 37-38, 21 (1966).

[85] B. Desplanques and J.Micheli, Phys. Lett. 68B, 339 (1977).

[86] D. Bailin, A. Love, D.V. Nanopoulos, and G.G. Ross, Nucl. Phys. B59, 177 (1973); M. Gari and J.H. Reid, Phys. Lett. 53B, 237 (1974); A. Andraši, J. Missimer, and D. Tadić, Phys. Rev. D11, 1331 (1975) and Interaction Studies in Nuclei, Ed. H. Jochim and B. Ziegler (North Holland, Amsterdam) 273 (1975); B. Desplanques and E. Hadjimichael, Nucl. Phys. B107, 125 (1976).

[87] V.K. Cung and C.W. Kim, Phys. Lett. 68B, 169 (1977); M.A. Peréz, preprint (1977).

[88] P.Sandars, talk at this Conference.

[89] A. Benvenutti et al., Phys. Rev. Lett. 36, 1478 (1976) and Phys.

Rev. Lett. 37, 1891 (1976); B.C. Barish et al., Phys. Rev. Lett. 38, 314 (1977).

[90] R.K. Ellis, Nucl. Phys. B106, 239 (1976); J.F. Donoghue, E. Golowich, and B.R. Holstein, Phys. Rev. D15, 1341 (1977).

[91] M.A. Ahmed and G.G. Ross, Phys. Lett. 61B, 287 (1976).

[92] B. Guberina and D. Tadić, IRB preprint.

[93] M.A. Box, A.J. Gabric, K.R. Lassey, and B.H.J. McKellar, J. Phys. G: Nucl.Phys. 2 (1976) L 107.

[94] B. Desplanques and J. Missimer, Carnegie-Mellon University Preprint COD-3066-82, Nucl.Phys. to be published.

[95] E.M. Henley, A.H. Huffman, and D.U.L. Yu, Phys. Rev. D7, 943 (1973); L. Wolfenstein, Nucl. Phys. B72, 111 (1974).

[96] B.H.J. McKellar and P. Pick, Nucl. Phys. B22, 625 (1970); E.M. Henley, Phys. Rev. Lett. 27, 542 (1971).

[97] E.M. Henley, Phys. Rev. C7, 1344 (1973); P. Herczeg and L. Wolfenstein, Phys. Rev. D8, 4051 (1973); P. Herczeg and P. Singer, Phys. Rev. D11 611 (1975); M.A. Peréz, Phys. Rev. D15, 704 (1977).

[98] P. Herczeg, Nucl. Phys. 48, 263 (1963).

[99] G.S. Danilov, Phys. Lett. 18, 40 (1965), 35B, 579 (1971); Sov. J. Nucl. Phys. 14, 443 (1972); B.H.J. McKellar, Nucl. Phys. A254, 349 (1975); J. Missimer, Phys. Rev. C14, 347 (1976).

[100] K. Kubodera, J. Delorme, and M. Rho, Phys. Rev. Lett. 38, 321 (1977).

[101] B.H.J. McKellar, Phys. Lett. B38, 401 (1972).

[102] G. Breit and M.L. Rustgi, Nucl. Phys. A161, 337 (1971).

[103] M. Gari, Interaction Studies in Nuclei, Eds. H. Jochim and B. Ziegler, (North-Holland, Amsterdam), 307 (1975); M. Gari and J.Schlitter, Phys. Lett. 59B, 118 (1975).

[104] J.L. Friar, Phys. Rev. C12, 2127 (1975); D.O. Riska, unpublished, J.L. Friar, Ann. Phys. (N.Y.) 104, 380 (1977).

[105] Corresponding references are listed in table 2. Additional information may be found in Refs. 93 and 94.

[106] E.M. Henley, Phys. Lett. 28B, 1 (1968); E. Bellotti, E. Fiorini, P. Negri, A. Pullia, and L. Zanotti, Interaction Studies in Nuclei, Eds. H. Jochim and B. Ziegler

(North Holland, Amsterdam) 411 (1975).
[107]M. Gari, J.B. McGrory, and R. Offermann, Phys. Lett. 55B, 277 (1975).
[108]B.H.J. McKellar and K.R. Lassey, University of Melbourne preprint UM-P-76/12 (1977).
[109]L. Stodolsky, Phys. Lett. 50B, 352 (1974).
[110]P.K. Kabir, G. Karl, and E. Obryk, Phys. Rev. D10, 1471 (1974); Can. J. Phys. 53, 2661 (1975).
[111]M. Forte, ILL Research Proposal 03-05-002 (1976).
[112]A. Barroso and D. Tadić, J. Phys. G: Nucl. Phys. 3, L 147 (1977); D. Tadić and A. Barroso, Nucl. Phys., to be published.
[113]J. Byrne et al., Research proposal to ILL Grenoble (1977).
[114]G. Karl and D. Tadić, Phys. Rev. C, to be published.
[115]E.M. Henley, Proceedings of the Summer Studies on High Energy Physics with Polarized Beams ANL/HEP 75-02, p. XII-1 (1974); E.M. Henley and F.R. Krejs, Phys. Rev. D11, 605 (1975).
[116]L.L. Frankfurt and V.B. Kopeliovick, Nucl. Phys. B103, 360 (1976); L.L. Frankfurt, JETP Lett. 12, 379 (1970); Sov.Phys. JETP 34, 23 (1972).
[117]M. Dobovoy, P. Langacker, and M. Suzuki, Phys. Rev. D4, 1474 (1971).
[118]N.N. Nikolaev, Sov. J. Nucl. Phys. 17, 62 (1973).
[119]K.H. Craig, Nucl. Phys. B109, 156 (1976).
[120]J. Missimer, L. Wolfenstein, and J. Gunion, Nucl. Phys. B111, 20 (1976).
[121]F. Halzen, Phys. Rev., to be published.
[122]E. Fischbach and G.W. Look, Phys. Rev. D13, 752 (1976); in High Energy Physics with Polarized Beams and Targets, ed. M.L. Marshak (AIP Conference Proceeding No 35, American Institute of Physics, New York) 258 (1976).
[123]E. Fischbach and G.W. Look, Purdue University preprint (1977).
[124]A. Barroso and D. Tadić, Sussex University preprint.
[125]B.H.J. McKellar, private communication.
[126]G.S. Danilov, Nucl. Phys. B24, 165 (1970).
[127]V.M. Lobashov et al., Sov. Phys. - JETP. Lett. 5, 59 (1967).
[128]N. Vinh-Mau, in Symposium on Nucleons and Weak Interactions, eds. B. Eman and D. Tadić (Institute "Rudjer Bošković",

Zagreb), p. 99 (1971).

[129] E. Kuphal et al., Nucl. Phys. A234, 308 (1974).

[130] E. Maqueda and R.J. Blin-Stoyle, Nucl. Phys. A91, 460 (1967); O. Dumitrescu, M. Gari, H. Kümmel, and J.G. Zabolitzky, Z. Naturforschung 27a, 733 (1972).

[131] W. Kummer and G. Segrè, Nucl. Phys. 64, 585 (1965); N. Christ, Phys. Rev. 176, 2086 (1968); G. Segrè, Springer Tracts in Modern Physics 52 (Springer-Verlag Berlin) 171 (1970).

[132] B. Desplanques, private communication.

[133] M. Simonius, Eidg. Technische Hochschule, Zürich preprint 1977.

[134] M. Simonius, Eidg. Technische Hochschule, Zürich preprint 1977.

[135] Yu. G. Abov, P.A. Krunchickii, Uspekhi Fizicheskih Nauk 118, 141 (1976).

[136] J. Missimer, private communication. Experiment is by the Wäffler group.

[137] E. Fiorini, private communication.

[138] M. Gari, J.B. McGrory, and R. Offermann, Phys. Lett. 55B, 277 (1975).

[139] M. Porrmann and M. Gari, Phys. Rev. Lett. 38, 947 (1977); M. Gari, private communication.

[140] J.D. Bowman et al., Phys. Rev. Lett. 34, 1184 (1975).

[141] S. Wahlborn, Phys. Rev. 138, B530 (1965).

[142] S. Weinberg, "Unified Gauge Theories", Invited talk presented at the VIIth International Conference on High-Energy Physics and Nuclear Structure, Zürich, 1977.

[143] P. Herczeg, Phys. Rev. D6 , 1934 (1972).

[144] B.A. Craver, A. Tubis, and Y.E. Kim, Purdue University preprint 1977.

[145] D.O. Riska and G.E. Brown, Phys. Lett. 38B, 193 (1972).

[146] H. Wäffler, private communication.

UPPER LIMIT FOR A SMALL PARITY-VIOLATING ASYMMETRY IN P-P SCATTERING AT 15 MeV

D. E. Nagle
Los Alamos Scientific Laboratory

Over the past five years LASL & Illinois have made a search for a dependence of the total pp scattering cross-section on the sign, or helicity, of a longitudinally polarized beam. Such an effect would indicate a parity violating piece of the pp interaction at low energy. During this time we have reduced the statistical errors and the estimated systematic errors from a level of 10^{-3} to about 10^{-7}. The central value in pp scattering is close to zero, but the present level of the errors does not allow us to rule out an effect of one part in 10^{-7}, which is the estimate of V. Brown, E. Henley and F. Krejs.[1]

The presence of a weak nucleon-nucleon force is inferred from a number of experiments which seem to show appreciable parity violating effects in nuclei, in medium and heavy nuclei.[2] Unfortunately it is difficult to extract the basic nucleon-nucleon PV interaction from these results because of the complexities of the many-nucleon problem. Hence the interest in the two-nucleon system. The only experiment reporting a non-zero effect with two nucleons is that of Lobashov et al.[3]

For these reasons we embarked on a study of pp scattering with a longitudinally polarized beam, using the Los Alamos Tandem Van de Graaff accelerator. Earlier reports of this effort are given by J. M. Potter et al.[4,5]

The basic idea of the experiment is to measure the integrated scattered beam over a solid angle of nearly four π, and divide it by the transmitted beam. This number depends on the integrated cross section, and a change in it with sense of the (longitudinal) polarization of the beam would be a direct manifestation of a parity violating piece in the scattering amplitude.

Figure 1 shows the general layout of the experiment. The beam spin direction from the polarized source is rapidly reversed at a frequency controlled by a reference oscillator. The 15 MeV beam scatters and the synchronous component of the cross section is stored in a computer. The scattering takes place when the beam passes along the axis of a cylindrical pressure vessel containing hydrogen gas at about 100 psi (Fig. 2). The outer part of the vessel contains four slabs of scintillating plastic, symmetrically arranged above, below, to the right, and to the left of the beam. Each slab is viewed by three photomultipliers. About 99% of the beam is transmitted through the scattering chamber to a set of five gold electrodes, arranged as center, up, down, and left and right with respect to the beam direction.

The transmitted current from the rear set of electrodes is divided into the scattered current, using an analog divider, and the result recorded in an on-line computer. The source is modulated to give a spin reversal rate of about 1 kHz, and the synchronous component of the divided signal from the detectors is found using a phase locked amplifier (P.L.A.). The result is digitized and stored in an on-line computer at one second intervals.

A typical intensity for the polarized beam out of the tandem is two or three hundred nanoamps, which implies that in a few days a 10^{-7} level can be reached. Most of the effort with the experiment is aimed at improving our understanding of the principal sources of systematic errors, reducing their magnitudes, measuring the residual amounts, and correcting the data for them to the extent required. This requires a very careful study of the polarized source and the reversal technique.[6] Figure 4 shows a schematic drawing of the Lamb Shift Source. Details of the several methods used for fast reversal are given in references 4-6. They are most important for understanding the systematic errors, most of which arise in the reversal region. The beam leaving the Lamb-Shift "spin filter" region of the source is a mixture of hydrogen atoms in the ground state (unpolarized) plus atoms in the (2s, $m_s = \frac{1}{2}$) state. The beam next passes into the so-called reversal region, where an alternating magnetic field precesses the spin as required. We enumerate some of the physical processes occurring in the reversal region giving rise to modulation of the beam; they include:

a) quenching of the 2s state in H, giving current modulation,
b) ionization of a small part of the beam in the reversal region, giving position modulation and transverse polarization modulation,
c) improper values of magnetic field gradients, giving residual transverse polarization modulation, and also giving rise to (C′) a spatial distribution of transverse polarization having a non-zero second moment at the detector.

Each of these effects has been extensively studied and brought under control to a substantial degree. Table I, taken from Potter et al,[5] illustrates the situation. Work on C′ has been done quite recently.

TABLE I TYPICAL SYSTEMATIC ERRORS

	Sensitivity	Amplitude	Correction to A_L
Current Modulation	$4x10^{-4}$	$6x10^{-7}$	$2.4x10^{-10}$
Position Modulation (one coordinate)	$3.7x10^{-4}$/mm	$9x10^{-5}$mm	$3.3x10^{-8}$
Transverse Polarization (one coordinate)	$2.7x10^{-5}$	$1x10^{-3}$	$2.7x10^{-8}$

Some of the other sources of systematic errors are ground loops, leading to electrical paths which can transmit a spurious signal at the reference or spin modulation frequency.

The general program for reducing the effects of these and other errors is as follows: first, understand the physical processes going on in the source which give rise to the unwanted beam modulation; second, modify the source to reduce that beam modulation component; third, adjust the detector symmetry to reject that component; fourth and fifth, measure the amount of residual modulation, and store the result in the computer along with the PV signal at the standard 1-sec interval. From time to time measure the residual sensitivity of the source to that component. Finally, make a numerical correction for that effect (off-line after the run).

As an example, consider the error reduction program as it has been applied to one beam variable, current modulation at the reference frequency. The principal causes of this are; 1) undesired quenching of the 2s state at frequency fo. and 2) deflection of a prematurely ionized component followed by interception of a portion of the beam at some aperture. The quenching is reduced by supplying a compensating electric field at right angles to the alternating magnetic field which precesses the spin. The magnitude of the E field is set so that for the atoms moving at the mean velocity of the beam, the sum of the Lorenz fields is zero. The reversal scheme is such that the fundamental component of residual current modulation is at $f_o \div 2$, not fo. The f_o component is much reduced.

The unwanted ionization is reduced by cryopumping the critical regions of the source, and with appropriate sweeping fields. The Lorenz balancing of the fields also means that for the ions, synchronous forces are suppressed.

In the detector electronics system, the ratio of scattered to transmitted beam in first order is independent of current modulation. The remaining current modulation on the transmitted (B) signal is used as the error signal on a feedback servo to the source, where the amplified signal is applied to an auxiliary quench electrode. Finally, the residual current modulation on the B signal is monitored by the computer. The detector sensitivity to current modulation is measured from time to time, and an attempt is made to correct for this residual, (see Table I). Similar techniques are used to suppress the other unwanted types of moduations on the beam.

Carrying out that program has brought us to the present level of sensitivity. It has meant that perhaps 95% of the running time has been devoted to development studies of the source, beam, detector, electronics, and only about 5% or so to actual data-taking.

At the present time, there are seven servo control channels applied to keeping the beam centered and unmodulated. When the servo feedback loops are closed, the tandem beam is very stable and quiet. We record into a computer the nominal PV signal, plus seven other synchronous channels. Various control tests are carried out during data taking.

It is most helpful for the PV measurement that hydrogen at 15 MeV has a very low integrated analyzing power for protons, about 0.1% or less.[7] This means the sensitivity to residual transverse polarization or polarization misalignment is small.

In order better to study the effect of residual transverse polarization, we use ^{4}He in place of H in the detector. ^{4}He averaged over our detector has an analyzing power of about 36%. So the effects are magnified over hydrogen by a factor of about 400.[8] This study, which is nearing completion, has revealed some interesting facets of the operation of the Lamb-Shift polarized source, and of our method of fast spin reversal. We think we are now about ready to resume work on hydrogen, with the aim of pushing the systematic and the statistical errors down to a level of a few parts in 10^8.

Table II gives the present state of our knowledge of the longitudinal asymmetry A_L for 15 MeV protons on ^{1}H and ^{2}H. Longitudinal asymmetry is defined as $[\sigma(+) - \sigma(-)] \div [\sigma(+) + \sigma(-)]$, where + and - denote the two helicity states of the beam. It is seen that A_L is consistent with zero at the 10^{-7} level in H and D.

TABLE II LONGITUDINAL ASYMMETRY

	D_2	H_2
$A_L \times 10^7$ uncorrected	$-1.2 \pm .85$	$+0.8 \pm 1.4$
Sum of systematic corrections	-0.81	0.75
$A_L \times 10^7$ corrected	-0.35 ± 0.85	$+0.05 \pm 1.4$

At the present time, therefore, there is no disagreement with the estimate of Brown et al.[1] Since the result of Lobashov et al.[3] is orders of magnitude larger than had been expected, this is most interesting.

The development of weak interaction theory, which presently is going rapidly and is of great topical interest, will be influenced by what we can learn experimentally about the weak coupling of mesons and nucleons, which gives rise to the weak nuclear force. Since the interpretation of the pp result is not obscured by the complexities arising from the structure of the heavier nuclei, it would be of great interest to push the statistical and systematic errors of the pp measurement still lower. We believe this will be possible with some modificaitons of present techniques.

During the course of this work we have been developing detectors suitable for a LAMPF experiment with 800 MeV protons, and a Lamb shift source has been developed by P. Chamberlain and R. Stevens embodying many of the ideas developed by J. McKibben and the rest of us. Next year the experiments are scheduled to begin with 800 MeV protons.

REFERENCES

1. V. R. Brown, E. M. Henley and F. R. Krejs, Phys. Rev. C9, 935 (1975).

2. F. Boehm, "High Energy Physics and Nuclear Structure," A.I.P. Conf. Proc. No. 25, Santa Fe and Los Alamos 1975, p.488. H. Frauenfelder, ibid., p.448.

3. V. M. Lobashov, D. M. Kaminker, G. I. Kharkevich, V. A. Kniazkov, N. A. Lozovoy, V. A. Nazarenko, L. E. Sayenko, L. M. Smotritsky, and A. I. Yegorov, Nucl. Phys. A197, 241 (1972).

4. J. M. Potter, J. D. Bowman, C. F. Hwang, J. L. McKibben, R. E. Mischke, D. E. Nagle, P. G. Debrunner, H. Frauenfelder, and L. B. Sorensen, "Test of Parity Conservation in p-p Scattering," Phys. Rev. Lett. 33, 1307 (1974).

5. J. M. Potter, "High Energy Physics with Polarized Beams and Targets," A.I.P. Conf. Proc. No. 35, (ANL 1976).

6. J. M. McKibben, bid., p. 375.

7. P. Lavoi, Thesis, Dept. of Physics, Univ. of New Mexico (1976).

8. D. C. Dodder et al., LA-6389-MS.

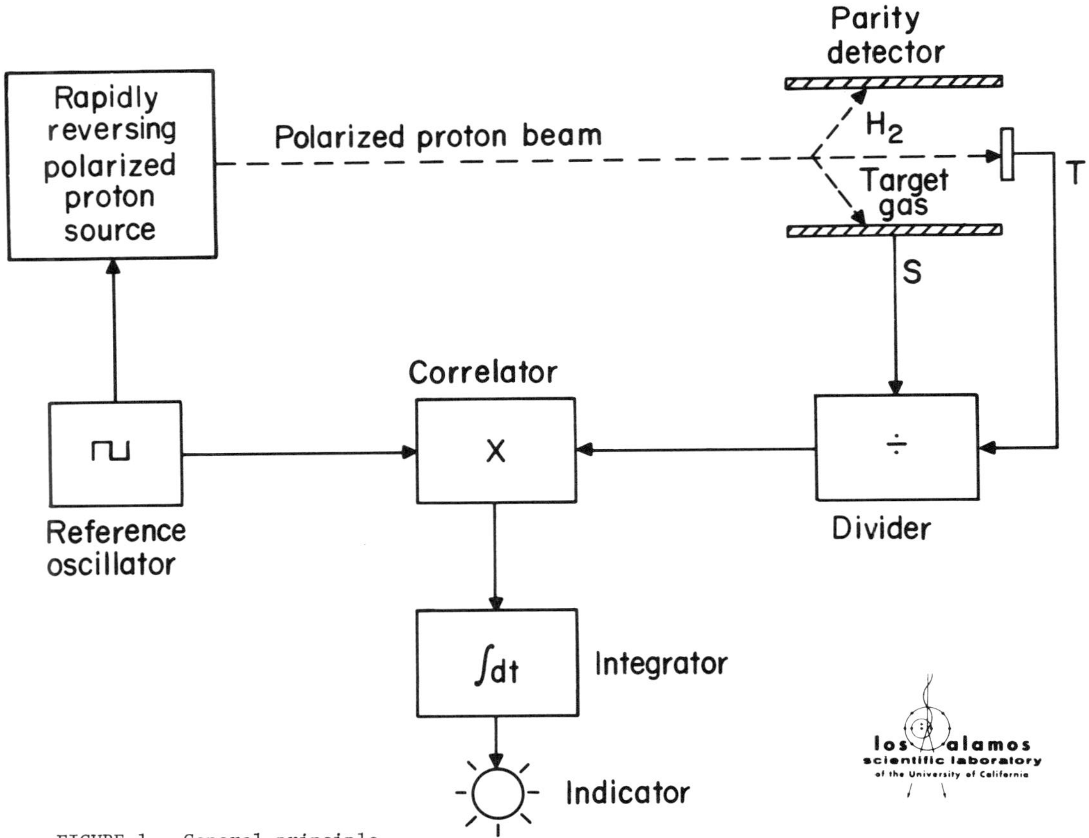

FIGURE 1 - General principle
of the experiment.

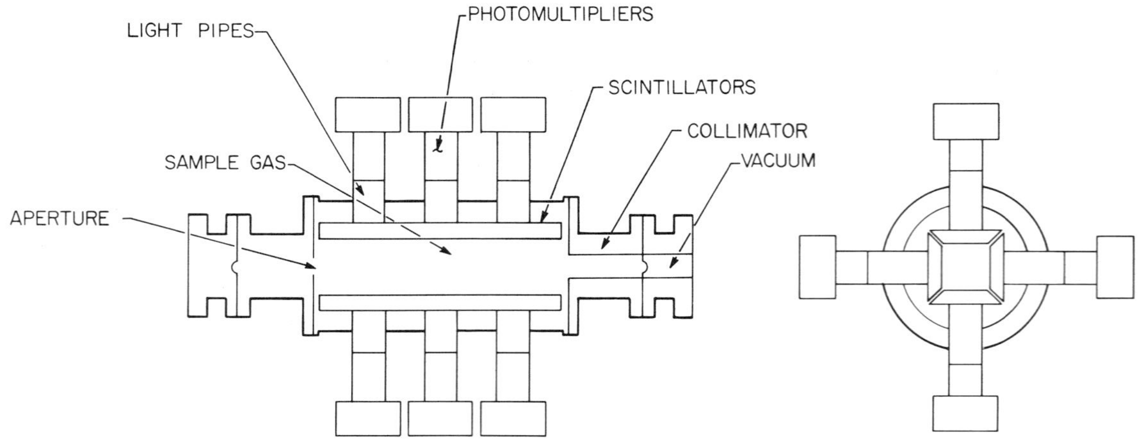

PNC SCATTERING CHAMBER

FIGURE 2 - Geometry of scattering chamber and detector. The scintillators are 380mm long and separated by 76mm.

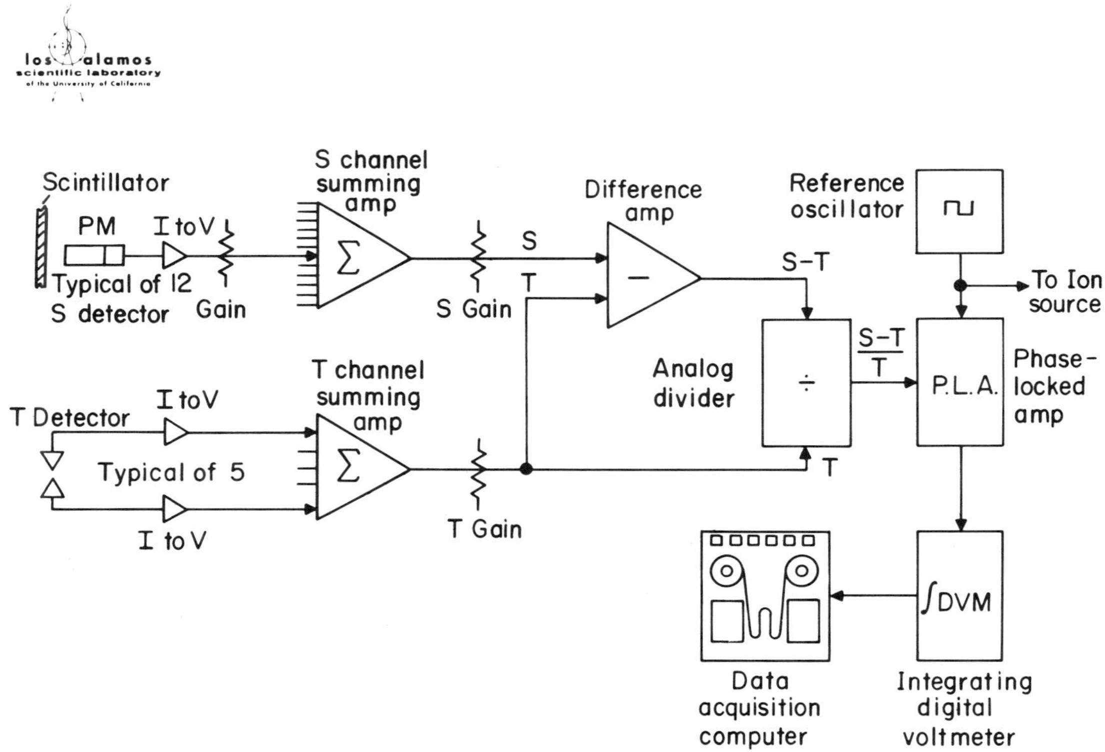

FIGURE 3 - Layout of electronics for parity experiment.

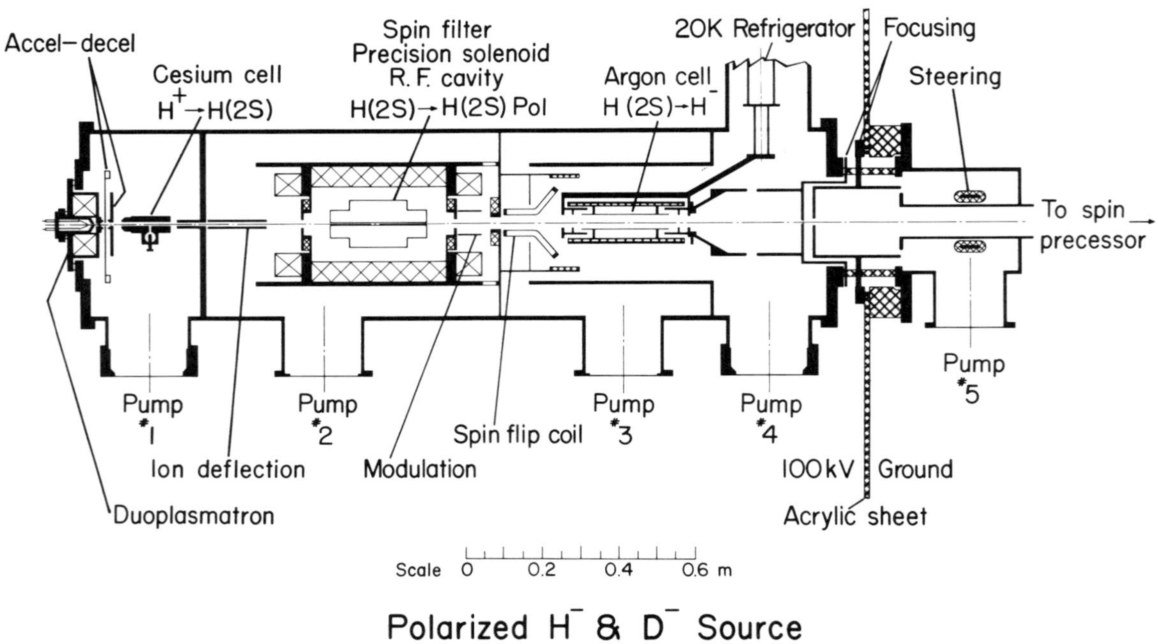

FIGURE 4 - Lamb shift polarized ion source.

Search for New Particles Beyond Charm

EVIDENCE FOR, AND PROPERTIES OF, THE τ LEPTON*

Martin L. Perl
Stanford Linear Accelerator Center
Stanford University, Stanford, California 94305

ABSTRACT

An outline of the evidence on the e^+e^- annihilation of a new charged lepton, the τ, is presented. Measured properties of the τ are summarized and some still open questions as to its properties are discussed.

TABLE OF CONTENTS

1. Introduction

I do not have the time in this short talk to present all the evidence for the existence of a new charged lepton or at least lepton-like particle of mass 1.9 ± 0.1 GeV/c, called the τ. Since the discovery at SPEAR[1] of the eμ events produced in e^+e^- annihilation, eight different experiments at SPEAR and DORIS have data either requiring such a new particle or consistent with the existence of such a particle. I recently reviewed all the data[2] and an earlier review was given by Flügge.[3] Therefore I will begin the talk with a general description of the signatures for charged heavy lepton production in

*Work supported by the Department of Energy

e^+e^- annihilation, and then I will concentrate on the $e\mu$ events as examples of one type of data. The larger portion of the talk will present the measured properties of the new particle. I will conclude with some open questions on the new particle.

2. Signatures for Charged Heavy Lepton Production in e^+e^- Annihilation

If a charged lepton $L^\pm$ is produced in e^+e^- annihilation we expect the following kinds of experimental signatures:

A. Pair Production

We expect single pair production

$$e^+ + e^- \rightarrow L^+ + L^- \tag{1}$$

but not

$$e^+ + e^- \rightarrow L^+ + L^- + \text{hadrons}. \tag{2}$$

Reaction (2) would mean that L has strong interactions and is not a lepton.

B. Point Particle Production Cross-Section

For a spin $\frac{1}{2}$ lepton we expect the production cross-section

$$\sigma_{e^+e^-\rightarrow L^+L^-} = \frac{2\pi\alpha^2\beta(3-\beta^2)}{3E^2_{c.m.}} \tag{3}$$

where β is the lepton's velocity in units of c, α is the fine structure constant and $E_{c.m.}$ is the total energy. I note that we do not expect a production form factor, whereas in hadron pair production we expect a form factor like

$$F(E_{c.m.}) = \text{constant}/E^n_{c.m.}, \quad n \geq 2 \tag{4}$$

to enter as the square in Eq. 3.

C. Assuming Radiative Decays are Suppressed or Prohibited, Purely Leptonic Decays Will Occur

Assuming the radiative decays

$$L^- \rightarrow e^- + \gamma \quad (5)$$
$$L^- \rightarrow \mu^- + \gamma$$

are suppressed or prohibited we expect the lepton to decay through the weak interaction into the purely leptonic final states

$$L^- \rightarrow \nu + e^- + \bar{\nu}_e \quad (6)$$
$$L^- \rightarrow \nu + \mu^- + \bar{\nu}_\mu \,.$$

At this time we need make no assumption as to the precise nature of the ν in Reactions (6). From Reactions (6) we expect the <u>$e^\pm\mu^\mp$ event signature</u>

$$\begin{array}{ccccc} e^+ + e^- \rightarrow & L^+ & + & L^- & \quad (7) \\ & \downarrow & & \downarrow & \\ & \bar{\nu} e^+ \nu_e & & \nu \mu^- \bar{\nu}_\mu & \end{array}$$

and its charge conjugate. Such $e\mu$ events have only two particles which can be detected by conventional apparatus and have substantial missing energy. Succinctly:

$$e^+ + e^- \rightarrow e^\pm + \mu^\mp + \text{missing energy} \quad (8)$$

D. <u>Momentum Spectrum of e or μ in $e\mu$ Events Will Be "Hard", But Not as "Hard" as A Two-Body Decay</u>

As shown in Fig. (1), the e or μ momentum spectrum in the laboratory frame will be "harder" than that of a weakly decaying hadron such as a singly-charmed meson.[4] But the spectrum will not be as "hard" as that from a two-body decay $M^- \rightarrow e^- + \bar{\nu}_e$.

E. <u>Semi-Leptonic Decays May Also Occur</u>

If the L mass is sufficiently large there may be semi-leptonic decays such as

$$L^- \rightarrow \nu^- + \pi^- \quad (9)$$
$$L^- \rightarrow \nu^- + K^-$$

$$L^- \rightarrow \nu^- + \rho^-$$ (9) (Continued)

$$L^- \rightarrow \nu + K^{*-}$$

$$L^- \rightarrow \nu + A_1^-$$

$$L^- \rightarrow \nu + \pi^- + \pi^+ + \pi^- + \pi^o$$

$$L^- \rightarrow \nu + \bar{p} + n$$

. .
. .
. .

F. Semi-Leptonic Decays Lead to e-Hadron and μ-Hadron Events

If one L decays purely leptonically and the other decays semi-leptonically, we expect e-hadron and μ-hadron events such as

$$\begin{array}{ccccc} e^+ + e^- \rightarrow & L^+ & + & L^- & \\ & \downarrow & & \downarrow & \\ & \nu\mu^+\nu_\mu & & \nu\rho^- & \\ & & & \downarrow & \\ & & & \pi^-\gamma\gamma & \end{array} \qquad (10)$$

G. Summary of Signatures

I want to emphasize that signatures A through F can provide very substantial evidence for the existence of a new lepton regardless of whether the new lepton fits all the requirements of a particular model. I have made this remark because although there is now substantial evidence that the τ is a lepton-like particle, we do not yet know whether the τ fits any of our usual lepton models. I shall briefly describe next, three of those models. In all of these models we require a mechanism to suppress the radiative decays, Eq. (5), because the radiative decays of the τ have not been found, and we at least want the model to fit that property of the τ.

3. Some Specific Lepton Models

A. Ortholepton

L^- might have the same lepton number as the e^- or μ^-. Llewellyn Smith[5,6] calls this an ortholepton where ortho is for orthodox. To supress the radiative decays, Eq. (5), we must construct a theory which suppresses the $L^- e^- \gamma$ vertex.

B. Paralepton

If L^- has the same lepton number as the e^+ or μ^+, the radiative decays, Eq. (5), are prohibited (if lepton number conservation is assumed). The para in paralepton is taken from paradoxical.[5]

C. Sequential Heavy Lepton

If there is a sequence of charged leptons each with its own unique, conserved, lepton number and its own unique neutrino:

charged lepton	associated neutrino
$e^\pm$	ν_e, $\bar{\nu}_e$
$\mu^\pm$	ν_μ, $\bar{\nu}_\mu$
$L^\pm$	ν_L, $\bar{\nu}_L$
.	.
.	. ,

(11)

and if ν_L is lighter than L, the purely leptonic decay modes

$$L^- \rightarrow \nu_L + e^- + \bar{\nu}_e \ , \ \nu_L + \mu^- + \bar{\nu}_\mu \qquad (12)$$

will occur. But the radiative decays, Eq. (5), will be prohibited. This is the model which motivated my interest in looking for new heavy leptons, but as you will see we don't yet know if the τ is a sequential heavy lepton.

If we assume that the lepton decays through the conventional weak interactions, that the neutrino mass is zero, and that the L-ν_L coupling is V-A, then the various decay rates can be calculated.[8,9] Table I gives the predicted branching ratios for a mass 1.9 GeV/c^2 lepton. Figures (2) and (3) show the branching ratios and lifetime for masses less than 10 GeV/c^2. We note that Table I predicts that decay modes of 1.9 GeV/c^2 lepton in this model will usually have only one charged track. (The specific prediction is about 85%.)

TABLE I

Predicted branching ratios for a τ^- sequential charged heavy lepton with a mass 1.9 GeV/c^2, an associated neutrino mass of 0.0, and V-A coupling.

Decay Mode	Branching Ratio	Number of Charged Particles in Final State
$\nu_\tau e^- \bar{\nu}_e$	.20	1
$\nu_\tau \mu^- \bar{\nu}_\mu$	.20	1
$\nu_\tau \pi^-$	.11	1
$\nu_\tau K^-$	.01	1
$\nu_\tau \rho^-$	.22	1
$\nu_\tau K^{*-}$	.01	1
$\nu_\tau A_1^-$	.07	1,3
ν_τ(hadron continuum)$^-$	.18	1,3,5

4. Outline of Evidence for the τ

There is now substantial evidence[2,3] for the existence of a charged lepton we call the τ with:

a) τ mass = 1.9 ± 0.1 GeV/c^2;

b) associated neutrino mass $\leq$ 0.6 GeV/c^2 with 90% confidence;

c) observed decay modes:

$$\tau^- \rightarrow \nu_\tau + e^- + \bar{\nu}_e$$

$$\tau^- \rightarrow \nu_\tau + \mu^- + \bar{\nu}_\mu$$

$$\tau^- \rightarrow \nu_\tau + h^- + \geq 0 \text{ photons}$$

$$\tau^- \rightarrow \nu_\tau + \rho^-$$

$$\tau^- \rightarrow \nu_\tau + h^- + h^+ + h^-$$

where h is a hadron.

From now on we shall refer to the neutrino associated with the τ as ν_τ for convenience, although we have not proven that the τ is a sequential-type lepton. Since this a brief presentation I will only discuss the $e^\pm\mu^\pm$ events in any detail.

A. $e^\pm\mu^\pm$ Events

The eμ events:

$$e^+ + e^- \rightarrow e^\pm + \mu^\pm + \text{no other detected particles,} \tag{13}$$

have been seen by four experiments, Table II. All experiments find the same production rate and same kinematic distribution properties for these events.

TABLE II

Data on eμ events. In addition to the lower limits on p_e and p_μ, all these sets of p_μ events have acoplanarity requirements such as 10° or 20°. The references should be consulted for details on the event selection criterion.

Experimental Group or Detector	$E_{c.m.}$ Range (GeV)	Lower Limits on p_e p_μ (GeV/c)	Total Number of eμ Events	Number of Back-ground Events	Comments	Ref.
M. Bernardini *et al.*	1.2 to 3.0				Early search at ADONE, lepton mass $\geq$ 1.0 GeV/c	10

(Table II, continued)

Experimental Group or Detector	$E_{c.m.}$ Range (GeV)	Lower Limits on p_e p_μ (GeV/c)	Total Number of eμ Events	Number of Background Events	Comments	Ref.
S. Orioto *et al.*	2.6 to 2.0				Early search at ADONE, lepton mass ≥ 1.15 GeV/c^2	11
SLAC-LBL Magnetic Detector	3.8 to 7.8	0.65 0.65	190	46	First evidence. Used to determine m_τ, m_{ν_τ}, τ-ν_τ coupling	1,2
PLUTO Group	3.6 to 5.0	0.3 1.0	23	1.9	Very clean. Strong argument against charm.	3, 12, 13, 14
LBL-SLAC Lead Glass Wall	3.7 to 7.4	0.4 0.65	22	0.4	Very clean. Low p_e cutoff.	15, 16
DASP Group	4.0 to 5.2	0.15 0.1	11	0.7	Good γ detection. Good hadron identification.	17

Figures (4) and (5) show two examples of the "hard" momentum spectrum of the e and μ. Thus the existence of the p_μ events and their momentum spectrum are in agreement with signature requirements C and D (Sec. 2) for a lepton. The eμ events are also consistent with signature A because even at the highest energies very few eμ events with hadrons have been found[1,3,12] relative to the number of pure eμ events, and these eμ events with hadrons are all accounted for as background from hadronic events in which the hadrons have been misidentified as electrons or muons. Indeed, an expected source of eμ events with hadrons - the joint semi-leptonic decay of a pair of singly charmed mesons - has not yet been found. This is because the lepton spectrum from singly-formed meson decays is relatively soft, and such events are eleminated by the lower momentum limits on the e and μ in eμ events.

B. <u>τ Production Cross-Section From eμ events</u>

It is convenient to eliminate the $E_{c.m.}^{-2}$ dependence in Eq. (3) by defining

$$R_{\tau^+\tau^-} = \frac{\sigma_{e^+e^- \to \tau^+\tau^-}}{\sigma_{e^+e^- \to \mu^+\mu^-}} = \frac{\beta(3-\beta^2)}{2} \quad (14)$$

As a preliminary we define

$$R_{e\mu} = \frac{\sigma_{e^+e^- \to e\mu}}{\sigma_{e^+e^- \to \mu^+\mu^-}} = \frac{\beta((3-\beta^2)}{2}\,(2B_e B_\mu A) \quad (15)$$

Here B_e and B_μ are the leptonic branching ratios of the τ, and A is the acceptance taking into account the geometry of the detector and the kinematic cuts. The acceptance A is almost independent of $E_{c.m.}$.

Figure (6) shows $R_{e\mu}$ and Fig. (7) shows $R_{\tau^+\tau^-}$. These figures are consistent with Eqs. (14) and (15), respectively, and are consistent with lepton masses in the range 1.9 ± 0.1 GeV/c^2. Thus the eμ events are consistent with signature A.

C. <u>e-Hadron and μ-Hadron Events</u>

We do not take the time to discuss it here, but events of the form

$$e^+ + e^- \to e^\pm + h^\mp + \geq 0 \text{ photons} \quad (16)$$

$$e^+ + e^- \to \mu^\pm + h^\mp + \geq 0 \text{ photons} \quad (17)$$

have been observed by the following experiments:

SLAC-LBL Magnetic Detector (SPEAR) [18]

PLUTO (DORIS)[12,13,14]

DASP (DORIS)[17]

LBL-SLAC "Lead Glass Wall" Experiment (SPEAR)[15,16]

DELCO (SPEAR)[19]

Maryland-Princeton-Pavia Experiment (SPEAR)[20, 21, 22]

All this data is consistent with the τ being a lepton.

5. Measured Properties of the τ

Table III lists measurements on the τ mass, m_τ .

TABLE III

Measurements of m_τ assuming V-A coupling and $m_{\nu_\tau} = 0.0$.

Experiment	Data Used	Method	τ Mass (GeV/c^2)	Comments	Ref.
SLAC-LBL Magnetic Detector	$e\mu$	p	1.91 ± .05	Statistical error	23, 24
		$\cos\theta_{coll}$	1.85 ± .10	Statistical error	
		r	1.88 ± .06	Statistical error	
		composite	1.90 ± .10	Statistical and systematic error	
PLUTO Group	μx	$\sigma_{\mu x}$	1.93 ± .05		14

B. ν_τ Mass

Two upper limits have been set on m_{ν_τ} . Using $e\mu$ events[23,24]: $m_{\nu_\tau} \leq 0.6$ GeV/c^2 with 95% CL. Using μx events[14]: $m_{\nu_\tau} \leq 0.54$ GeV/c^2 with 95% CL.

C. τ-ν_τ Coupling

The SLAC-LBL $e\mu$ events cannot be fit by V + A coupling of the τ to the ν_τ[23]. However V - A, pure V or pure A coupling is acceptable.

D. τ Lifetime

The PLUTO group finds the τ lifetime is less than 1.0×10^{-11} sec with 95% confidence. Using the SLAC-LBL $e\mu$ events I find the τ lifetime is less than 1.1×10^{-11} sec with 95% confidence.

E. Type of Lepton

Experiments using muon neutrinos rule out the τ being a muon-related ortholepton or paralepton with conventional coupling strengths.[26-29] Our measurements on anomalous ee and μμ events produced in e^+e^- annihilation have ruled out the τ being an electron-related paralepton.[30] All data is consistent with the τ being a sequential lepton or an electron-related ortholepton.

6. Observed Decay Modes of the τ

A. Purely Leptonic Decays

Table IV lists all measurements on the purely leptonic decay modes of the τ. The agreement is surprisingly good considering the difficulty of making these measurements and the variety of methods used. A safe overall value assuming $B_e = B_\mu$ is 18 ± 4%, which is in excellent agreement with Table I.

TABLE IV

The measured fractional decay rates B_e and B_μ. V - A coupling, m_τ = 1.9 GeV/c and m_{ν_τ} = 0.0 was used to calculate acceptances.

Experimental Group or Detector	Data Used	B_e or B_μ	Comments	Ref.
SLAC-LBL Magnetic Detector	eμ	0.186±.010±.028	Assume B_e=B_μ. First error is statistical, second is systematic.	23, 24
SLAC-LBL Magnetic Detector	μx	0.175±.027±.030	Assume B_x=0.85 First error is statistical, second is systematic.	23, 24
PLUTO Group	μx	B_μ = 0.14±.034		14
PLUTO Group	μx,eμ	B_e = 0.16±.06		14
LBL-SLAC Lead Glass Wall	eμ	0.224±.032±.044	Assume B_e=B_μ. First error is statistical, second is systematic.	15, 16
DASP Group	eμ	0.20±.03	Assume B_e=B_μ.	17

(Table IV, continued)

Experimental Group or Detector	Data Used	B_e or B_μ	Comments	Ref.
DELCO Group	ex	0.15	No error given.	19
Iron Ball	μμ	$0.22 {}^{+.07}_{-.08}$		31
Maryland-Princeton-Pavia	μx	0.20±.10		22

B. Semi-Leptonic Decay Modes

The semi-leptonic decay modes of the τ which have been observed are listed in Table V.

TABLE V

Observed semi-leptonic decay modes of the τ. V - A coupling, $m_\tau = 1.9$ GeV/c^2 and $m_{\nu_\tau} = 0.0$ was used to calculate acceptances. Here h means hadron.

Experimental Group or Detector	Decay Mode (for τ^-)	Branching Ratio	Ref.
LBL-SLAC Lead Glass Wall	$h^- + \nu_\tau + \geq 0\ \gamma$ s	0.45±0.19	15, 16
DASP Group	$\rho^- + \nu_\tau$	0.24±.09	17
PLUTO Group	$(3h)^- + \nu_\tau$		14
LBL-SLAC Lead Glass Wall and SLAC-LBL Magnetic Detector	$(3h)^- + \nu_\tau$		32

In Table V $(3h)^-$ means a resonant state containing three hadrons which are probably all pions and has a mass of about 1100 MeV/c^2 and a width of several hundred MeV/c^2. The motivation for looking for this state is that the τ should decay into an A_1.[33] The data on the $(3h)^-$ state is consistent with it being an A_1 but there is as yet no proof that it is an A_1 .

C. Upper Limits on Rare Decay Modes

Table VI gives upper limits on radiative and other rare decay modes of the τ. The limits in Table VI are the reason we have emphasized models in Sec. (3) in which radiative decays are suppressed.

TABLE VI

Upper limits on rare decay modes of the τ using V - A coupling, $m_\tau = 1.9$ GeV/c^2 , $m_{\nu_\tau} = 0.0$ for acceptance calculations.

Experimental Group or Detector	Mode	Upper Limit on Branching Ratio	C.L.	Ref.
PLUTO Group	$\tau^- \to$ (3 charged particles)$^-$	0.01	95%	3
PLUTO Group	$\tau^- \to$ (3 charged leptons)$^-$	0.01	95%	3
SLAC-LBL Magnetic Detector	$\tau^- \to$ (3 charged leptons)$^-$	0.006	90%	34
SLAC-LBL Magnetic Detector	$\tau^- \to \rho^- + \pi^0$	0.024	90%	35
PLUTO Group	$\tau^- \to e^- + \gamma$ $\tau^- \to \mu^- + \gamma$	0.12	90%	3
LBL-SLAC Lead Glass Wall	$\tau^- \to e^- + \gamma$	0.026	90%	36
LBL-SLAC Lead Glass Wall	$\tau^- \to \mu^- + \gamma$	0.013	90%	36

7. Open Questions

A. Precise Mass of the τ

We do not understand why the τ mass, 1.9 ± 0.1 GeV/c^2, is close to the mass of the singly-charmed D^+ meson, 1.865 GeV/c^2. Is it a coincidence, does it have some significance, or is it related to the relative closeness of the π and μ mesons? We have no answer to these questions. In spite of the very strong evidence that the τ is a lepton and is not the D^+ meson, it is still very desireable to have a precise measurement of the τ mass. If the τ mass is different from the D^+ mass this is a final proof that the particles are different. If the τ mass is the same as the D^+ mass, it means that we understand neither singly-charmed hadrons nor leptons. Therefore it is important to make a precise measurement of the τ mass.

B. The $\tau^- \rightarrow \nu_\tau + \pi^-$ Decay Mode

The sequential lepton model with conventional weak interactions and V - A coupling predicts about a 10% branching ratio to this mode (Table I). The DASP group[17] and the PLUTO group[25] have both looked for this mode. Their results are preliminary and their statistics are small. However, neither group has found this mode. Each result is null by about 2 or 3 standard deviations. If the $\tau^- \rightarrow \nu_\tau + \pi^-$ mode is absent then the τ does not fit any of the specific lepton models in Sec. (3). Nevertheless, all the other data still requires the τ to be a lepton. The τ would then be a lepton of a new and very strange kind, a kind stranger than we ever expected. Therefore it is important not to draw conclusions from these preliminary results, but to wait for higher statistics experiments. The continued data-taking at SPEAR by the DELCO group and the new experiment at SPEAR by the SLAC-LBL collaboration using the new MARK II magnetic detector will hopefully help to answer both this question and the mass question in the coming year.

REFERENCES

1. M. L. Perl et al., Phys. Rev. Lett 35, 1489 (1975); Phys. Lett. 63B, 117 (1976).
2. M. L. Perl in Proceedings of the 1977 International Symposium on Lepton and Photon Interactions at High Energies (Hamburg, West Germany, 1977, to be published); also issued as SLAC-PUB-2022.
3. G. Flügge in Proceedings of the 1977 Experimental Meson Spectroscopy Conference (Northeastern University, Boston, Mass., 1977); also issued as DESY 77/35.
4. For a discussion on this point and experimental illustrations see references (14), (16), and (17).
5. C. H. Llewellyn Smith, Oxford University preprint 33/76 (1976).
6. J. D. Bjorken and C. H. Llewellyn Smith, Phys. Rev. D7, 88 (1973).
7. A. Ali and T. C. Yang, Phys. Rev. D14, 3052 (1976).
8. H. B. Thacker and J. J. Sakurai, Phys. Lett. 36B, 103 (1971).
9. Y. S. Tsai, Phys. Rev. D4, 2821 (1971).
10. M. Bernardini et al., Nuovo Cimento 17, 383 (1973).
11. S. Orioto et al., Phys. Lett. 48B, 165 (1974).
12. H. Meyer in Proceedings of the Orbis Scientiae (Coral Gables, Fla., 1977).
13. J. Burmester et al., DESY 77/25.
14. G. Knies in Proceedings of the 1977 International Symposium on Lepton and Photon Interactions at High Energies (Hamburg, West Germany, 1977, to be published).
15. A. Barbaro-Galtieri et al., Phys. Rev. Lett. 39, 1058 (1977).

16. A. Barbaro-Galtieri in Proceedings of the 1977 International Symposium on Lepton and Photon Interactions at High Energies (Hamburg, West Germany, 1977, to be published).

17. S. Yamada in Proceedings of the 1977 International Symposium on Lepton and Photon Interactions at High Energies (Hamburg, West Germany, 1977, to be published).

18. G. J. Feldman *et al.*, Phys. Rev. Lett. **38**, 1977 (1976).

19. J. Kirkby in Proceedings of the 1977 International Symposium on Lepton and Photon Interactions at High Energies (Hamburg, West Germany, 1977, to be published).

20. M. Cavalli-Sforza *et al.*, Phys. Rev. Lett. **36**, 588 (1976).

21. G. Snow, Phys. Rev. Lett. **36**, 766 (1976).

22. H. Sadrozinski in Proceedings of the 1977 International Symposium on Lepton and Photon Interactions at High Energies (Hamburg, West Germany, 1977, to be published).

23. M. L. Perl *et al.*, Phys. Lett. (to be published); also issued as SLAC-PUB-1997 (1977).

24. M. L. Perl in Proceedings of the XII Rencontre de Moriond (Flaine, France, 1977), edited by Tran Thanh Van (R. M. I. E. M., ORSAY).

25. H. Meyer, talk given at the Stanford Linear Accelerator Center, Stanford, Calif. (September 1977).

26. B. C. Barish *et al.*, quoted by D. C. Cundy in Proceedings of the XVII International Conference on High Energy Physics (London, England, 1974), p. IV-147.

27. C. H. Albright, Phys. Rev. **D13**, 2508 (1970).

28. M. Murtagh in Proceedings of the 1977 International Symposium on Lepton and Photon Interactions at High Energies (Hamburg, West Germany, 1977, to be published).

29. C. Baltay, private communication.

30. F. B. Heile *et al.*, to be published.

31. U. Camarini *et al.*, paper submitted to the 1977 Symposium on Lepton and Photon Interactions at High Energies (Hamburg, West Germany, 1977, to be published).

32. J. Jaros *et al.*, to be published.

33. J. J. Sakurai in *Proceedings of the 1975 International Symposium on Lepton and Photon Interactions at High Energies* (Stanford University, Stanford, Calif. (1975), p. 353.

34. G. J. Feldman, private communication.

35. H. K. Nguyen, private communication.

36. J. Jaros, private communication.

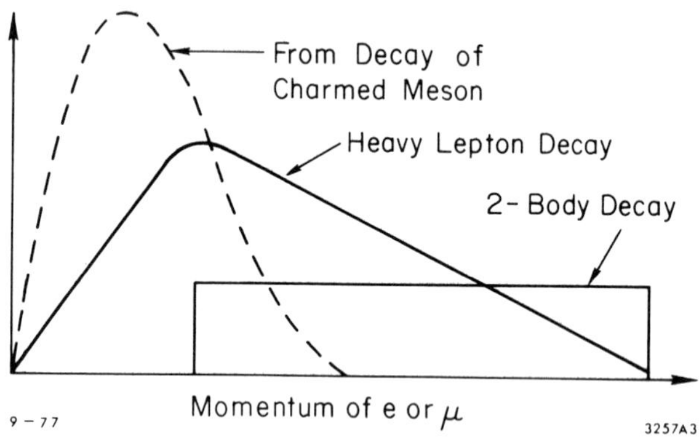

Fig. 1 Schematic comparison of the momentum spectrum for a lepton from a heavy lepton decay compared to the lepton spectrum from a charmed particle semileptonic decay or from a two-body decay.

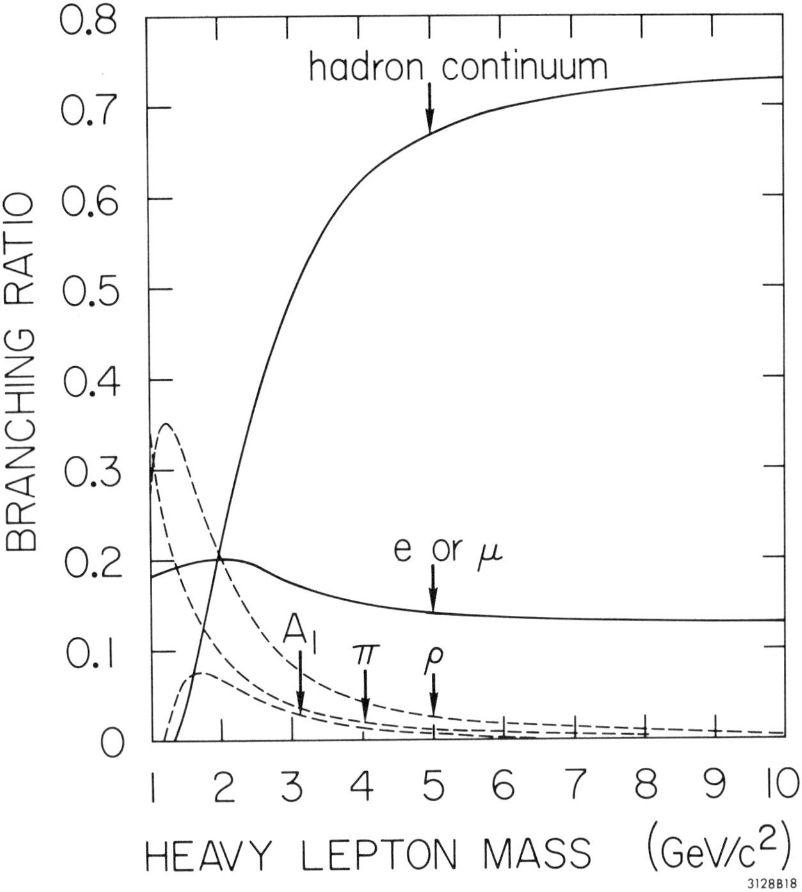

Fig. 2 Branching ratios versus mass for a charged lepton with conventional weak interactions, V - A coupling, and a massless neutrino.

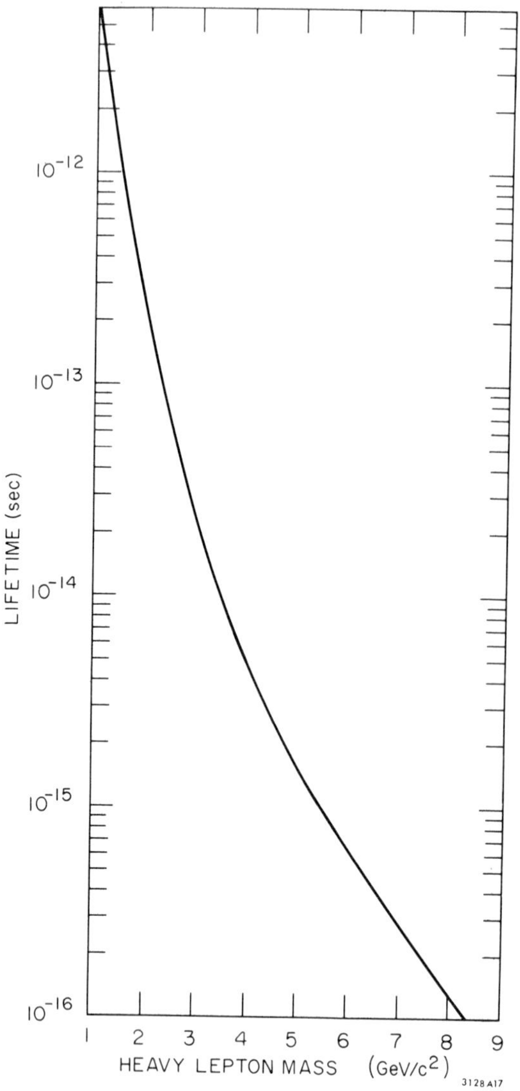

Fig. 3 Lifetime versus mass for a charged lepton with conventional weak interactions, V - A coupling, and a massless neutrino. The conventional weak coupling constants are assumed.

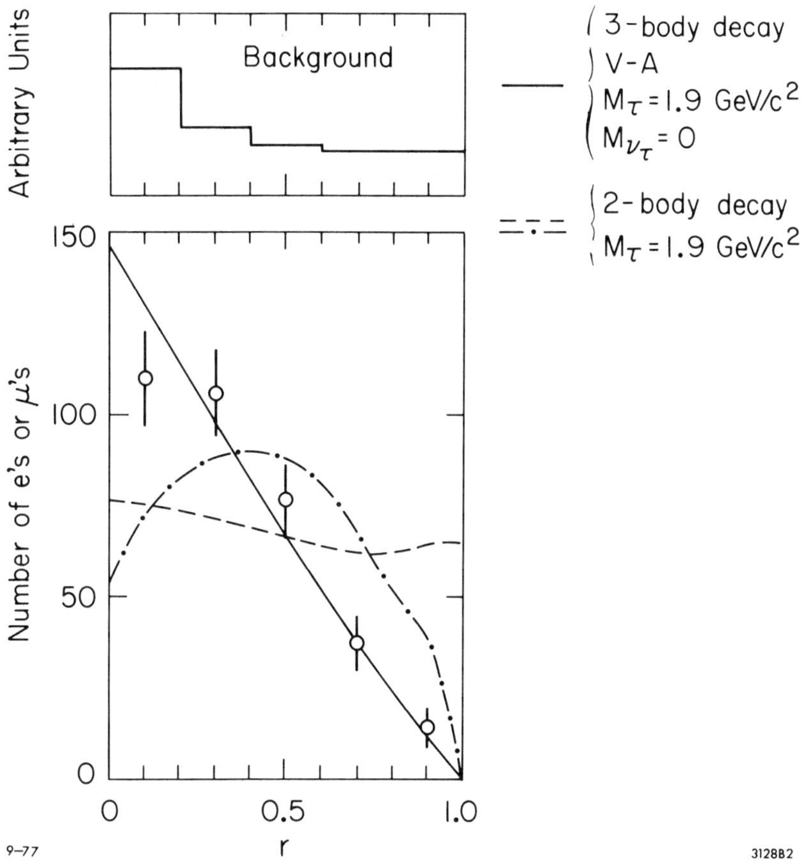

Fig. 4 The momentum spectrum for eμ events, with $3.8 \lesssim E_{c.m.} \leq 7.8$ GeV, from the SLAC-LBL Magnetic Detector Collaboration[1,2] corrected for background. Here $r = (p - 0.65)/(p_{max} - 0.65)$ where p is the e or μ momenta in GeV/c. The solid theoretical curve is for the three-body leptonic decay of a mass 1.9 GeV/c^2 τ; the dashed theoretical curve is for the two-body decay of an unpolarized boson; and the dash-dotted theoretical curve is for the two-body decay of a boson produced only in the helicity = 0 state.

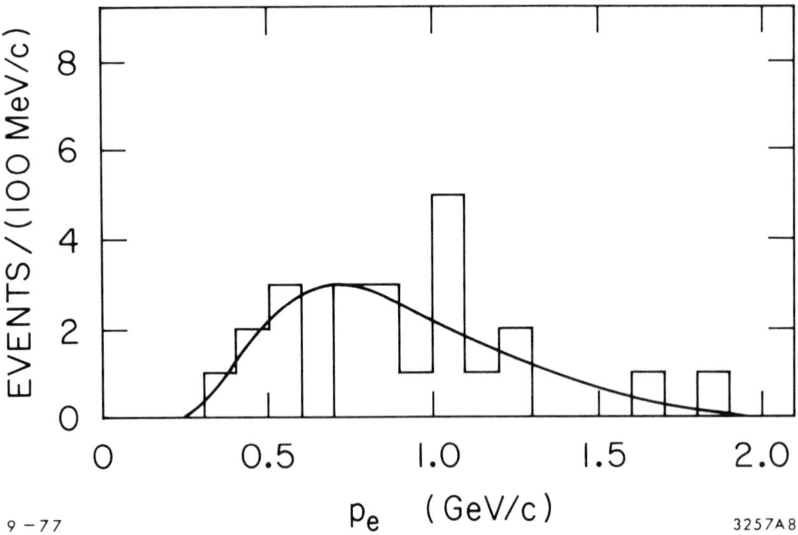

Fig. 5 The electron momentum spectrum for eμ events with $4.0 \leq E_{c.m.} \leq 5.0$ GeV from the PLUTO Group,[14] compared with the theoretical curve for the three-body leptonic decay of a mass 1.9 GeV/c^2 τ.

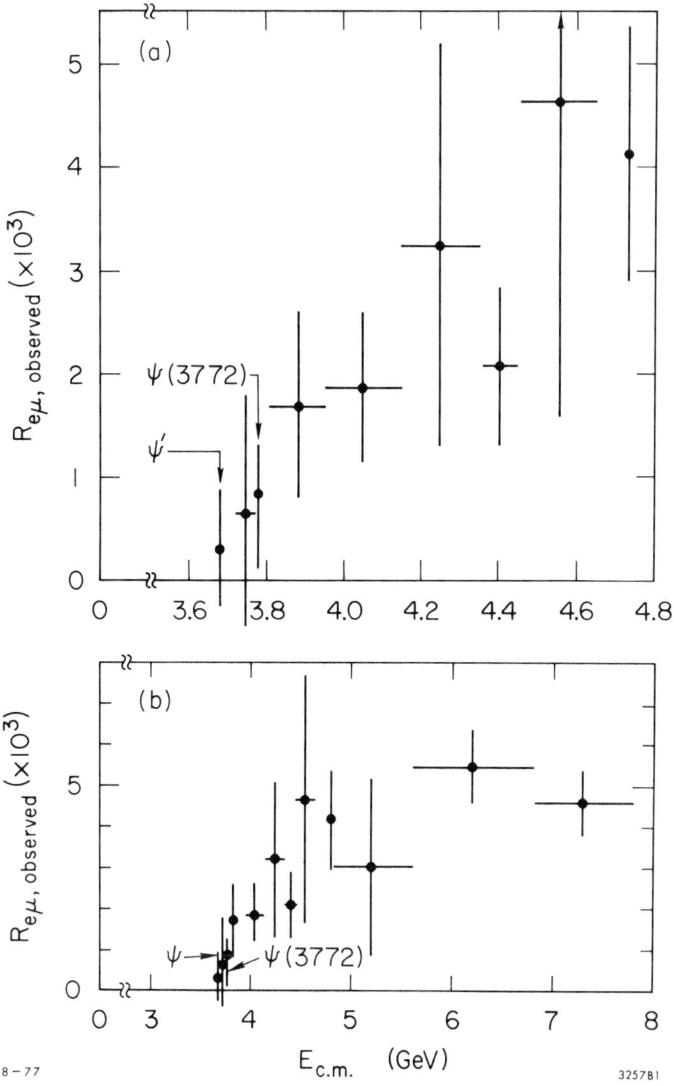

Fig. 6 $R_{e\mu,observed}$ for (a) $3.6 \leqslant E_{c.m.} \leqslant 4.8$ GeV and (b) $3.6 \leqslant E_{c.m.} \leqslant 7.8$ GeV.

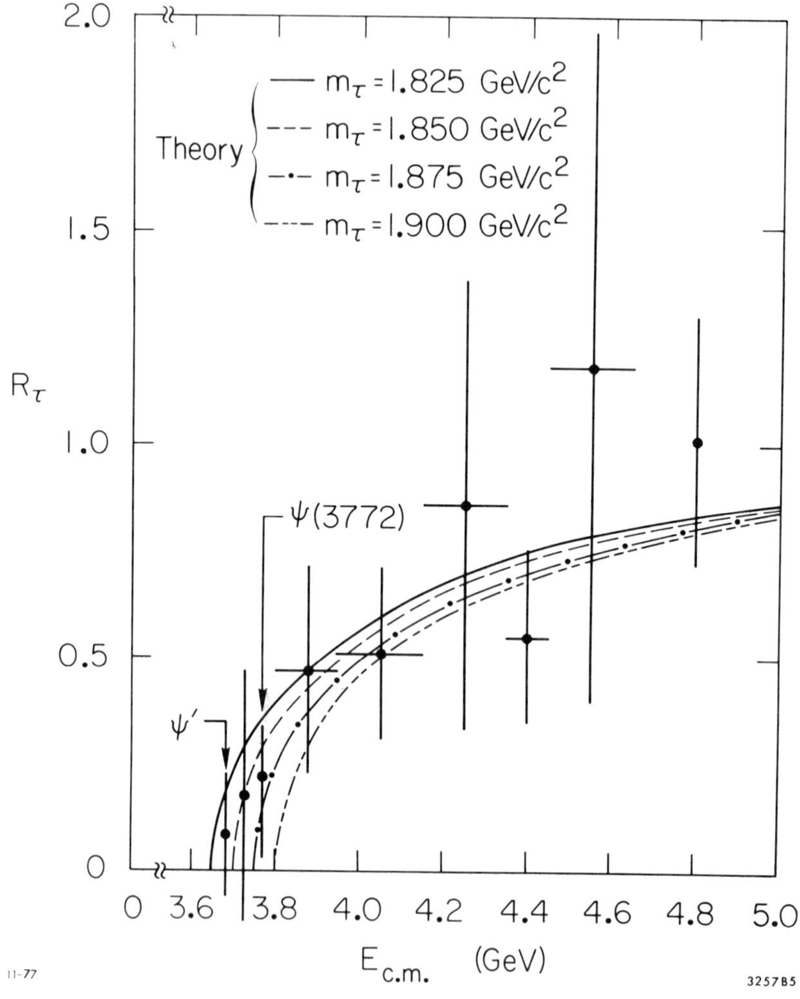

Fig. 7a R_τ for $3.6 \leq E_{c.m.} \leq 4.8$ GeV, compared with the theoretical R_τ for various τ masses.

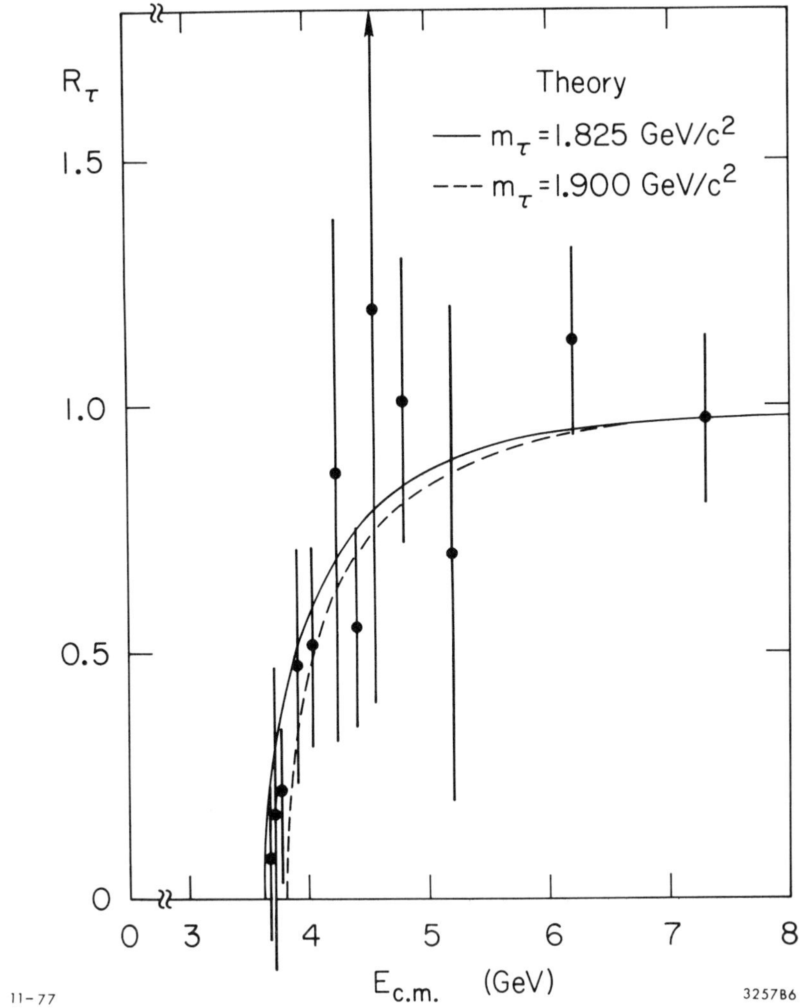

Fig. 7b R_τ for $3.6 \leq E_{c.m.} \leq 7.8$ GeV, compared with the theoretical R_τ for various τ masses.

New Quarks and Old Ones, Too

L. M. LEDERMAN

Columbia University, New York, N.Y. 10027

I. Introduction

This will be a brief review of the data taken by the Columbia-Fermilab-Stony Brook collaboration at Fermilab.*

---

* See Chart I.

The reaction is:

$$P + N \rightarrow \mu^+ + \mu^- + \text{practically anything}$$

in which a double arm spectrometer, Fig. 1, registers the muon pairs and reconstructs the effective mass and other kinematic properties of the dimuon. Primary protons are used, as many as the apparatus can stand. Details of the troubles and the tricks have appeared in the Physical Review Letters.[1,2] The results, at 400 GeV, are presented in Figs. 2 and 3 and show a yield with dilepton mass which falls rapidly ($\sim e^{-m}$) with dimuon mass from $\sim$ 5 GeV up to the highest mass recorded at 17 GeV. This fall is interrupted by a dramatic enhancement in the region of 9-10.5 GeV which is shown in detail in Figs. 2 and 3.

Again a very detailed study of the apparatus was carried out[3] in order to verify that the enhancement is real and not an artifact of acceptance, software etc. The thrice checked acceptance at 400 GeV is given in Fig. 4 and is seen to be rather flat above 5 GeV up to the kinematic limit (27.3 GeV). Another crucial aspect of the apparatus is the resolution (Fig. 5) which is tested on the J/ψ and ψ' on the one hand and also on the calculated and observed target reconstruction distribution.

The data presented here, 26,000 events in the mass range greater than 5 GeV, were obtained during a three-calendar months' run (May - August 1977) in the P-center lab. This represents a breakthrough in dilepton data rates (taken with a 2% resolution) by a factor of the order of 100 over any previous study. The basic ideas here are the use of Be as a hadron filter to minimize multiple coulomb scattering with meticulous attention to shielding details, the absence of any detectors upstream of the spectrometer magnets (these would limit the number of protons on target) and the use of a solid iron dipole to redetermine the muon momentum (this served to eliminate spurious tracks). The simultaneous registration of $\mu^+\mu^-$, $\mu^-\mu^+$, $\mu^+\mu^+$ and $\mu^-\mu^-$ pairs gave a continuous measurement of essentially all the backgrounds.

II. The Upsilon

It is qualitatively evident that the enhancement is resolved into at least two peaks and Table I summarizes a statistical study of the fitting of resonance peaks to a continuum background. We note that the peaks at 9.4 and 10.0 GeV (which we call Υ and Υ') are consistent with the resolution of the apparatus. This means that they are less than $\sim$ 100 MeV [σ(rms)] wide; they could of course be much, much narrower. This is the most important result since a 10 GeV object made of ordinary quarks would be expected to have a width of at least 1 GeV (think of the ρ, ρ' etc.). There is clearly a new selection rule which prohibits decay into the almost infinite number of possible hadronic final states. The intrigue is compounded by the proximity of two peaks. It is then understandable that a large number of theorists have proposed that the Υ represents a bound state of a new quark b to its antiquark $\bar{b}$. This would not only explain the narrowness but would account for a family of closely spaced levels due to radial excitations e.g. 1^3S, 2^3S, 3^3S, etc. It is well known that the possibility of heavier quarks has been widely anticipated for a variety of reasons - some esthetic (quark-lepton symmetry), some based upon no longer valid interpretations of neutrino data and some just to make trouble.

Alternative interpretations of the data have been considered by us as skeptical experimentalists. Appealing to existing theoretical structures there are Z^o's (weak vector boson) and Higgs particles (scaler bosons). Now a 10 GeV Higgs object, as we understand it, would decay largely into $\tau^+\tau^-$ with a very small branching[4] into $\mu^+\mu^-$ and a vanishing rate into e^+e^-. Our old e^+e^- data[5] contain a fairly clear cluster of $\sim$ 6 events near 9.5 GeV which made it unlikely that Upsilon = higgs. More recently, there are reports of clusters at 9.5 GeV in two ISR[6] experiments looking at e^+e^- final states. Thus we can rule out Higgs particles, at least of the garden variety.

The possibility of a highly premature Z^o can be investigated by searching for apparent C-violation effects, the simplest and cleanest for our experiment being the ratio,

$$R = \frac{N(E^+ > E^-) - N(E^- > E^+)}{N(E^+ > E^-) + N(E^- > E^+)} .$$

Energies are measured in the c.m. We have studied this for our data sample and present the results in Fig. 6. Here we make cuts designed to eliminate very slow (in the c.m.) dileptons which cannot contribute to the asymmetry. We do not have a measure of the sensitivity of this test since we would need a theory of the Z^0 and a model of its production.[7] Nevertheless, this result, taken with the fact that there are two peaks close together, makes Z^0 a dubious possibility.

A quantitative comparison of the data of Table I with predictions of the quark model with $\Upsilon = b\bar{b}$ runs into the following problem: the Υ'-Υ splitting of 600 MeV is close to that of the $\psi'\psi$ separation instead of a predicted[8] 450 MeV based on the linear confinement potential model. The model is constrained by much data in charmonium spectroscopy. There are apparently several ways to resolve this problem and these are discussed elsewhere.[9,10] However, one residual problem is the ratio of Υ'/Υ which we measure to be ~ 0.4. The $C\bar{C}$ corresponding ratio $\psi'/\psi \sim 0.03$ and the reasons for so large a difference are, to our knowledge, not yet understood.[10] Thus there is reason to leave the door to alternative interpretations somewhat ajar.

We conclude our discussion of Υ by several comments: (1) There is some evidence (see Fig. 11 below) that the production of Υ involves a process different from that of the neighboring continuum. This comes from our study of $\langle P_T \rangle$. (2) If someone asks about a third peak, Table I supplies the answer - there is weak indication of Υ'' at 10.4 GeV.* (3) The first confirmation of Υ outside of E-288 comes from a Northeastern experiment[11] whose data are shown in Fig. 7.** (4) Data at other energies (200 and 300 GeV) were recently taken in order to study the scaling properties of Υ and the continuum. Some of these data are shown in Fig. 8. We note that there is no sign of a resonance effect at 200 GeV. (If this had been a 200 GeV accelerator, as was originally planned, we would clearly not have found the Υ - we then owe its discovery to the FNAL Director!) Scaling of the Υ production is given in Fig. 9.

III. The Continuum

We have recently published[12] the results of an analysis of the continuum data between 5 and ~ 17 GeV. A somewhat updated version (small corrections and additional data) results in:

$$\left.\frac{d^2\sigma}{dm\,dy}\right|_o = (2.60 \pm 0.02)\ e^{-(0.987 \pm 0.006)m} \times 10^{-33}\ \frac{cm^2}{GeV/nucl} \quad (1)$$

We have plunged into the Drell-Yan model in order to test its fruitfulness in the new experimental regime. Recall that in this model, dileptons are produced via the annihilation of a quark from one of the colliding nucleons and an anti-quark from the other nucleon. To obtain the mass distribution, the probabilities for finding quarks and anti-quarks with fractional momenta X_1 and X_2 must be known:

---

* Note added "in proof": more data since this was written strengthens the Υ'' to ~ 3σ and gives $\Upsilon''/\Upsilon \sim 0.12 \pm 0.04$. (See Figure 12)

** ISR confirmation[6] came in later.

$$\frac{d^2\sigma}{dmdy} = \frac{8\pi\alpha^2}{9m^3} \sum_{k=u,d,s} e_k{}^2 x^2 \{q_k{}^p(x) q_{\bar{k}}{}^N(x) + q_{\bar{k}}{}^p(x) q_k{}^N(x)\} \quad . \tag{2}$$

Here we have set $X_1 = X_2 = X$ to correspond to the fact that, at 400 GeV, data are obtained very near $y = 0$ ($|y| < 0.3$). Taking Fermi motion, complex nuclei into account, we can use the data on deeply inelastic lepton-nucleon scattering[13] to supply the information on valence quark distributions and use our data to determine the antiquark distributions, presumed, in the model to come from the sea:

$$\left.\frac{d^2\sigma}{dmdy}\right|_o = \frac{8\pi\alpha^2}{9m^3}\ x\ S(x)[\nu W_2{}^{ep}(x) + \nu W_2{}^{eN}(x) - \tfrac{4}{3}\ x\ S(x)] \quad . \tag{3}$$

We find a very good fit:

$$xS(x) = 0.6(1-x)^{10} \quad . \tag{4}$$

Actually, νW_2 is written as a function of x and Q^2 with $Q_o{}^2 = 9$ being a reference. If we use $Q^2 = m_{\mu\mu}{}^2$, our result changes somewhat:

$$xS(x) = 0.5(1-x)^9 \quad . \tag{5}$$

This merely indicates that the observed scale breaking of νW_2 has a relatively minor effect above $x = 0.2$ over the Q^2 range we study. Extension to lower values of x require $E >$ 400 GeV since lowering the mass runs us into the resonance region.

IV. Scaling: Mass Spectrum

Our quantitative study of the behavior of dilepton production with s is still preliminary. There is a complication: In a fixed laboratory arrangement, the centroid of the acceptance in X_F or rapidity y moves a bit as shown in this table:

E_o	$\sqrt{s}$	$\langle y \rangle$
200	19.4	~ 0
300	23.7	0.15
400	27.3	0.30

Thus we must be careful not to confuse y-variations with $\sqrt{s}$ variations. These are fairly small effects* ($\leqslant$ 30%) but concern about them is the penalty of having good data. The preliminary conclusion is that $m^3 d\sigma/dm$ scaling is a good fit to the data in the interval:

$$0.2 \leq m/\sqrt{s} \leq 0.5 \quad .$$

If we ask whether this is in disagreement with the observed[13] scale breaking seen in μp scattering and which influences the valence quark distributions, the answer already given is no - in the interval studied, these effects are $\leqslant$ 20% and we don't see them at this stage in the analysis.

V. Scaling in P_T

Our final topic has to do with the transverse momentum distribution of the dileptons. The key issue here is that in the parton-antiparton annihilation model (the original paper is now generally referred to as Naive-Drell-Yan - and it is up to the reader to decide whether the first word modifies Drell, Yan the model or is the name of a new collaborator), the transverse momentum of the dilepton must be related to that of the annihilating quarks:

* Assuming gentle behavior with y. Newer analysis shows a rather significant variation of $d\sigma/dy$ and preliminary conclusions confirm the scaling behavior i.e. $m^2 d^2\sigma/d\tau\, dy = F(\tau, y)$.

$$\langle P_T^{\,2}\rangle_{\mu\mu} = \langle k_T^{\,2}\rangle_q + \langle k_T^{\,2}\rangle_{\bar{q}} \quad . \tag{6}$$

Thus we probe the transverse constituent motions by our measurement of P_T. I should comment here that there has been a literal explosion of theoretical interest in this problem. Figure 10 shows a sample of the data at 400 GeV. We find a very nice fit to the shapes with a purely empirical form

$$\left(\frac{1}{(bP_T)^2+1}\right)^6 \tag{7}$$

with b = 0.360 at $\sqrt{s}$ = 27.3 GeV. We find that the shape fits all masses from 5-14 GeV except in the region of Υ.

We have recently studied the same distributions at 300 and 200 GeV and a table of preliminary fit parameters is given:

$\sqrt{s}$ (GeV)	19.3	23.7	27.3
$b(\text{GeV}/c)^{-1}$	0.42	0.39	0.36

Thus we have a representation of the data. In Fig. 11, we present $\langle P_T\rangle$ vs. m for the three energies. We can make the following conclusions:

i) $\langle P_T\rangle$ is essentially independent of mass above 5 GeV, ruling out such "hadronic" behavior as $\exp b(-\sqrt{m^2+P_T^{\,2}})$.

ii) The striking flatness of $\langle P_T\rangle$ with respect to $m_{\mu\mu}$ is in contradiction with many QCD calculations.[14] These predict $\langle P_T\rangle^2 \sim m_{\mu\mu}^{\,2}/\log m_{\mu\mu}^{\,2}$ and predict much larger $\langle P_T\rangle$ at $m_{\mu\mu}$ > 10 GeV than is observed. Damping terms such as (1-x) are not enough.

iii) The CIM model[15] produces high P_T dileptons via quark scattering and quark annihilation; in both cases a cloud meson supplies the quark or anti-quark. This gives a good account of the average P_T's independence on mass but also a power law distribution which falls as P_T^{-4}, hence not in agreement with (7).

iv) Covariant parton models as published,[16] give poor representation of the data presented here. It would be helpful if theorists provided predictions for the distributions in P_T vs. m, $\sqrt{s}$ and y, so that more incisive confrontations can be made.

On a more speculative and very preliminary basis:

v) We can fit the data with

$$\langle P_T \rangle = 0.50 + 0.026\sqrt{s}$$

(preliminary)

or $$\langle P_T^2 \rangle = 0.750 + 0.00155\ s$$

which suggests that for large enough s (or large enough P_T), dimensional scaling holds - this is a consequence of several QCD calculations[17,18], one of which[18] explicitly predicted the above result with a very high degree of accuracy. The first term above is related to the "primordial" quark transverse momentum in the nucleon.

Looking at the data from the Naive et al point of view, the effective quark transverse momentum (assuming $\langle k_T \rangle_q = \langle k_T \rangle_{\bar{q}}$) is

$$\langle k_T^2 \rangle_q \sim 0.4 + 0.008\ s\ (\text{GeV/c})^2 \qquad \text{(preliminary)}\ .$$

In conclusion, the author takes personal responsibility for any incorrect statements and acknowledges his very great debt to the E-288 group for their efforts, especially in supplying the preliminary new data.

References

[1] S. Herb et al, Phys. Rev. Lett. 39, 252 (1977).

[2] W. Innes et al, Phys. Rev. Lett. 39, 1240 (1977).

[3] L.M. Lederman: Cargese Lectures July 1977.
W. Innes, Proceedings of the SLAC Topical Conference, Stanford, July 1977.

[4] J. Elles, M. Gaillard, D. Nanopoulos, Nucl. Phys. B106, 292 (1976).

[5] D. Hom et al, Phys. Rev. Lett. 36, 1236 (1976).
D. Snyder, Ph.D. Thesis, Columbia University (1978).

[6] J.H. Cobb et al, Phys. Lett. 72B, 273 (1977).
CERN-Columbia-Oxford-Rockefeller Collaboration, private communication (1978).

[7] R.W. Brown, K.O. Mikaelian, M.K. Gaillard, Nucl. Phys. (1974).

[8] E. Eichten, K. Gottfried, Phys. Lett. 66B, 286 (1977).

[9] C. Quigg, J. Rosner, Fermilab Pub - 77/82-THY (1977).

[10] K. Gottfried, Proceedings of the Conference on High Energy Photons and Electrons, Hamburg (1977).

[11] D.A. Garelick et al (Northeastern-Seattle-Tufts), submitted to Phys. Rev. Lett.

[12] D. Kaplan et al, to be published in Phys. Rev. Lett.

[13] W. Francis, DPF Meeting, Argonne National Lab (1977).
T. Kirk, private communication. See also:
H. Anderson et al, Phys. Rev. Lett. 38, 1450 (1977).

[14] J.B. Kogut, Phys. Lett. 65B, 377 (1976).
C.S. Lam, T.M. Yan, CLNS-365 (1977).

14 cont'd

J. Kogut, J. Shigemibu, CLNS-364 (1977).

J. Hinchliffe, C. Llewellyn-Smith, Phys. Lett. 66B, 281 (1977).

15 M. Duong-van, K.V. Vasavada, R. Blankenbecler, SLAC PUB 1882 (1977); M. Duong-van, R. Blankenbecler, submitted to Phys. Rev. 1977. Also M. Fontannaz, Phys. Rev. D14, 3127 (1976).

16 P.V. Landshoff, DAMTP (Cambridge) 76/23 (1976).

J.F. Gunion, UCD-76-V (1977).

J.C. Polkinghorne, Nucl. Phys. B108, 253; B116, 347 (1976).

17 K. Kajantie, R. Raitio, HU-TFT-77-21 (Helsinki) (1977).

H.D. Politzer, HUTP-77/A029 (Harvard) (1977).

18 H. Fritsch, P. Minkowski, TH-2400-CERN (1977).

Fermilab
E-288

Columbia *	Fermilab +	Stony Brook *
S.W. Herb	W.R. Innes	D.M. Kaplan
D.C. Hom	J.A. Appel	A.S. Ito
L.M. Lederman	B.C. Brown	H. Jostlein
J.C. Sens	C.N. Brown	R.D. Kephart
H.D. Snyder	K. Ueno	
J.K. Yoh	T. Yamanuchi	

* N.S.F.

\+ D.O.E.

Chart I

TABLE I

Resonance Fit Parameters[a]

		2 peak	3 peak	
Υ	M_1	9.38 ± 0.009	9.34 ± 0.001	GeV
	$B\frac{d\sigma}{dy}\Big\vert_{y=0}$	0.31 ± 0.01	0.30 ± 0.01	pb
Υ'	M_2	10.03 ± 0.03	9.938 ± 0.036	GeV
	$B\frac{d\sigma}{dy}\Big\vert_{y=0}$	0.130 ± 0.007	0.13 ± 0.007	pb
Υ''	M_3	---	10.32 ± 0.12	GeV
	$B\frac{d\sigma}{dy}\Big\vert_{y=0}$	---	0.037 ± 0.012	pb
	χ^2/DF	45/14	20/12	

[a] Table revised 1/20/78 in response to new data, systematic corrections. Results are preliminary.

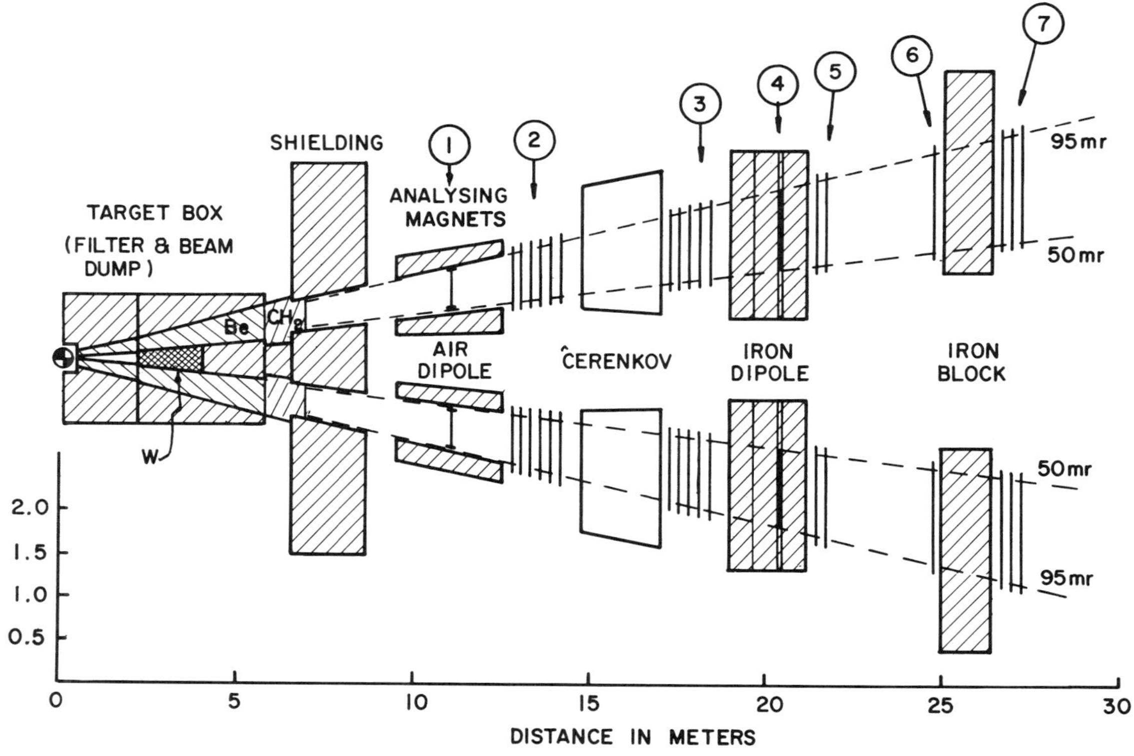

FIG. 1

μμ II Apparatus

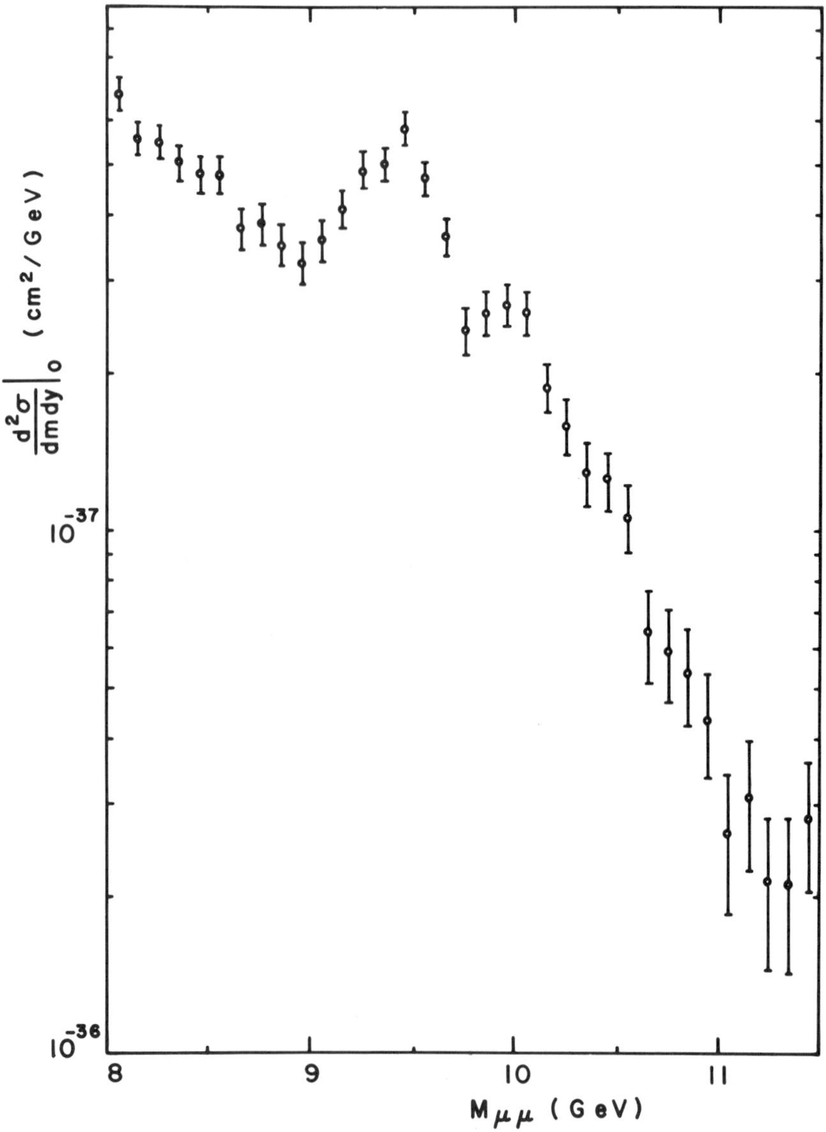

FIG. 2 Mass distribution in the Υ region

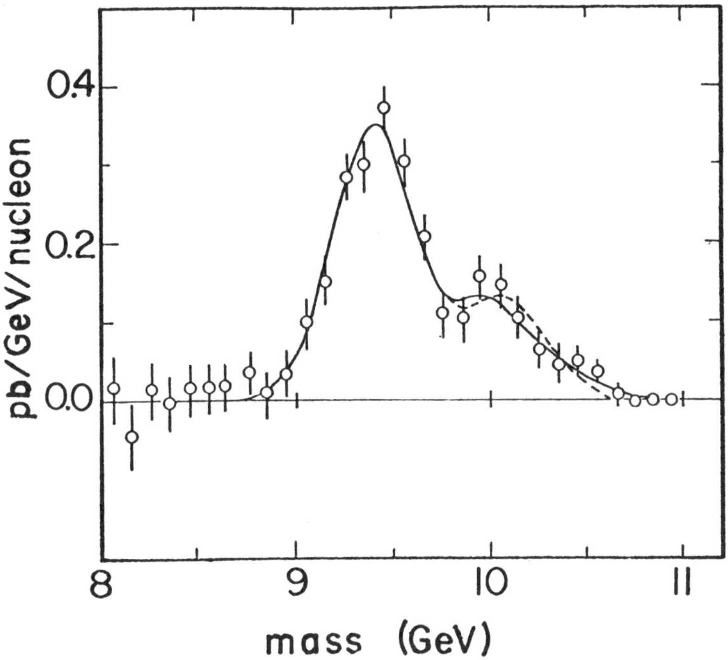

FIG. 3 Excess of the data over the continuum fit. Errors shown are statistical only. The solid curve is the three-peak fit; the dashed curve is the two-peak fit. Data as of 15 August 1977.

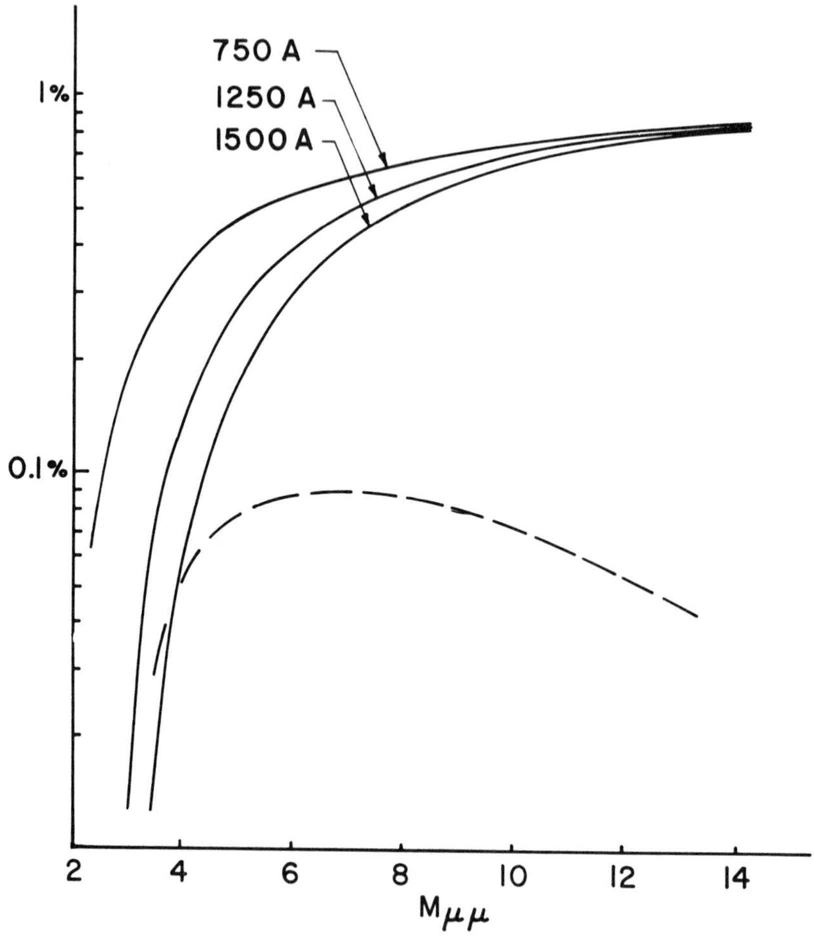

Fig. 4 Mass acceptance for various magnet currents. The dashed curve is the μμ I acceptance. This acceptance is calculated with isotropic decay. There are ~ 30% effects due to other distributions.

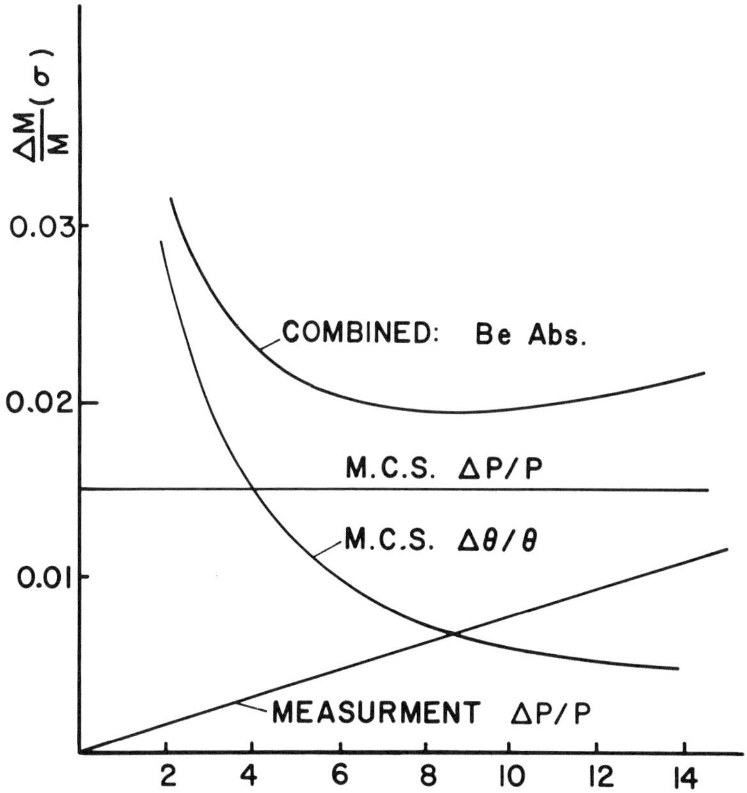

Fig. 5 Mass resolution with all Be absorber. Replacing Be by Cu increases the $\Delta\theta/\theta$ contribution only and raises the total curve to $\sim$ 2.1% above 6 GeV. Analytically $\Delta m/m = 0.11/M + 0.015 + 0.0009$ M for Cu. Additions are in quadrature.

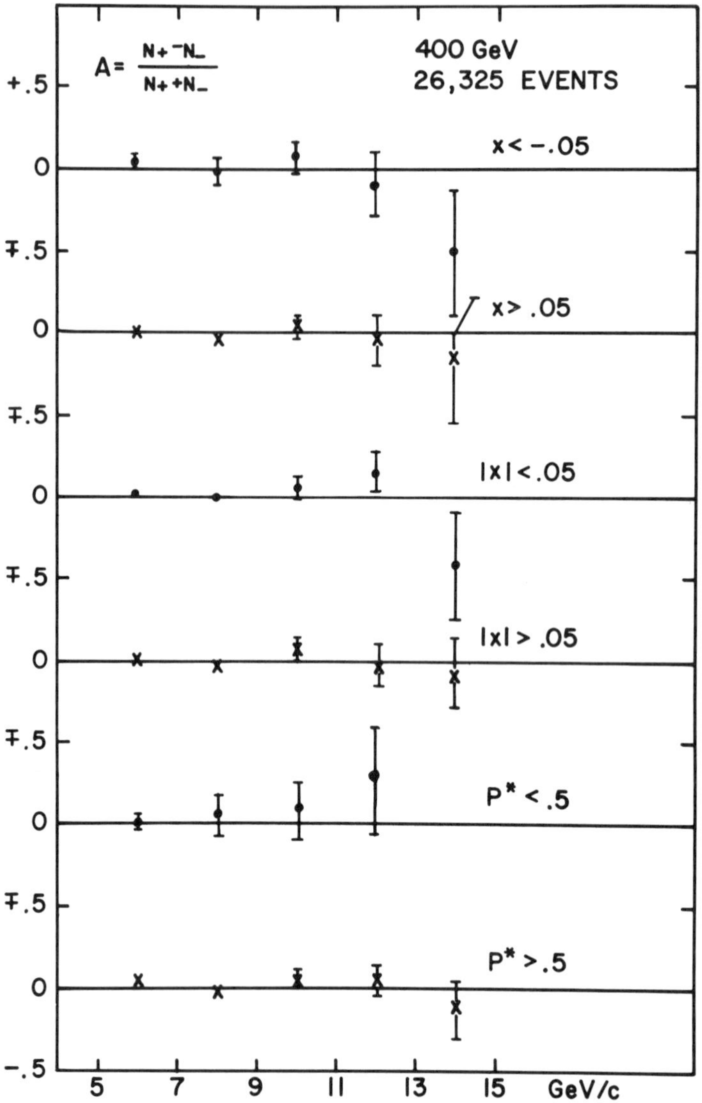

Fig. 6 Charge asymmetry vs. mass. N_+ is the number of events in which the μ^+ has higher energy in the c.m. than μ^-. x and P* refer to the dilepton longitudinal motion in the c.m.

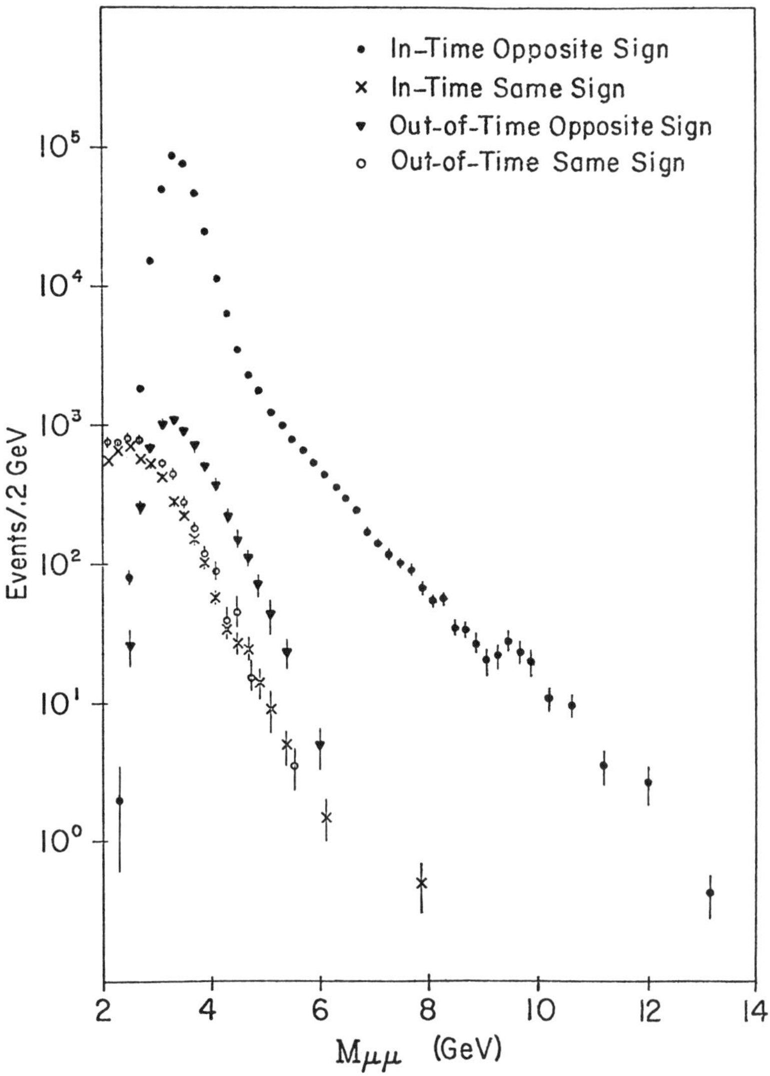

Fig. 7 Data from the Northeastern-Seattle-Tufts dimuon experiment at Fermilab.

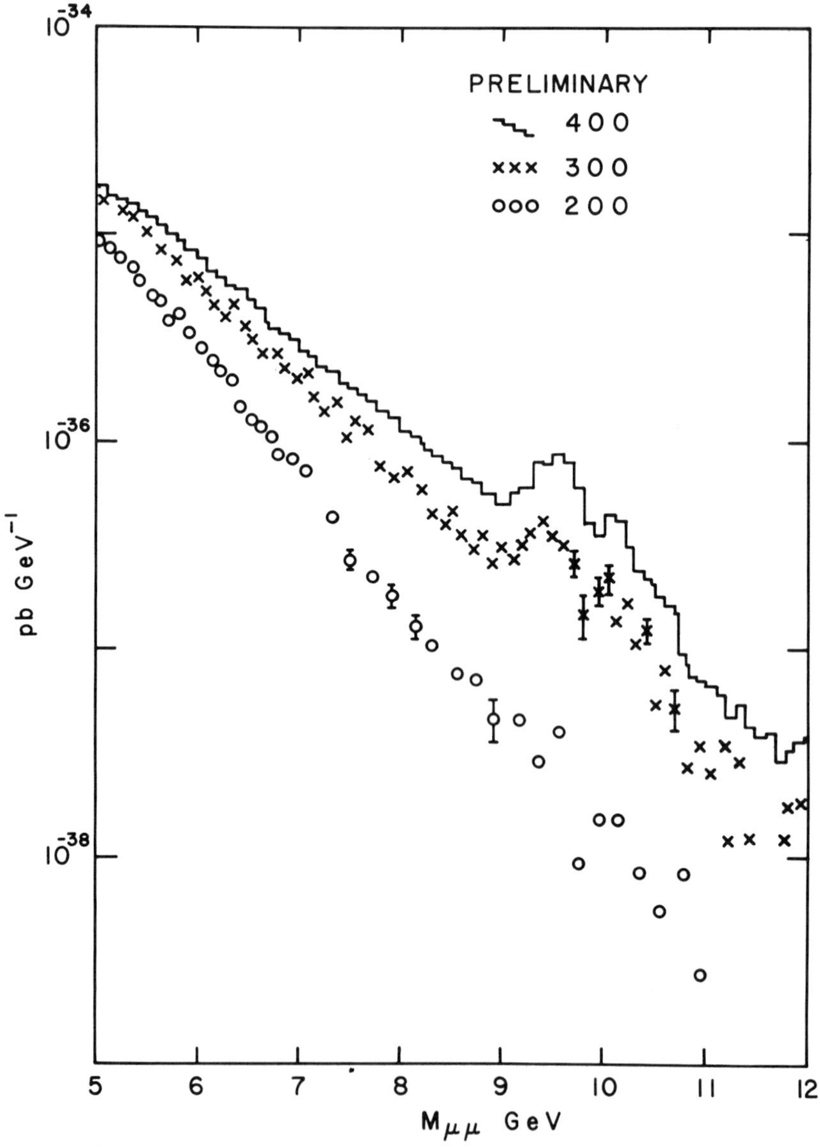

Fig. 8 Preliminary dimuon spectra at 200, 300 and 400 GeV

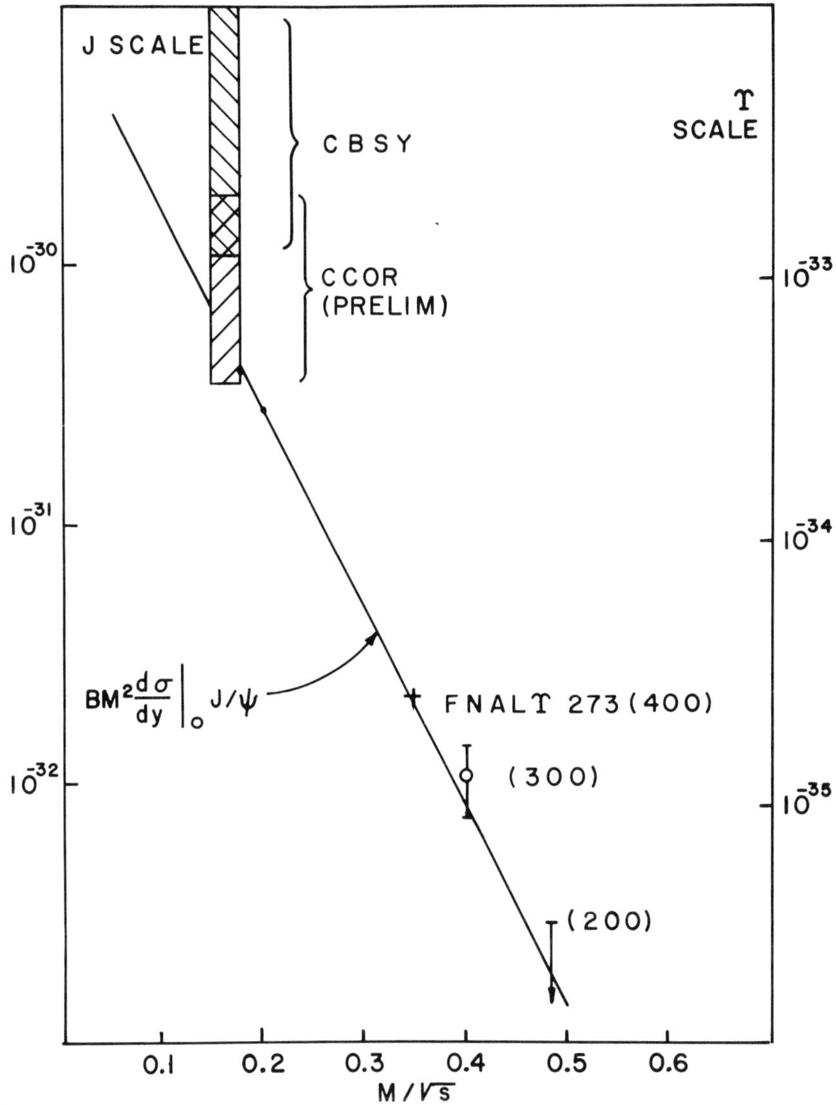

Fig. 9 Comparison of J/ψ and Υ excitation

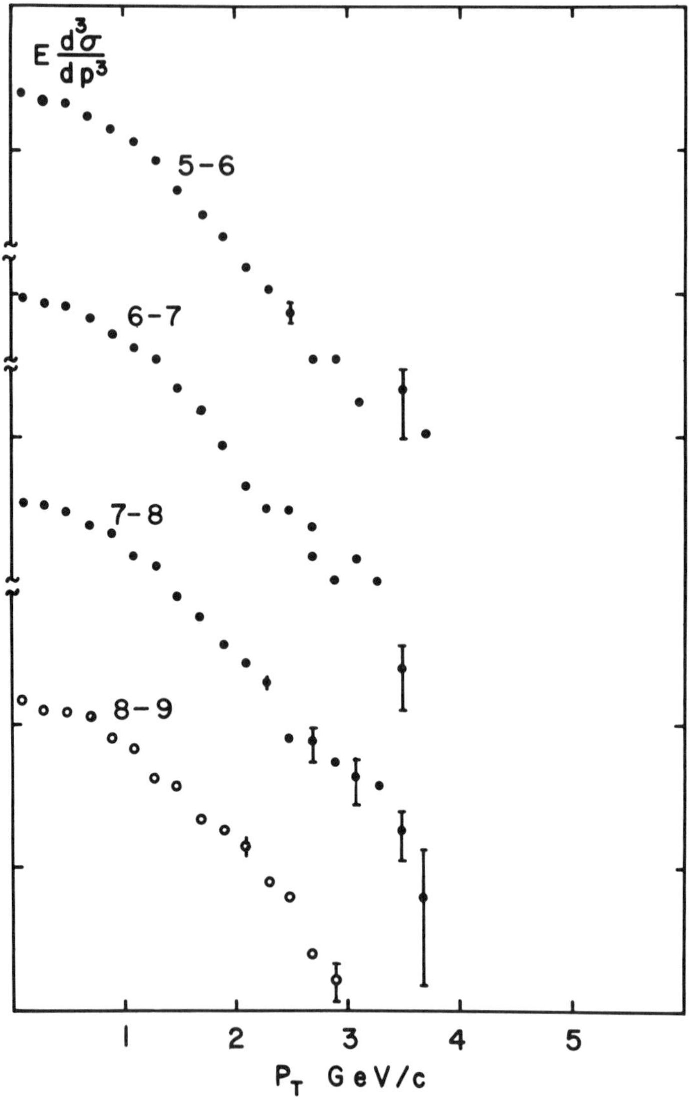

Fig. 10 Some typical P_T distributions

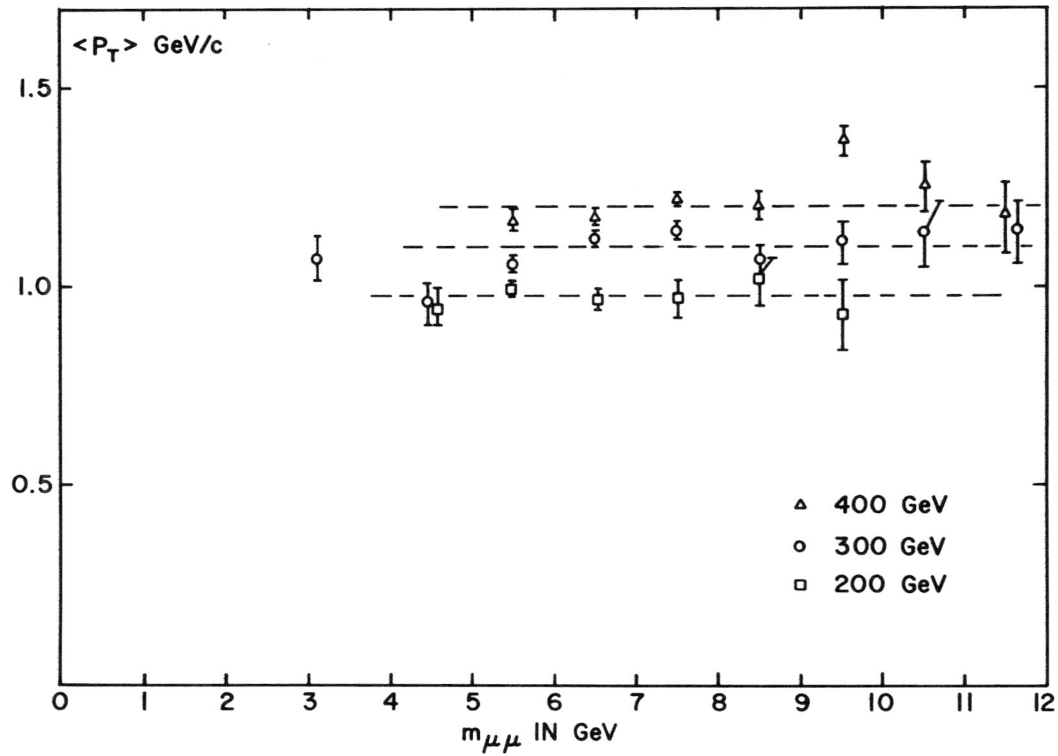

Fig. 11 Average P_T vs. mass at three energies. Preliminary data.

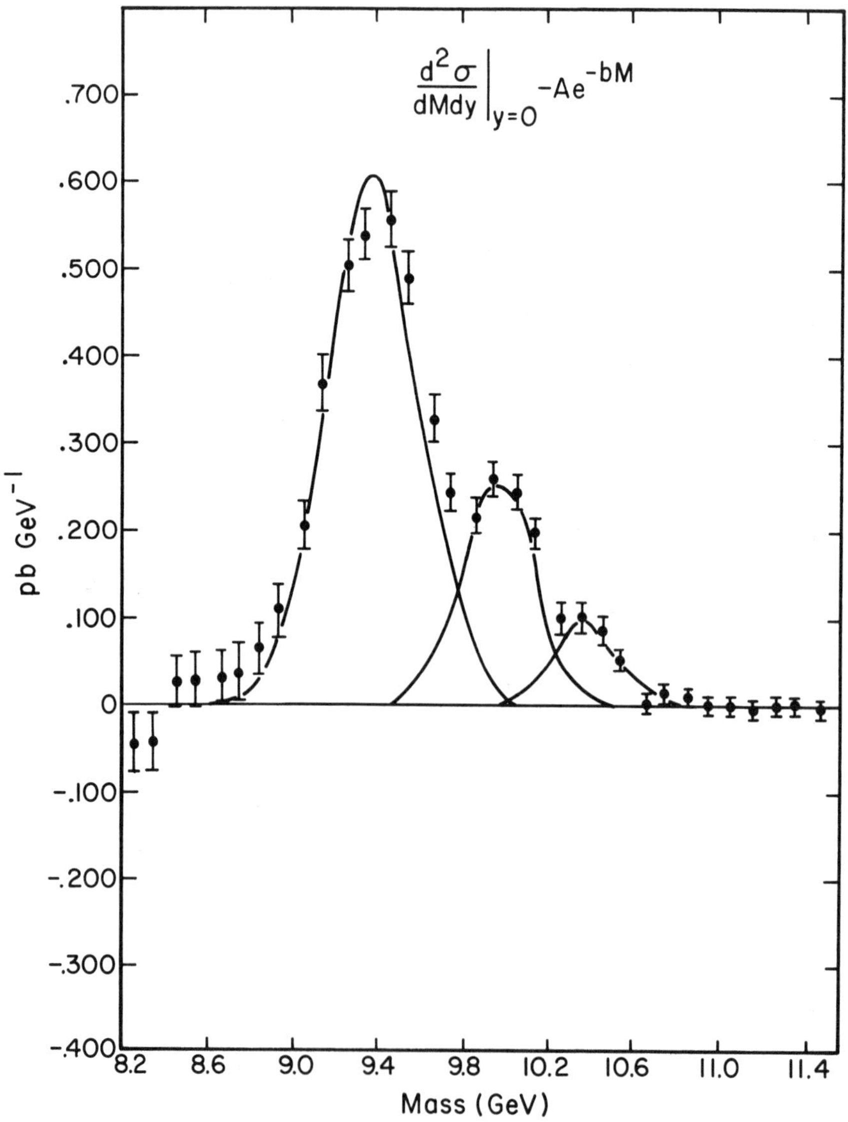

Fig. 12. Preliminary new upsilon results - resonance fits are sketched approximately.

MULTILEPTON PRODUCTION BY HIGH ENERGY NEUTRINOS AT FERMILAB

Don D. Reeder
University of Wisconsin

Implications of multilepton production

The voluminous and detailed studies of decay processes have clearly established the change in isospin and/or strangeness characteristic of the charged current weak interaction. Recent discoveries have confirmed the presence of another "flavor", viz. charm, and suggest still more possible hadronic degrees of freedom.[1] One might expect the weak interaction does not conserve these new hadronic degrees of freedom and thus particles with these quantum numbers could be singly produced by neutrinos, and subsequently decay. This flavor changing process would produce $\mu\mu$ and μe events in the ν_μ beam.

Another source of multileptons would be the production of pairs of these flavored particles in the hadronic cascade produced in deep inelastic ν_μ interactions. Trilepton ($\mu\mu\mu$, μee, etc.) events as well as dileptons could result from this process.

New leptons, if they exist, might also couple to ν_μ and could be produced, decay back into the old leptons and result in a multilepton final state.

The search for multilepton production in neutrino interactions provides a sensitive and relatively model independent test of new particle production. And who knows what other surprises may appear in the guise of multileptons?

Of course, in order to establish these new particles, the production of leptons by old particles or known processes must be identified and eliminated. More specifically these are: 1) the known flavor changing decays $\pi \to \nu\mu$ and $K \to \nu\mu$, $K \to e\nu\pi$ etc. 2) the production of electron or muon pairs by electromagnetic interactions.

Dilepton Production

The first hint of possible new phenomena was provided by the HPWF collaboration at the International Conference on High Energy Physics held at London in July, 1974.[2] Two possible examples of

$$\nu_\mu + N \to \mu^-\mu^+ X(\nu)$$

were reported. The electromagnetic background could be eliminated, assuming conservation of lepton number, by the absence of a third muon. However, the rejection of meson decay could not be conclusively argued because of the small statistics. However, this possibility would be eliminated if μ^-e^+X final states were observed.

In November 1974 the question of the existence of new particles was conclusively and dramatically answered by the simultaneous observation at BNL and SLAC of hidden charm (ψ/J particle).[3] Subsequently the SPEAR group detected the D mesons which carried the new charm quantum number.

Nonetheless the investigations using neutrino beams continued until today the observation of dilepton final states is well established. In Table I I summarize the number of events detected to date in the three major experiments at FNAL.[3]

Table I

Cumulative Dilepton events detected at FNAL

Reaction	FHPWR	CRF	BC
$\nu_\mu + N \rightarrow \mu^-\mu^+X$	∿300	∿70	∿100
$\bar{\nu}_\mu + N \rightarrow \mu^+\mu^-X$	∿ 80	∿30	∿ 10

Among the general conclusions that can be drawn from these data are:

i) that muon neutrino induced events producing a μ^- and another charged lepton exist and have a non-trivial origin

ii) that the rate of μ^-e^+ events is comparable to that of $\mu^-\mu^+$ events implying no large violation of μe universality in the decay of the new particles.

iii) that most of these events arise from the production by both ν and $\bar{\nu}$ and the decay of singly produced charmed particles.

Within the quark parton model of neutrino-nucleon interactions the distributions in $x = Q^2/2m\nu$ and $y = \nu/E_\nu$ for these events can be calculated. Here Q^2 is the four momentum transfer from ν_μ to μ^-, ν is the energy transfer and E_ν is the laboratory neutrino energy.

Let d, s, c be the down, strange, and charmed quarks respectively and $\bar{d}$, $\bar{s}$, $\bar{c}$ their antiparticles. Then the possible reactions are:

a) $\nu_\mu + d \rightarrow \mu^- + c + X$. Although this reaction can proceed via valence quarks and thus at large x, it is inhibited by a Cabibbo angle $\sin^2\theta_c \sim 0.5$.

b) $\nu_\mu + s \rightarrow \mu^- + c + X$

$\bar{\nu}_\mu + \bar{s} \rightarrow \mu^+ + \bar{c} + X$ Here no Cabibbo suppression is present but the s and $\bar{s}$ quarks are not valence quarks but come from the sea of virtual quarks. These events appear at small x.

As evidence for this interpretation I present the dimuon data obtained by the FHPWR collaboration. In Fig. 1 the distribution of visible energy for dimuons is presented. The absence of events with energies greater than 150 GeV for the $\bar{\nu}$ beam is characteristic of the much softer spectrum of antineutrinos. The sharp x distribution of $\bar{\nu}$ induced events indicative of production on the sea is shown in Fig. 2. Also shown is the x-distribution of ν induced events showing production at large x (valence quarks) in addition to the small x sea. Finally, Fig. 3 shows the flat y distribution for both ν and $\bar{\nu}$, again consistent with the predictions of the model.

In Fig. 4 the correlation of the two muon momenta is shown. Charm production is indicated by the bands of events near the axis.

The rate of dilepton production compared to single muon deep inelastic scattering is measured by several experiments is shown in Table II.

Table II

Published rates for dilepton production

Experiment	Limit of $E_{\ell 2}$(GeV)	Lepton Type	Corrections	$R(\ell^+\mu^-)/R(\mu^-)$ ν (%)	$\bar{\nu}$ (%)	E_{VIS} (GeV)
HPWF[4] (1975)	4	$\mu^+\mu^-$	GEOM	0.8 ±0.3	2. 1.	> 40
CTFR[5] (1977)	2.4	$\mu^+\mu^-$	GEOM	0.8 ±0.5	1.4 ±0.7	>150 ∿
BCHW (76)	0.8	$e^+\mu^-$	GEOM	0.8 ±0.3		> 10
COL-BNL[7]	0.3	$e^+\mu^-$	GEOM	0.8 ±0.2		> 40
CDHSB[8] (1977)	4.5	$\mu^+\mu^-$	NONE	>0.5 ±0.1	>0.5 ±0.1	> 50
BHW[9] (1977)	0.8	$e\mu$	GEOM	0.34 +0.23 -0.07	0.10 +0.13 -0.07	$<E_\nu>=47$ $<E_{\bar{\nu}}>=30$

There is good agreement among the various experiments that about .5 to 1.0 percent of neutrino interactions have an additional charged lepton.

One tantalizing question remains open. The BCHW group who first reported μ^-e^+ events observed an anomalously large number $(1.8^{+.6}_{-.5})$ of K° mesons produced with these events. Subsequently the Columbia-BNL experiment did not confirm this result, instead they obtained a smaller number (.6 ± .2) of K° mesons. However, recent results from BEBC at CERN appear to corroborate the large value of BCHW.(1.7 ± .5)[10]

Thus by 1977 the opposite sign dimuons provide a new trivial source of multileptons--charm. The game now moves to a new arena--Are there events which are not explained by charm?

Beyond Charm

A close examination of Fig. 4 reveals about 5-15 percent of dimuons are in the symmetric region defined as $\frac{P}{2} \leq P_+ \leq P_-$. The number of events to be expected assuming only charm production is model dependent and poorly known. So the opposite sign dimuons might suggest other phenomena but not in a compelling way.

Another indication of raw phenomena may be like sign dimuon production of non-trivial origin. An example is shown in Fig. 5 wherein the event is shown superimposed on the FHPWR detector. Both tracks originate in the fiducial volume of an iron scintillator calorimeter and pass through at least 5.6 m of iron. The measured momenta of both of the tracks exceed 10 GeV and it is extremely improbable that meson decay could be the source.

Several candidates for like sign dimuons have been recorded by FHPWR. To indicate the validity of this sample, the like sign and opposite sign dimuons are compared in Fig. 6 where the distance between the tracks at the vertex (Δ) is plotted. Although there is more background in the like sign sample there is a distinct signal. To determine if meson decay could be responsible the rate in targets of different density can be compared. In Fig. 7 the ratio of $\mu^-\mu^-$ to $\mu^-\mu^+$ events is plotted as a function of inverse density. Although the errors are large, qualitatively the data are consistent with direct production.

In Fig. 8 the correlation of the two momenta can be seen. As was the case for $\mu^+\mu^-$ about 5-15% are in the symmetric region. Although not yet conclusive, the data strongly suggest a non-trivial source of direct $\mu^-\mu^-$ events. Possible origins could be associated production of charmed pairs, new leptons, or something unforseen.

Trimuon Production

In March 1977 the CFR group published another hint of new phenomena.[11] Two examples of apparent trimuon production were reported.

$$\nu + N \rightarrow \mu^-\mu^?\mu^?X$$

$$\nu + N \rightarrow \mu^-\mu^+\mu^?X$$

The signs of all the muons could not be measured in part because one of the accompanying μ was very soft (< 10 GeV).

By May 1977 the FHPWR group had confirmed trimuons and perhaps found something new.[12] They now have 13 examples of

$$\nu + N \rightarrow \mu^-\mu^+\mu^-$$

and one example of

$$\bar{\nu} + N \rightarrow \mu^+\mu^-\mu^?$$

The distribution of these events in visible energy is shown in Fig. 9. One of two particularly interesting events is shown in Fig. 10 superimposed on the detector. Three very high energy muons and no hadronic energy. The estimated backgrounds for this event and another are presented in Table III. The only possible background of importance is the direct pair production from the lepton vertex. Initial calculations have indicated that its contribution is small, but detailed calculations by a number of theorist are in progress and the result should be known soon.

TABLE III

Background Estimates for Exceptional Trimuon Events

TYPE	Event Number 119	Event Number 281
Random Space-Time Coincidence	$<10^{-3}$ event	$<10^{-3}$ event
i) 3 single muons (ν events)		
ii) 1 single muon, 1 dimuon (ν)		
iii) 1 dimuon event, 1 stray muon		
iv) 3 stray muons		
Decay of π/K from dimuons	$<10^{-1}$	$<10^{-1}$ event
Electromagnetic Processes		
i) $\nu + N \rightarrow \mu + X$		
$\mu + N_Z \rightarrow \mu\mu\mu + N_Z$	$<10^{-2}$	$<10^{-2}$ event
ii) dimuon from hadron vertex	$<10^{-2}$	$<10^{-2}$ event
iii) direct pair from lepton vertex	?	?

In summary, the direct production of $\mu^-\mu^-\mu^+X$ induced by ν has been observed at an estimated cross section $\sigma \sim 10^{-40}cm^2$ at $E_\nu \gtrsim 100$ GeV. There seem to be qualitatively at least two types of trimuons; i) moderate energies with one or two soft muons ($p \lesssim 10$ GeV/c), ii) high energy with all muons ($p \gtrsim 10$ GeV) and large total muon energy. The production of trimuons by antineutrinos remains an open question. The current data suggest $\sigma(\bar{\nu}\to3\mu)/\sigma(\nu\to3\mu)\lesssim1$, but the process is not established. A search for more than three muons has resulted in a limit to the rate of four or more muons is about one-fifth the rate for trimuons.

It is probable that several sources of trimuons exist. The details of some of the events suggest that at least one source is concerned with the lepton vertex. All three neutrino facilities are now in operation at Fermilab and one may expect progress in this field in the near future.

I would like to acknowledge the work of all the members of these neutrino facilities: the FHPWR group A. Benvenuti, F. Bobisut, D. Cline, P. Cooper, M. Heagy, R. Imlay, M. Johnson, T. Y. Ling, R. Lundy, A. Mann, P. McIntyre, S. Mori, D. D. Reeder, J. Rich, C. Rubbia, R. Stefanski and D. Winn; the CFR Group, B. Barish, J. Bartlett, A. Bodek, K. Brown, D. Buchholz, Y. Chu, F. Scuilli, E. Siskind, L. Stutte, H. Fisk, G. Krafezyk, D. Nease, and O. Fackler; and the myriad of bubble chamber users. On behalf of all of us who have had the opportunity to explore the physics of neutrinos at Fermilab, I acknowledge the advice, support and enthusiastic encouragement of Dr. Benjamin Lee.

References and Footnotes

*Supported in part by the U. S. Department of Energy.

1. A comprehensive and detailed survey of the charm situation is available in the Proceedings of the Summer Institute on Particle Physics (Aug. 1976) SLAC reprint 198 edited by M. C. Zipf. Also Herb et al., Phys. Rev. Letters 39, 252, 1977.

2. C. Rubbia et al., Proceedings of the 17th International Conference on High Energy Physics, London, July 1974.

3. The groups involved are the Fermilab, Harvard, Pennsylvania, Wisconsin, Rutgers collaboration (FHPRW); the Cal Tech, Fermilab, Rockefeller collaboration (CFR) and the 15' Bubble Chamber program (BC) in which the groups at the University of California at Berkeley, Lawrence Berkeley Laboratory, University of Hawaii, University of Wisconsin, Columbia University, Brookhaven National Laboratory, the University of Washington and Fermilab have participated.

4. A. Benvenuti et al., Phys. Rev. Letters 35, 1199 (1975).
A. Benvenuti et al., Phys. Rev. Letters 35, 1249 (1975).

5. B. Barish et al., Phys. Rev. Letters 39, 981 (1977).

6. J. von Krogh et al., Phys. Rev. Letters 36 710 (1976).
P. Bosetti et al., Phys. Rev. Letters 38, 1248 (1977).

7. C. Baltay et al., Phys. Rev. Letters 39, 62 (1977).

8. P. Block, Proceedings of the 1977 International Symposium on Lepton and Photon Interactions at High Energies, Hamburg, 1977.

9. T. Burnett, Proceedings of the 1977 International Symposium on Lepton and Photon Interactions at High Energies, Hamburg, 1977.

10. K. Schulgi, Proceedings of the 1977 International Symposium on Lepton and Photon Interactions at High Energies, Hamburg, 1977. W. F. Fry, CERN private communication.

11. B. C. Barish et al., Phys. Rev. Letters 38, 577, 1977.

12. A. Benvenuti et al., Phys. Rev. Letters 38, 1110, 1977 and D. Reeder, Proceedings of the 1977 International Symposium on Lepton and Photon Interactions at High Energies, Hamburg, 1977.

Figure 1

The distribution in visible energy ($E_\nu \geq E_{vis}$) of the dimuon events observed.

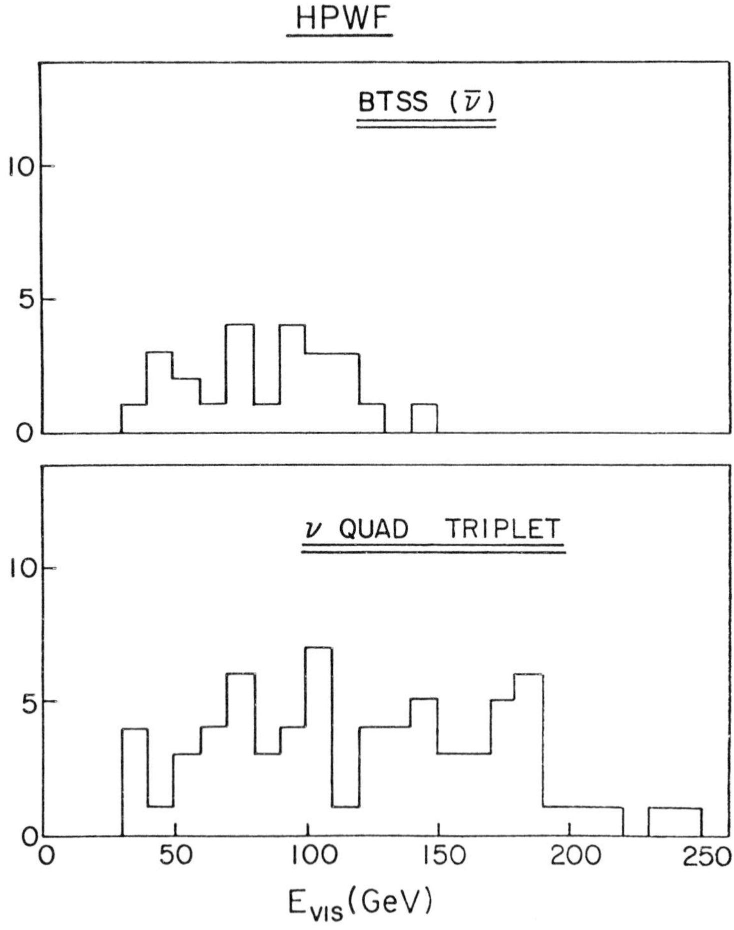

Figure 2

Distribution in $x = Q^2/2m\nu$ for dimuon events.

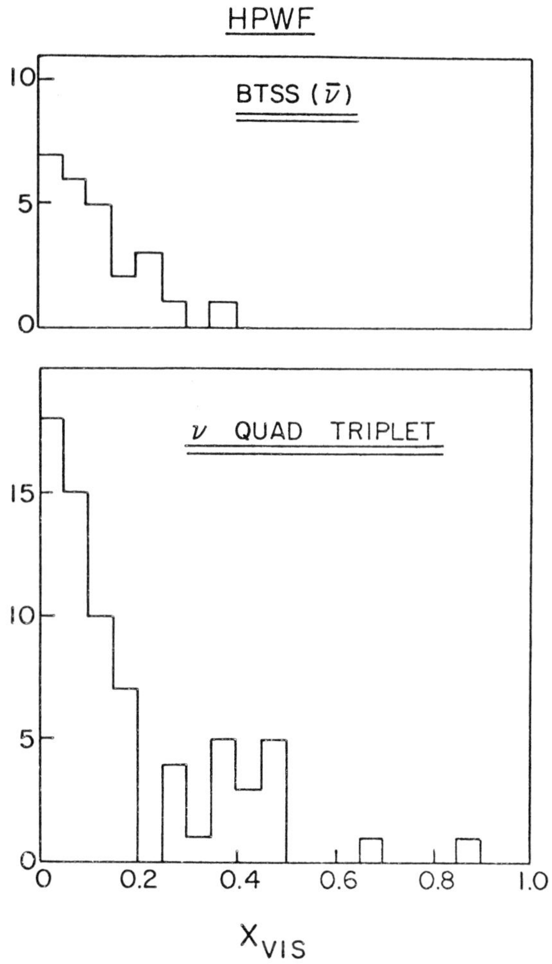

Figure 3

The distribution in $y = \nu/E_{vis}$ for dimuon events.

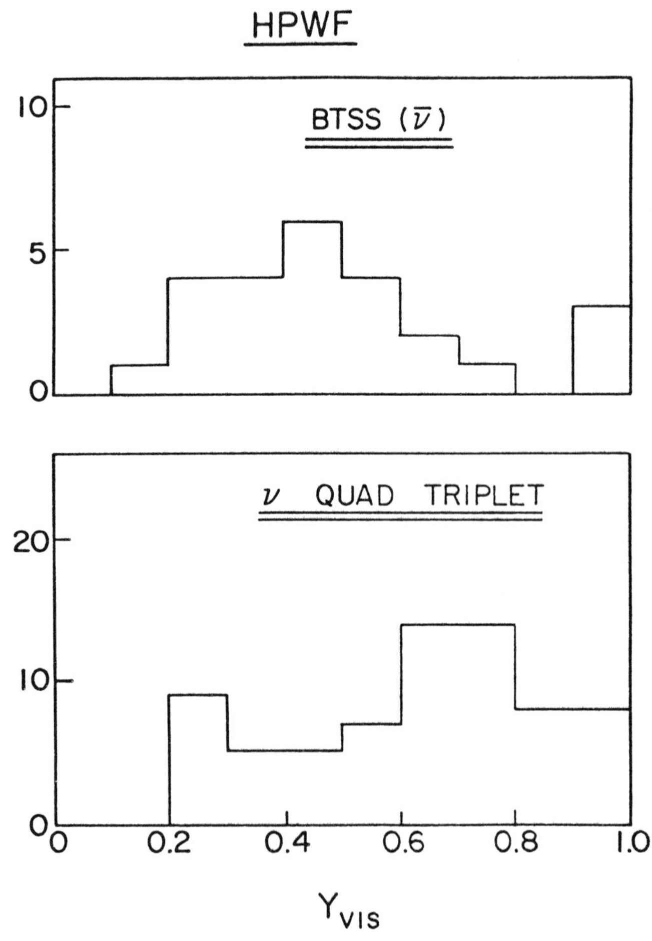

Figure 4

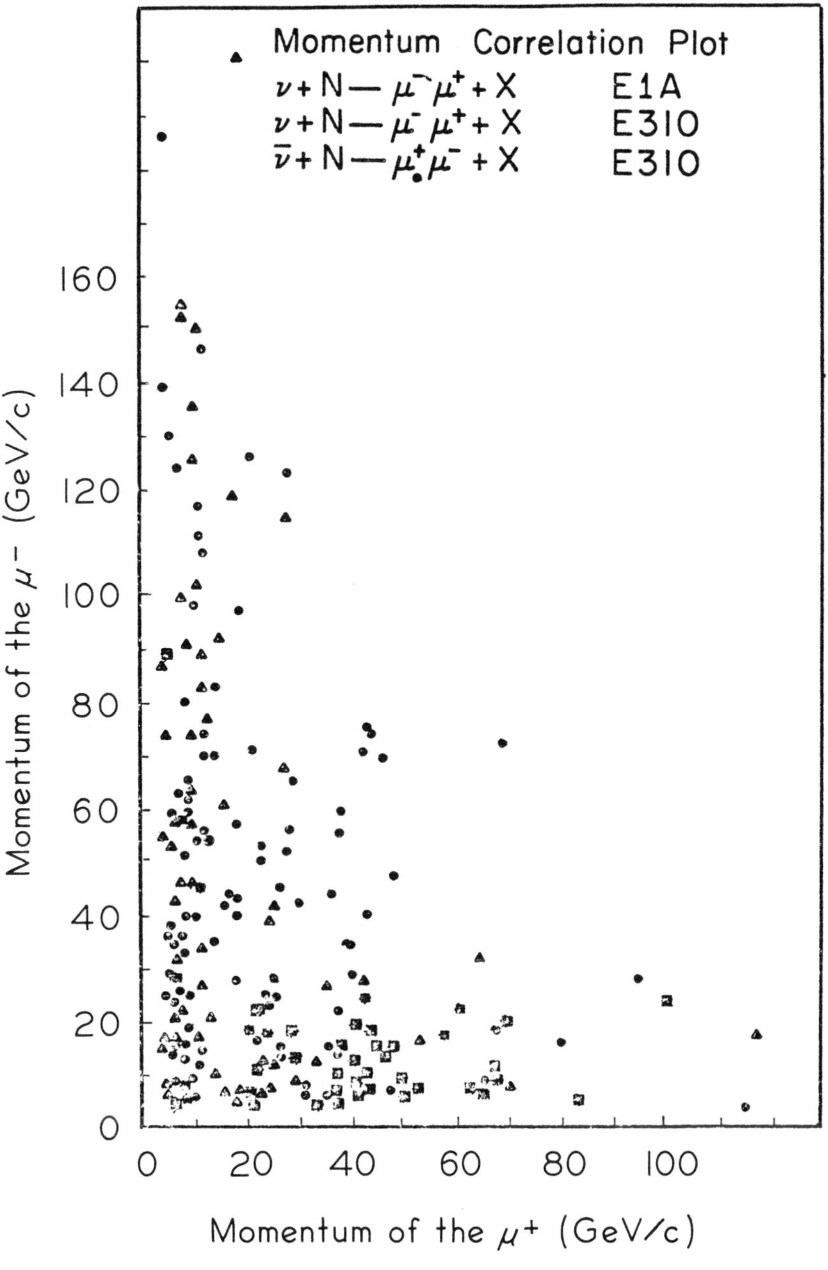
Momentum Correlation Plot
ν + N — μ⁻ μ⁺ + X E1A
ν + N — μ⁻ μ⁺ + X E310
ν̄ + N — μ⁺ μ⁻ + X E310
Momentum of the μ⁻ (GeV/c)
Momentum of the μ⁺ (GeV/c)
0
20
40
60
80
100
120
140
160

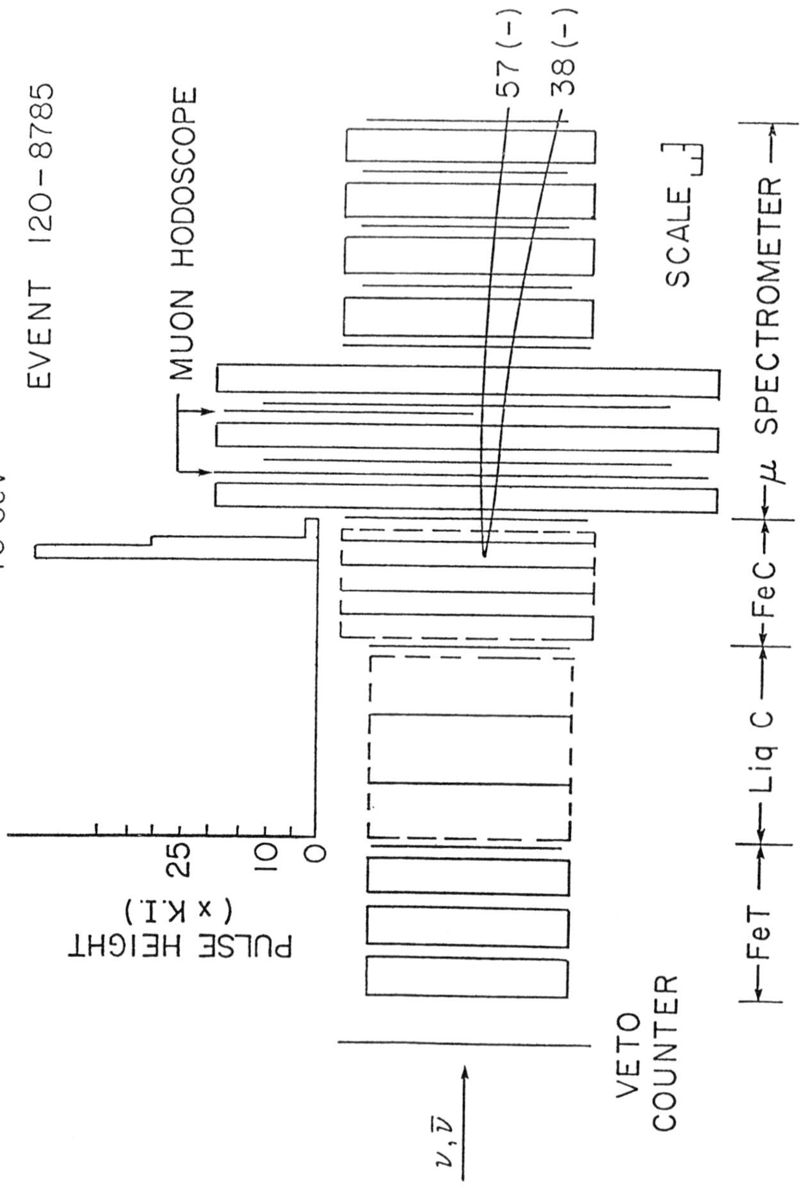
EVENT 120-8785
MUON HODOSCOPE
15 GeV
PULSE HEIGHT
(x K.I.)
25
10
0
57(-)
38(-)
SCALE
VETO
COUNTER
$\nu,\bar{\nu}$
FeT
Liq C
FeC
μ SPECTROMETER

Figure 5

Figure 6
Distribution of the distance between the muon tracks at the vertex for dimuons of opposite and like charges.

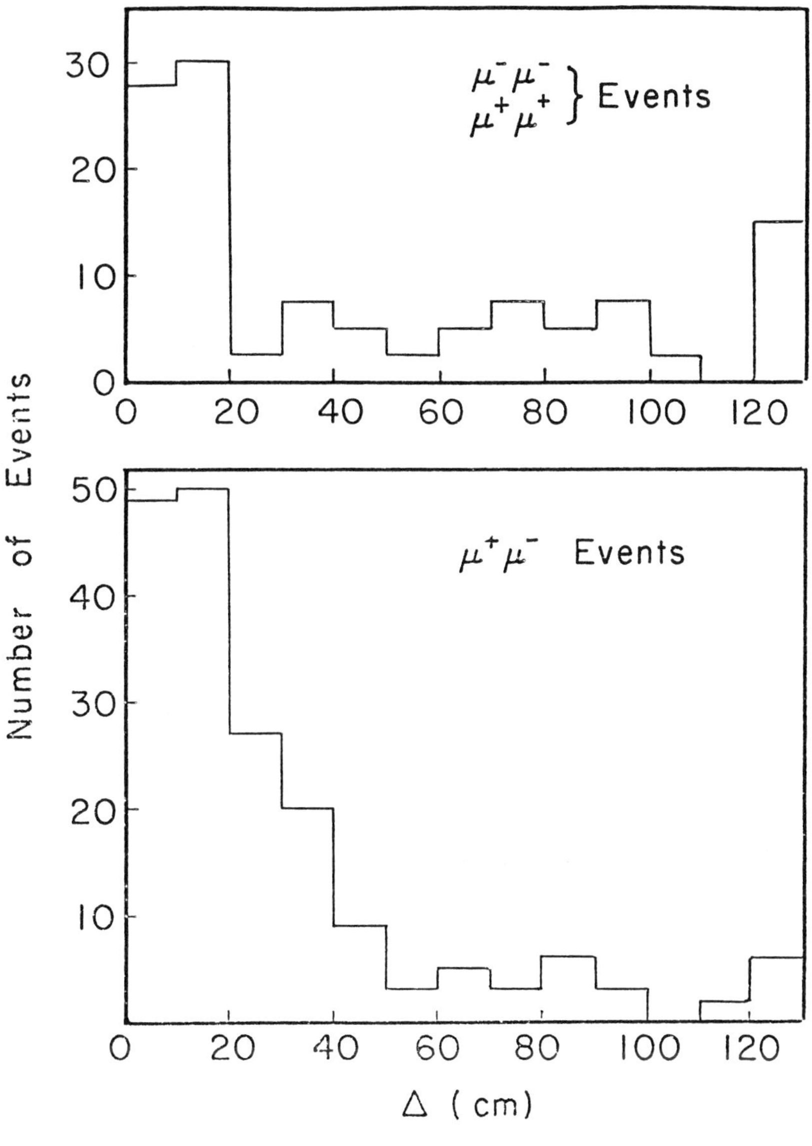

Figure 7

The rate of like sign to unlike sign dimuons

as a function of reciprocal density.

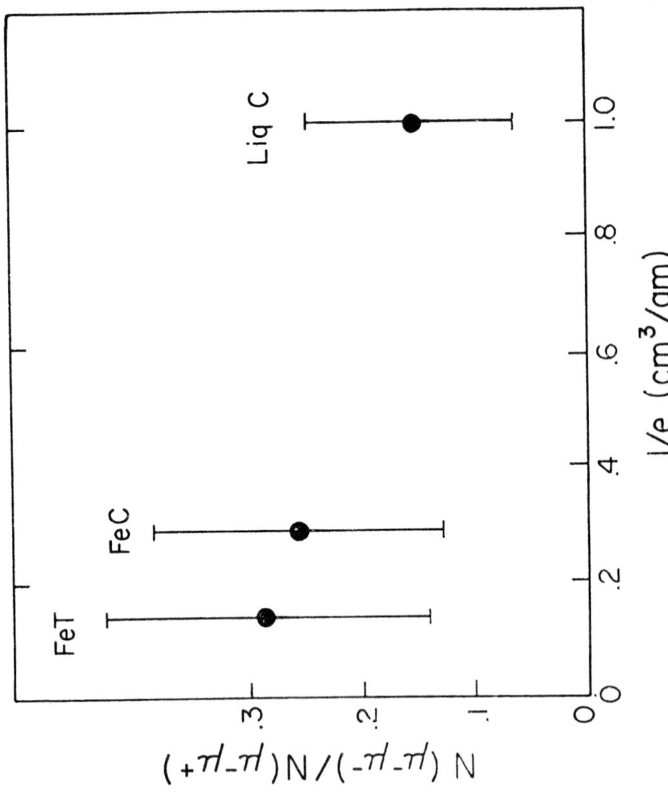

Figure 8

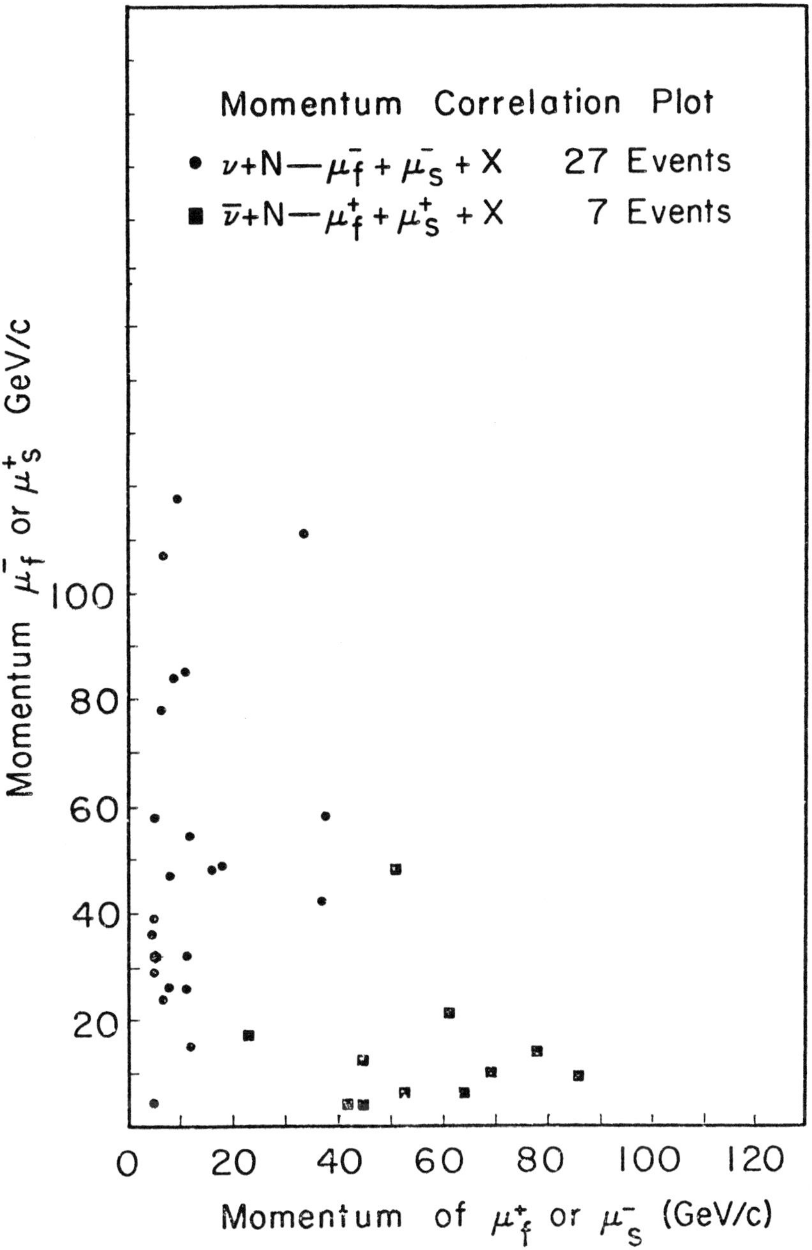

Momentum Correlation Plot
● $\nu + N \rightarrow \mu_f^- + \mu_s^- + X$ 27 Events
■ $\bar{\nu} + N \rightarrow \mu_f^+ + \mu_s^+ + X$ 7 Events
Momentum μ_f^- or μ_s^+ GeV/c
100
80
60
40
20
0
20
40
60
80
100
120
Momentum of μ_f^+ or μ_s^- (GeV/c)

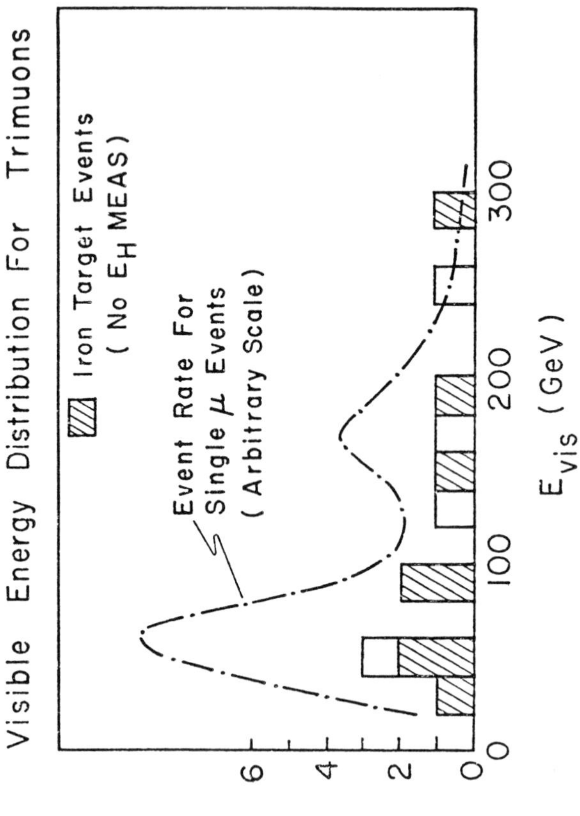
Visible Energy Distribution For Trimuons
Iron Target Events
(No E_H MEAS)
Event Rate For
Single μ Events
(Arbitrary Scale)
0
2
4
6
0
100
200
300
E_{vis} (GeV)

Figure 9

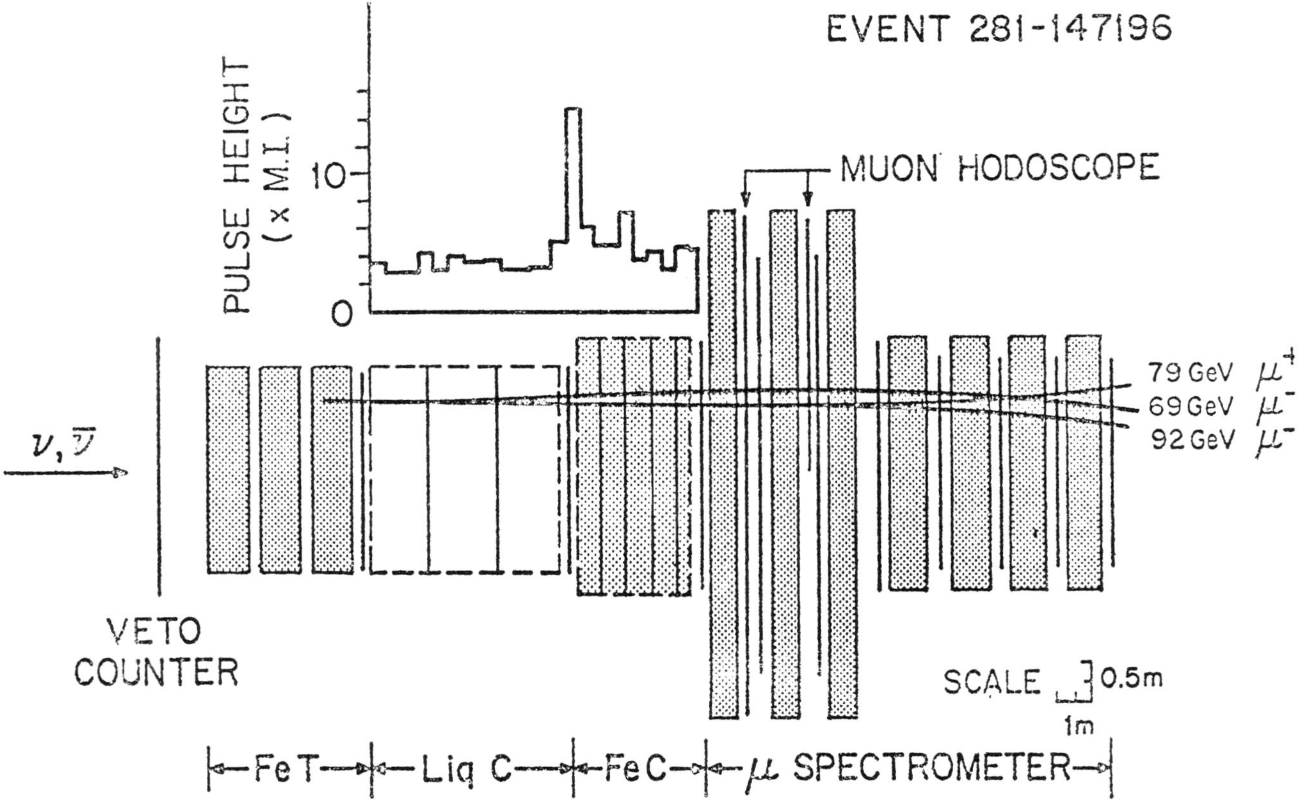
EVENT 281-147196
PULSE HEIGHT (x M.I.)
10
0
MUON HODOSCOPE
$\nu, \bar{\nu}$
VETO COUNTER
79 GeV μ^+
69 GeV μ^-
92 GeV μ^-
SCALE 0.5m 1m
FeT
Liq C
FeC
μ SPECTROMETER

Figure 10

Observation of muon-electron pair production by neutrinos with a visible energy between 3 and 13 GeV[1)]

E. Radermacher

III. Physikalisches Institut, Technische Hochschule Aachen, Aachen, Germany

Abstract

Muon-electron pairs, which are not accompanied by other particles, except recoil protons, were detected in the Aachen-Padova multiplate spark chamber, exposed to the neutrino beam from the CERN PS. The background comes mainly from muonic π^0 production with one decay gamma lost. It was determined from empirical data, together with the small contribution from inelastic ν_e reactions. For electron energies above 2 GeV the expected background is 4 ± 2 events, whereas 12 (μe)-candidates have been observed. Hence, statistically the effect is established, the rate being some 10^{-4}, if one normalizes to all muonic reactions with a visible energy > 3 GeV. From the absence of hadrons, and from the small angles of the two leptons and the high energy of the electron, charm creation as origin of the (μe)-production process can be excluded. It resembles rather the production and decay of a neutral heavy lepton.

The search for μe pairs in neutrino experiments is old. First indications for such pairs were reported by the CERN spark chamber group in 1963[2]. There, the relative frequency, around 10^{-3}, was used to derive a limit of $\geqq$ 1.8 GeV for the mass of the indermediate boson [3,4], which then was thought to be the only source of such lepton pairs.

Today, the production of (μe)-pairs at low energies has been established by the GARGAMELLE Collaboration, both, with[5] and without[6] accompanying V^0. Like the concurrent observations at high energy[7], these ($\mu^- e^+$)-pairs have been interpreted in terms of the production and decay of charmed hadrons[8].

The (μe)-pairs we have observed[9], appear to be of a different character. They have been selected excluding hadrons altogether (save low energy proton recoils). The geometry of the spark chamber, as well as the acceptance criteria, favour small lepton angles with respect to the neutrino direction. Besides, the effect is most pronounced at electron energies $\geqq$ 2 GeV. These properties are not at all typical for charm. They resemble rather the production and decay kinematics of a heavy lepton.

The (μe)-events were detected in the 20 to (fid. volume) multi-plate spark chamber of the Aachen-Padova Group[10]. It was exposed to the CERN PS wide band neutrino beam with a peak energy of 1.5 GeV (see fig. 1). The experimental set-up is shown in fig. 2. 258 Al plates, 2 × 2 m^2 in size, 1 cm thick and spaced by 1 cm, form the "target".

This detector is well suited for the detection of electromagnetic showers, since the radiation length X_0 is only 22 cm in space. Because of its large angular acceptance, and its sizeable length (22 X_0), it is suited as well for the detection of both gammas from a π^0-decay. On the other hand, there is no possibility to recognize from the structure of the

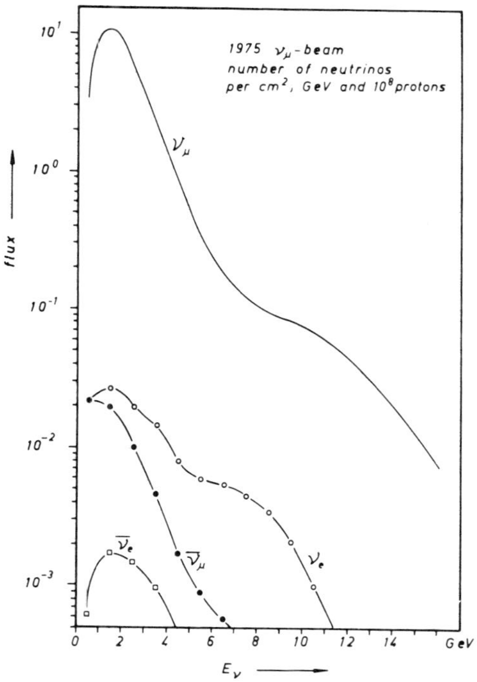

Fig. 1: The CERN PS neutrino spectrum.

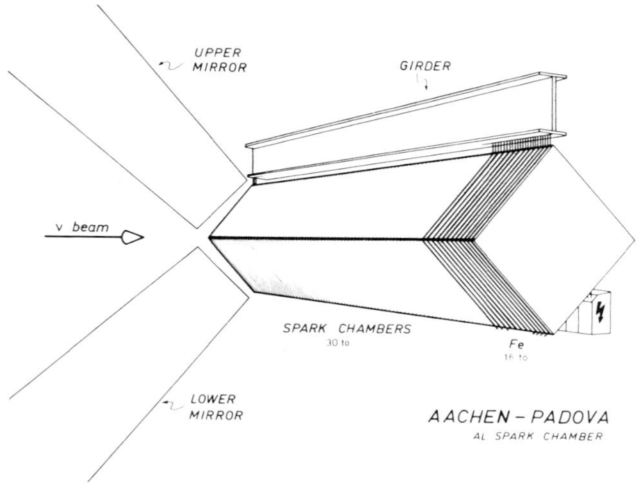

Fig. 2: The Aachen-Padova 30 to Al spark chamber. The 4 cm iron plate sections at the end (IMI) serve to identify muons.

electromagnetic shower, whether it was induced by electrons or positrons, or rather by gamma rays.

The muons are identified from the absence of nuclear interactions over a certain path-length ℓ. The discrimination between muons and hadrons is easy, since the Al part of the chamber is already 10 nuclear interaction lengths Λ_o deep. Another 6 Λ_o are concentrated in the Internal Muon Identifier (IMI) at the end of the set-up (fig. 2) which consists of 4 cm thick iron plates, interleaved with the standard Al-spark chambers.

373000 pictures (60% of the total statistics) from neutrino exposures have been analysed for "naked" (μe)-pairs. A (μe)-candidate was defined as follows:

1) There must be one track with a length $\ell > 4\ \Lambda_o$, showing neither interaction nor scattering in the chamber.

2) There must be an electromagnetic shower with an energy $E_e > 300$ MeV, starting immediately at the vertex, i.e. at the origin of the track.

3) Additional recoils at the common vertex are accepted, provided their length is $\leqq 1\ \Lambda_o$.

There are two types of background to worry about:

A. Charged current ν_μ events with an accompanying π^o, where one decay gamma converts close to the vertex, whereas the other one is lost.

B. Inelastic ν_e-events resulting in a genuine high-energy electron and a pion.

As it turns out, background A is more important. In order to get an experimental handle on it, a control sample was collected which satisfies the same criteria, except that the shower starts at a distance from the vertex.

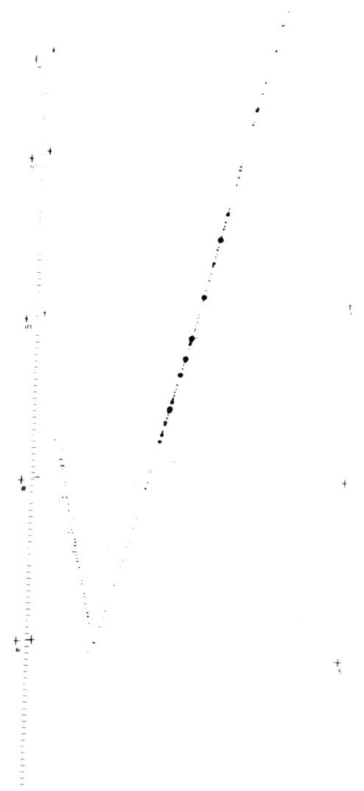

ig. 3: Typical (μe)-candidate in the Aachen-Padova spark chamber.

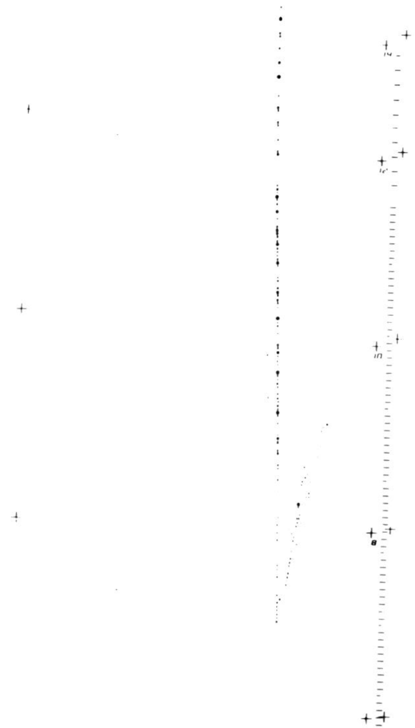

Fig. 4: Single gamma ray, converted at a distance from the muon origin. (A second gamma was probably lost)

Pictures of a (μe)-candidate and a clear ($\mu\gamma$)-background event are shown in fig. 3 and fig. 4, respectively.

The background sample yields a conversion curve, which is shown in fig. 5. Already at this stage of selection the candidate sample, which fills the first bin of the curve, is larger than what one would have expected from extrapolating the background. In fact, the case for an excess in the first bin is stronger than it appears, since sparks may be lost in the first gaps of genuine electron tracks.

In order to test if the experimental ($1\mu + 1\gamma$)-background can really be explained by charged current π^0-production with only one decay-gamma observed, also a sample of events was collected, where the track fulfils selection criteria 1) and where the complete π^0, i.e. two showers are observed. Using this selected events as input for a Monte-Carlo-calculation, the π^0-decay-kinematics was simulated, as well as the γ-conversion-probability, the geometry of the apparatus, and the experimental measurement uncertainties. The result of this ($\mu\gamma$)-background-calculation is displayed in fig. 5 as the smooth curve. It agrees well with the experimental data, if one disregards the first bin (χ^2/ndf = 0.97). Subtraction of this curve and of the Dalitz pairs from the first bin leads to an excess of 15.5 ± 8.4 events, presumably electrons.

The energy distribution of the 62 showers in the first bin of the conversion curve is given in Fig. 6. As compared to the corresponding background curve (dashed line) one recognizes an enhancement at energies above 2 GeV. This observation, together with the knowledge that a contribution of the decay of charmed hadrons in this kinematical region is small[11)], suggests to investigate the conversion curve of events with a shower-energy above 2 GeV. The result is shown in Fig. 7: There are 12 events in the first bin. The Monte-Carlo-calculation (smooth curve) instead predicts a background in the first bin of 3.0 ± 1.9 events. This comprises

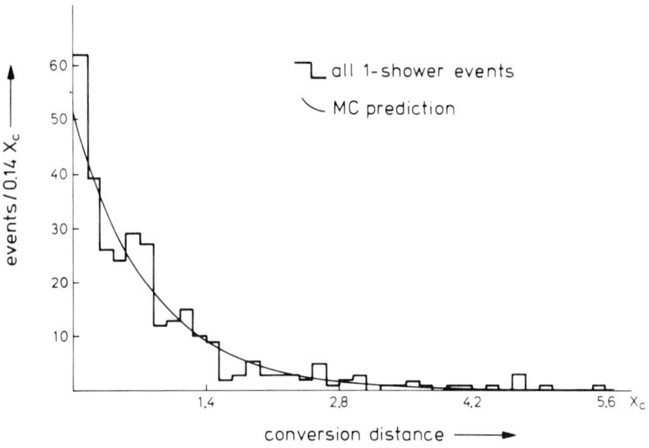

Fig. 5: Conversion curve of all single showers associated to a muon track.

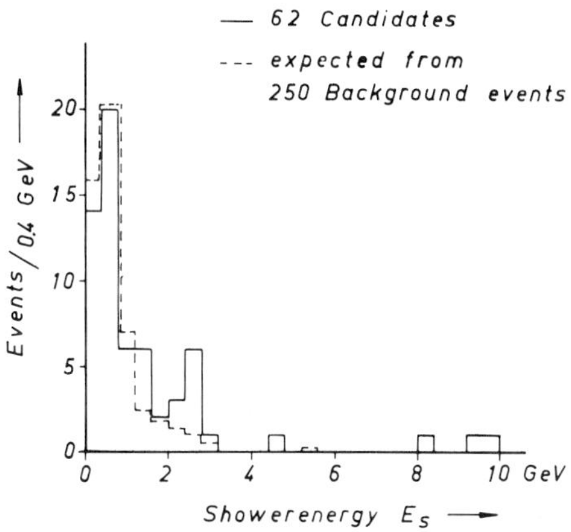

Fig. 6: Energy distribution of showers having no missing shower-spark at the vertex, i.e. of the candidates.

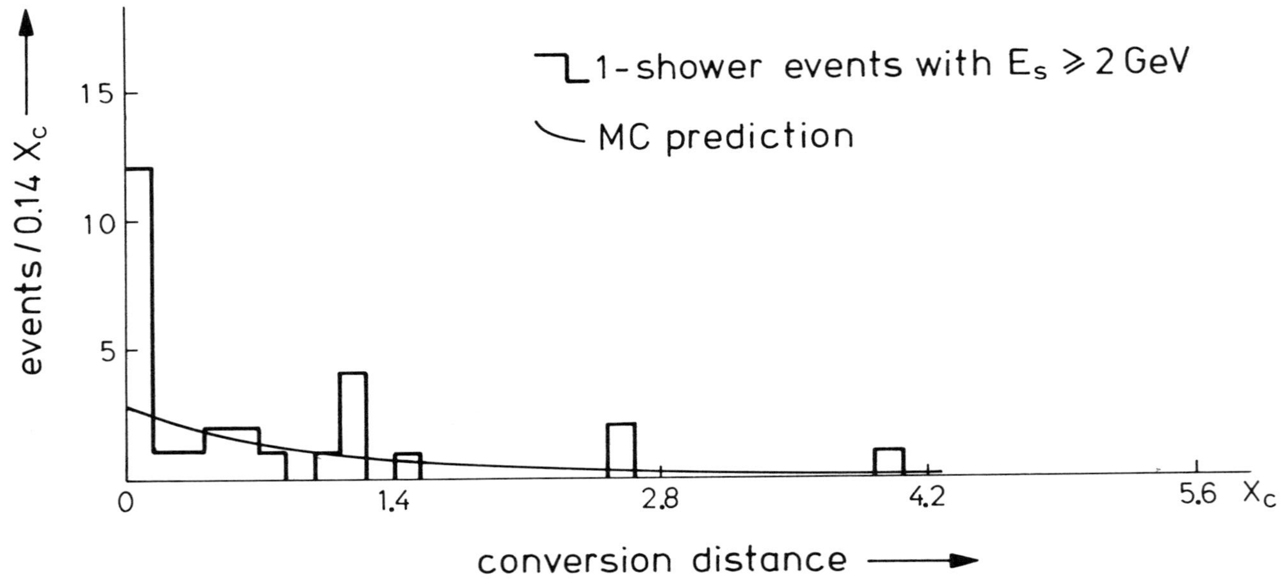

Fig. 7: Conversion curve of single showers, with an energy larger than 2 GeV, associated to a muon track.

the contribution from Dalitz pairs, and combined with the expectation for overlapping gammas from π^o-decay, gives a total background A of 3.0 ± 1.9 events.

In addition, the ν_e contribution, background B, was computed from known production and interaction cross sections. With a typical ν_e event rate of 2% relative to the ν_μ rate, requiring a non-interacting π of $\ell \geqq 4\ \Lambda_o$, and considering the π decay in flight, the expected ν_e background is 1.0 ± 0.5 events.

This leads to a total background of 4.0 ± 2.0 events (s. Table I). The chance that the observed 12 candidates are caused by a statistical fluctuation of this (well known) background is < 1%.

To our knowledge, there are two possible explanations for the remaining μe pairs:

1) Production of charmed hadrons decaying semileptonically. The known production cross sections[12] and branching ratios[8,11,12], as well as the requirement of at least 2 GeV for the positron energy[11] give an expectation of 0.1 ± 0.1 events.

2) Weak trident production. At our low neutrino energy ($<E_\nu>$ = 2.3 GeV) one would expect[13] $\simeq 10^{-2}$ events.

Hence, there are 8 ± 4 genuine neutrino produced dileptons (μe). Thus, they are manifestations of some "new physics". The observed rate corresponds to a raw ratio of (μe)'s to all produced μ + X of 5.7×10^{-5}. Since the visible energy of the observed (μe)-events is > 3 GeV, a more significant figure may be obtained by normalizing to the (computed) number of muonic events above 3 GeV:

$$R_{\mu e}\ (> 3\ \text{GeV}) = 1.6 \times 10^{-4} \quad .$$

Table I: Background to Dilepton-Production

Reaction	Number of events expected	Comment
$\nu_\mu N \rightarrow \mu^- P \pi^o$, $\pi^o \rightarrow \gamma(+\gamma)$	3.0 ± 1.9	Including Dalitz decay and over-lapping 2γ-events
$\nu_e N \rightarrow e^- N n\pi^\pm$ $(n \geq 1)$	1.0 ± 0.5	conservative estimation
$\nu_\mu N \rightarrow \mu^-$ N'X X = Strange Particle decaying via a π^o into γ's	-	excluded by conversion-analysis
total Background	4.0 ± 2.0	

Since we see only part of the muon energy, the (μe)-events observed have probably all total energies > 5 GeV, and hence the ratio

$$R_{\mu e} \; (> 5 \text{ GeV}) = 4.4 \times 10^{-4}$$

is presumably the most meaningful number to be compared with theory.

It seems premature to speculate too much about the origin of the μe pairs. However, as has been emphasized above, they cannot come from charm production on kinematical grounds. Instead, the small momentum transfer to the nucleon, their forward emission with high energies, and the apparent symmetry between electron and muon[14] - all hint at the production of a massive particle, which subsequently decays into the observed leptons. Of the possible schemes, one seems to be particularly fitting, namely the transformation of the primary muon-neutrino, by a neutral current coupling, into a muon-type neutral "ortho-lepton" M^o. This, in turn, decays in the usual charged current induced way into μ^- and a ($e^+\nu_e$)-pair:

$$\nu_\mu + \mathcal{N} \rightarrow M^o + \mathcal{N}' \qquad (1)$$
$$\hookrightarrow \mu^- + e^+ + \nu_e$$

The existence of such "heavy neutrinos" is postulated in several recent gauge models[15]. Moreover, two examples of reaction (1) have apparently been observed, at Serpukhov, in the heavy liquid bubble chamber SKAT[16].

A calculation[17] for reaction (1) was made, assuming a mass of 2 GeV for the M^o. Its result was compared to several kinematical distributions (see fig. 8, 9), where the known measured ($\mu\gamma$) background was properly subtracted from the μe signal. The similarity between measurement and prediction is remarkable.

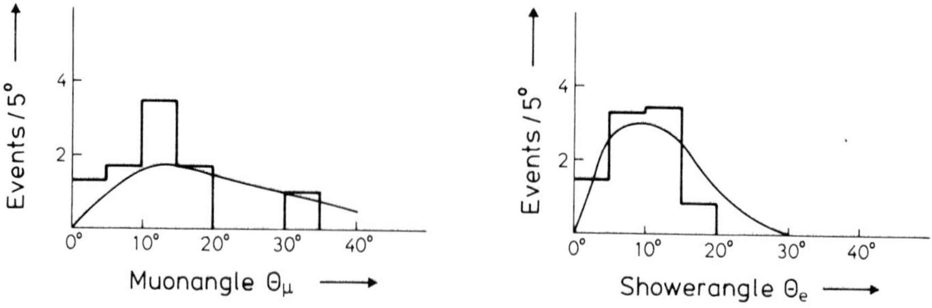

Fig. 8: Muonangle θ_μ and showerangle θ_e, both to ν-direction, of the (μe) signal after subtraction of the ($\mu\gamma$)-background. The curve shows the M^0 prediction (area normalized).

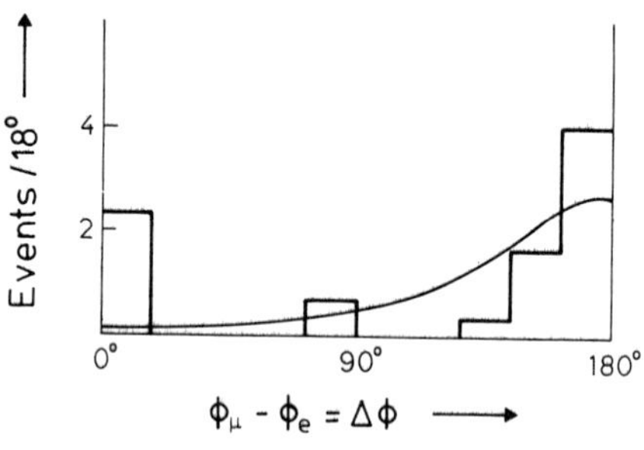

Fig. 9: Difference in azimuthal angle $\Delta\phi$ between muon and electron of the (μe)-signal after subtraction of the ($\mu\gamma$)-background. The curve shows the M^0 prediction (area normalized)

References

1) The experiment was performed by the Aachen-Padova collaboration:
H. Faissner, E. Frenzel, T. Hansl, D. Hoffmann, E. Radermacher, H. Reithler, W. Thomé, H. de Witt, III. Physikalisches Institut, Technische Hochschule Aachen, Aachen, Germany
M. Baldo-Ceolin, F. Bobisut, H. Huzita, M. Loreti, G. Puglierin, Istituto di Fisica Galileo Galilei dell'Universita di Padova, Istituto Nazionale di Fisica Nucleare, Sezione di Padova, Padova, Italy

2) G. Bernardini et al., Proc. Internat. Conf. High Energy Physics Sienna (1963) Vol. 1, p. 571
see also H. Faissner, Acta Physica Austriaca, Supplement I, (1964) p. 189

3) G. Bernardini et al., Phys. Letters 13, 86 (1964)

4) G. Bernardini et al., Nuovo Cimento 38, 608 (1965)

5) J. Blietschau et al., Phys. Letters 60B, 207 (1976)

6) H. Deden et al., Phys. Letters 67B, 474 (1977)

7) J. von Krogh et al., Phys. Rev. Letters 36, 710 (1976)
J. von Krogh et al., Proc. International Neutrino Conference Aachen (1976) p. 41
C. Baltay et al., Proc. International Neutrino Conference Aachen (1976) p. 60

8) see M.K. Gaillard, Proc. International Neutrino Conference Aachen (1976) p. 553 and
M.K. Gaillard, B.W. Lee, and J.L. Rosner, Rev. Mod. Phys. 47, 277 (1975)
L.M. Sehgal and P. Zerwas, Nucl. Phys. B108, 483 (1976)

9) H. Faissner et al., Proc. Internat. Neutrino Conference Elbrus (1977), to be published

10) H. Faissner, Proc. European Phys. Soc. Conf. on High Energy Physics Palermo (1975) p. 291
H. Faissner et al., presented by F. Bobisut, Proc. Internat. Neutrino Conference Aachen (1976) p. 223
H. Reithler, Proc. Internat. Symposium on Lepton and Photon Interactions at High Energies, Hamburg (1977), in press

11) A.J. Buras, CERN TH Preprint 2142 and Nucl. Phys. B109, 373 (1976)
G. Köpp, L.M. Sehgal, P.M. Zerwas, Nucl. Phys. B123, 61 (1977)
D. Rein, L.M. Sehgal, P.M. Zerwas, to be published

12) R.E. Shrock and B.W. Lee, Fermilab-Pub-75/80-THY (November 1975)

13) K. Fujikawa, Annals of Physics 68, 102 (1971)
R.W. Brown, R.H. Hobbs and J. Smith, Phys. Rev. D3, 794 (1971)

14) A. Pais and S.B. Treiman, Phys. Rev. Letters 35, 1206 (1975)

15) see H. Fritzsch, Proc. Int. Summer School Bielefeld (1977), in press

16) D.S. Baranov et al., Phys. Letters 70B, 269 (1977) and Proc. Internat. Neutrino Conference Elbrus (1977), to be published
L.D. Soloviev, Proc. Internat. Symposium on Lepton and Photon Interactions at High Energies, Hamburg (1977), to be published

17) D. Rein, L.M. Sehgal, P.M. Zerwas, to be published

PHENOMENOLOGY OF HEAVY QUARK AND HEAVY LEPTON PRODUCTION BY NEUTRINOS*†

V. Barger
Physics Department, University of Wisconsin, Madison, WI 53706

Charm: State of the Art

The first experimental indication for naked charm was the observation at Fermilab[1,2] of dimuon production by neutrinos

$$\nu N \rightarrow \mu^- \mu^+ X$$

$$\bar{\nu} N \rightarrow \mu^+ \mu^- X$$

with the extra muon signaling the semileptonic decay of the charm quark. Subsequent accumulations of opposite sign $\mu\mu$ and μe events[3-6] by-and-large confirm charm production predictions of the standard quark-parton model for the x and y distributions of the primary muon

$$\frac{d\sigma_c^{\nu N}}{dxdy} \simeq x[2s(x)] + \sin^2\theta_C x[u(x)+d(x)]$$

$$\frac{d\sigma_c^{\bar{\nu} N}}{dxdy} \simeq x[2\bar{s}(x)]$$

in units G^2ME/π. For $\bar{\nu}N$ dimuons, the x_{VIS} distribution is concentrated at small x, as expected for charm production off sea quarks, whereas the νN dimuon x_{VIS} distribution has a strong valence quark component.

The predicted charm dimuon rates at energies well above threshold are

$$\left[\frac{\sigma(2\mu)}{\sigma(\mu)}\right]^{\nu N} \simeq [f_s + \sin^2\theta_C]\ B(c\to\mu)$$

$$\left[\frac{\sigma(2\mu)}{\sigma(\mu)}\right]^{\bar{\nu} N} \simeq 3f_s\ \ B(c\to\mu)$$

where f_s is the momentum fraction carried by the strange sea relative to that carried by valence quarks. For typical estimates $f_s \simeq 0.025$ and branching fraction $B(c\to\mu) \simeq 0.2$, we find

$$\left[\frac{\sigma(2\mu)}{\sigma(\mu)}\right]^{\nu N} \simeq \left[\frac{\sigma(2\mu)}{\sigma(\mu)}\right]^{\bar{\nu} N} \simeq 1.5 \times 10^{-2}$$

These rates are in the range of the measured values, where there are sizeable rate uncertainties associated with undetected secondary leptons below energy acceptance cut-offs. Typical detection efficiency losses are 30-50% in counter experiments.[7]

Calculation of distributions for the secondary lepton requires further input about the fragmentation function $D(z = E_D/\nu)$ of the charm quark into D (or F) mesons and about the semileptonic modes of decay, e.g., $D \to K\ell\nu$, $D \to K^*\ell\nu$. With reasonable inputs about these, the secondary lepton distributions and the primary-secondary lepton correlations are well-reproduced.[7-10]

This success with charm calculations inspires confidence that neutrino production of further new quarks or new leptons can also be estimated reliably.

Trimuons: New Physics Possibilities

The discovery at Fermilab of neutrino produced trimuons[11-13]

$$\nu N \to \mu^-\mu^+\mu^-\ X$$

triggered a wave of theoretical response in the anticipation that new quarks or further new leptons are being produced.[14-33] Ben Lee was deeply involved in these investigations in the months preceding his death.

Most of the new physics proposals to explain trimuons invoke double weak decays. Three categories of these mechanisms are: (i) heavy lepton cascade (LC), (ii) quark cascade (QC), and (iii) lepton-hadron (LH), as illustrated in Fig. 1. The possibility exists that more than one contributes. Assumed masses for the participating new leptons or quarks are guessed from comparisons of predictions with observed invariant masses and p_t-distributions in 3μ events. An exception is the mass of the b-quark, which is identified as the constituent of the Υ(9.5) meson. Typical masses used for the particles in Fig. 1 are

(LC)	M^-	7	GeV
	M^o	2.5	
(QC)	t	8	
	b	4.75	
(LH)	M^o	5	
	b	4.75	

Since the total sample of 3μ events on which these mass estimates are based is small, we must at present be content to explore the general characteristics of the mechanisms; detailed tests will require more data.

Fortunately, the distributions are insensitive to the left or right-handed nature of the decay couplings. Since neutrino to heavy lepton couplings are necessarily left-handed, only the quark couplings (which specify production y-distributions) are essential to calculations.

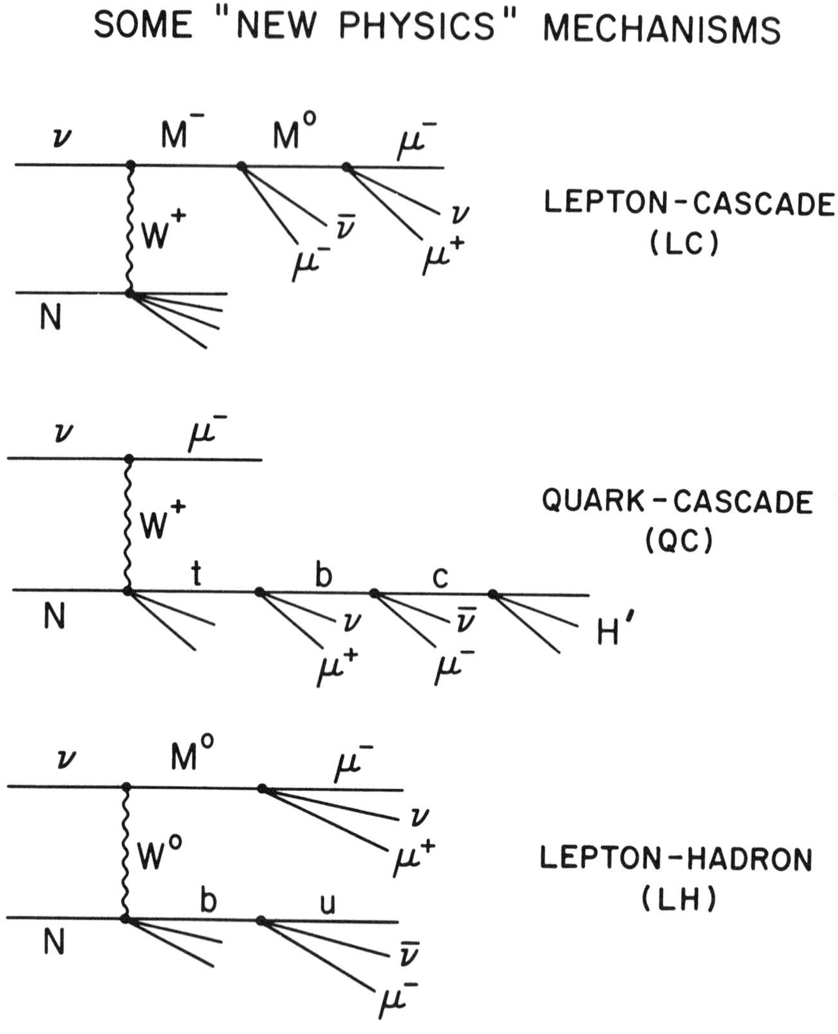
SOME "NEW PHYSICS" MECHANISMS
ν
M⁻
M⁰
μ⁻
W⁺
ν̄
μ⁻
ν
μ⁺
N
LEPTON-CASCADE
(LC)
ν
μ⁻
W⁺
t
b
c
N
ν
μ⁺
ν̄
μ⁻
H′
QUARK-CASCADE
(QC)
ν
M⁰
μ⁻
ν
μ⁺
W⁰
b
u
N
ν̄
μ⁻
LEPTON-HADRON
(LH)

Figure 1

In the following comparisons we assume the quark production transitions

$$
\begin{array}{lll}
\text{(LC)} & d \rightarrow u & L \\
\text{(QC)} & d,s \rightarrow t & L \\
\text{(LH)} & d \rightarrow b & R.
\end{array}
$$

Our QC mechanism is based on the left-handed 6 quark model. For the fragmentation of heavy quarks to new hadrons, we adopt Bjorken's conjecture that[34,35]

$$\langle z \rangle \simeq 1 - \frac{1\ \text{GeV}}{m(\text{quark})} .$$

With this minimal set of assumptions, we can calculate distributions for trimuon events.

Invariant Mass and Energy Distributions

Figure 2 shows results for the $m(3\mu)$, $m(\mu^-\mu^+)$, and $m(\mu^-\mu^-)$ invariant mass distributions. In the QC and LH mechanisms negative muons originate from both lepton and hadron verticies. This causes their invariant mass distributions to be quite broad; 3μ events to populate these high mass regions have not yet been observed.

Figure 3 compares predictions for $E(3\mu)$, the total energy of the three-muon system, based on the Fermilab quadrupole triplet spectrum. Average energies are rather similar in all cases, with

$$
\begin{array}{ll}
\langle E(3\mu) \rangle & \sim 100\ \text{GeV} \\
\langle E(\text{HADRONS}) \rangle & \sim 50\ \text{GeV} \\
\langle E(\text{MISSING}) \rangle & \sim 50\ \text{GeV}
\end{array}
$$

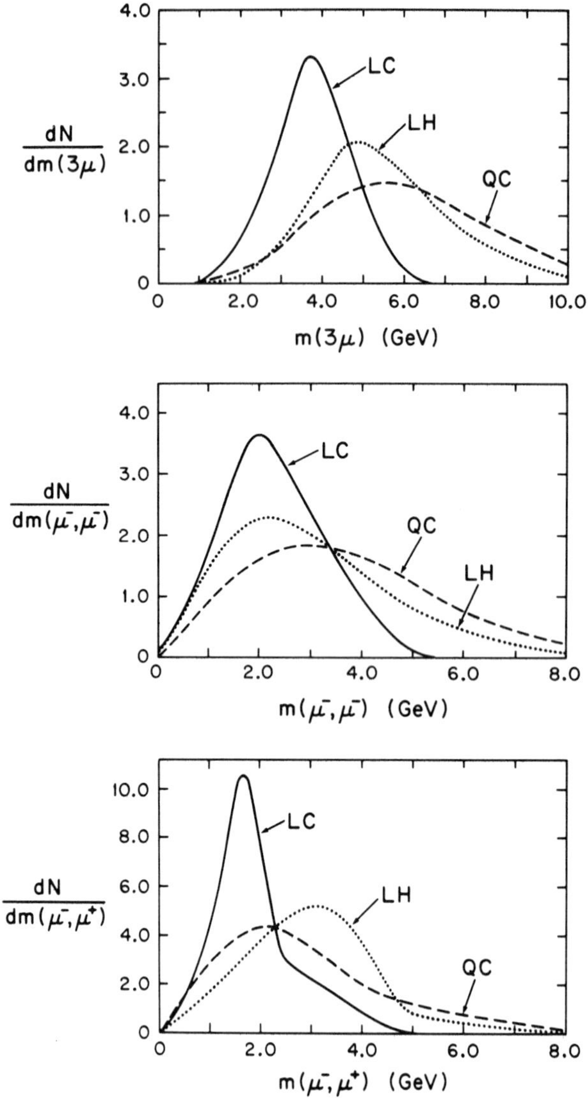
4.0
3.0
2.0
1.0
0
dN
dm(3μ)
LC
LH
QC
0
2.0
4.0
6.0
8.0
10.0
m(3μ) (GeV)
4.0
3.0
2.0
1.0
0
dN
dm(μ⁻,μ⁻)
LC
QC
LH
0
2.0
4.0
6.0
8.0
m(μ⁻,μ⁻) (GeV)
10.0
8.0
6.0
4.0
2.0
0
dN
dm(μ⁻,μ⁺)
LC
LH
QC
0
2.0
4.0
6.0
8.0
m(μ⁻,μ⁺) (GeV)

Figure 2

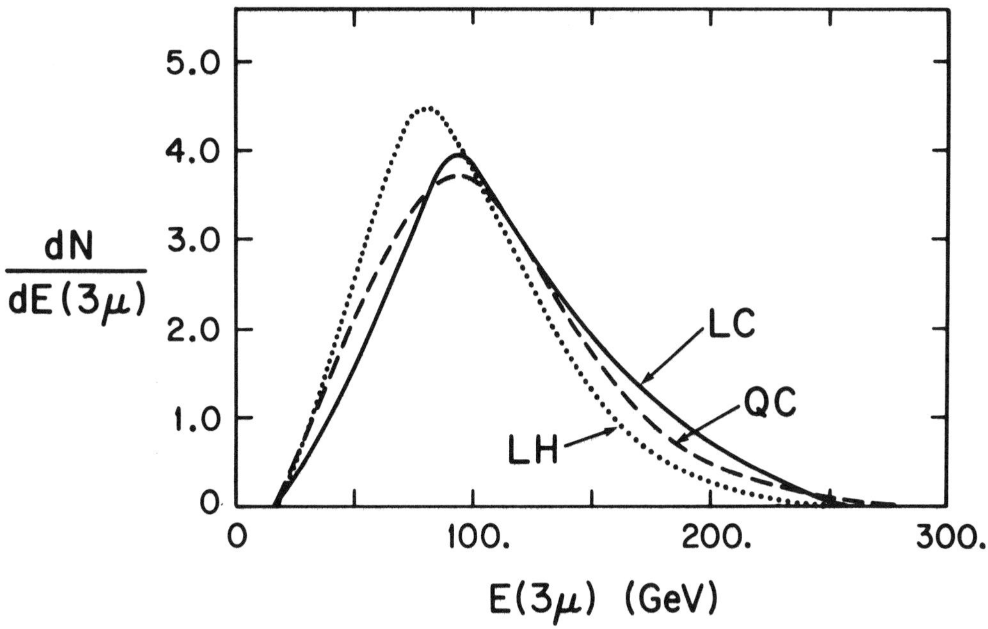

Figure 3

Substantial energy is carried off by decay neutrinos. The $E(3\mu)$ distribution dies off rapidly above 200 GeV, so these mechanisms are improbable explanations of the two exceptional trimuon events with $E(3\mu) \sim 250$ GeV.

In the QC model the $E(\mu^+)$ distribution reflects the t - b mass difference: see Fig. 4. For $m_t = 6$ GeV, the μ^+ is relatively slow. If the t and b quarks are nearly degenerate in mass, the low Q-value for $t \to b$ decay would give a μ^+-energy spectrum far softer than indicated by the current trimuon data.

Trimuon Production Rates

In the double weak decay models the 3μ rate

$$\frac{\sigma(3\mu)}{\sigma(\mu)} = \frac{\sigma_{prod}}{\sigma(\mu)} B(3\mu)$$

is controlled by three factors:

(i) the net branching fraction of the produced leptons or quarks into three muons, typically estimated at

$$B(3\mu) \sim (0.2)^2 \sim 4 \times 10^{-2}$$

[Exception: by a clever arrangement of multiplets and masses so that $M^- \to \mu^o \mu^- \bar{\mu}^o$ and $\mu^o \to \mu^- X$ are the relevant LC decay channels, a branching fraction $B(3\mu) \sim 0.25$ can be achieved.[22]]

(ii) the kinematic threshold suppression for the production of massive particles, typically of order 10^{-1} for the examples under consideration here.

(iii) the weak coupling strengths at the production verticies. In $SU(2) \times U(1)$ gauge models the strengths of the couplings are tightly constrained by experimental limits on the validity of μ - e and

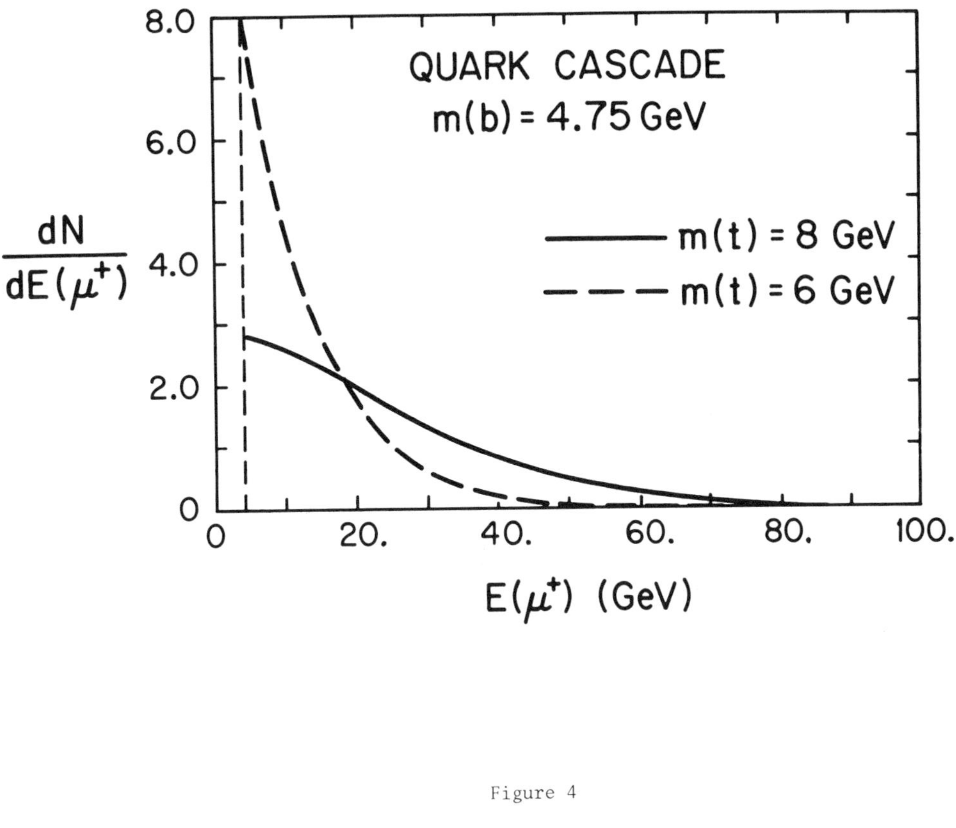
QUARK CASCADE
m(b) = 4.75 GeV
m(t) = 8 GeV
m(t) = 6 GeV
dN/dE(μ+)
8.0
6.0
4.0
2.0
0
0
20.
40.
60.
80.
100.
E(μ+) (GeV)

Figure 4

Cabibbo universality.

For the LC mechanism, the asymptotic ratio of M^- to μ^- production in SU(2) × U(1) models is bounded by

$$\frac{\sigma(\nu \to M^-)}{\sigma(\nu \to \mu^-)} \lesssim 0.04$$

In a QC mechanism based on the 6-quark model with left-hand couplings

$$\begin{pmatrix} u \\ d' \end{pmatrix}_L \qquad \begin{pmatrix} c \\ s' \end{pmatrix}_L \qquad \begin{pmatrix} t \\ b' \end{pmatrix}_L$$

the asymptotic ratio of t to c quark production by νN is given by

$$\frac{\sigma_t}{\sigma_c} \simeq \sin^2\theta_2$$

where θ_2 is one of three Euler angles generalizing the Cabbibo angle. To maintain the original successful estimate of the charm quark mass from the K_L - K_S mass difference calculation, $\sin^2\theta_2$ is bounded by[23]

$$\sin^2\theta_2 \lesssim 0.12 \quad \text{for} \quad m_t = 8 \text{ GeV}$$

Figure 5 illustrates the upper bound on σ_t/σ_c versus E; due to threshold suppression of t-production, the approach to the asymptotic bound is slow.

Translation of the preceding constraints gives typical SU(2) × U(1) rate bounds of

$$\text{(LC)} \qquad \frac{\sigma(3\mu)}{\sigma(\mu)} \lesssim 2 \times 10^{-4} \qquad M^- = 7 \text{ GeV}$$

$$\text{(QC)} \qquad \frac{\sigma(3\mu)}{\sigma(\mu)} \lesssim \begin{matrix} 6 \times 10^{-5} & t = 6 \text{ GeV} \\ 3 \times 10^{-5} & t = 8 \text{ GeV} \end{matrix}$$

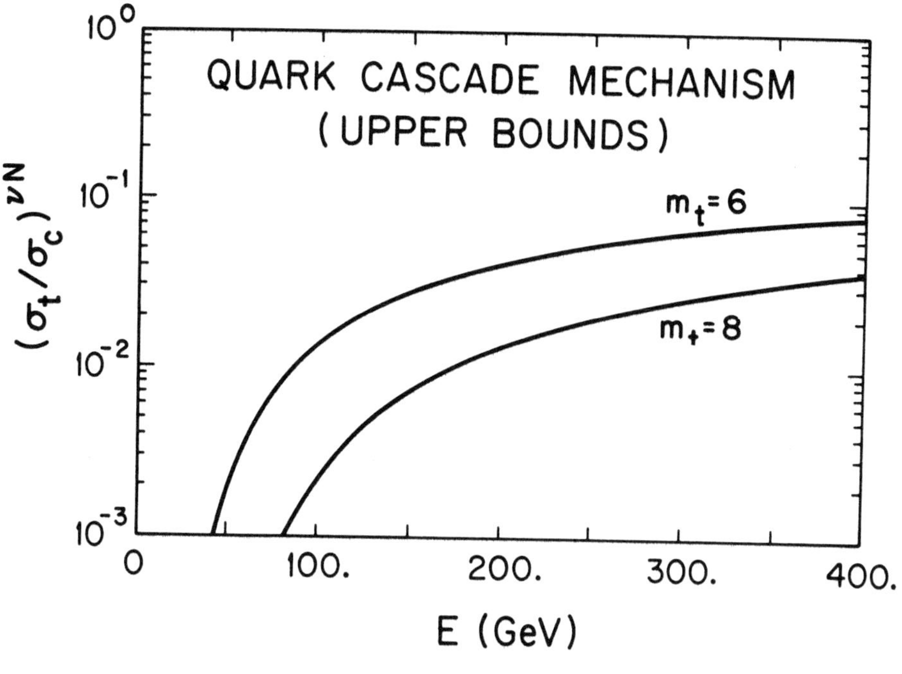
QUARK CASCADE MECHANISM
(UPPER BOUNDS)
$m_t = 6$
$m_t = 8$
10^0
10^{-1}
10^{-2}
10^{-3}
$(\sigma_t/\sigma_c)^{\nu N}$
0
100.
200.
300.
400.
E (GeV)

Figure 5

averaged over the Fermilab quadtriplet spectrum for $E > 100$ GeV and with acceptance cuts $E_\mu > 4$ GeV. However, these constraints can be avoided by going to larger gauge groups, or by clever multiplet choices, as in the LC exception mentioned previously.

The HPWF estimate of their observed rate is

$$\frac{\sigma(3\mu)}{\sigma(\mu)} \sim 5 \times 10^{-4}$$

for $E > 100$ GeV with $E_\mu > 4$ GeV cuts.

Like-Sign Muons and Multimuons

The physics of same-sign dileptons is likely to be closely linked to that of trimuons. Most trimuon mechanisms predict $\mu^-\mu^-$ (and μ^-e^- events) at a rate

$$\frac{\sigma(\mu^-\mu^-)}{\sigma(3\mu)} \sim 2 \text{ to } 4$$

These are obtained in the diagrams of Fig. 1 by replacing the $\mu^+\nu$ vertex by $e^+\nu$ or hadrons.

Tetra and pentalepton events would be an exciting possibility. Theoretical rates of order

$$\frac{\sigma(4\mu)}{\sigma(3\mu)} \sim \frac{1}{5} \qquad \frac{\sigma(5\mu)}{\sigma(3\mu)} \sim \frac{1}{25}$$

can be generated with a cascade mechanism, but lower predictions for these rates are more common.

Antineutrino Trimuons

Knowledge of antineutrino trimuon production rates will help to discriminate among mechanisms. For HPWF spectra with E > 100, the relative rates (without E_μ cuts) for different models are listed below:

	$[\sigma(3\mu)/\sigma(\mu)]^{\bar{\nu}N}/[\sigma(3\mu)/\sigma(\mu)]^{\nu N}$	
LC(d $\rightarrow$ u)	0.9	
LC(d $\rightarrow$ t,V+A)	0.005	(sea/valence suppression)
QC(m_t = 6)	0.8	
QC(m_t = 8)	0.3	
LH	0.9	
RADIATIVE	0.9	

We note that these relative rates are independent of (i) experimental flux normalizations and (ii) any overall theoretical weak current coupling factors. The charges of the muons produced by $\bar{\nu}N$ are mechanism-dependent.

Old-Physics Mechanisms

The diagrams of Fig. 6 yield trimuon final states without recourse to new particle production.

Associated Charm

This contribution to triumon production has recently been calculated from QCD diagrams with a parton emitting a timelike color gluon that subsequently decays into a $\bar{c}c$ pair.[36] The resulting trimuon rate is estimated to be $\sim 3 \times 10^{-6}$. The like sign dimuon rate from associated charm is $\sigma(\mu^-\mu^-) = \sigma(3\mu)/B(c\rightarrow\mu) \sim 2 \times 10^{-5}$. Thus associated charm

SOME "OLD PHYSICS" MECHANISMS

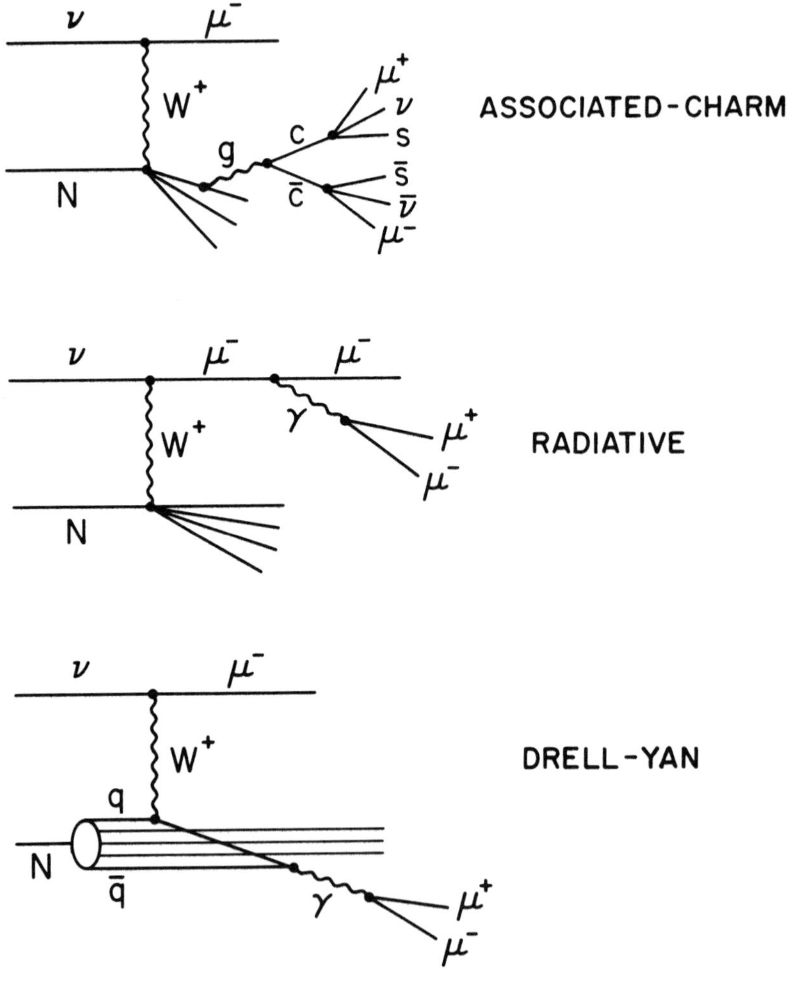

Figure 6

production is unlikely to play a significant role in explaining the trimuon observations. If associated charm had been a dominant source of trimuons, we would have expected a relatively low threshold, which does not seem to be the case experimentally.

Radiative Muon Pairs

Attention is now being focused on radiative muon-pairs as a background source of trimuons.[37,38] For gauge invariance, diagrams with the pair originating from both the primary muon and the parton lines must be included. The radiative calculations indicate non-negligible trimuon rates; for E > 100 GeV and with E_μ > 4 GeV cuts the rate is

$$\frac{\sigma(3\mu)}{\sigma(\mu)} \simeq 0.3 \times 10^{-4}$$

The lower of the two $\mu^+\mu^-$ mass combinations in each event has a distribution that is sharply peaked at very low mass. By restricting attention to trimuon events with $m(\mu^+,\mu^-)_{LOWER} \gtrsim 1.3$ GeV, the radiative background contribution can be largely eliminated, independent of rate considerations. Several of the reported HPWF events survive such a cut. The two exceptionally energetic trimuon events are improbable with the radiative mechanism.

It is worth noting that the radiative process does not lead to $\mu^-\mu^-$ events, except when the μ^+ energy is below the acceptance cutoff. This occurs at the level of 17% of the observed 3μ rate for a E_μ > 4 GeV cutoff, which gives $\sigma(\mu^-\mu^-)/\sigma(\mu) \sim 10^{-5}$ from the radiative mechanism.

The internal Drell-Yan process in Fig. 6 is another electromagnetic source of muon pairs.[39] This is basically a double-scattering process that can be estimated by analogy with Glauber theory. Preliminary estimates[37] indicate a rate $\sigma(3\mu)/\sigma(\mu) \sim 10^{-5}$.

Other Novel Possibilities

The two exceptionally energetic trimuon events with

$$E(3\mu) \sim 250 \text{ GeV}$$

$$E(\text{HADRONS}) < 30 \text{ GeV}$$

are improbable in almost all the mechanisms considered thus far. These energies are at the extreme high energy tail of the incident ν-spectrum, and any mechanism with missing neutrinos in the final state has a difficult time feeding most of the incident energy into the 3μ system. To account for the exceptional events, several novel possibilities have been entertained which have no or little missing energy. One possibility is the internal Drell-Yan process that puts nearly all the beam energy into the trimuons. An investigation of this mechanism is in progress.[37] Other possibilities are illustrated in Fig. 7.

The neutral current decay of a heavy lepton would yield a unique 3μ invariant mass. The present exceptional events have different $m(3\mu)$.

The second possibility in Fig. 7 is a flavor-changing neutral current decay of a massive hadron t. For $m_t \gtrsim 10$ GeV, this diagram leads to muon energies comparable to those observed in the exceptional events. However the neutral current decay is highly unfavored, which makes the 3μ rate from this unpromising.[40]

OTHER NOVEL 3μ MECHANISMS

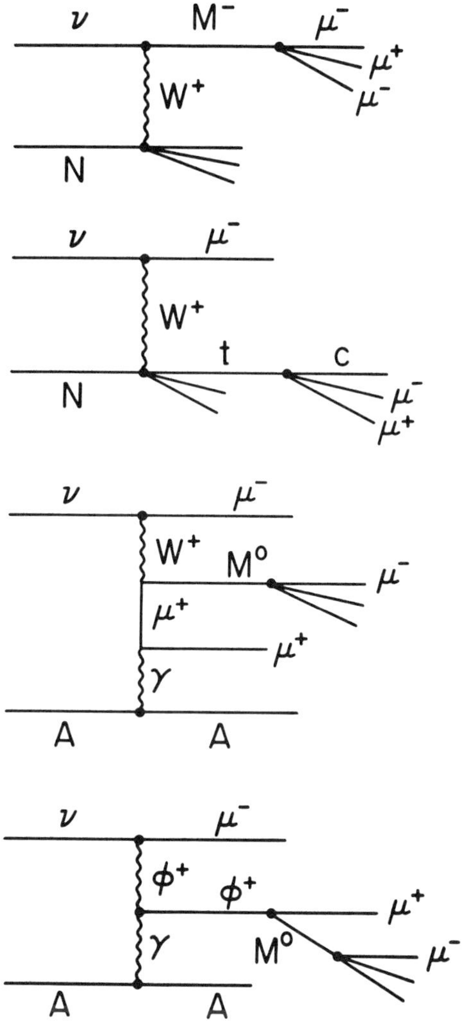

Figure 7

The third speculation[40] in Fig. 7 is coherent electromagnetic production of a heavy neutral lepton with right-hand coupling to μ^-. For $m_{M^o} \sim 1/2$ to 1 GeV the rate is $\sigma(3\mu)/\sigma(\mu) \sim 1.5 \times 10^{-5}$.

The final diagram in Fig. 7 is electromagnetic production of a changed Higgs meson which decays via a neutral heavy lepton.[32]

Summary

The potential that neutrino produced trimuons will yield information about new leptons or quarks remains high. However, with a variety of new and old sources possible, the situation is more complex than at first realized; higher statistics trimuon data (as well as $\mu^-\mu^-$ and possibly multimuon data) are needed for the sifting and winnowing process. More attention will have to be devoted to the trident background from radiative muon pairs. Essentially all mechanisms but lepton cascade predict high invariant masses which have not yet been observed. The two exceptionally energetic trimuon events are not very probable in most of the mechanisms considered, and these may be the key to the most exciting physics to be learned.

Acknowledgement

I am indebted to Tom Gottschalk for his assistance in preparing the material reported in this talk.

References

*Work supported in part by the University of Wisconsin Research Committee with funds granted by the Wisconsin Alumni Research Foundation, and in part by the Energy Research and Development Administration under contract EY-76-C-02-0881, COO-881-7.

1. A. Benvenuti et al., Phys. Rev. Lett. 34, 419 (1975); 35, 1199, 1203, 1249 (1975).
2. B. C. Barish et al., Phys. Rev. Lett. 36, 939 (1976); 39, 981 (1977).
3. J. Von Krogh et al., Phys. Rev. Lett. 36, 710 (1976);
   P. Bosetti et al., Phys. Rev. Lett. 38, 742 (1977).
4. J. Blietschau et al., Phys. Lett. 60B, 207 (1976).
5. C. Baltay et al., Phys. Rev. Lett. 39, 62 (1977).
6. M. Holder et al., Phys. Lett. 69B, 377 (1977).
7. V. Barger, T. Gottschalk, and R. J. N. Phillips, Phys. Rev. D16, 746 (1977); Phys. Lett. 70B, 51 (1977).
8. R. Odorico, CERN Report TH-2360 (1977).
9. L. M. Sehgal and P. M. Zerwas, Nucl. Phys. B108, 483 (1976).
10. E. Derman, Nucl. Phys. B110, 40 (1976).
11. A. Benvenuti et al., Phys. Rev. Lett. 38, 1110, 1183 (1977); report to Int. Symp. on Lepton and Photon Interactions at High Energy, Hamburg (1977).
12. B. C. Barish et al., Phys. Rev. Lett. 38, 577 (1977).
13. M. Holder et al., Phys. Lett. (to be published).
14. C. H. Albright, J. Smith and J. Vermaseran, Phys. Rev. Lett. 38, 1187 (1977); Stony Brook Reports ITP-SB-77-25,32.

15. V. Barger, T. Gottschalk, D. Nanopoulos, J. Abad, and R. Phillips, Phys. Rev. Lett. 38, 1190 (1977); Phys. Rev. D16, 214 (1977); UW-Madison Reports 598, 600, 602, 603, 604 (1977).

16. R. M. Barnett and L. M. Chang, SLAC-PUB-1932 (1977).

17. M. A. B. Beg, R. N. Mohapatra, A. Sirlin, and H.-S. Tsao, Rockefeller Univ. Report COO-22328-133 (1977).

18. F. Bletzacker, H. T. Nieh, and A. Soni, Phys. Rev. Lett. 38, 1241 (1977), E 39, 306 (1977); F. Bletzacker and A. Soni, Stony Brook Report ITP-SB-77-34; F. Bletzacker and H. T. Nieh, ITP-SB-77-42; A. Soni, UCSB Report Th-4-77.

19. G. C. Branco, S. Nandi, H. P. Nilles, and V. Rittenberg, Bonn Report (1977).

20. P. H. Cox and A. Yilding, Harvard Report HUTP-77/A033.

21. V. K. Cung and C. W. Kim, Phys. Lett. 69B, 359 (1977).

22. A. De Rujula, H. Georgi, and S. L. Glashow, Harvard Report HUTP-77/A037.

23. J. Ellis, M. K. Gaillard, D. V. Nanopoulos, and S. Rudaz, CERN Report TH.2346 (1977).

24. H. Harari, SLAC Summer School Lectures (1977).

25. D. Horn and G. G. Ross, Phys. Lett. 69B, 364 (1977).

26. J. Kandaswamy, J. Schechter, and M. Singer, Syracuse Report SU-4210-96 (1977).

27. P. Langacker and G. Segre, Phys. Rev. Lett. 39, 259 (1977); Univ. Penn. Reports UPR-0072T; G. Segre and M. Golshani, UPR-0075T; P. Langacker, G. Segre, and M. Golshani, UPR-0079T.

28. B. W. Lee and S. Weinberg, Phys. Rev. Lett. 38, 1237 (1977).

29. E. L. Lipmanov, JETP Letters 23, 327 (1976).

30. S. Pakvasa, H. Sugawara, and M. Suzuki, LBL Report 6419 (1977).

31. L. M. Sehgal and P. M. Zerwas, Aachen preprint (1977).

32. Y. Tomozawa, Univ. Michigan Reports UM HE 77-20,22.

33. A. Zee, F. Wilczek and S. B. Treiman, Phys. Lett. 68B, 369 (1977).

34. J. D. Bjorken, SLAC-PUB-1992 (1977).

35. M. Suzuki, Phys. Lett. 68B, 164 (1977) and CERN Report TH.2369 (1977).

36. H. Goldberg, SLAC-PUB-2007 (1977); T. Walsh, et al. (to be published).

37. V. Barger, T. Gottschalk, and R. J. N. Phillips (to be published).

38. J. Smith and J. A. M. Vermaseren (to be published).

39. D. Winn (unpublished).

40. V. Barger, T. Gottschalk, P. Langacker, R. J. N. Phillips, and G. Segre (unpublished).

Weak Interactions in Astrophysics

WEAK INTERACTIONS IN ASTROPHYSICS: SUPERNOVAE AND GRAVITATIONAL COLLAPSE

David N. Schramm
Enrico Fermi Institute and
Department of Astronomy and Astrophysics
University of Chicago

Introduction

The purpose of this talk is to review the role that weak interactions play in the final stages of stellar evolution, gravitational collapse and supernovae. We will see that weak interactions are important in determining the composition and structure of massive stars as well as in the dynamics of the final collapse to make a neutron star (or black hole). It should be remembered that the bulk of the released binding energy in making a neutron star or a black hole comes off as neutrinos and that these neutrinos may be important in ejecting the outer part of the star while allowing the central regions to collapse. These points will be reviewed in more detail later in this paper.

For the sake of having a common discussion ground we will talk primarily in terms of the standard "onion-skin" massive star model[1,2,3]. Figure 1 shows a cross section of such a star just before the final core collapse. We will come back to this model in more detail shortly. Let us first briefly mention the various roles that the weak interaction can play in the problems associated with the evolution and explosion of such a star. We will go back and examine each of these roles in more detail.

1. There is a need for a pure leptonic ν-emission (e.g. $e^+ e^- \to \nu \bar{\nu}$) to obtain pre-supernova configurations for massive

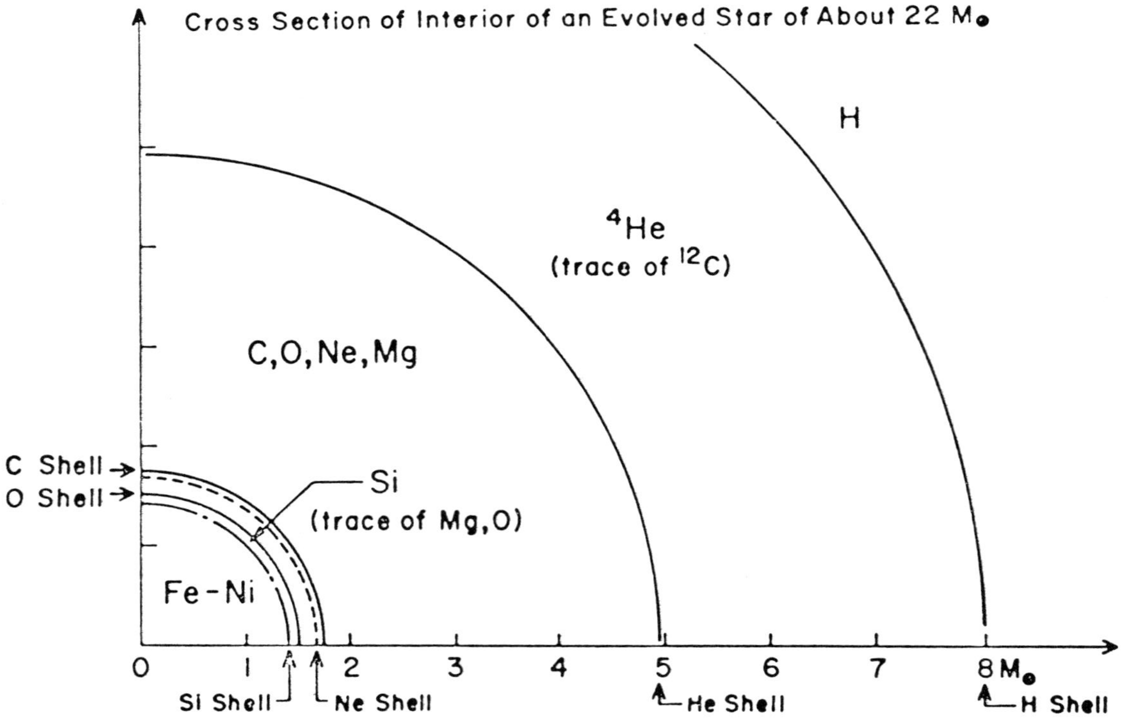

Figure 1. Cross section of an evolved 22 $M_\odot$ star showing the 1.4 $M_\odot$ Fe-Ni core with the surrounding shells of Si, O, Ne, C, He and H envelope.

stars ($M \gtrsim 6\ M_\Theta$) which have thick shells outside the central core. The ejection of these shells is the way in which heavy elements are added to the interstellar medium. For example, if it were not for pure leptonic neutrino emission one could not understand how a supernova such as the Crab Nebula could have ejected so much helium.

2. The ability of neutrinos, through neutral currents, to coherently scatter [4,5] with a cross section $\sigma \propto A^2$ (where A is the atomic mass of the target) is important to momentum transfer during the final stages of gravitational collapse. In particular such scattering may aid in the ejection of the heavy element enriched outer part of the star while enabling the neutronized central region to collapse. This possibility stimulated much recent research [6,7,8] but has yet to come to any decisive conclusion (c.f. ref. 9). It should be noted that coherent scattering depends on the isoscalar component of the neutral current.

3. In addition to its role in coherent scattering, it is also extremely important that the neutral current exists because of simple neutrino-nucleon scattering. This scattering can significantly increase the neutrino opacity and trapping in the neutronized central part of the collapsing core and is important in causing the filling of neutrino phase space [5,10,11,12].

4. The number of neutrino types is also quite important for gravitational collapse, since the more neutrino types, the greater the neutrino flux out of the core and the greater the possible momentum transfer to the mantle [13,9].

5. One extremely important aspect of gravitational collapse is the rate of electron capture. Particularly the competetion of electron capture with photodisintegration for complex nuclei. This competetion

determines the composition and thus the equation of state for the matter during gravitational collapse. This equation of state determines the rate of collapse and the strength and propagation of the hydrodynamic bounce. The nature of this bounce is critical to the ejection of matter and the formation of neutron stars or black holes.

6. One final point which we will merely mention here and not expand upon later, is the possibility that neutrinos mix. That is perhaps ν_e can change to ν_μ to ν_τ to ν_x and back to ν_e on some timescale. This idea has been proposed by Pontecorro (c.f. ref. 14) to explain the solar neutrino situation. In addition, some weak interaction theories with finite mass neutrinos have such mixing. If such mixing does occur, it can effect gravitational collapse, in that an emitted ν_e might change into a ν_μ or ν_τ which has a lower probability of interacting. However, it should be remembered that all types of ν's probably undergo neutral current scattering. Thus it would be difficult to have such mixing make tremendous differences in the mean free paths in the core.

Established "Facts"

Let us now review the "established" results of theoretical calculations regarding supernovae and the advanced stages of stellar evolution. The following points have been obtained in several independent calculations and seem to be agreed on by most of the workers in the field.

A. Stars with $M \gtrsim 8\ M_\odot$ evolve to the "onion-skin" model mentioned earlier (see Figure 1). Just prior to final collapse the star consists of an Fe-Ni core surrounded by concentric shells of ^{28}Si, ^{16}O, ^{20}Ne, ^{12}C, ^{4}He and H with some admixture of ^{24}Mg and ^{32}S in the inner zones. This basic configuration has been found in the calculations of Arnett and Schramm [1], Weaver [15], Iben [16].

It is also found in these calculations that different mass stars will have different mass shells (see Figure 2). All the massive stars have essentially the same 1.4 $M_{\odot}$ Fe-Ni core just prior to final collapse and all have each of the above mentioned shells. However, the lower mass stars (say 15 $M_{\odot}$) have a very rapid density fall-off and therefore have very thin Si, O, Ne and C shells and a thick He shell. Whereas, the more massive stars (say $\sim$ 25 $M_{\odot}$) have much thicker Si, O, Ne and C zones. Since the mass averaged ejecta of these stars will be more like the massive stars[1], one can then easily understand the bulk of the heavy element abundances. Likewise, since the more typical supernova will be at the lower end, we can understand why the Crab Nebula shows such a large He dominance.

B. It is also known that if by some mechanism such a star as described in (A) blows up to become a supernova with the observed energetics then the following 2 things occur:

1. The shock propagating out through the envelope will produce an external light output which looks like that from a type II supernova. This has been shown in calculations by Falk [19], Chevalier [18], Lasher [19].
2. Such a shock will heat up the various shells of the mantle material and cause nuclear reactions. (In addition the high temperatures in the shells just prior to the shock will cause similar effects.)

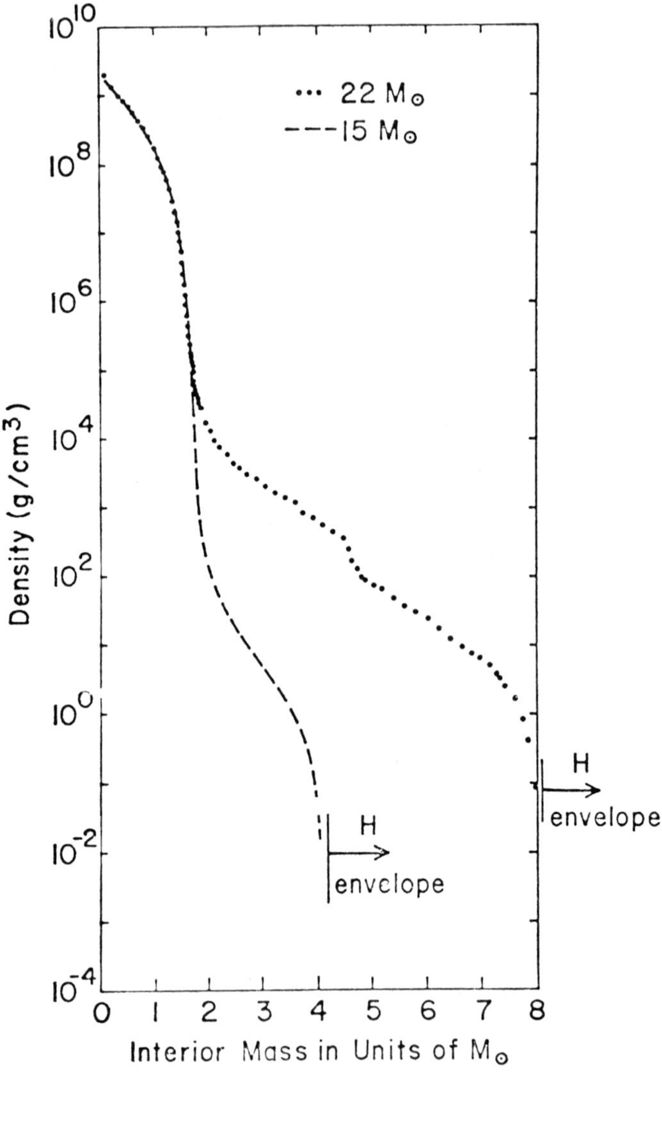

Figure 2. The density profile of a $\sim$ 22 $M_\odot$ and a $\sim$ 15 $M_\odot$ star.

These reactions will process some of the predominantly alpha - particle nuclei into their non - alpha - particle neighbors. This will mean that the ejecta will not only consist of things such as ^{24}Mg and ^{28}Si, but also ^{25}Mg, ^{26}Mg, ^{27}Al and even ^{26}Al. This type of ejection has been described in ref. 1, 3, 15, 20.

C. The bulk of the binding energy of the remnant neutron stars is emitted as neutrinos. It has been well established by several Soviet[38] and American groups [6,7] that neutrinos are the prime way in which energy is lost from the collapsing central core on the way to its becoming a neutron star or black hole. In fact, it has been shown that even for non-radial collapses the neutrinos seem to damp out perturbations faster than other dissipative mechanisms [21]. In particular neutrino radiation seems to be much more important than gravitational radiation [22] for realistic collapse.

Need for Pure Leptonic ν-Emission

It was mentioned earlier that pure leptonic processes such as $e^+ e^- \rightarrow \nu \bar{\nu}$ (or any other ν-pair process - see table 1) are needed in order to get thick shells in the onion-skin model. Let us examine this point in a little more detail.

The important point here is that from the carbon burning stage of evolution to iron core neutronization the prime mechanism by which the core loses energy is via leptonic neutrino emission at each burning stage. The newly formed products build up in the center until they reach some critical mass where they can no longer support themselves and they gravitationally collapse until the core reaches sufficiently high temperatures that it begins to burn.

During the early stages of stellar evolution, before the neutrino losses dominates the critical collapse mass is the maximum mass tha supported isothermally at the temperature of the burning shell. However, once neutrino emission is possible, the energy is able to be extracted from the core and a hydrostatic isothermal core is not able to be maintained. The critical mass is then the maximum mass supported by degenerate electron pressure - the Chandrasekhar mass [23]. For the massive stars, this mass is much less than the mass of an isothermal core. Because of the ability of neutrinos to carry energy out of the core and because of the smaller critical mass, the core is able to evolve much more rapidly. Thus there is still much unburned material in the burning shell when the core begins its collapse to the next core burning stage. This is not the case if ν-emission does not occur. Then the surrounding burning shells are extremely thin and the core is much larger [24].

It is known that the Crab Nebula has a large amount of helium. A massive star seems to only have a thick helium shell if neutrino emission is the prime cooling mechanism.

TABLE 1

Emission	Reference
Fermi Processes	
Electron capture	
$e^- + p \xrightarrow{(W)} n + \nu_e$	Bludman and Van Riper (1977) [24]
Positron capture	
$e^+ + n \xrightarrow{(W)} p + \bar{\nu}_e$	Bludman and Van Riper (1977) [24]
Pair Processes	
Pair annihilation	
$e^+ + e^- \xrightarrow{(W,Z)} \nu + \nu$	Dicus (1972) [25]
Plasma	
$\text{plasmon} \xrightarrow{(W,Z)} \nu + \bar{\nu}$	Dicus (1972) [25]
Photo	
$\gamma + e \xrightarrow{(W,Z)} e + \nu + \bar{\nu}$	Dicus (1972) [25]
Bremsstrahlung	
$e + z \xrightarrow{(W,Z)} e + z + \nu + \bar{\nu}$	Dicus et al. (1976) [26]
Nuclear de-excitation	
$A^* \xrightarrow{(Z)} A + \nu + \bar{\nu}$	Bahcall et al. (1975) [27]

Role of Neutrinos in Final Gravitational Collapse

To see the role neutrinos might play in collapse let us first review what happens to the Fe-Ni core of these massive stars. Let us look at the evolution of the core once it reaches the point of being an iron-nickel core of about 1.4 M_Θ. All of the cores tend to pass through a point with the temperature around 10^{10} °K and the density around 10^{10} g/cm^3. They are supported by degenerate electrons with an adiabatic index $\gamma \approx 4/3$. Around this point four things begin to happen:

A. Electron captures occur, electrons being captured due to the high chemical potential of the electrons. Each electron capture causes a neutrino to be emitted and capturing electrons removes the electrons which are supporting the core.

B. Photodisintegration occurs, breaking the iron and nickel down into alphas and then eventually down into free neutrons and protons.

C. The surrounding shell of silicon continues to burn and thus add more iron and nickel to the core, increasing the core mass to greater than 1.4 $M_{\odot}$.

D. The core is extremely hot, so neutrino pair emission takes place, carrying further energy out of the core.

All four of these processes yield the net effect of the core collapsing. Eventually the collapsing core will switch over from its adiabatic index being dominated by the electrons degeneracy pressure to the pressure being dominated by a thermal neutron gas with a γ closer to 5/3. This transition from a γ near to 4/3 to a stiffer γ will cause a hydrodynamic bounce [33]. It's interesting to note that the mass of the core inside of the bounce region seems to always be greater than about 1.1 $M_{\odot}$. This implies that one can never eject that inner 1.1 $M_{\odot}$, so at least 1.1 $M_{\odot}$ would always go to form a remnant. Thus we could say that the mass of any remnant left must be greater than 1.1 $M_{\odot}$ as a result of the core collapse of these massive stars. This is in good agreement with the observations of the masses of neutron stars [35].

What happens next in the collapse is a matter that is currently under much uncertainty. It is clear that the manner in which the core releases the bulk of its binding energy is via neutrinos. Thus in order to get down to a neutron star, it must release $\sim 10^{53}$ ergs of neutrinos. Whether or not these neutrinos are what cause the ejection of the outer part of the star is a matter of some debate. To see whether this ν-emission is able to have any further effects than merely being an energy sink, we need to review the possible neutrino opacity sources. These are summarized with the exchange current labeled in table 2 both for the core region with atomic weight wear unity and for the mantle with $A > 1$.

TABLE 2
(See review in ref. 29)

A = 1	A > 1
$\nu_e + n \xrightarrow{(W)} p + e^-$	$\nu_e + (N,Z) \xrightarrow{(W)} (N-1, Z+1) + e^-$
$\nu + e \xrightarrow{(W,Z)} \nu + e$	$\nu + e \xrightarrow{(W,Z)} \nu + e$
$\nu + n \xrightarrow{(Z)} \nu + n$	$\nu + A \xrightarrow{(Z)} \nu + A$
$\nu + p \xrightarrow{(Z)} \nu + p$	
$\nu + \nu \overset{(Z)}{\quad} \nu + \nu$	$\nu + \nu \xrightarrow{(Z)} \nu + \nu$

The opacity sources divide themselves into those that go via a charged or a charged and a neutral current, such as a neutrino electron scattering and neutrino absorption by neutrons, and the pure neutral current process, which is neutrino scattering off nucleons. It has been pointed out by Freedman [4] (see ref. 5 or 29 for a complete derivation including non-isoscalar effects) that the cross section for neutrino scattering off complex nuclei scales like atomic weight squared at low energies and thus this mechanism may provide a way to eject the heavy-element-rich mantle while allowing the central neutronized core to collapse on down to make a neutron star. This would explain why the outer part of the star is ejected and the inner part falls in to make a very dense remnant. However, there have been a number of complications in this seemingly nice scenario. One is the fact that ν-nucleon scattering is not pure isoscalar nor pure vector and has a cross section $\sim$ factor 6 or 8 times the A = 1 limit of coherent scattering [5]. This also means that to know the neutrino opacity in the core one needs to know the νn cross section at low energies. Another important complication is the effect of degeneracy [8,10], both the degeneracy of electrons and the degeneracy of neutrinos in the very central core. The electron degeneracy inhibits all reactions that have electrons in the final state, such as the neutrino-electron scattering and the neutrino absorption by neutrons going to a proton plus and an electron. The neutrino degeneracy also causes an inhibition in reactions that have neutrinos in the final state but in scattering processes there is a neutrino in the initial state as well as the final state. Thus if it's completely conservative scattering, the effect can cancel. This has been pointed out in ref. 30, 31. That is, the enhancement in the scattering caused by the initial neutrino state driving the reaction in a forward direction balances out the inhibition caused by neutrinos in the final state. This cancellation of the inhibition effects due to the neutrino being both in the forward and backward channel occurs only when one has conservative scattering; that is, scattering where there is no energy change or only a very small energy change for the neutrino. This is the case for the neutrino scattering off nuclei and off nucleons. However, in the neutrino electron scattering, one has very nonconservative scattering where the neutrino can gain or lose large amounts of energy in the scattering [5,12]. Therefore the cancellation of the degeneracy effects does not occur and the situation is much more complex. However, in the central core regions of the collapsing stars, the conservative neutrino scattering off nucleons dominates. The role of degeneracy is extremely important because it determines the energy spectrum of neutrinos coming out of the core as well as their ability to come out of the core. Since neutrino cross sections go roughly like E^2, the energy of the neutrinos coming out of the core is extremely important.

The effect of all of these processes is to make the following of the detailed neutrino transport in the core a very difficult problem. It is currently being attacked by a variety of methods from simple diffusion to flux-limited diffusion to complete S_n diffusion by Yueh and Buchler [32] to complete Monte Carlo treatment by Tubbs [12]. It does seem that the central core has extremely large neutrino opacities which would make some form of diffusion a very good approximation. However,the region just outside of the central core where the neutrinos have a mean free path comparable to the size of the region is the region where, if the neutrinos are going to do significant transfer, they must do it and it is that region where it is extremely complicated to follow the neutrino transport [7]. However, in the past few years it has been noted that the equally important question is the core where the luminosity is determined. It should also be noted that while neutrino flow might not be important in the central core, the bulk neutrino pressure may have an important effect on the equation of state in that region.

Bruenn, Arnett and Schramm [33] have described the criterion for the three standard mechanisms for mass ejection in collapsing non-rotating stellar cores. These mechanisms are:

1. Neutrino Energy Transport
2. Neutrino Momentum Transport
3. Hydrodynamic Bounce

The first of these, neutrino energy transport, was originally used by Colgate and White [34], however, it tends to require more energetic neutrinos to be effective, thus the probable mechanisms reduce in the current models to #2 and #3.

It has been shown [34,35] that the critical question as to whether the neutrinos are able to drive off the outer material via momentum transport is whether or not the neutrino luminosity coming out of the central core region is greater than the Eddington Limit for neutrinos. We can estimate the Eddington Limit Luminosity in the following way [35]

$$L_{\nu}^{Edd} \approx \frac{4\pi GMc}{\kappa_{\nu} \langle 1-\cos\Theta \rangle}$$

where M is the interior mass, κ_{ν} is the neutrino opacity in the shell and $\langle 1-\cos\Theta \rangle$ is the average scattering angle.

The neutrino opacity in the shell is dominated by the Freedman coherent scattering process. If the shell has a composition of heavy elements it is very easy to estimate its opacity and thus calculate the neutrino luminosity necessary to push it off. This luminosity is approximately 10^{54} ergs/sec. If one merely estimates the number of neutrinos coming from electron capture alone, we can see that that number is the order of 10^{57} neutrinos. The typical energy of these neutrinos is the order of 15-20 MeV. If they were emitted with the freefall time for a core at 2×10^{11} g/cm^3, then the neutrino luminosity would indeed be greater than the Eddington Limit. However, we know that they are not emitted over a freefall time since the core is collapsing somewhat slower than freefall In fact, due to the diffusion out of the core taking a finite time, the timescale for which the neutrinos are escaping is quite a bit longer than freefall. The question is the timescale over which it takes the neutrinos to get out. If they escape rapidly then in fact we will have a luminosity near the Eddington Limit and the neutrinos can cause ejection.

Another way to augment the luminosity is via neutrinos from heavy leptons since these neutrinos will be produced in equal numbers with muon neutrinos in the pair processes. The more types of neutrinos there are, the greater the neutrino luminosity would be.

The relative importance of each of the above effects varies from one calculation to the next depending primarily on what equation of state is used.

Figure 3 shows the neutrino time spectrum of typical gravitational collapse calculations run by Wilson [13]. Note that initially there is a burst of the electron neutrinos which gradually diffuse out of the core followed by the emission of the pair neutrinos, the muon neutrinos and the heavy lepton neutrinos. The anti-electron neutrinos are suppressed for a while due to the fact that the electron-neutrino phase space in the core has been filled by the neutrinos from electron capture. Only after these neutrinos from electron capture have been able to diffuse out can the pair processes which

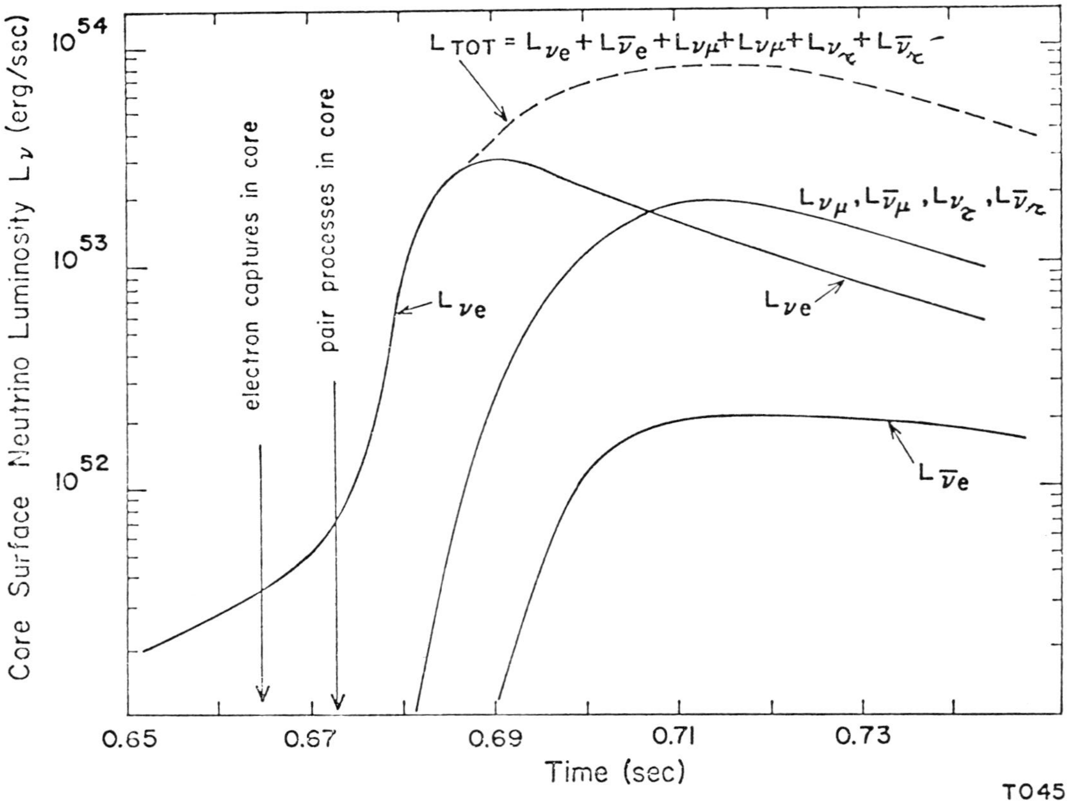

Figure 3. The neutrino luminosity versus time for a typical gravitational collapse calculation. This is taken from the work of Wilson [13].

produce electron-anti-neutrinos in pairs with electron neutrinos be successful in producing anti-electron neutrinos. The typical neutrino energies are ∿ 10 MeV.

Figure 4 shows the trajectories of mass zones as a function of time for a particular model of Wilson's [13] which had mass ejection. Notice how the zones fall to a bounce point and undergo a hydrodynamic bounce. In this particular model, as the bounce material reached its peak and was about to fall back in, it was hit by the neutrino flux which had just then diffused out and was at the Eddington Limit and was able to drive off the material before it fell back onto the core. There was thus ejection and a supernova occurred. Due to the large uncertainties in the equation of state which is what determines the strength and location of the bounce and due to some uncertainties still present in the neutrino transport out of the core, it is clear that the calculations are by no means definitive at the present time.

Jim Wilson's [13] best current calculations are now showing ejection via a hydrodynamic bounce with the neutrinos serving no major role. Because of the sensitivity of mass ejection results to the equation of state, Ken Van Riper of Chicago has carried out a series of hydrodynamic calculations using a parameterized equation of state. He uses an effective adiabatic index γ_{min} to describe the equation of state at sub-bounce densities and γ_{max} at the higher densities.

Figure 5 shows the trajectories with $\gamma_{min} = 1.24$ and $\gamma_{max} = 1.38$. Notice that the shock damps itself out. Figure 4 shows the trajectories $\gamma_{min} = 1.33$ and $\gamma_{max} = 1.38$. In this latter case a strong shock develops and ejects the outer part of the star via hydrodynamic bounce with no neutrino aid. Van Riper finds that if γ_{min} is greater than about 1.28, the

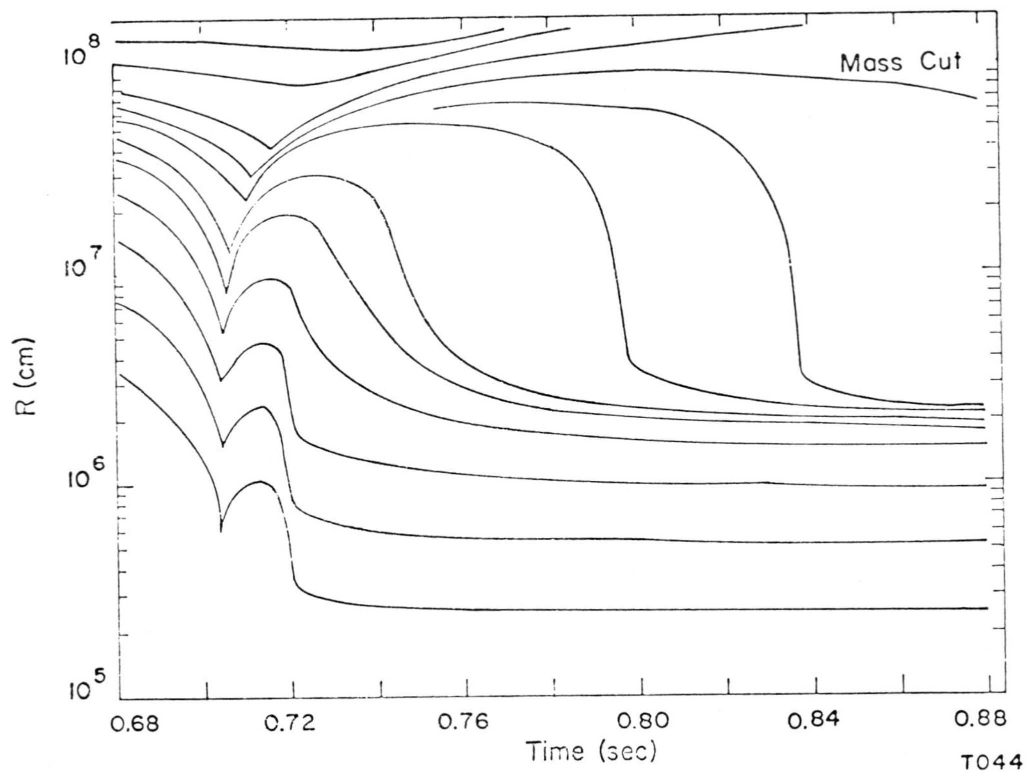

Figure 4. The trajectories of various mass zones in a calculation of Wilson's with the coherent scattering opacity arbitrarily increased by a factor of ~3. The mass ejection occurs when the shocked mantle material is hit by the neutrino flux (compare times with Fig. 3) before it can fall back on the core.

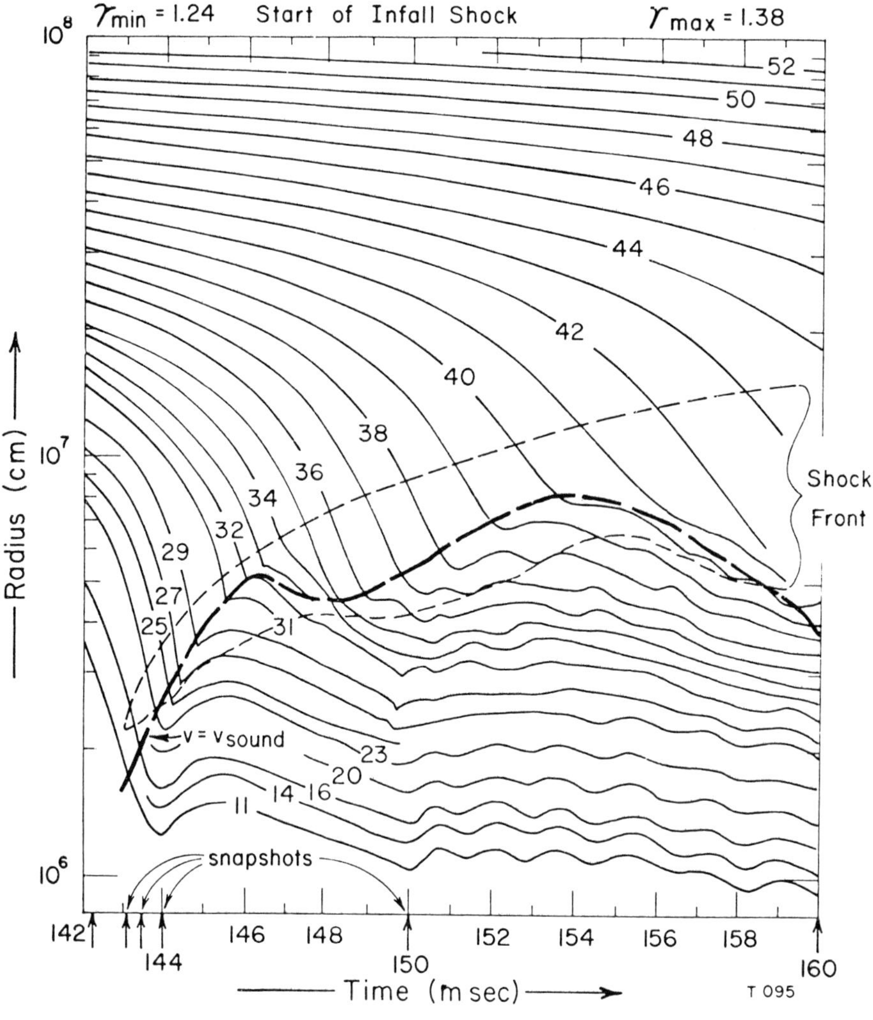

Figure 5. Van Riper's calculations with γ_{min} = 1.24 and γ_{max} = 1.38. Notice that the shock damps itself out.

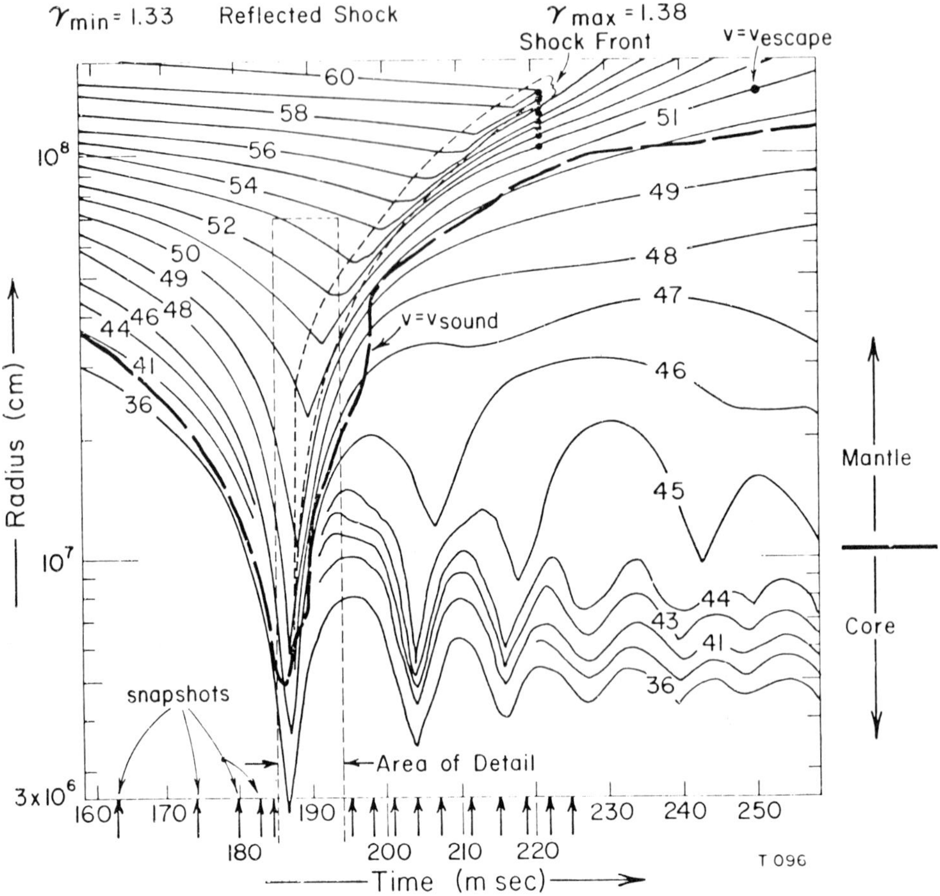

Figure 6. Van Riper's calculations for γ_{min} = 1.33 and γ_{max} = 1.38. Note the strong reflected shock.

shock tends to damp out as it tries to traverse this region. (Van Riper's calculations do not include neutrino damping, so his cores tend to oscillate a good deal after the initial bounce. However, even without neutrino damping, he finds the oscillations tend to damp out, due to acoustical damping prior to gravitational radiation becoming more important).

At the present time there exist many equation of state uncertainties, so the true effective value of γ_{min} and γ_{max} are unknown. Therefore at the present time a strong statement cannot be made one way or the other with regard to whether or not ejection occurs via neutrinos and/or via a bounce or via some other mechanism in the core.

The major problem with collapse calculations is the equation of state of the high density (10^{10} - 10^{15} g/cm^3) - high temperature (10^{10} -10^{12} K) matter during the collapse. In particular, what is the composition as a function of time in the core. To determine this composition one needs to examine photodisintegration and electron capture rates for the collapsing material. One worry is that very exotic neutron-rich nuclei may get created by electron capture before photodisintegration occurs. If that were to be the case then the solution of the problem may be a long way off because so little is understood about the physics far from β-stability, particularly for low mass nuclei where the droplet models break down.

In addition to the nuclear composition effects on the equation of state, the neutrinos themselves can have significant effects on the pressure due to neutrino trapping in the core.

Although there are many uncertainties, there are some recent rays of hope. One of these comes from Willy Fowler, who points out that the nuclear partition functions used in most of the calculations have not adequately treated the continuum. In fact the simple partition function

$$\sum_i (2J_i + 1)\, e^{-E_i/kT}$$

which is used in most nuclear statistical equilibrium calculations, diverges at high T for standard density of states assumptions. Whereas an accurate partition function should go to zero as $T \to \infty$, since a complex nucleus should breakdown to free nucleons. The proper treatment should,in analogy to atoms, treat the continuum in such a way as to have its contribution dominate over the bound states at high temperature. The direction of such a correction will be:

1. To have more free particles and thus have a stiffer equation of state and thus have better shock propagation and

2. The breakdown may be fast enough to prevent getting too far from stability. Thus we may have a better chance of actually determining the equation of state.

Another ray of hope comes from the Monte Carlo neutrino transport calculations of Dave Tubbs [12], where it was found that some of the worries about having to use complicated neutrino transport treatments are unnecessary. In particular Tubbs has shown that

1. The central core [out to within a few mean free paths of the surface ($\sim 10^{11}$ g/cm^3)] can be well approximated by a Fermi-Dirac "black body" neutrino radiator. However, in the important ν-photosphere the ν-distribution is not fully degenerate.

2. The high energy part of the ν-distribution equilibrates first.

3. The νn scattering can be as important as the νe scattering in energy redistribution. This occurs because there are many more νn

scatterings then νe scatterings thus even though each one has a small energy change, the total effect can be large.

4. In the ν-photosphere, neutronization and it's inverse ν-absorption dominate. This makes the transport easier to approximate than if scattering dominated.

Summary and Conclusions

We can summarize the problems in understanding core collapse as follows:

0. The problem of understanding the Presupernova Evolution; that is, the relative composition of the different zones in the star and the density profile of the core and the material just outside of the core. This can be strongly influenced by how convection is calculated.

1. The Equation of State, including the composition as the core collapses, in particular taking into account the rate of electron captures on complex nuclei and the rate of photodisintegration of these complex nuclei. This can make very large changes since, if one finds large amounts of heavy nuclei in the central core then there would be a significant increase in the neutrino opacity in the central core and this would significantly change the equation of state and the pressure balance in the central core. A very important point here with regard to the equation of state is the density at which the core bounce takes place and the temperature of the material at the core bounce. This is extremely sensitive to the equation of state and composition.

2. The second point is the Weak Interaction Theory, not just the strength of the isoscalar component of the neutral current which determines the strength of the coherent scattering opacity, but also the number of types of neutrinos. As far as the coherent scattering is concerned it doesn't matter what type of neutrino comes out since they all interact via a neutral current.

3. The Neutrino Transport. This must include the effects of both neutrino and electron degeneracy and the effect of conservative and non-conservative scattering to be able to follow the neutrinos coming out of the core both in their flux and in their energy. The energy is extremely important due to the sensitive dependence of neutrino cross sections upon the energy of the neutrino.

4. The Effects of Rotation and Magnetic Fields. All the calculations reported in this paper have been with spherically symmetric, non-rotating collapsing stars. No rotation or magnetic fields have been mentioned. It seems clear that if rotation and magnetic fields are added, one could change the nature of the problem. In fact, in calculations by LeBlanc and Wilson [36] and a follow-up on that by Meier et al.[37], an ejection of a very different nature was obtained due to the magnetic field winding up and ejecting jets out of the poles during a collapse. The effects of rotation and magnetic fields on the pressure balance, etc., have not been explored in any sort of detail in any of the present calculations.

So we see that there are many problems in the present core collapse, but these problems, although they affect the final stages of the collapse and whether or not mass is ejected, they are of relatively little importance to the main fact that neutrinos are emitted in copious numbers from collapse.

The basic fact is that the weak interaction is extremely important to the final stages of stellar evolution. Without a pure leptonic weak interaction with some neutral current component our calculations would be quite different. However, the reverse situation of using the astrophysics of collapse to make constraints on the weak interaction is a long way off because there are still too many other uncertainties in the calculations which could mask most effects.

Acknowledgments

I would like to thank Dave Arnett, Ken Van Riper, Jim Wilson, Dave Tubbs and Demos Kazanas for valuable discussions. This research was supported in part by NSF grant AST 76-21707 at the University of Chicago.

References

1. Arnett, W.D. and Schramm, D.N. 1973a, Astrophys. J. 184, L47

2. Arnett, W.D. 1973a, in Explosive Nucleosynthesis ed. by D. N. Schramm and D. W. Arnett (University of Texas Press: Austin)

3. Schramm, D.N. and Arnett, W.D. 1975a, Mercury, May/June, 16

4. Freedman, D.Z. 1974, Phys. Rev. D9, 1389

5. Tubbs, D. and Schramm, D.N. 1975, Astrophys. J. 201, 467

6. Wilson, J.R. 1974, Phys. Rev. Lett. 32, 849

7. Schramm, D.N. and Arnett, W.D. 1975b, Astrophys. J. 198, 628

8. Schramm, D.N. and Arnett, W.D. 1975c, Phys. Rev. Lett. 34, 113

9. Schramm, D.N. 1978, Proc. in "Neutrino-77", Caucausus, U.S.S.R., Sov. Acad. of Sci.

10. Mazurek, T. 1975, Astrophys. and Space Sci.

11. Arnett, W.D. 1978, Astrophys. J. In press

12. Tubbs, D. 1977, Fermi Institute preprint

13. Wilson, J.R. 1976, Private communication and talk at Aspen Astrophysics Workshop

14. Pontecorro, B. 1977, Proc. "Neutrino-77", Caucausus, U.S.S.R., Sov. Acad. of Sci.

15. Weaver, T. 1977, presented at Santa Cruz Supernova Workshop. Preprint

16. Iben, I. 1977, in Advanced Stages of Stellar Evolution, Iben, Renzini and Schramm (Observatoire de Geneve, Switzerland)

17. Falk, S. and Arnett, W.D. 1977, Astrophys. J.

18. Chevalier, R. 1977, in Supernovae, ed. D. N. Schramm (Reidel: Dordrecht, Holland)

19. Lasher, G. 1977, in Supernovae, loc. cit.

20. Hainebach, K., Norman, E.G. and Schramm, D.N. 1976, Astrophys. J. 203, 245

21. Kazanas, D. and Schramm, D.N. 1976, Nature 267, 671

22. Shapiro, S. 1977, Cornell University preprint

23. Chandrasekhar, S. 1931, Astrophys. J.

24. Sugimoto, R. 1973, Proc. IAU Symp. on Advanced Stages of Stellar Evolution, Warsaw, Poland

25. Dicus, D. 1972, Phys. Rev. 133B, 1046

26. Dicus, D., Kolb, F., Schramm, D.N. and Tubbs, D. 1976, Astrophys. J.

27. Bahcall, J., Trieman, S. and Zee, A. 1975, Phys. Lett.

28. Schramm , D.N. 1975, Ann. N.Y. Academy Sci. 262, 65

29. Freedman, D.Z., Schramm, D.N. and Tubbs, D. 1977, Ann. Rev. Nuc. Sci. 27

30. Bludman, , S. 1976, Talk at Chicago workshop on Weak Interaction in Astrophysics, April 3-4, 1976

31. Pethick, C. and Lamb, D. 1976, preprint

32. Yueh, W.R. and Buchler, R. 1977, Astrophys. J.

33. Bruenn, S., Arnett, W.D. and Schamm, D.N. 1977, Astrophys. J. 213, 213

34. Colgate, S. and White, R. 1966, Astrophys. J. 143, 626

35. Joss, P.C. and Rappaport, S.A. 1977, Astrophys. J.

36. Le Blanc, J. and Wilson, J. 1970, Astrophys. J. 161, 541

37. Meir, D.L., Epstein, R.I., Arnett, W.D. and Schramm, D.N. 1976, Astrophys. J. 204, 869

38. Chechetkin, V.M., Imshennik, V.S., Ivanova, L.N. and Nadyozhin, D.K., in Supernovae, ed. D. N. Schramm (Reidel: Dordrecht, Holland)

WEAK INTERACTIONS IN ASTROPHYSICS - THE EARLY UNIVERSE

Gary Steigman
Yale University Observatory

Introduction

The exciting developments in weak interaction physics have not passed unnoticed by those of us concerned with astrophysics and cosmology. Indeed, the experimental discoveries and the proposed theoretical models have already made a significant impact on the directions and even the results of current astrophysical research. The flow of information is not entirely one-way however, since certain theoretical studies coupled with astronomical observations can lead to important constraints capable of limiting or eliminating some of the models proposed for the weak interaction. The Universe, in some circumstances, can provide an alternative to the accelerator as a laboratory in which new theories may be tested.

Neutrino astrophysics provides an example of the relationship between particle physics and astrophysics. The discussion here will be limited to the role of neutrinos in cosmology. Before proceeding, it should be noted (and few in this audience will be surprised to learn) that much of what I will report on is research stimulated by work done by Ben Lee (with Steve Weinberg) shortly before his untimely death.

Why Neutrinos?

There are basically two distinct epochs during which neutrinos can influence the evolution of the Universe. The early evolution is affected by the number of "flavors" of massless or "light" (in this context, "light" means $m_\nu c^2 << 1MeV$) neutrinos. The later stages of evolution, including the present epoch, can reflect the role played by massive neutrinos (for these epochs, neutrinos with masses in the range 1eV - 10GeV may be important). Light neutral leptons ($1eV \lesssim m_\nu c^2 \lesssim 1$ MeV) can make their presence felt during both epochs; in contrast, heavy neutral leptons ($1MeV \lesssim m_\nu c^2 \lesssim 10GeV$) will play little role early on but, may well dominate the Universe today. Before dealing with these effects in some detail, it is worthwhile to raise and answer the following question.

Neutrinos are notoriously difficult to detect and there is, at present, no evidence for a universal background of primordial neutrinos (in contrast, for example, with the relic photons of the microwave background). On what basis then is it believed that neutrinos are indeed present and in sufficient abundance to have influenced the evolution of the Universe? The answer to this question comes from cosmology and from statistical mechanics.

Consider first the statistical mechanics. In thermodynamic equilibrium (at some temperature T) the probability a particle of mass m and momentum p is present is described by a distribution function $f_\pm(pc/kT)$. For $x = pc/kT$ and $z = mc^2/kT$ and for fermions (+) or bosons (-), the appropriate distribution function is

$$f_\pm(x,z) = \{\exp[(x^2 + z^2)^{1/2}] \pm 1\}^{-1}. \qquad (1)$$

The macroscopic quantities of interest (e.g.: number density, energy density, pressure, etc.) are obtained from various weighted integrals (d^3p) of $f_\mp$ (see, for example, reference 1). It is important to emphasize that, in equilibrium, such quantities will not depend on the details of the interactions nor on the specific coupling constants. Provided thermodynamic equilibrium can be established (and, it will be shown that it can be during the early Universe), all details become irrelevant and the interesting physical quantities depend only on the temperature, the mass, the number of particle states (e.g.: spin states, antiparticles, flavors, etc.) and, the statistics (e.g.:bosons or fermions). It is the simplification provided by this generality that permits the conclusion that neutrinos may, indeed, play an important role in cosmology. Before turning to the cosmology, recall the following simple results; they will be of value in the subsequent discussion.

In equilibrium at temperature T, the number density and the energy density of a particle of mass m and multiplicity g are

$$n(T;m) = n(z) = g\left(\frac{1}{2\pi^2}\right)\left(\frac{kT}{\hbar c}\right)^3 \int_0^\infty f_\pm(x,z)x^2dx, \tag{2}$$

$$c^2\rho(T;m) = c^2\rho(z) = g\left(\frac{1}{2\pi^2}\right)\left(\frac{kT}{\hbar c}\right)^3 kT \int_0^\infty f_\pm(x,z)x^2(x^2+z^2)^{1/2}dx. \tag{3}$$

For example, for photons (bosons with $m = 0$, $g = 2$)

$$n_\gamma = \frac{2\zeta(3)}{\pi^2}\left(\frac{kT}{\hbar c}\right)^3 \approx 20\ T^3\ (\text{cm}^{-3}), \tag{4a}$$

$$\rho_\gamma = \frac{3\zeta(4)}{\zeta(3)}\left(\frac{kT}{c^2}\right)n_\gamma \approx 2.7\left(\frac{kT}{c^2}\right) n_\gamma \approx 8.4 \times 10^{-36}T^4(\text{gcm}^{-3}). \tag{4b}$$

In (4) the temperature is in °K; occasionally, kT will be written in energy units as, for example, T_{eV}, T_{MeV}, T_{GeV}, etc.

For neutrinos, in the low mass (or, high temperature) limit ($m_\nu c^2 << kT$; $z_\nu << 1$)

$$n_\nu = \frac{3}{8} g_\nu n_\gamma \quad ; \qquad \rho_\nu = \frac{7}{16} g_\nu \rho_\gamma \tag{5}$$

The neutrino multiplicity, g_ν, is being considered a free (but, in principle, determinable) parameter. The motivation here is that g_ν may be different for massless and massive neutrinos and that, through g_ν, different neutrino flavors may be considered together. The importance of (5) is that it shows that, in equilibrium, relativistic neutrinos ($z_\nu << 1$) are as important as photons. In contrast, for heavy neutral leptons (or, low temperatures; $z_\nu^{-1} << 1$) these same ratios are

$$\frac{n_\nu}{n_\gamma} \approx g_\nu \left(\frac{\pi}{2}\right)^{1/2} \frac{z_\nu^{3/2} e^{-z_\nu}}{4\zeta(3)} \approx 0.26\, g_\nu z_\nu^{3/2} e^{-z_\nu} << 1, \tag{6a}$$

$$\frac{\rho_\nu}{\rho_\gamma} \approx \frac{m_\nu n_\nu}{\rho_\gamma} \approx \left(\frac{z_\nu}{2.7}\right) \frac{n_\nu}{n_\gamma} \approx 0.10\, g_\nu z_\nu^{5/2} e^{-z_\nu} << 1. \tag{6b}$$

The exponential factor reflects the cost ($\gtrsim 2m_\nu c^2$) of making massive $\nu\bar{\nu}$ pairs. When the average thermal energy is small compared to the neutrino mass, there are simply too few particles present with sufficient energy to make very many $\nu\bar{\nu}$ pairs.

To make the connection with cosmology it is necessary to verify that, even for weakly interacting particles, thermodynamic equilibrium prevails during the early Universe. For a homogeneous, isotropic Universe, the early expansion is analogous to (but the time reversal of) a free fall collapse (see, for example, reference 2). That is, for early epochs the age (t) and the total density (ρ) are related by

$$\rho t^2 = \text{constant} \tag{7}$$

Now, recall that for a relativistic gas (radiation), $\rho \sim V^{-4/3}$. In contrast, for a non-relativistic gas of conserved particles (matter), $\rho \sim V^{-1}$. Although it is likely that the present Universe is matter dominated ($\rho \sim V^{-1} \sim T^3$), the early Universe ($V \to 0$) will have been radiation dominated ($\rho \sim V^{-4/3} \sim T^4$). During the early stages then, the age vs. temperature relation must have been of the form

$$t(T) \sim T^{-2}. \tag{8}$$

For weakly interacting particles to be in equilibrium it is necessary that the weak interaction rate ($\Lambda = n\sigma v$) exceed the expansion rate (t^{-1}). Now, for temperatures above $T_{MeV} \gtrsim 1$, electron-positron pairs are abundant ($n_e \approx 1.5\, n_\gamma \sim T^3$); further, at such temperatures $\sigma v \sim T^2$ so that $\Lambda \sim T^5$ and

$$\Lambda(T)\; t\,(T) \sim T^3 \tag{9}$$

Then, for sufficiently early times $\Lambda t \gg 1$ and the weakly interacting particles will be in equilibrium.

During the early evolution, neutrinos are present with their abundances given by (5) and/or (6). For the early Universe (say, for $T_{MeV} \gtrsim 1$), only relatively light neutral leptons ($m_\nu c^2 < 1$MeV) are important (see equations (5) and (6)). As the Universe expands and cools ($T \sim V^{-1/3}$), an epoch is reached such that equilibrium can no longer be maintained ($\Lambda t < 1$). For heavy neutrinos ($m_\nu c^2 > 1$MeV), however, departure from equilibrium occurs earlier. The astrophysical consequences of the surviving primordial neutrinos depend crucially on when this departure from equilibrium occurs. Thermodynamic equilibrium in the early Universe ensures that neutrino pairs were once copiously produced; during this stage the details of the weak interaction are irrelevant. Neutrino survival is determined by the departure from equilibrium; during this stage the details of the weak interaction are crucial, as are the details of the cosmology. It is this interconnection, this independence and interdependence which provides the basis for the symbiotic relation between weak interaction physics and astrophysics.

Neutrino Survival

Although the basic physics is the same, there are qualitative as well as quantitative differences between the survival of light ($m_\nu c^2 \lesssim 1$ MeV) and heavy ($m_\nu c^2 \gtrsim 1$ MeV) neutral leptons. First, consider the light neutrinos.

Light Neutrinos ($m_\nu c^2 \lesssim 1$ MeV)

For $m_\nu c^2 \lesssim 1$ MeV, thermodynamic equilibrium is maintained for $T_{MeV} \gtrsim 1$ so that for such neutrinos: $n_\nu/n_\gamma = 3g_\nu/8$. For $T_{MeV} \lesssim 1$, the weak interaction can no longer be in equilibrium (the transition, of course, is gradual, not sharp) and, crudely speaking, $\nu\bar{\nu}$ pairs are neither created nor annihilated; the $\nu\bar{\nu}$ pairs which were present at $T_{MeV} \approx 1$ are "frozen-out". Thus, subsequent to the time ($\approx$1 sec.) when $T_{MeV} \approx 1$, neutrino pairs (with, recall, $m_\nu c^2 \lesssim 1$ MeV) are conserved. If no new photons were produced, then at all later epochs, the ratio of neutrinos to photons would be given by (5). For $T_{MeV} \lesssim 1/2$, however, the electron-positron pairs annihilate, dumping their energy into the photon background (the neutrinos, which have already decoupled, cannot share this energy). The photon temperature is thus raised over that of the neutrinos (after decoupling the neutrinos are cooled by expansion, $T_\nu \sim V^{-1/3}$). For $T_{MeV} \ll 1$, $T_\gamma \approx (11/4)^{1/3} T_\nu$; for details see, for example, reference 1. After $e^\pm$ annihilation is complete,

$$\frac{n_\nu}{n_\gamma} = \frac{3}{8}\, g_\nu \left(\frac{T_\nu}{T_\gamma}\right)^3 \approx \frac{3}{22}\, g_\nu, \tag{10a}$$

$$\frac{\rho_\nu}{\rho_\gamma} = \frac{7}{16}\, g_\nu \left(\frac{T_\nu}{T_\gamma}\right)^4 \approx 0.11\, g_\nu. \tag{10b}$$

The above analysis, first presented by Cowsik and McClelland[3] shows that light neutrinos were very abundant in the early Universe ($T_{MeV} \gtrsim 1$; $t \lesssim 1$ sec). Indeed, for the known neutrinos (ν_e, $\overline{\nu}_e$; ν_μ, $\overline{\nu}_\mu$), $g_\nu = 4$ so that $\rho_\nu = (7/4)\, \rho_\gamma$; for $g_\nu > 6$, neutrinos would dominate that total density ($\rho_\nu > \rho_\gamma + \rho_{e\pm}$).

It should be noted that equation (10a), which reflects neutrino-pair conservation, is valid for all temperatures $\lesssim 0.1$ MeV. In contrast, equation (10b) is only valid while the neutrinos are relativistic ($kT_\nu \gtrsim m_\nu c^2$). For later times (and, lower temperatures; $kT_\nu \lesssim m_\nu c^2$) when the neutrinos are nonrelativistic,

$$\rho_\nu = m_\nu n_\nu \approx \frac{3}{22}\, g_\nu m_\nu n_\gamma. \tag{11}$$

Recalling that $n_\gamma \approx 20 T_\gamma^3$ and that, at present, $T_{\gamma o} \approx 2.7$ °K, it follows that

$$\rho_\nu \approx 5 \times 10^{-33} g_\nu m_{eV} T_\gamma^3 \ (\mathrm{gcm}^{-3}), \tag{12a}$$

$$\rho_{\nu o} \approx 1 \times 10^{-31} g_\nu m_{eV} (\mathrm{gcm}^{-3}). \tag{12b}$$

In (12), the neutrino mass is in eV (i.e.: 1 eV $\Leftrightarrow$ 1.8×10^{-33}g). The importance of such neutrinos at present is measured by the density parameter Ω_ν which is the ratio of the present mass density in neutrinos ($\rho_{\nu o}$) to the critical density (ρ_{co}) where

$$\rho_{co} = 2 \times 10^{-29} h_o^2 \ (\mathrm{gcm}^{-3}). \tag{13}$$

In (13), h_o is the present value of the Hubble parameter in units of 100 $kms^{-1}Mpc^{-1}$. Comparing (12b) and (13),

$$g_\nu m_{eV} \approx 200\Omega_\nu h_o^2 \tag{14}$$

The corresponding quantity for ordinary matter (nucleons) is not well determined bu. it is likely that[4] $\Omega_N h_o^2 \gtrsim 5 \times 10^{-3}$; neutrinos may dominate nucleons for $g_\nu m_{eV} \gtrsim 1$. In constrast, the absence of evidence for strong deceleration in the expansion of the Universe suggests that $\Omega_\nu h_o^2 \lesssim 1/2$ so that $g_\nu m_{eV} \lesssim 100$. As Cowsik and McClelland[3] noted, this last inequality provides the best limit to the mass of the e-neutrino and the μ-neutrino.

$$m_{\nu e} + m_{\nu\mu} \lesssim 50 \text{ eV}. \tag{15}$$

Before considering the effects of such light neutrinos, let us consider the survival of the heavy neutral leptons.

<u>Heavy Neutrinos ($m_\nu c^2 \gtrsim 1$ MeV)</u>

Pairs of heavy ($m_\nu c^2 \gtrsim 1$ MeV) neutral leptons are copiously produced only for $kT \gtrsim m_\nu c^2$. For $kT \lesssim m_\nu c^2$ fewer and fewer ν-pairs are produced and n_ν/n_γ decreases rapidly (see equation (6)). The exponential decrease does not continue indefinitely since the decreasing ν-density makes annihilation more and more difficult. Annihilation ceases when the neutrino density is so low that $\nu\bar{\nu}$ collisions do not occur in the age of the Universe. Thus, for some temperatures $\lesssim m_\nu c^2$ the heavy neutral lepton pairs are in equilibrium and their densities are given in (6). Eventually, however, a temperature T_f is reached such that $\nu\bar{\nu}$ pairs are neither created nor annihilated; equilibrium can no longer be maintained and the neutrinos are "frozen-out". As before, of course, this transition away from equilibrium is gradual and there is no special significance to a very precise value of T_f. The rapidly decreasing exponential factors do, in fact, make this departure from equilibrium relatively sharp and this exponential behavior permits a rather simple and accurate determination of T_f and of the numbers of surviving heavy neutrino pairs.

The analysis to be outlined below is due to Lee and Weinberg[5] (LW). It was their insight that the problem posed and solved by Cowsik and McClelland[3] was, in fact, only half a problem. Although the physics is the same (i.e.: the weak interaction rate must be compared with the cosmological expansion rate), the details are very different when considering the survival of heavy neutrinos.

At temperatures above T_f the neutrino pairs are in equilibrium and their density is given by (6) for $kT_f < m_\nu c^2$. For temperatures below T_f equilibrium is no longer maintained so that the evolution of the neutrino density due to pair creation, pair annihilation and cosmological expansion must be considered. LW have solved the appropriate equations numerically and, also, found a good analytic fit to their results; for details the reader is referred to their paper. The following considerations reproduce the LW results rather accurately; they have the further advantage of clearly displaying the physics and exposing the sources of possible uncertainties in the final results.

For $kT_f < m_\nu c^2$, pair production may be neglected and only annihilation and expansion are important. First consider the expansion rate (i.e.: the t vs. T relation). It has already been noted that for early epochs, $t \sim \rho^{-1/2}$ and that ρ is dominated by relativistic particles.

For $1 \lesssim T_{MeV} \lesssim 100$, the known particles which will contribute to ρ are γ, $e^{\pm}$, $\nu_e(\bar{\nu}_e)$, $\nu_\mu(\bar{\nu}_\mu)$ so that

$$\rho = \rho_\gamma + \rho_e + \rho_{\nu e} + \rho_{\nu_\mu} = 4.5\rho_\gamma \tag{16}$$

It then follows that[1]

$$t \approx 0.81\ T_{MeV}^{-2}. \tag{17}$$

If, however, there were, as yet undiscovered, particles with $mc^2 \lesssim 100$ MeV, their presence would modify (16) and (17) for $kT \gtrsim mc^2$. In that case (17) will become

$$t \approx 0.81(1 + \frac{\Delta N}{4.5})^{-1/2} T_{MeV}^{-2}, \tag{18}$$

where $\Delta N = 1/2(7/16)$ for each additional boson (fermion) species and spin state. Indeed, it should be noted that in obtaining the coefficient 4.5 in equation (16) it was assumed that ν_e and ν_μ were massless so that $g_{\nu e} = g_{\nu\mu} = 2$; the modification suggested in (18) may be due additional states of the known leptons or, to the presence of new leptons such as, perhaps, the τ-neutrino (if $m_{\nu\tau} \lesssim 1$MeV). To keep track of the effects of this uncertainty, write $t = AT^{-2}$.

The expansion rate must be compared to the annihilation rate $\Lambda_a = n v\sigma_a$. The annihilation cross section is, of course, not known; σ_a is model dependent. For $kT < m_\nu c^2$, LW assume that $\sigma_a v = BM^2$; they use a specific model to evaluate B. In particular, B depends on the number of final channels available in annihilation. Finally, from (4) and (6) it follows that $n_\nu(T) = T^3 F(z_\nu)$ where F is dominated by e^{-z_ν}. Defining the "freeze-out" epoch by $\Lambda_a(T_f) \approx t^{-1}(T_f)$ leads to an equation for $z_{\nu f}$

$$F(z_{\nu f}) = \frac{z_{\nu f}}{ABM^3} \quad . \tag{19}$$

The exponential behavior of F permits an easy estimate of z_f; LW find for $M_{GeV} \approx 1 - 10$, $z_{\nu f} \approx 20$ ($T_{f,MeV} \approx 50\ M_{GeV}$). The density of surviving neutrinos is

$$\frac{n_\nu}{T_\nu^3} = \frac{n_{\nu f}}{T_{\nu f}^3} = F(z_{\nu f}). \qquad (20)$$

For the standard (LW) values of the parameters and, for $T_{MeV} << 1$,

$$n_\nu = n_{\bar{\nu}} \approx 10^{-6} M_{GeV}^{-3}\ T_\gamma^3\ (\mathrm{cm}^{-3}). \qquad (21)$$

Compared to photons, surviving massive neutrinos are rare.

$$\frac{n_\nu}{n_\gamma} \approx 10^{-7} M_{GeV}^{-3}. \qquad (22)$$

However, such neutrinos may contribute significantly to the total density of the Universe; their present density would be

$$\rho_{\nu o} = m_\nu n_{\nu o} \approx 8 \times 10^{-29} M_{GeV}^{-2}\ (\mathrm{gcm}^{-3}), \qquad (23)$$

which corresponds to a density parameter

$$\Omega_\nu \approx 4(M_{GeV} h_o)^{-2}. \qquad (24)$$

For $2 \lesssim M_{GeV} \lesssim 10$, heavy neutral leptons may dominate the dynamical evolution of the Universe.

The above estimates depend not only on the mass of the neutral lepton but also on the uncertain parameters A and B. Fortunately, the exponential which dominates $F(z_\nu)$ leads to a value of T_f which is only logarithmically dependent on A and B. The results expressed in equations (21)-(24) or, the more precise results of LW, are relatively insensitive to these parameters. If, however, a neutral lepton with a mass of a few GeV should be discovered, it would be very important to determine A and B as accurately as possible; it is through these parameters that the particle physics can constrain the cosmology and vice-versa.

Implicit in the previous analysis there were two unstated but rather important assumptions. The massive neutrinos were assumed stable and nondegenerate. It is worth spending a moment discussing the reasonableness of these assumptions and calling attention to the limitations they impose.

Neutrino Degeneracy

In our discussion of primordial neutrinos, only those pairs produced thermally in the early Universe were considered. Indeed, equal numbers of neutrinos and antineutrinos were guaranteed by the absence of a chemical potential in (1). But, is it known that the Universe is, in fact, symmetric between neutrinos and antineutrinos? Not at all; at present, the relic neutrinos are undetectable. It is, however, virtually certain that the Universe is not baryon symmetric.[1,6] A measure of the baryon asymmetry is the dimensionless entropy per nucleon (equivalent to the photon to nucleon ratio: $n_\gamma/n_N \approx 10^8 - 10^{10}$). By analogy, a neutrino asymmetry might seem reasonable. Such an asymmetry would, however, introduce into cosmology another free parameter (initial condition) for each neutrino flavor. The arbitrariness of these parameters would destroy the simple connection between physics and cosmology. If, for example, there is an excess of heavy neutrinos over antineutrinos (or vice-versa), the direct connection between the neutrino mass and the present mass density is broken. In such a case there is no a priori way of knowing whether or not the neutrinos are dynamically important.

In the absence of experimental evidence to the contrary and, for simplicity, it is natural to assume equal numbers of neutrinos and antineutrinos. At the same time it is wise to keep in mind that the same reasonable assumption is clearly violated for nucleons.

The choice of a zero chemical potential ensures that the neutrinos are nondegenerate. But, what of the momentum space distribution of the relic neutrinos; is it that of a degenerate Fermi gas? The early history of the neutrinos ensures that it is not. Early on, when thermodynamic equilibrium is established, the phase space distribution of the neutrinos is given by f_+ (see equation (1)). Kinetic equilibrium is maintained through collisions with electrons and positrons until $T_{MeV} \lesssim 1$. Afterwards the neutrinos are collisionless and their phase space distribution remains nondegenerate.

Neutrino Stability

In the analyses of Cowsik and McClelland[3] and Lee and Weinberg[5] it is assumed that the massive neutral lepton is stable. This is quite a restrictive assumption; in the present context it requires that the neutrino lifetime be greater than the age of the Universe ($t_o \approx 1 - 2 \times 10^{10}$yr $\approx 3 - 6 \times 10^{17}$ sec). Such stability would seem to require a new, conserved quantum number associated with the massive neutrino (LW).

In contrast, Dicus *et al*[7] have considered the cosmological consequences of an unstable, massive neutral lepton. Although perhaps more reasonable, this assumption opens a Pandora's Box. The effects of such an unstable neutrino depend not only on its mass and lifetime, but also on its decay scheme. In particular, the restrictions on the mass of a stable neutrino imposed by limits to the present density of the Universe no longer apply. With the lifetime as a free parameter, entire regions of the mass vs. lifetime plane are consistent with the cosmological constraints.[7] Without restricting the mass and/or the lifetime, nothing very definite can be said regarding the cosmological consequences of such neutrinos. The valuable connection between physics and cosmology is lost.

In the following it is assumed that the neutrino is stable. Although perhaps not the most reasonable assumption, it does have the advantage that the problem posed is more narrowly defined. It should, however, be noted that there are astrophysical constraints capable of excluding large regions of the mass vs. lifetime plane for unstable neutrinos.[7,8] For details the reader is referred to the papers of references 7 and 8; below is presented a brief summary of some of the relevant arguments.

The constraints depend on the decay scheme of the massive neutral lepton; more severe restrictions follow if one of the decay products (carrying a non-negligible fraction of the energy) is a photon. Independent of the decay scheme it is clear that the density of the Universe at present is not dominated by the decay products of an unstable, massive, neutral lepton. That is, there is good evidence that the Universe, during recent epochs, has not been radiation dominated.[4] For example, for a radiation dominated Universe the relation among the age (t_o), the Hubble parameter (H_o) and the density (as measured by Ω_o) is

$$H_o t_o = (1 + \Omega_o^{1/2})^{-1}. \tag{25}$$

For $\Omega_o = 1$ and $H_o \gtrsim 50$ kms^{-1}Mpc^{-1} the age of the Universe would be $t_o \lesssim 10 \times 10^9$ yr which is in conflict with the ages of globular clusters.[9] Futhermore, matter perturbations do not grow in a radiation dominated

Universe[10], so that galaxy formation could not occur. Considerations of galaxy formation via gravitational instability suggest that $\Omega_{rad} \lesssim \Omega^2_{matter}$; from the results of reference 4 this leads to $\Omega_{rad} \lesssim 3 \times 10^{-3}$.

Much more restrictive constraints follow[8] if the heavy neutrino decays include photons ($\nu \rightarrow \gamma + ---$). If $t_\nu \gtrsim 10^{12}$ sec (but, $t_\nu < t_o$), the photons from such decays would dominate x-ray, γ-ray, ... backgrounds. Observations of, or, limits to, these background radiations lead to the conclusion that $t_\nu < 10^{12}$ for virtually all neutrino masses of interest. Furthermore, for $t_\nu < 10^{12}$ sec the decay photons would distort the microwave background. The absence of distortions suggests that $t_\nu \lesssim 10^3$ sec for $m_\nu c^2 \lesssim 100$ MeV($t_\nu \lesssim 10^{12}$ sec for $m_\nu c^2 \lesssim 100$ GeV). For $t \lesssim 10^3$ sec no cosmological imprint remains; massive neutrinos with $t_\nu \lesssim 10^3$ sec are possible since they do not affect the cosmology in any detectable way.[8]

Cosmological Consequences

The "new" neutrinos have consequences for cosmology which fall into two distinct categories. Early on, during the radiation dominated era, relativistic neutrinos ($m_\nu c^2 \lesssim kT$) may contribute significantly to (and even dominate) the total density. These neutrinos can influence the early evolution of the Universe. The quantitative effects of such neutrinos depend on the neutrino multiplicity g_ν (i.e.: on the number of distinct neutrino flavors and spin states). The abundances of the light elements produced in primordial nucleosynthesis reflect the effects of these "light" ($m_\nu c^2 \lesssim 1$ MeV) neutrinos.[11,12] During more recent epochs both light ($\lesssim 1$ MeV) and heavy ($\gtrsim 1$ MeV) stable massive neutrinos may be dynamically important.[3,5,8,13] Indeed, it has already been seen that estimates of the present mass density (or, equivalently, limits to the deceleration of the expansion of the Universe) restrict the mass of a stable neutrino to be $\lesssim 50$ eV[3] or, $\gtrsim 2$ GeV.[5] Consistent with these restrictions, massive neutrinos may still dominate the recent evolution of the Universe.[8,13]

It is the purpose of the following discussion to present a brief summary of the aforementioned effects of neutrinos in cosmology. The details regarding primordial nucleosynthesis can be found in the papers of references 11 and 12; for neutrino effects during more recent epochs see references 8 and 13.

Primordial Nucleosynthesis

It is worth reviewing the physics of primordial nucleosynthesis in the "standard" model before considering the effects of "new" neutrinos (see, for example, the review article by Schramm and Wagoner[14]).

For $T_{MeV} > 1$, photons, electrons, e-neutrinos and μ-neutrinos are in equilibrium ($T_\gamma = T_e = T_\nu$) and, for $g_\nu = 4$, the age vs. temperature relation is $t(\text{sec}) \approx 0.81\ T_{MeV}^{-2}$. The weak interaction, during this epoch, is sufficiently rapid to maintain equilibrium between neutrons and protons through the reactions

$$p + e^- \rightleftarrows n + \nu_e\ ,\quad p + \bar{\nu}_e \rightleftarrows n + e^+,\ n \rightleftarrows p + e^- + \bar{\nu}_e. \tag{26}$$

In equilibrium the neutron-to-proton ratio is

$$\frac{n}{p} = \exp\left(\frac{-\Delta Mc^2}{kT}\right). \tag{27}$$

Which elements (if any) can be built up and their abundances clearly will depend on this ratio. For $T_{MeV} \lesssim 1$, the weak interaction freezes-out leaving a ratio $n/p \approx 0.22$ (or, a neutron abundance by mass $X_n = n(n + p)^{-1} \approx 0.18$). All the while, two body collisions between nucleons are occurring.

The most important of these reactions is $p + n \rightleftarrows D + \gamma$. For sufficiently high temperatures ($T_{MeV} \gtrsim 0.2$) there are large numbers of photons present with sufficient energy to photodissociate the deuteron. At such temperatures then, the deuterium abundance is very low. The negligible deuterium abundance provides a bottleneck to further nucleosynthesis. For $T_{MeV} \lesssim 0.1$, photodissociation of deuterium is no longer rapid, the deuterium abundance increases and this permits further build up of the other light elements by reactions such as

$$p + n \rightarrow D + \gamma, \text{---}, D + D \rightarrow {}^4He + \gamma, \text{---}. \qquad (28)$$

There is a gap (no stable nucleus) at mass 5 which provides a bottleneck to the build up of the heavier elements. Since ^4He is the most tightly bound light nucleus, virtually all the neutrons end up in ^4He (occasionally, the neutron decays). As the temperature and density drop the increasing coulomb barrier and the decreasing density combine to quickly shut off further nuclear reactions; by the time $T_{MeV} \approx 0.03$ most of the action is over. Recall, also, that the electron-positron pairs have annihilated leading to $T_\gamma \approx (11/4)^{1/3} T_\nu$; the age vs. temperature (T_γ) relation has become $t(sec) \approx 1.4\ T_{MeV}^{-2}$. The evolution just described is illustrated in Figure 1 taken from reference 15. The mass fraction of the light elements is plotted versus the age (upper horizontal axis) and the temperature (in units of 10^9 °K on the lower horizontal axis).

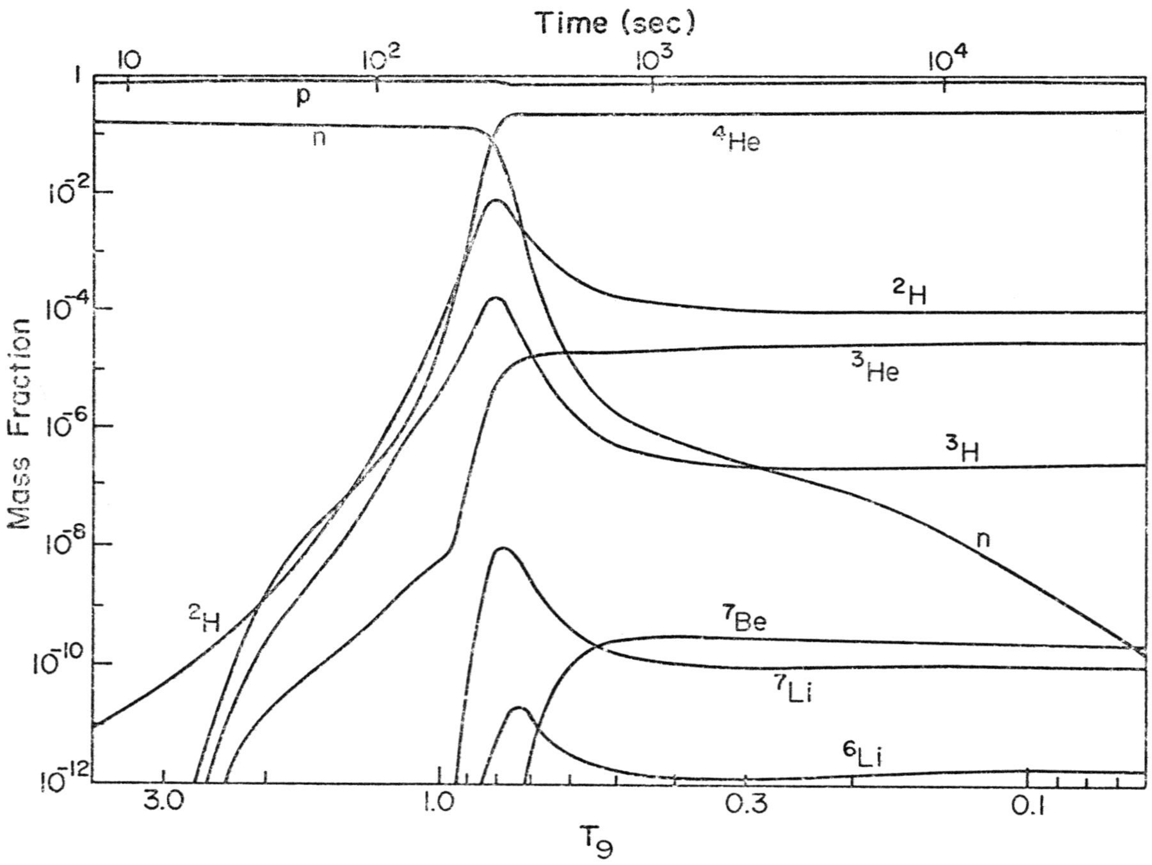

FIGURE 1: Evolution of the abundances of the light elements during nucleosynthesis.

The final abundances depend on two effects which may be considered separately. Since virtually all neutrons end up in ^{4}He, the helium abundance is most strongly dependent on the neutron to proton ratio at freeze-out. In contrast, the abundances of the other light elements depend critically on competing two-body reaction rates; these abundances are very sensitive to the nucleon density at the time of nucleosynthesis. Since nucleons are conserved, the nucleon density during the epoch of nucleosynthesis may be related to the present nucleon density

$$\rho_N = \rho_{NO}\left(\frac{T}{T_o}\right)^3 \gtrsim 10^{-30}\ \Omega_N h_o^2\ T^3 \quad (\mathrm{gcm}^{-3}). \tag{29}$$

Figure 2 (also from reference 15) illustrates the final abundances as a function of the present nucleon density ρ_N or, equivalently, the nucleon density parameter Ω_N. As Figure 2 clearly shows, the observed abundance of deuterium can lead to constraints on the present nucleon density. The helium abundance, in contrast, is relatively independent of nucleon density. It is the frozen-out neutron to proton ratio which determines the helium abundance. Since the n/p ratio is determined by the competition between the weak interaction rate (known from beta decay) and the expansion rate (dependent on the numbers of relativistic particles present at $T_{MeV} \gtrsim 1$), the helium abundance can be used to constrain g_ν[11].

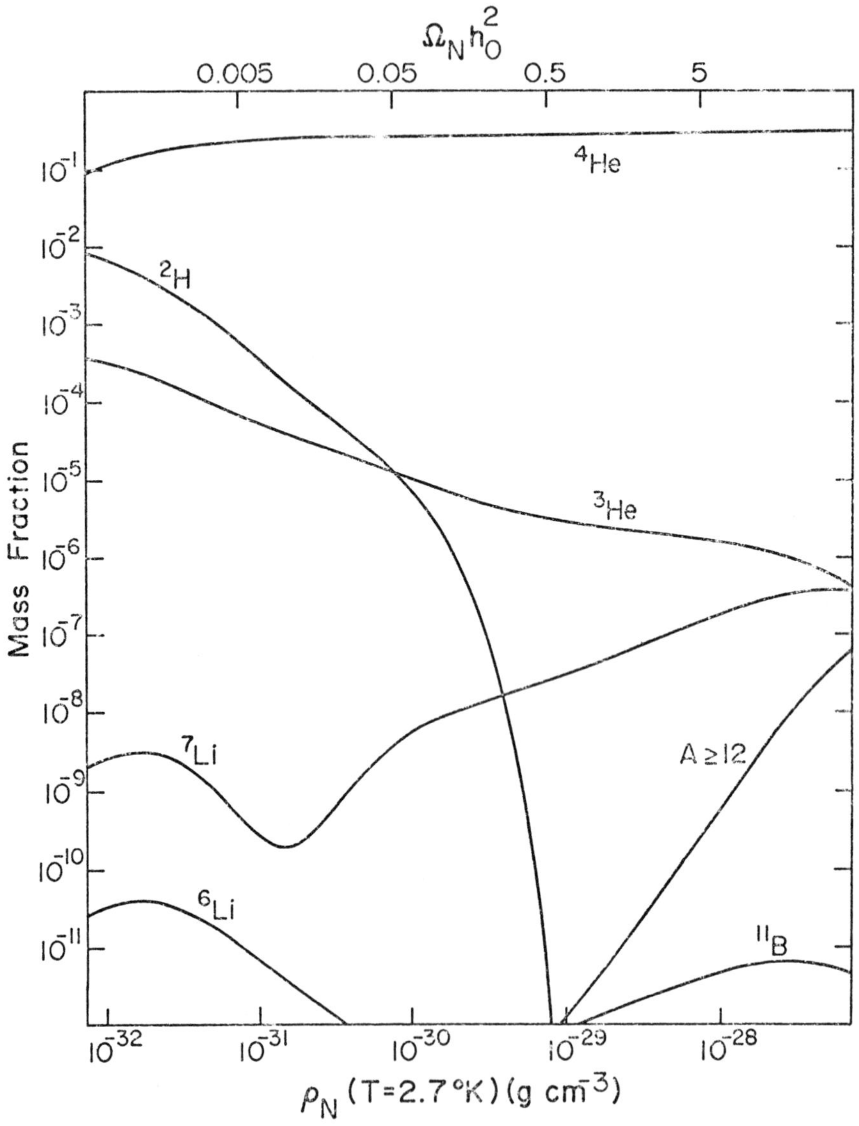

FIGURE 2: Primordial abundances as a function of the present nucleon density.

Let us now consider the effect of "new" neutrinos on the previous analysis. In the above discussion only two component e-neutrinos and μ-neutrinos were considered . In this case (g_ν =4) the total density at early epochs is

$$\rho = \rho_\gamma + \rho_e + \rho_{\nu e} + \rho_{\nu_\mu} = 4.5\rho_\gamma \tag{30}$$

With allowance for new neutrinos, $g_\nu \to g'_\nu = 4 + \Delta g_\nu$ so that

$$\rho \to \rho' = (1 + \frac{7\Delta g_\nu}{72})\rho. \tag{31}$$

The expansion rate is then changed to $t \to t' \equiv \xi^{-1}t$ where

$$\xi^2 \equiv \rho'/\rho = 1 + 7\Delta g_\nu/72. \tag{32}$$

The frozen out neutron to proton ratio becomes

$$\frac{n'}{p'} \approx \exp[-(\frac{\Delta Mc^2}{kT})\xi^{-1/3}]. \tag{33}$$

For $\Delta g_\nu > 0$, $\xi > 1$ so that $(n/p)' > n/p$. This leads to a new helium abundance in excess of the "standard" estimate. The effect is illustrated in Figure 3 (from reference 12) where the "standard" helium abundance is compared with that for $\Delta g_\nu = 10$ (i.e.: five new two component neutrinos). According to Steigman et al[11], a good estimate of the change in the helium abundance by mass (Y) with g_ν is provided by

$$\Delta Y \approx 0.008\Delta g_\nu \tag{34}$$

[In comparing reference 11 with the discussion here, note that $\Delta g_\nu \Leftrightarrow 2\Delta N_L$].

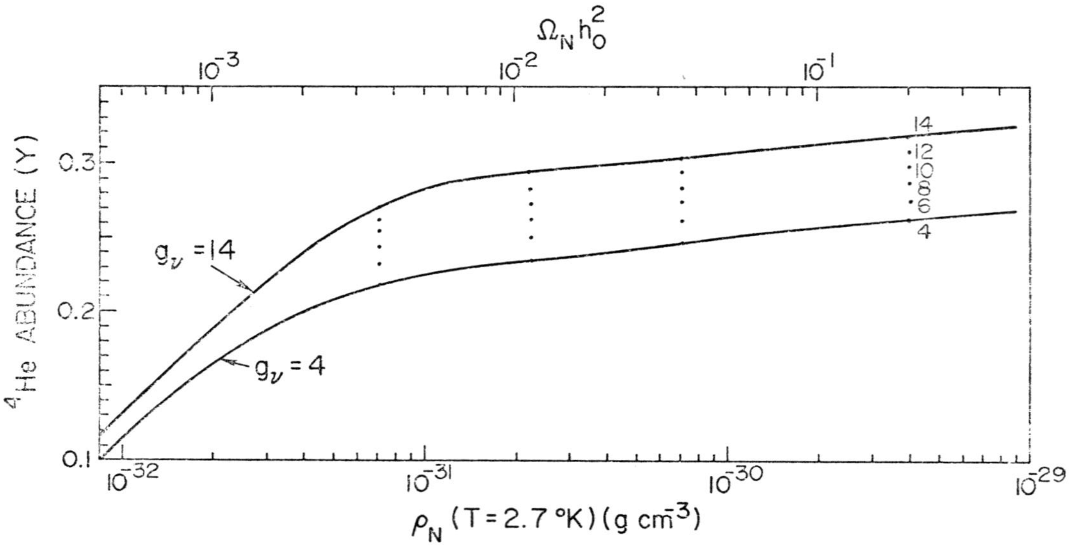

FIGURE 3: Helium abundance as a function of the present nucleon density and the neutrino multiplicity.

The present helium abundance is $Y \lesssim 0.29$ which, combined with an estimate of the lower limit to ρ_{No} of $2 \times 10^{-31} gcm^{-3}$ (reference 4), leads to the constraint $\Delta g_{\nu} \lesssim 5$ (reference 11). It is likely, however, that some of the presently observed helium was processed in stars during the course of galactic evolution. Estimates of the truly primordial helium abundance are uncertain but it is not unreasonable that $Y \lesssim 0.26$.[12] In this case, $\Delta g \lesssim 4$ for $\rho_{No} \gtrsim 2 \times 10^{-31} gcm^{-3}$. Indeed, even for $\Delta g_{\nu} = 0$, the standard model predicts $Y \gtrsim 0.26$ for $\rho_{No} \gtrsim 3 \times 10^{-30} gcm^{-3}$; this suggests the Universe is not closed by nucleons.

As a timely illustration of these considerations note that if the τ-neutrino is distinct from the e-neutrino and if $m_{\nu\tau} < 1MeV$ (see, Perl, this conference) then $\Delta g_{\nu} \geq 2$. For $Y \lesssim 0.26$ this implies (see Figure 3), $\rho_{No} \lesssim 7 \times 10^{-31}$.

The deuterium abundance as a function of the density (ρ_{No}) and the neutrino multiplicity (g_{ν}) is shown in Figure 4. The strong density dependence (deuterium is destroyed by two body collisions so that higher nucleon density corresponds to lower deuterium abundance) means that the primordial abundance of deuterium (X_D) can lead to severe constraints on the present nucleon density.[4] In contrast, X_D does not constrain Δg_{ν}. The point is the following. For $X_D \gtrsim 3 \times 10^{-5}$, $\rho_{No} \lesssim 4 \times 10^{-31}$ gcm^{-3} if $\Delta g_{\nu} = 0$ (see Figure 4). For $\Delta g_{\nu} = 10$, $X_D \gtrsim 3 \times 10^{-5}$ for $\rho_{No} \lesssim 7 \times 10^{-31} gcm^{-3}$. Now, if ρ_{No} were precisely known then Δg_{ν} could be determined. But, ρ_{No} is not known accurately and, as Figure 4 shows, any uncertaintly in ρ_{No} can be compensated for by Δg_{ν} and, vice versa. The deuterium abundance does, however, confirm the conclusions based on helium that $\Delta g_{\nu} \lesssim 10$ and $\rho_{No} \lesssim 7 \times 10^{-31} gcm^{-3}$ ($\Omega_N h_o^2 \lesssim 0.035$).

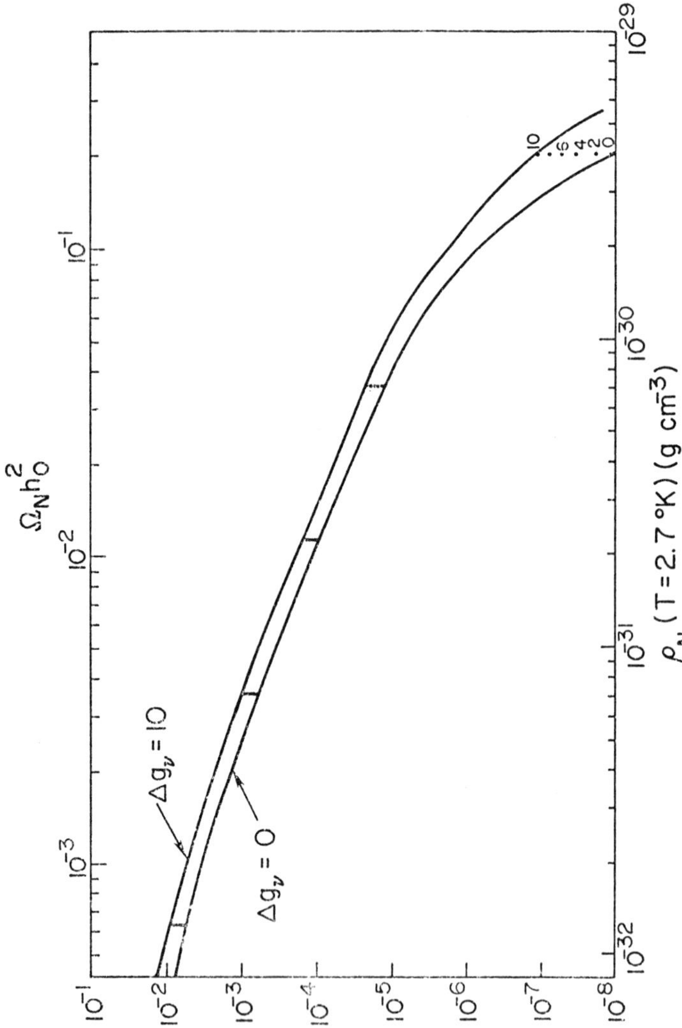

FIGURE 4: Deuterium abundance as a function of the present nucleon density and the neutrino multiplicity.

Recent Epochs

Having decoupled in the early Universe, a smooth background of relic neutrinos would remain collisionless. The neutrinos do, however, interact gravitationally. The gravitational effects of the relic neutrinos, if they are massive and stable, may well dominate the cosmological expansion at present. In Figure 5 (from reference 8) is shown the present density in relic neutrinos as a function of their mass. Current observations suggest $\rho_o \lesssim 2 \times 10^{-29} gcm^{-3}$ so that $m_\nu c^2 \gtrsim$ 2 GeV and $m_\nu c^2 \lesssim$ 50 eV. But, for $m_\nu c^2 \lesssim$ 10 GeV or $m_\nu c^2 \gtrsim$1 eV, relic neutrinos may dominate the density. What other effects could be due to such neutrinos? Are they detectable?

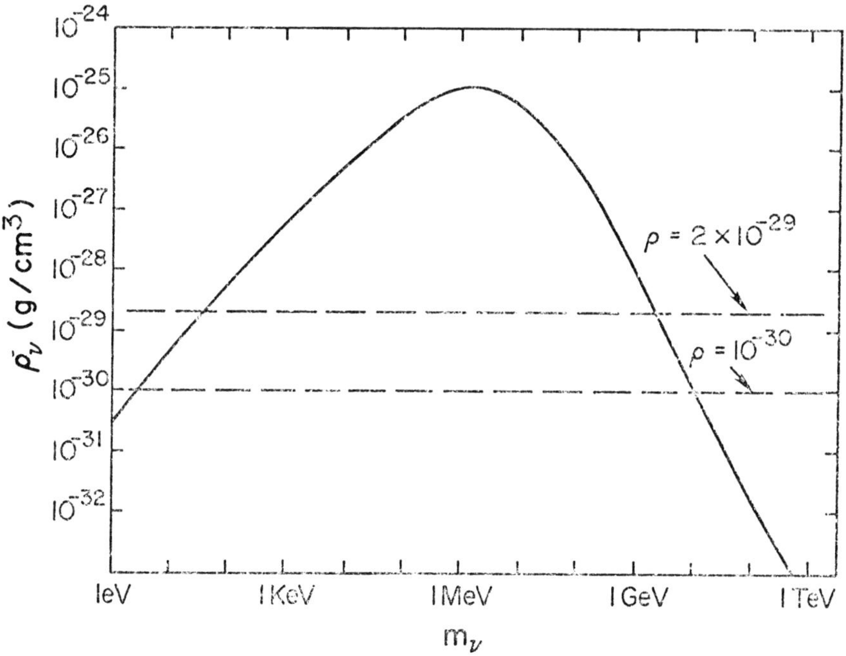

FIGURE 5: The present density in stable relic neutrinos as a function of their mass.

Since the neutrinos interact only gravitationally and via the weak interaction it might seem at first sight that the relic neutrinos are unobservable. This pessimistic view is not entirely justified. Neutrino detectability and, indeed, the astrophysical consequences of such neutrinos depend crucially on whether or not the relic neutrinos are smoothly distributed or clustered. For example, although neutrino pair annihilation ceased in the early Universe, condensations of neutrinos could lead to the resumption of annihilation; the neutrinos could be detected by observing their annihilation products. Indeed, large condensations of massive, stable neutrinos could provide the "missing mass" for galaxies and clusters of galaxies as well as the Universe. Could gravitational instability, believed responsible for the development of structure in an otherwise homogeneous Universe, lead to clustering of neutrinos and initiate a new "neutrino era"? The work of Gunn et al[8] and Steigman et al[13] attempts to answer this question.

Since the neutrinos are a collisionless gas the exchange of energy between ordinary matter and heavy neutrinos is very difficult. The neutrinos, therefore, are effectively isolated from energy changing processes such as shocks, dissipation, etc. The gravitational interaction, weak though it may be, prevents the complete isolation of the neutrinos. Neutrinos which find themselves in a bound perturbation will collapse along with the ordinary matter turning potential energy into kinetic energy. In a classic paper, Lynden-Bell[16] showed that the violently changing gravitational field of a collapsing system will be effective in changing the statistics of particle orbits. Within a few collapse times this violent relaxation produces an isothermal distribution with the temperature proportional to the mass.[16,17,18] Through their gravitational interactions then, the neutrinos will relax violently and be virialized. Since the resulting velocity dispersion is independent of mass, the neutrinos, the nucleons, the stars, etc. will all have the same velocity dispersion. For collapse without dissipation the neutrinos and the ordinary matter fall together. In any structures formed by such collapse (e.g.:elliptical galaxies, clusters of galaxies) the neutrino distribution and the matter distribution are the same.

Some structures such as stars, spiral galaxies, etc. required the dissipation of internal energy for their formation. The neutrinos cannot get rid of their internal energy and, it would seem cannot participate in such collapse. For practical purposes this is correct but, it is worth noting that it is not strictly true that the neutrinos cannot participate in dissipative collapse.[13] The crucial point is that energy is not conserved in a time-dependent potential. Then, for example, neutrinos which fall down a potential well may find, when they try to climb out, that the well has deepened and they are trapped. Steigman _et al_[13] show that neutrinos may, in this manner, be captured by collapsing perturbations. However, the captured neutrinos are weakly bound and at large radii; they will be easily stripped from such objects. For an initially bound region (e.g.: a galaxy), some part of which undergoes further collapse (e.g.: to from a star or star cluster), the neutrinos lag behind the ordinary matter. For slow collapse (collapse time >>free fall time), Steigman _et al_[13] find $R_\nu \sim R_{matter}^{3/4}$; the neutrinos soon get left behind and are easily stripped from the perturbation. For fast collapse (collapse time $\gtrsim$ free fall time) the neutrinos are left behind even more rapidly[13] ($R_\nu \sim R^{1/2}_{matter}$). We return then to our initial conclusion that there will be negligible numbers of neutrinos in objects, such as stars, formed by dissipative collapse. This, we will see, is a fortunate result.

In the initial, non-dissipative collapse to form some galaxies (ellipticals) and clusters of galaxies, the neutrinos and the ordinary matter collapse together and virialize. The neutrino mass fraction in such objects may, therefore, be large. The "missing mass" in elliptical galaxies and in galaxy clusters may be in massive neutrinos. In virial equilibrium there will be a "halo" of neutrinos such that $(\Delta v_\nu)^2 \gtrsim GM/R$; in convenient units

$$\frac{M}{M_\odot} \approx 10^{11} \left(\frac{\Delta v_{100}}{2}\right)^2 \left(\frac{R_{kpc}}{10}\right) \tag{35}$$

$M_\odot$ is the solar mass (2×10^{33}g), Δv_{100} is the velocity dispersion in units of 100 kms^{-1} and, R_{kpc} is the distance in kiloparsecs (1 kpc $\gtrsim 3 \times 10^3$ light years). A typical galactic luminosity is[19] $L \approx L_* \gtrsim 3 \times 10^{10} L_\odot \gtrsim$ (where the solar luminosity $L_\odot \approx 4 \times 10^{33}$ergs^{-1}) so that the mass-to-light ratio associated with a halo distribution of neutrinos plus ordinary matter is

$$\frac{M/L}{(M/L)_\odot} \approx \frac{1}{12} (\Delta v_{100})^2 R_{kpc} \tag{36}$$

Consider now, some of the consequences.

If our galaxy has a neutrino halo then, for $R_{kpc} \approx 10$, $\Delta v_{100} \approx 2$ so that $M(<10kpc) \gtrsim 10^{11} M_\odot$. Such a mass (in addition to the mass in luminous objects) is required to stabilize the galactic disk against bar instabilities.[20] The missing mass may, but need not necessarily, be contributed by massive neutrinos.

For typical sizes of galaxies ($R_{kpc} \gtrsim 15$) and typical rotational velocities ($\Delta v_{100} \gtrsim 2-3$), a neutrino halo component would contribute $M/L \gtrsim 5-10(M/L)_{\odot}$ in agreement with observations. There is, indeed, some evidence for a halo component around some galaxies. For M31, Roberts and Whitehurst[21] find a flat rotation curve (Δv = constant) out to 30 kpc implying $M \sim R$.

Since some dissipation is required in the formation of spiral galaxies, the optical size of a spiral galaxy may be smaller than that of its accompanying neutrino halo. Dissipation seems to be ruled out for ellipticals[22], so that if most of the mass is in neutrino halos, one should find $(M/L)_E \gtrsim (M/L)_S$; this seems to be the case.

Finally, for clusters of galaxies, $\Delta v_{100} \sim 10$ and $L \sim 10^2 L_*$ so that the large M/L ratios observed for many clusters can be due to neutrinos.

It must be emphasized that the above results are consistent with, but do not require the existence of, a background of massive, stable, relic neutrinos. The only evidence, albeit indirect, that could require such a background might come from the gamma rays (if any) which result from $\nu\bar{\nu}$ annihilation. Gunn et al[8] consider the constraints imposed by the observed γ-ray background.

Suppose a significant fraction of the annihilation energy is carried off in f^{-1} γ-rays; that is, in a typical annihilation there are f^{-1} photons each with energy (on average) $E_\gamma \gtrsim f m_\nu c^2$. Such annihilation photons would never dominate the observed isotropic γ-ray background. On the other hand, were there a neutrino halo around our galaxy, the flux from annihilation would exceed that observed unless $f \lesssim 0.1$. If theory and/or experiment can constrain f, the γ-ray observations can test the hypothesis of a background of massive relic neutrinos.

Conclusions

It was the purpose of the previous discussion to support the claim that there is a symbiotic relationship between particle physics and cosmology. The role of neutrinos in the Universe was chosen to illustrate the point. Although many of the details presented here may change with our changing ideas about the weak interaction and cosmology, the basic connection between particle physics and astrophysics will remain. The cosmological consequences depend on the details of the particle theories and, in turn, the cosmology can lead to constraints which restrict those theories.

In the present context it was shown that the weak interaction is in equilibrium during the early Universe. During this stage, neutrinos of all flavors and masses would have been present. As the Universe expands the weak interaction drops out of equilibrium leaving a remnant of relic neutrinos. These neutrinos may have a profound influence on processes occurring during the evolution of the Universe. Light or massless neutrinos can influence the expansion rate early on and effect the abundances of the elements produced during primordial nucleosynthesis. Limits to the primordial abundance of helium can constrain the numbers of "new" light neutrinos. On the other hand, if such new neutrinos should be discovered, the helium abundance then provides a constraint on the present density of nucleons, a poorly known but valuable cosmological parameter. If massive, stable neutrinos exist, their gravitational effects can influence the evolution of the Universe during recent epochs (and, at present). Indeed, limits to the deceleration of the expansion of the Universe place severe constraints on the mass such a stable neutrino may have. Neutrinos consistent with these constraints may dominate the masses of galaxies, clusters of galaxies and, even the Universe. Massive neutrinos are a fine hiding place for the "missing mass". Such neutrinos, however, are potentially detectable. Although, at low densities, neutrino pair annihilations are rare, the gamma ray flux from such annihilations may be detectable. Current, or future, gamma ray observations may confirm (or, exclude) the presence of a background of relic, massive neutrinos.

The subject of neutrino astrophysics is in a phase of rapid and exciting development. It is reassuring that the connection between particle physics and cosmology offers the possibility that laboratory experiments and astronomical observations can constrain our euphoric speculations.

References

(1) Steigman, G. 1973. Cargese Lect. Phys. 6, 505.

(2) Weinberg, S. 1972. Gravitation and Cosmology. New York: Wiley Interscience.

(3) Cowsik, R. and McClelland, J. 1972. Phys.Rev.Lett. 29, 669.

(4) Gott, J.R., Gunn, J.E., Schramm, D.N. and Tinsley, B.M. 1974. Ap.J. 194, 543.

(5) Lee, B.W. and Weinberg, S. 1977. Phys.Rev. Lett. 39, 165.

(6) Steigman, G. 1976. Ann. Rev. Astron. & Astrophys. 14, 339.

(7) Dicus, D.A., Kolb, E.N. and Teplitz, V.L. 1977, Phys.Rev. Lett. 39,168.

(8) Gunn, J.E., Lee, B.W., Lerche, I., Schramm, D.N., and Steigman, G. Ap.J. (Submitted November 1977).

(9) Demarque, P. and McClure, R. 1977. The Evolution of Galaxies and Stellar Populations (ed. B.M.Tinsley and R.B.Larson), 199.

(10) Meszaros, P. 1974. Astron. and Astrophys. 37, 225.

(11) Steigman, G., Schramm, D.N. and Gunn, J.E. 1977. Phys.Lett.66B, 202.

(12) Yang, J., Steigman, G., Schramm, D.N. and Rood,R.T. 1977, in preparation.

(13) Steigman, G., Sarazin, C., Quintana, H. and Faulkner, J. 1978, (Submitted to A.J., January 1978)

(14) Schramm, D.N. and Wagoner, R.V. 1977. Ann. Rev. Nucl. Sci., 27, 37.

(15) Wagoner, R.V. 1973. Ap.J. 179, 343.

(16) Lynden-Bell, D. 1967, M.N.R.A.S. 136, 101.

(17) Peebles, P.J.E. 1970. Astron.J. 75, 13.

(18) Gott, J.R. 1973, Ap.J. 186, 481.

(19) Schecter, P. 1976, Ap.J. 203, 297.

(20) Ostriker, J.P. and Peebles, P.J.E., 1973. Ap.J. 186, 467.

(21) Roberts, M.S. and Whitehurst, R.N. 1975. Ap.J. 201, 327.

(22) Gott, J.R. 1977. Ann. Rev. Astron. & Astrophys. 15, 235.

QUANTUM BLACK HOLES

M. J. Perry,
Department of Applied Mathematics and Theoretical Physics,
University of Cambridge,
Silver Street,
Cambridge,
England.

Gravitation is the weakest of all the interactions known to physics. In microscopic systems, for all intents it may be neglected. For example, in the hydrogen atom the contribution to the ground state energy is roughly 2.3×10^{-30}eV, then $E_{grav}/E_{tot} \sim 1.7 \times 10^{-31}$. Gravitation would remain too weak to be significant except for one important factor. The principle of Equivalence states that gravitation affects the motion of all bodies in the same way. This was first noticed by Galileo who performed experiments with an inclined plane and rolling bodies. He concluded that all bodies were attracted to the earth at the same rate. The equivalence principle has since been tested in many experiments. The most accurate experiment to date, performed by Dicke and his co-workers, enables one to conclude that Aluminium and Gold fall towards the Sun with the same acceleration to a precision of one part in 10^{11}.[1] The interpretation of the principle of equivalence is that gravitation couples directly to the energy-momentum tensor of matter. One knows from the requirements of quantum mechanics that energy densities in any system must always be positive. Consequently, one concludes that gravitation must be a purely attractive force. It is the fact that unlike electromagnetism, gravitation is always attractive, that makes it such an important force on large distance scales. These ideas were synthesized into Einstein's General Theory of Relativity which provides a unified description of the geometry of spacetime and the phenomenon of gravitation. Gravitational forces are seen to be a consequence of the curvature of space-time.

During the 1960's, Hawking and Penrose proved a series of theorems which govern the evolution of singularities in space-time.[2] These theorems show that once a "trapped surface" has formed in some region of space-time, the occurence of a singularity in the future is inevitable. A singularity is a region of space-time where the curvature has become infinite, and consequently all laws of physics must break down. A trapped surface is formed as a consequence of a relatively high concentration of matter such that its gravitational field is sufficiently large to prevent the escape of light from a particular region. Trapped surfaces are those surfaces from which light cannot escape.

An appropriate concentration of matter seems to be inevitable under certain circumstances. For example, gravitational collapse can only be prevented either by thermal pressure, or degeneracy pressure. Consider a large gas cloud, at first gravitational collapse will occur and be halted as the gas becomes sufficiently hot for nuclear reactions to begin, thus forming a star. Eventually, the stars' nuclear fuel will be exhausted and gravitational collapse will continue. It may be halted by electron degeneracy pressure. In this case a white dwarf star is formed. It is impossible for degeneracy pressure in such an object to prevent gravitational collapse if its mass is greater than the Chandrasekhar mass $M = 1.2M_{\odot}$.[3] For more massive objects, gravitational collapse can be halted by the degeneracy pressure of nucleons. The maximum equilibrium mass of a neutron star is not known exactly owing to various uncertainties in the nature of nucleon-nucleon strong forces. The most recent estimates seem to give estimates of about $1.8M_{\odot}$.[4] There is ample evidence for the existence of rapidly rotating neutron stars; these are pulsars which are frequently found in supernova remnants.[5]

For more massive objects nothing can halt the collapse. Furthermore a trapped surface will form once there is a surface of radius r such that the amount of matter inside m, is given by $m > r/2$. We use units such that $G = c = \hbar = k = 1$ throughout. For one solar mass, the corresponding radius is 3 km. Provided there is no catastrophic explosive process which causes massive stars to disrupt into small fragments during their collapse, the collapse must inevitably continue until this stage is reached.[6]

Once this has happened, a singularity will inevitably form. There are two possibilities. The first is that the singularity will remain surrounded by trapped surfaces. The outermost of these trapped surfaces is called the event horizon. This configuration is termed a black hole. The other possibility is that the singularity is not surrounded by a trapped surface. This is termed a naked singularity. A naked singularity is an unpleasant object. It is visible to observers at large distances : under such circumstances one is unable to make predictions about the evolution of physical systems. The unpalatibility of this possibility led Penrose to propose the Cosmic Censorship conjecture[7], which basically forbids the occurence of naked singularities. For objects which are not rotating, it seems certain a black hole will form as a consequence of collapse. For objects with angular momentum,the situation is less certain. A black hole cannot have an angular momentum J greater than M^2, where M is the mass of the hole. This means that in order for most stars to evolve into black holes, they would have to lose much of their angular momentum in the collapse process. Whilst it has not been shown that this must be the case, no counter-examples to Cosmic Censorship are known.

Next we consider where one might find black holes in the universe. Roughly speaking, the types of black hole which we might find are those which were generated in the early universe, and those which are the consequence of stellar collapse. Primordial black holes [8] may well turn out to be the most interesting to particle physicists. These black holes may have been formed in the early universe as a result of density fluctuations. Fluctuations occur in a random fashion: as a result of having a spontaneously overdense region, gravitational collapse can occur and a black hole results. The mass spectrum of such black holes depends on the precise nature of matter at very high densities ($\rho > 10^{15}$ g.cm^{-3}). However, one may expect a continuous range of masses from 10^{-5} gms up to 10^{43} gms ($10^{10} M_{\odot}$) to have been formed. Those with masses of less than 10^{15} gms will turn out to be the most interesting.

Once a black hole forms as a result of collapse, it will settle down to an equilibrium state. The time scale for this process is the light travel time across the event horizon.[9] All of the equilibrium states of black holes have their geometries described by the Kerr-Newman family.[10] These are a set of axisymmetric spacetimes. The only degrees of freedom in the Kerr-Newman family relate to the gravitational field - the mass of the black hole M, and its angular momentum J - and to vector gauge fields whose symmetry remains unbroken. The only known long-range vector field is electromagnetism. The only possible degrees of freedom that a black hole may have associated with electromagnetism, are the electric charge, Q, and the magnetic monopole charge. We also believe that colour is an exact symmetry of strong interactions, being described as a set of SU(3) gauge fields whose symmetry remains unbroken. Under these circumstances, a black hole can act as a source of this gauge field. The effect of this on geometry is the same as electric charge. Thus, in principle, a black hole can act as a source of colour.[11] Black holes cannot behave as a source of any other field. Since there are no gauge fields that couple to things like baryon number, electron number or muon number, one cannot determine the corresponding quantity associated with the hole. Thus, it is impossible to determine many details about the formation of a particular black hole. As an example, one could not tell whether a given black hole was formed from matter, anti-matter, or radiation since one cannot define its baryon number. A black hole is consequently described completely by its mass, angular momentum, electric charge, magnetic monopole moment and colour. In what follows, we shall ignore the magnetic monopole moment and colour.

There are a number of theorems which govern the equilibrium properties of black holes. They were first formulated by Bardeen, Carter and Hawking.[12] The surface gravity of a black hole, κ, defined as the gravitational acceleration evaluated on the event horizon. The zeroth law states that κ is constant on the event horizon. For the Kerr-Newman family

$$\kappa = \frac{(M^2 - Q^2 - J^2/M^2)^{\frac{1}{2}}}{2M^2 - Q^2 + 2M\,(M^2 - Q^2 - J^2/M^2)^{\frac{1}{2}}} \tag{1}$$

The first law relates to infinitesimal changes in the parameters describing the black hole

$$dM = \frac{\kappa dA}{8\pi} + \Omega dJ + \Phi dQ. \tag{2}$$

A is the surface area of the event horizon.

$$A = 4\pi\,[2M^2 - Q^2 + 2(M^4 - M^2Q^2 - J^2)^{\frac{1}{2}}] \tag{3}$$

Ω is the angular velocity of the hole

$$\Omega = \frac{J}{M(2M^2 - Q^2 + 2(M^4 - M^2Q^2 - J^2)^{\frac{1}{2}}} \tag{4}$$

and Φ is the electrostatic potential of the hole

$$\Phi = \frac{Q[M + (M^2 - Q^2 - J^2/M^2)^{\frac{1}{2}}]}{2M^2 - Q^2 + 2(M^4 - M^2Q^2 - J^2)^{\frac{1}{2}}} \tag{5}$$

The second law tells us how the surface area of the event horizon changes under any classical variation

$$\Delta A \geq 0. \tag{6}$$

Finally, there is the third law. This should rank as a conjecture rather than a law. If one believes the cosmic censorship hypothesis, then one cannot reduce κ to zero. The reason being that κ must be positive and real, or else one has a naked singularity and not a black hole. The Kerr-Newman solutions describe naked singularities if $J^2/M^2 + Q^2 \geq M^2$.

Classically, a black hole is regarded as an entity that is at a temperature of absolute zero since a black hole can only absorb things and never emit things. This is clearly at variance with the third law of thermodynamics. Hawking[13] showed that if one treated quantized fields on a classical geometry representing collapse to form a black hole, the resultant black hole behaves as if it has a temperature proportional to κ.

Here we give a treatment of a quantized field in equilibrium with a black hole enclosed in a box. Outside a black hole (with J = Q = 0), the geometry is described as the Schwarzschild metric

$$ds^2 = -\left(1 - \frac{2M}{\gamma}\right) dt^2 + \left(1 - \frac{2M}{\gamma}\right)^{-1} d\gamma^2 + \gamma^2(d\theta^2 + \sin^2\theta\, d\phi^2) \qquad (7)$$

γ is a radial co-ordinate, θ and ϕ are polar co-ordinates, and t is a time co-ordinate. This can be extended to give the entire Kruskal extension which is the maximally extended manifold related to Schwarzschild.

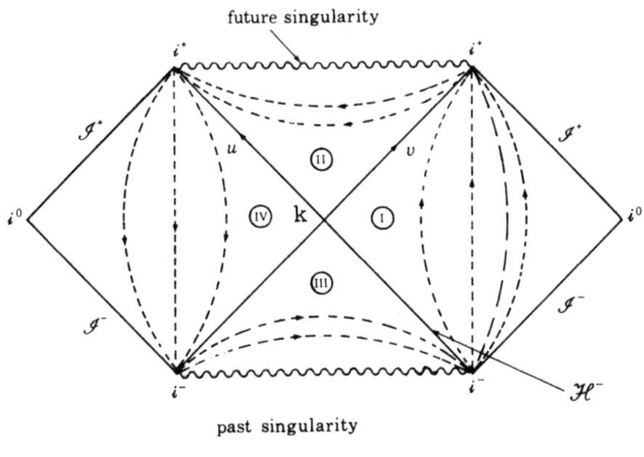

Fig. 1

Region I is the region exterior to the hole. Regions II and III represent the interior of the hole. Region IV is another region exterior to a related hole. H^+ and H^- are the past and future event horizons. The dashed lines represent the walls of the enclosing box. The metric over all of this manifold is given by

$$ds^2 = -\frac{32M^3}{\gamma}\exp\left(-\frac{\gamma}{2M}\right) dUdV + \gamma^2 d\theta^2 + \gamma^2 \sin^2\theta \, d\phi^2 \qquad (8)$$

where

$$UV = \left(1 - \frac{\gamma}{2M}\right)\exp\left(\frac{\gamma}{2M}\right) \qquad (9)$$

and

$$\frac{U}{V} = -\exp\left(\frac{t}{2M}\right) \qquad (10)$$

γ, t being defined by (7) in region I, where $U < 0$ and $V > 0$. One should not regard the Kruskal manifold as a physically attainable situation since the black hole represented here has always existed, and was not formed as the result of a collapse. This is why regions III and IV and H^- appear in this diagram. For realistic black holes they would not exist.

The scalar wave equation on this manifold is

$$\left(\Box - m^2\right)\phi = 0. \tag{11}$$

The propagator G(x,x') is defined on this manifold as a solution of the inhomogeneous equation

$$\left(\Box - m^2\right)G(x,x') = \delta(x,x') \tag{12}$$

As in flat space, there is no unique solution to this equation. The physically relevant Green'sFunction for propagation of fluctuations about the ground state is obtained by the Euclidicity postulate. For this situation we may adopt the same procedure.[14, 15] The result is that the propagator G(x,x') has the property of periodicity in imaginary time

$$G(\underline{x}, t; \underline{x}', t') = G(\underline{x}, t; \underline{x}', t' + i\beta) \tag{13}$$

where

$$\beta = \frac{2\pi}{k} = 8\pi M. \tag{14}$$

This is a direct consequence of the symmetry of the spacetime about the Kruskal crossover point, (k in figure 1). This symmetry is expressed through transformations of U and V, which leave the geometry invariant

$$U \to Ue^{2i\pi}$$
$$V \to Ve^{-2i\pi} \tag{15}$$

The property of periodicity in imaginary time indicates that one is dealing with a system in thermal equilibrium at a temperature of

$$T = \frac{1}{\beta} \tag{16}$$

Since the material must be in equilibrium with the black hole, we deduce that the temperature of the black hole must also be given by

$$T = \frac{1}{\beta} . \tag{17}$$

This result may be generalized to deal with situations where the black hole has $Q \neq 0$ and $J \neq 0$, and to the case of any other quantum field. The result is always that

$$T = \frac{1}{\beta} = \frac{k}{2\pi} . \tag{18}$$

For black holes where J = Q = 0,

$$T = 10^{-7}\left(\frac{M_{\odot}}{M}\right)\,^{\circ}K. \qquad (19)$$

hence for massive black holes the effect is rather unimportant.

We now see that the classical black hole theorems are nothing more than ordinary laws of thermodynamics, provided we identify the area of the event horizon with the entropy of the black hole, namely

$$S = \tfrac{1}{4}A \qquad (20)$$

The first law then becomes

$$dM = Tds + \Omega dJ + \Phi dQ. \qquad (21)$$

We can relate the entropy of the black hole to the information theoretic view of entropy. The entropy of a system is defined in terms of the number of ways it can be constructed from microscopic objects with the same macroscopic result.[16] The entropy of a hole of mass M, angular momentum $\underline{J}$, and charge Q is given by

$$S = \ln N, \qquad (22)$$

where N is the number of ways of forming such a black hole. We see the entropy of a black hole arising as a direct consequence of the uniqueness theorems for black holes which demonstrate one cannot determine how they were originally formed.

Since a black hole behaves like a hot body, it will emit thermal radiation at the appropriate temperature. The radiation will consist of all possible fundamental particles. At low temperatures it will emit only zero rest mass particles - photons, neutrinos and gravitons. Its luminosity is given by Stefans law

$$\begin{aligned} L &= \left(\frac{\pi^2}{15} n_b + \frac{7\pi^2}{120} n_f\right) \times T^4 \times A_{geom} \\ &= 3^2 (5\pi)^{-1} 2^{-15} (8n_b + 7n_f) M^{-2} \end{aligned} \tag{23}$$

n_b and n_f are the number of boson and fermion species respectively. A_{geom} is the geometric cross section of the black hole for radiation, which is $27\pi M^2$. We have put $Q = J = 0$ in this expression, the reason being that any black hole formed with charge or spin will rapidly lose it.[17] Since energy is conserved, the black hole will lose mass. A characteristic of gravitationally bound states is that they have negative specific heats. Thus the black hole will get hotter. As it gets hotter, extra particle species, mass m, will begin to be emitted when

$$m \sim T \sim \frac{1}{8\pi M}\,. \tag{24}$$

Roughly, the time evolution of the black hole's mass is given by

$$L = -\dot{M} \sim \frac{1}{M^2} \tag{25}$$

so

$$M(t) = (M_o^3 - 3t)^{1/3}. \tag{26}$$

Estimates have been made of the lifetime of black holes.[8, 18, 19] They yield, for a black hole of one solar mass, a lifetime of ~ 10^{58} Hubble times. For a black hole of 10^{14}gms, the lifetime is ~ 1 Hubble time. Any black hole in the present universe would be undetectable if its temperature was less than the temperature of the cosmic microwave background. Thus, only black holes of mass less than 6 x 10^{25}gms could possibly be visible. Even if small black holes were formed in the early universe, they would be difficult to detect. We could reasonably only expect to see black holes with masses of less than ~ 10^{14}gms.

Firstly, there is an upper limit on the possible number of such black holes which is derived from the limit on the hard gamma ray background. It turns out that this is roughly 10^4 black holes per cubic parsec. If, optimistically, these primordial black holes were clustered in galaxies, then the nearest one to Earth would be expected to be about one light-day away. On this basis, one reasonable way to observe them would be to look for the final burst of radiation as the black hole is vanishing. There are two possibilities here - either one believes in the quark soup type model,[20] or in Hagedorn type model,[21] of how matter behaves at very high densities. In the quark soup model one expects to be able to see very high energy gamma rays during the final phases. In contrast, if the Hagedorn model holds, the black hole terminates in a rapid explosion on a timescale of ~ 10^{-7} secs, giving a burst of gamma rays peaked at around 200 MeV. The energy in such an explosion would be about 10^{34} ergs . In either case, the phenomena would be very difficult to detect. For further details, see references 8, 18 and 19.

Perhaps a more fruitful approach is to search for characteristic bursts of electromagnetic radiation in the radio region of the spectrum.[22] These bursts would occur as a consequence of a sudden release of charged particles from the black hole into the interstellar magnetic field. In the Hagedorn model such a release does occur in the final burst, but the efficiency of the process is so small that the resultant pulse would be very difficult to detect.

There are other phenomena which could trigger the rapid ejection of charged particles from a black hole. It turns out that the phase transition in the Salam-Weinberg model at which the gauge symmetry of the W-bosons is restored could well cause such an event.[23, 24] This takes place at a temperature of

$$T = \frac{M_W}{e} \sim 600 \text{ GeV}. \tag{27}$$

It may well lead to a burst of radiation at millimetre wavelengths. Such radio pulses would be very conspicuous indeed.

A completely different way in which one might verify ones ideas on quantum gravity in relation to black holes, exploits some recent work on gravitational instantons.[25] We wish to consider the gravitational vacuum. Consider a timelike tube of radius r_o. One may determine the nature of what passes through the walls of the tube, but not what is inside the tube. The reason for this is that fluctuations inside may give rise to black holes, and their prescence will destroy information about what is in the tube. One may consider the density of states in the tube between energies E and E + dE. In terms of the vacuum persistance amplitude, Z

$$N(E) = \frac{1}{2\pi}\int Z(r)e^{-iEt}\,dt, \tag{28}$$

Z being given by

$$Z = \sum_n \langle n|\, e^{-iHt} |n\rangle \tag{29}$$

$|n\rangle$ being a complete set of states generated by the gravitational Hamiltonian H. In the path integral approach to quantum gravity [26]

$$Z = \int D[g]\,\exp(iI), \tag{30}$$

where I is the gravitational action. The integral is taken over all metrics which obey appropriate boundary conditions. Adopting the Euclidicity postulate to define this field theory, Z may be thought of as the partition function at a temperature $T = 1/\beta$, where

$$Z = \sum_n \langle n|\, e^{-\beta H} |n\rangle . \tag{31}$$

The integral in (30) is now taken over all fields which are periodic in imaginary time with period β. The dominant contribution to Z will come from metrics close to classical solutions that have the correct periodicity. The simplest possibility is flat space where I = 0. The next possibility is the Schwarzschild solution where

$$I = \frac{i\beta^2}{16\pi} = 4\pi i M^2. \tag{32}$$

One may evaluate the one loop corrections to this [27] and find that

$$\log Z = -4\pi M^2 + \frac{\pi^4 \gamma_0^3 \beta^{-3}}{17280} + \frac{212}{45} \ln\left(\frac{\beta}{\beta_0}\right) + \frac{i\pi}{2} + \text{lower order terms} \tag{33}$$

β_0 is a renormalization parameter, which is expected to be of order unity. The factor $\frac{i\pi}{2}$ arises because of the lack of positivity of the gravitational action. We thus find the density of states to be

$$N(E) = -\frac{1}{2\pi} \int_{-i\infty}^{+i\infty} iZ(\beta) e^{+\beta E} d\beta \sim 2e^{+4\pi E^2} \tag{34}$$

Thus we see that the microcanonical ensemble for a black hole is well-behaved.

The gravitational vacuum will be the microcanonical ensemble for E = 0. For this situation one must consider the possibility of having n black holes. Although there is no exact solution corresponding to this, one can consider the dilute gas approximation. The action then appears to be

$$I = \frac{in\beta^2}{16\pi} + \frac{212n}{45}\ln\left(\frac{\beta}{\beta_0}\right) \tag{35}$$

However, this action is not quite correct. We must introduce a set of collective co-ordinates so as to describe the possible translations of the black holes within the tube. This will introduce an overall factor of V^n, where V is the volume of the box which the tube describes. Also, each of our black holes will be indistinguishable; this introduces a factor of $n!$ Thus the action will be

$$I = \frac{V^n}{n!}\left[\frac{in\beta^2}{16\pi} + \frac{212n}{45}\ln\left(\frac{\beta}{\beta_0}\right)\right]. \tag{36}$$

This action is now extremised by the choice $\beta \sim \beta_0$ and $n \sim V$. Thus we expect the gravitational vacuum to be a collection of Planck mass black holes separated from each other by distances of the order of the Planck length.

This idea leads to certain effects that otherwise could not take place. For example, consider a simplified model of a proton as being a bag of radius 10^{-13}cm containing three pointlike quarks. If one quark encounters a black hole, then by conservation of colour and charge, we expect it to emerge as the same object. However, if all three quarks fell into a black hole, the only constraint would be that the final state be a colour singlet and have a charge of +1. Thus the final state could be a positron and a photon. The baryon number violating decays of a proton have an estimated partial lifetime of 10^{68} years. If the muon were a pointlike object, it too could have a gravitational decay mode

$$\mu^{+} \rightarrow e^{+} + \gamma,$$

and this is suspected to have a very much shorter lifetime. The calculation of this is presently being pursued.

Another possible way to observe might be in interference experiments over long path lengths. The phase of a particle going into a black hole will not be correlated with the phase of the outgoing particle. Thus, interference experiments over long distances may lead to the experimental confirmation of some of these ideas.

REFERENCES

1. Roll, P.J., Krotkov, R., and Dicke, R.H., Ann.Phys. (NY), 26, 442, 1964.
2. Hawking, S.W., and Ellis, G.F.R. "The Large Scale Structure of Space-time", C.U.P., London, 1973.
3. Chandrasekhar, S., Ap.J, 74, 81, 1931.
4. Anderson, P.W., personal communication.
5. Smith,F.G., "Pulsars", C.U.P., London, 1977.
6. Misner, C.W., Thorne, K.S., and Wheeler, J.A., "Gravitation", W.H.Freeman and Co., San Francisco, 1973.
7. Penrose, R.
8. Carr, B.J., Ap.J. 201, 1, 1975., Ap.J. 206, 8, 1976.
9. Doroshkevich, A.G., Zel'dovich, Ya.B., and Novikov, I.D., Sov.Phys.J.E.T.P., 22, 122, 1966.

10. Carter, B., "Black Hole Uniqueness" in Proceedings of the Marcel Grossman Meeting, Trieste, 1975, ed. R. Ruffini, North-Holland, Amsterdam, 1977.

11. Perry, M.J., Phys.Lett.B71, 234, 1977.

12. Bardeen, J.M., Carter, B., and Hawking, S.W., Comm.Math.Phys. 31, 161, 1973.

13. Hawking, S.W., Comm.Math.Phys, 43, 199, 1975.

14. Hartle, J.B., and Hawking, S.W., Phys.Rev.D 13, 2188, 1976.

15. Gibbons, G.W., and Perry, M.J., Phys.Rev.Lett. 36, 985, 1976,Proc.Roy.Soc.A., To be published.

16. Hawking, S.W., Phys.Rev.D. 13, 191, 1976.

17. Gibbons, G.W., Comm.Math.Phys. 45, 191, 1975.

18. Page, D.N., and Hawking, S.W., Ap.J., 206, 1, 1976.

19. Carter, B., Gibbons, G.W., Perry, M.J., Lin, D.N.C., Astronomy and Astrophysics, 52, 427, 1976.

20. Collins, J.C., and Perry, M.J., Phys.Rev.Lett. 34, 1353, 1975.
Perry, M.J., Ph.D. dissertation, University of Cambridge, unpublished.

21. Hagedorn, R., "The Thermodynamics of Strong Interactions" in Cargese Lectures in Physics, Vol.6 ed. E. Schatzman, Gordon and Breach, New York, 1973.

22. Rees, M., Nature, 266, 333, 1977.

23. Weinberg, S., Phys.Rev. D9, 3357, 1974.

24. Perry, M.J., in preparation.

25. Hawking, S.W., The Gravitational Vacuum" in preparation.

26. Gibbons, G.W., and Hawking, S.W., Phys.Rev. D15, 2452, 1977.

27. Gibbons, G.W., Hawking, S.W., and Perry, M.J., "Path Integrals and the Indefiniteness of the Gravitational Action", preprint, 1977.

BLACK-HOLE CHEMICAL POTENTIAL AND PARTICLE CONSERVATION

John Archibald Wheeler

Center for Theoretical Physics
The University of Texas at Austin
Austin, Texas 78712

Whether the laws of conservation of baryon and lepton numbers are upheld is a question of black-hole physics as well as a question of particle physics. In the Bekenstein-Hawking process of black-hole radiance, as presently considered, equal numbers of particles and antiparticles are evaporated by a small black hole ($m<10^{16}$g) as it disappears; this equality is unaffected by any inequality, no matter how great, between the numbers of particles and antiparticles that went to make the black hole in the first place. Contrary to earlier considerations[1] it is no objection to this problem of principle that it would take unbelievable energy and technology[2] to crush so small a mass to a black hole.

Two very different positions suggest themselves.

(1) The conservation laws are violated.

(2) The conservation laws are upheld, and upheld by the existence of black hole chemical potential[3].

Consider the standard expression for the rate of evaporation

$$N = \frac{v\ dk}{2\pi} \sum_{j,m,\pi} \Gamma_{\omega,j,m,\pi} \frac{1 \text{ or } \hbar\omega}{e^{C/T} \pm 1} \qquad (1)$$

where N is number of quanta or amount of energy emitted per unit time in specified interval dk, with total angular momentum

j, angular momentum m about some axis, and polarization π. Here Γ, normally in the range of 0 to 1, is the "access factor" and the temperature T is the multiple $(\hbar/2\pi)$ of the surface gravity. Present theory gives for the "cost of emission" C

$$C = \hbar(\omega - m\Omega) - e_i\Phi \qquad (2)$$

where $m\hbar$ is the azimuthal angular momentum of the emitted quantum or particle, e_i is its charge, Ω is the angular velocity of the black hole, and Φ is the electrical potential at its surface. No one sees how any chemical potential can arise that could subtract from the "cost of emission" C a chemical potential μ_i, as happens elsewhere in physics[4]. However, it may be appropriate to reword the remark of Eddington, when it was objected to him that thermonuclear reactions cannot take place in stellar interiors, "We do not argue with the critic who urges that the stars are not hot enough for this process; we tell him to go and find *a hotter place*"[5]. If no mechanism is now envisaged that would produce such a chemical potential, it may be necessary to look further afield. Piling up of particles at the horizon of the black hole, such as would bring the Pauli exclusion principle into play, is excluded by the theory of the horizon as presently understood. But particle couplings are described in terms of the Yang-Mills vacuum; the geometry around a black hole is described in terms of the Einstein vacuum; and these are not too distinct vacuums, but one and the same vacuum. Nowhere more than in this circumstance does it seem appropriate to look for the origin of a black hole chemical potential.

References

1. Wheeler, John A. "Conference Summary: More Results Than Ever in Gravitation Physics and Relativity," in General Relativity and Gravitation (G. Shavin and J. Rosen, ed.), John Wiley and Sons, N. Y. 1975.
2. Harrison, B. K., Thorne, K. S., Wakano, M. and Wheeler, J. A. Gravitation Theory and Gravitational Collapse, University of Chicago Press, Chicago, Illinois, p. 81. 1965.
3. Wheeler, John Archibald. "The Black Hole," in Proceedings of the Sixteenth Solvay Conference at the University of Brussels, September, 1973, Astrophysics and Gravitation, Université de Bruxelles, p. 303, 1974.
4. Landau, L. D. and Lifshitz, E. M. Statistical Physics (Translated from Russian by E. Peierls and R. F. Peierls), Pergamon, London, and Addison-Wesley, Reading, MA., 1958.
5. Eddington, A. S. The Internal Constitution of the Stars, Cambridge University Press, London, p. 301, 1926.

THE SOLAR NEUTRINO PROBLEM

William A. Fowler
California Institute of Technology, Pasadena, California 91125 USA

It has been almost 20 years since I wrote to Raymond Davis, Jr. of Brookhaven in January, 1958 suggesting that the few counts he was observing in his $^{37}Cl/^{37}Ar$ neutrino-detector below the Savannah River reactor might be due to solar neutrinos from the decay of ^{8}B in the sun.[1,2] The few counts that Davis[3,4] observed were less than expected from fission-product anti-neutrinos from the reactor if $\overline{\nu}$ were identical to ν. It has become clear that most of the counts he observed at that time were due to now understood backgrounds.

Twenty years later, some say the solar neutrino problem is still with us. Others say there never was a problem. My purpose today is to present this ambiguous situation to you in such a way that you can make your own judgment. I must confess at the outset that I will not be able to conceal my own pre-judice. Also, to give a 20 min talk on a 20 year old problem is well nigh impossible.

The situation is this: the observations by Davis and his collaborators yield at most one third the detectable neutrino flux at the earth which is predicted using the standard solar model with the best nuclear and atomic input data. The ambiguity arises because there are uncertainties in both the observational background and in the experimental input into the theoretical models. The whole question boils down to: how many standard deviations separate observation and prediction?

The Production of Neutrinos in the Sun

Let us start at the beginning. How are neutrinos produced in the sun? ill discuss only the proton-proton chain. As John Bahcall first pointed , if the carbon-nitrogen cycle dominated solar energy production, then the ar neutrino flux would be simply calculable on the basis of two neutrinos ı time the conversion of four protons into one helium nucleus produced ՎeV or 4×10^{-5} ergs of energy. The known energy flux from the sun at the th is equal to 2.0 cal min^{-1} cm^{-2} (solar constant) and translates into $\times 10^{10}$ neutrinos sec^{-1} cm^{-2}. The resultant detectable flux would be times that permitted by the present observations. Experimental reaction es confirm that the proton-proton chain dominates over the CNO cycle.

The proton-proton chain, Table I, produces detectable neutrinos mainly the result of several temperature sensitive branchings in nuclear proces- In our current view, it is a solar thermometer. It indicates that the tral temperature of the sun is less than that necessary to support the in hydrostatic equilibrium against gravitational collapse on the basis standard solar models to be described later. In Table I note the four trinos from the primary pp-reaction, from the pep-reaction, from electron ture by ^{7}Be and from the decay of ^{8}B. As we shall see, only the last ee have energies above the threshold for detection by Davis.

The pep-reaction is relatively infrequent and a straightforward calcu- ion[5] indicates that it occurs once for 400 primary pp-reactions. The x of neutrinos from ^{7}Be depends on the ratio of rate of the $^{3}He(\alpha,\gamma)^{7}Be$ ction to that of the $^{3}He(^{3}He,2p)^{4}He$ reaction. The flux from ^{8}B depends that branching ratio as well as on the ratio of the rate of the $(p,\gamma)^{8}B$ reaction to that of $^{7}Be(e^{-},\nu)^{7}Li$. Those ratios depend on experi- tal measurements in the nuclear laboratory. The measurements and their

Table I. The pp tri-chain

$H^1 + H^1 \rightarrow D^2 + e^+ + \nu$	$Be^7 + e^- \rightarrow Li^7 + \nu$
$H^1 + e^- + H^1 \rightarrow D^2 + \nu$ (pep)	$Li^7 + H^1 \rightarrow He^4 + He^4$
$D^2 + H^1 \rightarrow He^3 + \gamma$	OR
$He^3 + He^3 \rightarrow He^4 + H^1 + H^1$	$Be^7 + H^1 \rightarrow B^8 + \gamma$
OR	$B^8 \rightarrow Be^{8*} + e^+ + \nu$
$He^3 + He^4 \rightarrow Be^7 + \gamma$	$Be^{8*} \rightarrow He^4 + He^4$

OVERALL:

$$4p \rightarrow \alpha + 2e^+ + 2\nu$$
OR
$$4p + e^- \rightarrow \alpha + e^+ + 2\nu$$
NUCLEAR

$$4H^1 \rightarrow He^4 + 2\nu$$
ATOMIC

analysis have been reviewed by Kavanagh.[6] In Fowler[7] I suggested that a low energy resonance in $^{3}He(^{3}He,2p)^{4}He$ would reduce the production of ^{7}Be and ^{8}B. Dwarakanath[8] extended his earlier measurements on this reaction down to 30 keV which is only slightly above the most probable reaction energy in the sun. Figure 1 shows that he was able to detect a cross section as low as 0.06 nanobarns = 6×10^{-35} cm^{2}. He found no sign of resonance and confirmed this with thick target measurements. Furthermore, there is no evidence for an excited state in ^{6}Be near the relevant energies.

All laboratory measurements are subject to error which can be specified as a standard deviation. It must be emphasized that the errors translate into an uncertainty, also expressed as a standard deviation, in the neutrino flux calculated on the basis of any solar model, standard or otherwise. I shall return to this in what follows.

At this point let me comment on a recent "development." There was a flurry of astonishment when Slobodrian et al.[9-11] claimed evidence for the reaction $^{3}He(^{3}He,e^{+}\nu)^{2}D + {}^{4}He$ and suggested that the rate for $^{1}H(p,e^{+}\nu)^{2}D$ in the sun might have been underestimated by a factor of 10^{8} with the consequence that ^{7}Be and ^{8}B are not produced in the sun. Newman and Fowler[12] showed that an increase by 10^{8} in the primary reaction of the pp-chain was completely incompatible with current models of solar structure and with other empirical determinations of the strength of the weak interaction, particularly from the decay rate of $^{10}C(e^{+}\nu)^{10}B$ in which two alpha particles are spectators while one of the other of two "valence" protons transforms to a neutron just as in $^{1}H(p,e^{+}\nu)^{2}D$. The flurry subsided when Davies et al.[13] were unable to confirm the results of Slobodrian and his co-workers and suggested particle misidentification as the explanation for Slobodrian's results.

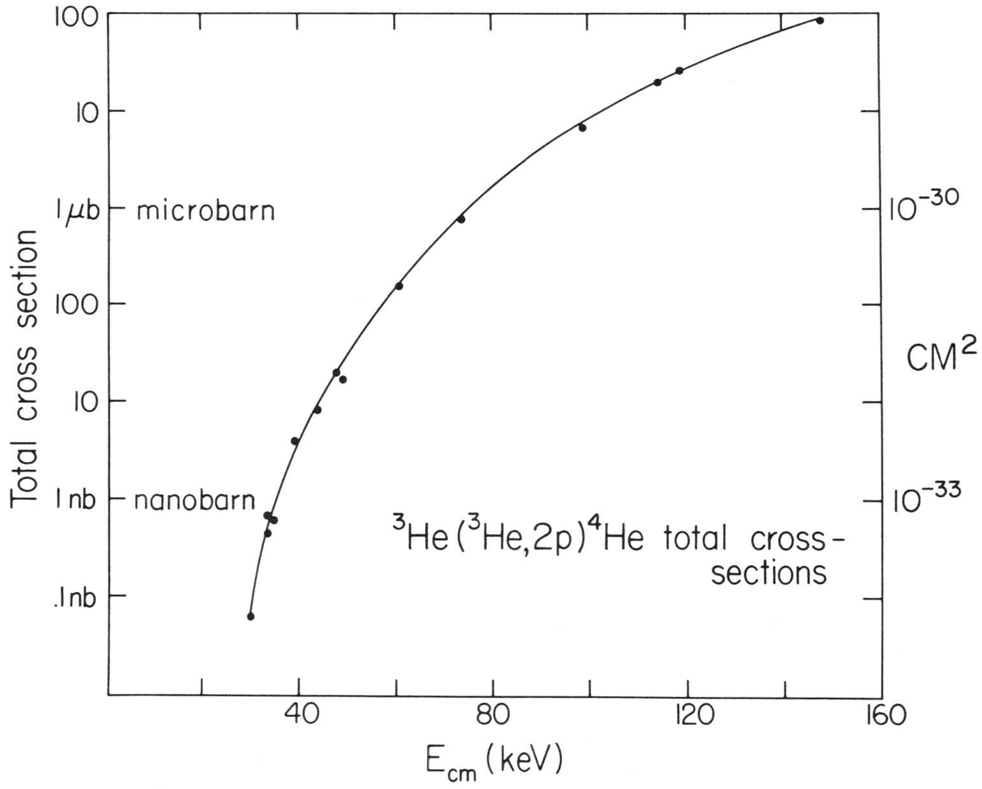

Fig. 1. Cross Section for the $^3He(^3He,2p)^4He$ reaction. See text for discussion.

The Detection of Solar Neutrinos

Table II lists the numerous neutrino absorption reactions which have been suggested as detectors of solar neutrinos. Neutrino electron scattering is not included but is another possible detector once charged current contributions versus neutral current contributions are clarified (see F. Reines, these Proceedings). The table also lists solar neutrino energies for comparison with the thresholds for the detection reactions. This discussion will be limited to the Davis results using the Pontecorvo-Alvarez technique based on $^{37}Cl(\nu,e^-)^{37}Ar(e^-,\nu)^{37}Cl^*(e^-_{AUGER})^{37}Cl$ in which the 2.8 keV Auger electrons emitted in electron capture by ^{37}Ar are detected. (The asterisk on ^{37}Cl indicates atomic excitation.) Note, however, that only ^{71}Ga, ^{115}In and ^{205}Tl have thresholds below the maximum energy of the primary pp-reaction neutrinos. There is a general consensus that a detector sensitive to these neutrinos must eventually be employed to reach a "solution" of the solar neutrino problem. The difficulties are formidable. For example, it has been estimated that 50 metric tons of ^{71}Ga are needed to make a significant observation on the pp-neutrinos. The present cost of the gallium alone is 15 million dollars. On an encouraging note, Davis (private communication) has found it possible to extract small amounts of germanium from 200 kilogram samples of gallium. For additional discussion of neutrino detectors see Davis and Evans.[14]

The absorption of neutrinos by ^{37}Cl to produce the ground state of ^{37}Ar has a threshold at 0.814 MeV neutrino energy and so is sensitive only to the monoenergetic (1.442 MeV) pep-neutrinos, to the monoenergetic ^{7}Be-neutrinos (90% at 0.862 MeV) and to most of the ^{8}B-neutrinos which have a continuous spread in energy from zero up to a broad maximum at 14.0 MeV. It is important to note that the cross section for detection varies approximately as the

Table II. Reactions for solar neutrino detection

REACTION	DETECT	HALF LIFE	THRESHOLD (MeV)
$^{37}C\ell(\nu,e^-)^{37}Ar$	2.8 keV e^-_{AUGER}	34.8 d	0.814
$^{37}C\ell(\nu,e^-)^{37}Ar^a(\gamma)$	"	"	5.803
$^{7}Li(\nu,e^-)^{7}Be$	0.478 MeV γ	53.4 d	0.862
$^{71}Ga(\nu,e^-)^{71}Ge$	9.3 keV e^-_{AUGER}	11.0 d	0.235
$^{115}In(\nu,e^-)^{115}Sn$	e^- + 0.116 MeV γ 0.498 MeV γ	3.3 μs PROMPT CASCADE	0.128
$^{205}T\ell(\nu,e^-)^{205}Pb^m(\gamma)$	^{205}Pb (MASS SPECTRA)	14 My (OLD Tℓ ORES)	0.046

SOLAR NEUTRINO ENERGIES

SOURCE	ENERGY (MeV)	SOURCE	ENERGY (MeV)
pp	0.420 MAX	^{8}B	14.0 BROAD MAX
pep	1.442	^{13}N	1.20 MAX
^{7}Be	0.862 (0.9) 0.384 (0.1)	^{15}O	1.73 MAX

square of the neutrino energy above the threshold. In addition, the analog state of the ground state of ^{37}Cl in ^{37}Ar at 4.989 MeV can be produced by neutrinos with energy greater than 5.803 MeV.

Table III lists the recent neutrino absorption cross sections for ^{37}Cl calculated by Bahcall[15] who discusses in detail his original calculations and those of others. In broad brush the neutrino capture rate to the ground state of ^{37}Ar is straightforwardly calculated from the measured ft-value for the electron capture by that state while the capture rate for ^{8}B neutrinos to the analog state, $^{37}Ar^{a}$, and other excited states of ^{37}Ar can be calculated in an empirical fashion using measured branching ratios in the isotopic analog decay $^{37}Ca(e^{+}\nu)^{37}K$. The importance of the analog state in ^{37}Ar lies in the fact that the transition to this state is super-allowed.

Table III also lists the flux of the various pp-chain and CN-cycle neutrinos from the sun at the earth calculated on the basis of the standard model and the so-called low Z model. By the standard model is meant, *inter alia*, a spherically symmetric solar model with support against gravity due to ionized gas pressure, plus a small amount of radiation pressure, using experimentally known or semiempirical-theoretical nuclear reaction rates and atomic opacities plus radiative heat transfer in the solar interior and convective heat transfer in the outer convective layer. For full details of the constraints on the standard model see Ulrich.[16] The heavy element mass fraction, Z = 0.02, is taken to be that observed in the solar photosphere. There is a standard model for each active investigator, differing slightly in input data and in computational techniques, but there is a remarkable consensus on the neutrino flux from practically all standard models. For the low Z model the heavy element abundance in the convective outer layer of the sun is taken to be 0.02 but in the radiative interior it is taken to be

Table III. Capture rates for solar models

(Bahcall et al., 1973; Bahcall, 1977)

ν	σ	STANDARD (Z = .02)		LOW Z(.002)	
		$\varphi(\oplus)$	$\sigma\varphi$	$\varphi(\oplus)$	$\sigma\varphi$
	10^{-46} cm^2	cm^{-2} s^{-1}	SNU	cm^{-2} s^{-1}	SNU
pp	0	6.1×10^{10}	0	6.3×10^{10}	0
pep	15.4	1.5×10^{8}	0.23	1.6×10^{8}	0.25
^{7}Be	2.35	3.8×10^{9}	0.90	1.4×10^{9}	0.33
^{8}B	1.08×10^{4}	3.2×10^{6}	3.45	5.4×10^{5}	0.58
^{13}N	1.63	2.6×10^{8}	0.04	1.2×10^{7}	0.002
^{15}O	6.59	1.8×10^{8}	0.12	4.3×10^{6}	0.003
TOTAL			4.74 SNU		1.16 SNU

one-tenth of this value.

Finally, for the two models, Table III lists the product of flux times cross section in solar neutrino units, SNU, defined as one capture per second per 10^{36} targets (^{37}Cl). Divide the number of SNU by 5.24 to obtain the number of counts per day expected in the Davis detector after correction for finite exposure and detection periods. (The half-life of ^{37}Ar is 34.8 d.)

As mentioned previously there are uncertainties in all of the empirical input data. These can be translated into uncertainties (standard deviation) of the calculated SNU-values using the methods of Bahcall et al.[17] A detailed analysis has been made by Ulrich[16] and is reproduced in Table IV. Careful study of Tables III and IV will reveal that the calculated values and standard deviations when rounded off become 4.7 ± 1.6 SNU for the standard model and 1.2 ± 0.3 SNU. These will be discussed after the observational data are presented.

Neutrino Observations

The Brookhaven Neutrino Observatory built by Davis consists of a 100,000 gallon tank of perchlorethylene, $C_2{}^{35}Cl_3{}^{37}Cl_1$, situated 4850 feet deep in the Homestake Gold Mine in Lead, South Dakota. The chamber in the mine in which the tank is located can be flooded with water to reduce background from neutrons produced by radioactive α,n-reactions in the rock walls. The known ^{37}Ar backgrounds are due to μ and ν produced by cosmic rays in the atmosphere. This background has been extrapolated from measurements at higher levels in the mine. After a suitable exposure time Davis flushes out the noble gas, ^{36}Ar or ^{38}Ar tracers plus neutrino produced ^{37}Ar, with another noble gas helium, then separates the argon from the helium in a cold trap and introduces the argon into a small proportional counter. By a number of tests he

Table IV. Percentage errors in solar neutrino flux computations

R. G. Ulrich, NEUTRINOS 1974 (Philadelphia), p. 270

Quantity	Standard Model		Low Z Model	
	Error in quantity	Error in $\Sigma\sigma\varphi$(SNU)	Error in quantity	Error in $\Sigma\sigma\varphi$(SNU)
1. Luminosity	1%	6%	1%	4%
2. Nuclear cross sections				
$S_{1,1}$	5	13	5	10
$S_{3,3}$	11	3	11	1.5
$S_{3,4}$	17	14	17	11
$S_{1,7}$	9	7	9	4
"$S_{e,7}$"	10	8	10	5
3. Opacity between T_1 and T_2 (10^6 K)				
10 15	10	13	5	7
7 10	15	9	5	3
5 7	15	9	5	3
3.5 5	15	6	5	3
4. Equation of state				
near $T = 8 \times 10^6$ K	1	8	1	8
Elsewhere	1	8	1	8
5. ν capture cross sections	10	10	10	10
RMS error in $\Sigma\sigma\varphi$(SNU)		33%		24%

has shown that he can quantitatively recover (> 90%) ^{37}Ar which he introduces into the tank or produces in the tank with a neutron source. Up until the present there has remained the problem whether the ^{37}Ar produced by neutrino capture would recoil out of the perchlorethylene molecule. He has solved this by quantitatively recovering ^{36}Ar from the decay of ^{36}Cl introduced into the perchlorethylene (private communication). The recoil given to ^{36}Ar in the decay $^{36}Cl(e^{-}\nu)^{36}Ar$ is less than that involved in the ^{37}Ar production.

Fig. 2 illustrates a sample of the data obtained by Davis and his collaborators. ^{37}Ar decays are isolated from background by a double signature. The energy spectrum (pulse height) of ^{37}Ar introduced into the proportional counters can be measured before exposure and only those events which fall within the full width at half-maximum are counted. The ions and electrons produced in the counter gas by the Auger electrons from ^{37}Ar are localized to a region small compared to the dimensions of the counter. Thus the ^{37}Ar pulses show a fast rise time and a large amplitude when differentiated. Such pulses fall between the two lines at roughly 45° as a function of energy in Fig. 2. It will be clear that valid ^{37}Ar events fall in the trapezoid indicated in the figure. Empirical corrections can be made for the few events which fall outside this trapezoid. The figure tells its own story regarding the exclusion of unwanted background.

Table V summarizes the results of Davis and his collaborators (private communication) in their Runs 18 through 47 over the period 1970 to 1977. Runs 1 through 17 only used energy discrimination in identifying ^{37}Ar events and suffered from large background to signal ratios. The runs have been systematically and consistently analyzed by Bruce Cleveland (private communication) by the method of maximum likelihood and the final result and standard deviation as given in Table V is 1.6 ± 0.3 SNU.

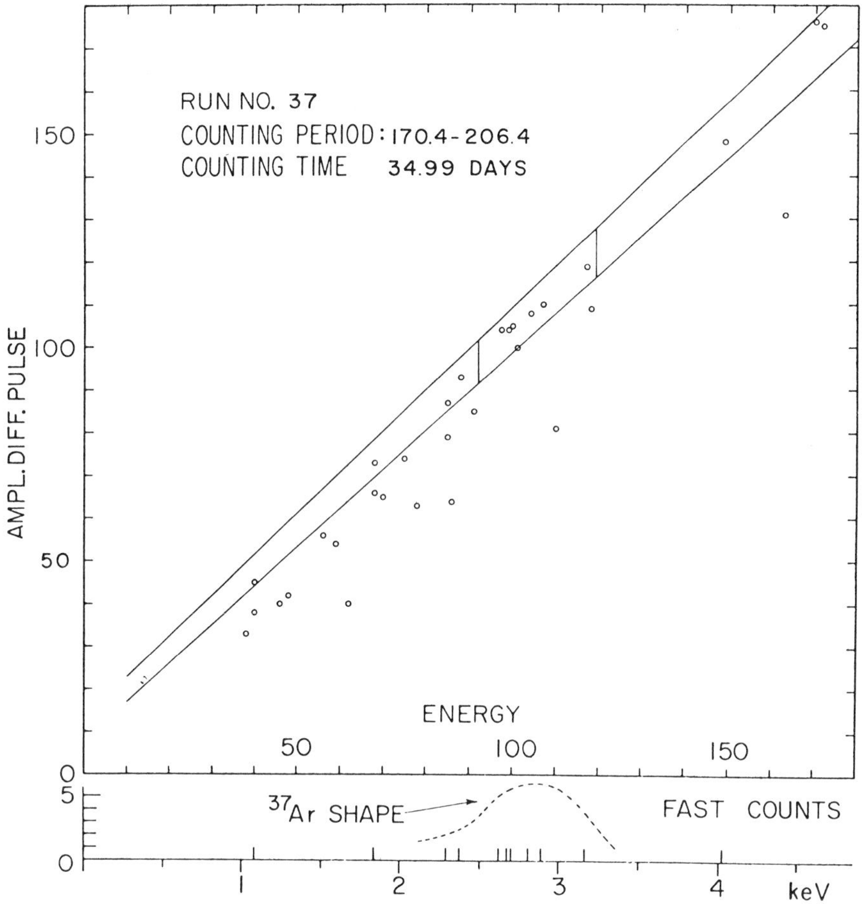

Fig. 2. Data from Run No. 37 obtained by Davis and his coworkers. See text for discussion.

Table V. Solar neutrino flux

Davis, Cleveland, et al. (1977)

RUNS 18 THRU 47	1970 to 1977
AVERAGE ^{37}Ar PRODUCTION RATE	= 0.38 ± 0.06
COSMIC RAY BACKGROUND ($\mu + \nu$)	= 0.08 ± 0.024
^{37}Ar FROM SOLAR NEUTRINOS (?)	= 0.30 ± 0.065 D^{-1}
NEUTRINO FLUX ABOVE KNOWN BACKGROUND → (5.24x)	= 1.6 ± 0.3 SNU
NO TIE TO THE SUN!	
STANDARD SOLAR MODELS:	
BAHCALL, et al. (1973); ULRICH (1974)	= 5.6 ± 1.8 SNU
BOLTON (1975)	= 5.3 ± 1.7
NEWMAN & FOWLER (1975)	= 5.1 ± 1.7
BAHCALL (1977)	= 4.7 ± 1.6
DIFFERENCE ≈ 2 STAN. DEVIATIONS	

Davis and Cleveland emphasize that the observation of 1.6 ± 0.3 SNU is that above known background and that there is no tie to the sun. When the 100,000 gallon tank detector was planned, the predicted count rate corresponded to 30 SNU. It was contemplated that careful measurements might just detect the 7% difference (maximum) when the earth is at perihelion in its orbit around the sun relative to aphelion. No chance, now!

Table V also lists a series of standard solar model calculations (1973 to 1977) culminating in the most recent ones by Bahcall.[15] Bahcall gives details of the changes in mass-37 empirical data which account for the change from 5.6 ± 1.8 SNU in 1973 to 4.7 ± 1.6 SNU in 1977. It is worth noting that this change is only one-half a standard deviation. What is most important is that the value calculated for the standard model is approximately two standard deviations (almost entirely arising in the calculations) from the maximum solar neutrino value permitted by the observations. The value calculated with the low Z model is very close to that observed taken at face value.

Discussion

The ambiguity in individual reactions to the solar neutrino problem lies in the two standard deviations of the standard model predictions from the observations. Some say "so what"; others frown and shake their heads. A number of people have suggested variations from the standard model. Almost all of these suggestions attempt to lower the interior temperature of the sun in order to eliminate the production of ^7Be and ^8B which decreases rapidly with decreasing temperature due to Coulomb barrier effects in the nuclear reactions involved. Lowering the central temperature of the sun by approximately 10% will do the trick. These suggestions will not be discussed here. Some are discussed in Bahcall.[15] In my book, all of them, including my own, meet with severe difficulties and raise more problems than they solve.

As an example, consider the low Z model. The opacity of solar material depends largely on its heavy element content. Thus, lowering the heavy element mass fraction by a factor of ten in the radiative interior lowers the temperature gradient from the surface to the center and thus lowers the central temperature. The ^{7}Be neutrino flux decreases to 37% of that in the standard model and the ^{8}B neutrino flux to 17%.

There is a price to pay, however. In order to achieve a solar model with all the properties of the sun (mass, luminosity, radius, surface temperature) the low Z model requires a shallow convective layer with M(Conv. Layer) = 7.5×10^{-4} M(⊙) [M(⊙) = total mass of the sun]. The standard model requires a much thicker layer with M(Conv. Layer) = 5.0×10^{-2} M(⊙) (Ulrich et al.[18]). At first sight this seems an advantage for the low Z model since it is then easier to accrete the heavy element mass fraction, Z = 0.02, observed in the convective layer as discussed by Talbot and Newman[19]; see earlier references therein. However, Michaud[20] shows that thin convective layers lose their heavy element content by diffusion into the radiative interior much more rapidly than thick layers. Moreover, Ulrich et al.[18] show that their observations on the 5 minute nonradial p-mode oscillations on the sun require M(Conv. Layer) > 10^{-2} $M_{\odot}$ and that the observed surface abundance of lithium requires M(Conv. Layer) < 10^{-1} $M_{\odot}$. The standard model convective layer falls within these limits; that of the low Z model does not!

It is worth mentioning that there may be a difficulty even with the standard model, as pointed out by Rood et al.[21] It is well known that main sequence stars like the sun produce substantial quantities, 3×10^{-3} by mass, of ^{3}He in the mass range from 0.5 to 0.7 M(⊙) where hydrogen burning does not go to completion to ^{4}He. During the red giant stage this is mixed into the rest of the envelope which then contains $\sim 10^{-3}$ of ^{3}He by mass. This is lost slowly to the interstellar medium or eventually ejected into it in a supernova explosion. The main sequence stars under consideration have been evolving into red giants only since the sun formed so there should have been a substantial enrichment of ^{3}He in the interstellar medium in the last 5×10^9 years. Observations on the interstellar medium and on planetary nebulae do not seem to show this (Predmore et al.[22]). There may be other explanations such as the continuing infall of primordial gas (^{3}He free) into the Galaxy but there are ways to avoid the ^{3}He ejection which also suggest a solution of the solar neutrino problem. Continuous or transient mixing in the solar interior brings the ^{3}He to the central regions of the sun. The ^{3}He is an energy rich, rapidly burning fuel and the central temperature is reduced accordingly. Bahcall[15] finds 1.5 SNU for a continuously mixed model. Transient mixing can lead to less than 0.1 SNU for several million years. There are many problems but the basic one is that all calculations indicate that the radiative interior of the sun is very stable against convective mixing. There is the additional possibility that the excess ^{3}He diffuses into the central region of the sun and enhances $^3He(^3He,2p)^4He$ relative to $^3He(\alpha,\gamma)^7Be$ with consequent reduction in the production of ^{7}Be and ^{8}B. It remains to be seen if the diffusion is fast enough.

Finally, it is impossible to conclude without mentioning at this Conference the possibility that the sun may be trying to teach us something about the properties of neutrinos as elementary particles. Bahcall, Cabibbo, and Yahil[23] suggested that the electron neutrino, with mass up to the 60 eV/c^2 limit from observations on the tritium beta-spectrum, might decay in the 8 minute passage from sun to earth to some other type of neutrino, such as the muon neutrino, plus some other particle such as a massless scalar or pseudoscalar. Earlier, Pontecorvo[24] had suggested oscillations between ν_e and ν_μ as a way to reduce the expected solar neutrino counting rate by a factor of two. Make it three with tau neutrinos! (See Mann and Primakoff,[25] for a recent discussion.) Neutrino oscillations! Will wonders never cease? Maybe there is something *neu* under the sun after all.

I am most grateful to Raymond Davis, Jr. and to Bruce Cleveland for permitting me to present their most recent observational results. This work was supported in part by the National Science Foundation [PHY76-83685 and AST76-80801].

References

1. W. A. Fowler, Astrophys. J. 127, 551 (1958).
2. A. G. W. Cameron, Ann. Rev. Nucl. Sci. 8, 299 (1958).
3. R. Davis, Jr., Phys. Rev. 97, 766 (1955).
4. R. Davis, Jr., Radioisotopes in Scientific Research, Vol. I (Pergamon Press, 1958).
5. J. N. Bahcall, W. F. Huebner, N. H. Magee, Jr., A. L. Merts, and R. K. Ulrich, Astrophys. J. 184, 1 (1973).
6. R. W. Kavanagh, in Cosmology, Fusion & Other Matters, ed. F. Reines (Colorado Associated University Press, 1972).
7. W. A. Fowler, Nature 238, 24 (1972).
8. M. R. Dwarakanath, Phys. Rev. C9, 805 (1974).
9. R. J. Slobodrian, R. Pigeon, and M. Irshad, Phys. Rev. Lett. 35, 19 (1975).
10. R. Pigeon, M. Irshad, and R. J. Slobodrian, Bull. Am. Phys. Soc. 21, 774 (1976).
11. R. J. Slobodrian, R. Pigeon, M. Irshad, S. Sen, and J. Asai, in Few Body

Dynamics, ed. A. N. Mitra et al. (North-Holland Publishing Co., 1976) p. 236.

12. M. J. Newman and W. A. Fowler, Phys. Rev. Lett. 36, 895 (1976).

13. W. G. Davies, G. C. Ball, A. J. Ferguson, J. S. Forster, D. Horn, and R. E. Warner, Phys. Rev. Lett. 38, 1119 (1977).

14. R. Davis, Jr., and J. C. Evans, Jr., Brookhaven National Laboratory Report (1976)

15. J. N. Bahcall, Astrophys. J. 216, 2115 (1977).

16. R. G. Ulrich, Neutrinos 1974 (Philadelphia) American Institute of Physics.

17. J. N. Bahcall, N. A. Bahcall, and R. G. Ulrich, Astrophys. J. 156, 559 (1969).

18. R. K. Ulrich and E. J. Rhodes, Jr.; E. J. Rhodes, Jr., R. K. Ulrich, and G. W. Simon, UCLA Astronomy & Astrophysics Preprints No. 11 & 12 (1977).

19. R. J. Talbot, Jr. and M. J. Newman, Astrophys. J. 34, 295 (1977).

20. G. Michaud, Nature 266, 433 (1977).

21. R. T. Rood, G. Steigman, and B. M. Tinsley, Astrophys. J. 207, L57 (1976).

22. C. G. Predmore, H. C. Goldwire, and G. K. Walters, Astrophys. J. 168, L125 (1971).

23. J. N. Bahcall, N. Cabibbo, and A. Yahil, Phys. Rev. Lett. 28, 316 (1972).

24. B. Pontecorvo, Zh. Eksp. Teor. Fiz. 53, 1717 (1967) [Sov. Phys.—JETP 26, 984 (1968)]. Also see S. M. Bilenky and B. Pontecorvo, Comments Nucl. Part. Phys. 7, 149 (1977).

25. A. K. Mann and H. Primakoff, Phys. Rev. D15, 655 (1977).

Theory

STATUS OF CONSERVED QUANTUM NUMBERS

Maurice Goldhaber
Brookhaven National Laboratory*
Upton, New York 11973

The phrase "conserved quantum numbers" has a rather definitive ring to it; it might be more cautious to talk of them as quantum numbers for which nonconservation has not been established.

Though the limits of quantitative measurements have been considerably extended in recent years, the status of these quantum numbers has not changed qualitatively. There is much activity, especially concerning the laws of conservation of baryons and of muon number, as if change were just around the corner.

If we look at the historical development of the conservation laws, we find one or more of the following conditions fulfilled:

1. No breakdown has been reported for the property being considered. (Early generalizations from somewhat qualitative evidence.)
2. A breakdown has been explicitly searched for and not found. (This usually leads to a quantitative limit.)
3. A particular conservation law is a consequence of a successful, general theory which explains many phenomena that appear at first sight to be unrelated (though the conservation law also may have been originally assumed in formulating the theory), the foremost example being conservation of electric charge.

*Under the auspices of the U.S. Department of Energy.

Two kinds of conservation law exist:

A. Restricted conservation law: A quantity is conserved with respect to some but not all interactions (e.g., strangeness, which is conserved in strong and electromagnetic, but not in weak, interactions).

B. Absolute conservation law: This is believed to hold with respect to all known interactions because of a fundamental symmetry, (e.g., conservation of electric charge follows from gauge invariance). An absolute conservation law cannot be proved empirically. This would require measuring quantities that are either zero or infinite, values which can be approached by successive improvements in experimentation but not reached. An absolute conservation law is therefore a theoretical statement.

From an operational point of view, a conservation law being absolute means that no known interaction leads to its breakdown; e.g., experimental limits on conservation of electric charge (discussed below) are good enough to conclude that if this conservation law were to break down it would be with an effective strength weaker than that of any of the known interactions. We should remember that from an empirical point of view such a possibility cannot be excluded.

The conservation laws that have withstood empirical tests are shown in Table I. Although the first three laws suffice to describe the empirical evidence, theoreticians like to hold on to the older fourth law, lepton number conservation, as a general framework for the discussion of a number of potential reactions, such as $\mu \to e + \gamma$.

TABLE I

CONSERVED QUANTUM NUMBERS

1) Q (ELECTRIC CHARGE)

2) B (BARYON NUMBER)

3) L_X (X = e, μ, (τ) ; SPECIFIC LEPTON NUMBER).

4) L (LEPTON NUMBER)

It may be useful to look back at the history of the oldest conservation law, which is still being hotly debated today, that of baryons, because this history provides some general lessons for other conservation laws. However, I shall not make an attempt at completeness here.

In an essay written in honor of Dirac for his 70th birthday, Pais[1)] wrote (p. 86): "Much speculation arose at that time to the effect that the negative energy states might be associated with the proton. For this and for another reason, a comment by Weyl [Z. Physik 56, 330 (1929)] is of interest: 'It is plausible to anticipate that, of the two pairs of components of the Dirac quantity, one belongs to the electron, the other to the proton. Further, two conservation laws of electricity will have to appear, which state (after quantization) that the number of electrons as well as of protons remain constant. To these conservation laws must correspond a twofold gauge invariance, involving two arbitrary functions.' Here then is the earliest version of the conservation law of baryons, involving all baryons known in 1929."

Weyl's genius led him to guess something important. But had he set himself the task of stating how the proton might decay (conserving electric charge), he would not at that time have known any particles it could decay into.

It did not take long for Weyl and others to realize that Dirac's positive particle could not be the proton. The conservation law for nucleons was explicitly postulated ten years later by Stueckelberg[2)] as a conservation law of a different kind from that of electricity.

He said (p. 317): "Besides the conservation law of electric charge, which follows from Maxwell's theory, there clearly (offenbar) exists a further conservation law: For all observed transformations of matter, no transformations of heavy particles (neutron and proton) into light particles (electron and neutrino) have yet been observed. We therefore wish to postulate a conservation law of the heavy charge (schwere Ladung)."

Ten years later Wigner[3)] rediscovered the conservation law of nucleons, saying (p. 525, footnote): "It is conceivable, for instance, that a conservation law for the number of heavy particles (protons and neutrons) is responsible for the stability of the protons in the same way as the conservation law for charges is responsible for the stability of the electron. Without the conservation law in question, the proton could disintegrate, under emission of a light quantum, into a positron, just as the electron could disintegrate, were it not for the conservation law for the electric charge, into a light quantum and a neutrino."

Why did these three learned gentlemen, Weyl, Stueckelberg, and Wigner, feel so sure that baryons are conserved? Well, you might say that it's very simple; they felt it in their bones. Had their bones been irradiated by the decays of nucleons, they would have noticed effects considerably exceeding "permissible radiological limits" if the nucleon lifetime were $<10^{16}$ years and if at least 10% of the nucleon rest mass were to appear as radiation absorbable in the body. That is a fairly sensitive measurement, but one can do much better by a deliberate experiment.

A quarter century after Weyl's conjecture the first attempts were made to test the assumption of nucleon stability explicitly[4)]. From simple considerations and by simple experiments we could conclude

a) $\tau_N > 10^{20}$ years, no matter how the nucleon disappears,

b) $\tau_{p_f} > 10^{21}$ years and $\tau_{N_b} > 10^{22}$ years,

if ionizing particles $\gtrsim$ 100 MeV are emitted in the decay of free protons (p_f) or bound nucleons (N_b). I shall return to details shortly.

Summing up, we said (p. 1158): "We cannot conceive of an experiment which would prove the absolute stability of nucleons, but judging from the demonstrated 'practical' stability of nucleons we conclude that the law of conservation of nucleons can be used with considerable confidence in discussions of 'practically observable' nuclear reactions. It proves very useful, for example, for hyperon reactions where it permits the conclusion that particles observed to decay into nucleons must be made from pre-existing nucleons or be produced in pairs (particles plus antiparticles). It also follows that nucleons must be found among the ultimate decay products of such hyperons; otherwise the decay of nucleons via virtual hyperon states would be observable. If nucleon pair production processes should be observed, the number of nucleons would only be conserved in an algebraic sense."

Pais[5)] suggested the collective name "baryon" for nucleons and "heavy" V particles. Since then one speaks of a conservation law of baryons. But of course, for practical purposes, tests of this law are confined to nucleons: free or bound protons, and bound neutrons.

During the next decade the spirit of testing conservation laws led to remarkable successes: observation of the breakdown of the symmetries P, C, and T.

To give some perspective on the lifetime limits with which we are dealing when we discuss nucleon decay, Table II shows some "long times" that play an important role in physics, geophysics, and astrophysics.

Unlike Weyl, who couldn't have said in 1929 how the proton might decay, we are now confronted by a multiplicity of possibilities. Figure 1 shows schematically all the particles which are lighter than a nucleon in mass and into which a nucleon might conceivably decay. If one counts particles and antiparticles as well as members of multiplets separately, one comes up with

TABLE II

SOME "LONG TIMES"

$T_{1/2}$ OF ^{238}U	:	4.5×10^9 YR
AGE OF EARTH	:	4.6×10^9 YR
TIME SINCE "BIG BANG"	:	$\sim 1.5 \times 10^{10}$ YR
$T_{1/2}$(SPONT.FISS.) OF ^{238}U	:	10^{16} YR
$T_{1/2}(\beta\beta)$ OF ^{130}Te	:	1.4×10^{21} YR

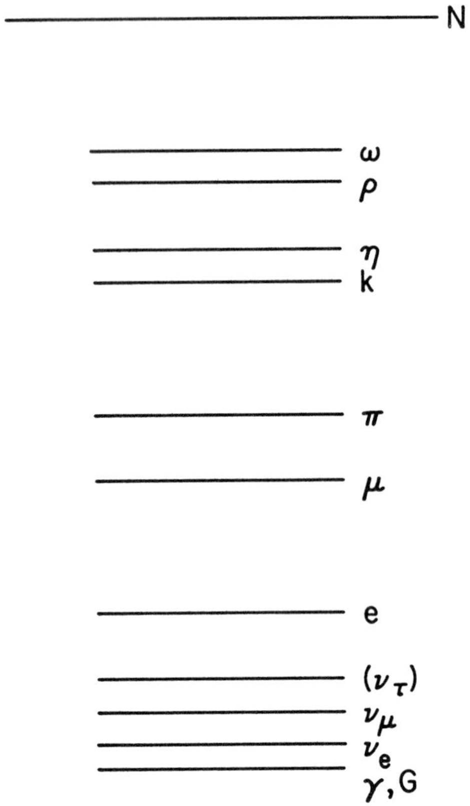

FIG. 1

Schematic of particles of mass lower than that of a nucleon.

some 20-odd known particles into which a nucleon could decay. Because of the many possibilities confronting an experimenter, I should like to urge those who are formulating theories of nucleon decay to try to predict, as far as possible, specific decay modes and specific branching ratios, because for each specific decay mode one can design a deliberate experiment — we could call it a dedicated experiment — not just take a free ride on various and sundry neutrino experiments. As an example, if there were a decay mode $p \to \pi^+ + \nu$, the π^+ would have a well-defined energy, equivalent to nearly half the proton mass, and a characteristic decay chain $\pi^+ \to \mu^+ \to e^+$. We might thus be able to see a π peak on top of the continuum induced by atmospheric neutrinos, and in principle this would allow us to measure very long lifetimes. Roughly, one can say that for measuring a lifetime of the order of 10^{31} years (which is a little beyond the limit given by Reines for a specific decay mode leading to μ^+ emission) one needs about 100 tons of matter to get of the order of 10 events per year, and proportionately more for longer times. Just to obtain a measurable signal one needs very large quantities of matter, and therefore it is natural to combine nucleon decay searches with neutrino searches, but by emphasizing as well as we can the detection of specific modes, we can probably push the search for certain partial decay modes to much longer times than would be otherwise possible.

Table III updates a 1958 paper with Gary Feinberg[6]. Here the symbol $\overset{\odot}{=}$ means "empirically equal" (see ref. 7). Again, let us remember that a measurement cannot yield an exact value zero, but the atomic beam measurements yield charge differences, shown in Table III, which are compatible with zero (see Table IV). If one makes the assumption that the neutron has charge zero, one can obtain better limits from experiments on bulk matter[8] (where different proton-neutron ratios, e.g., different isotopes, have so far not

TABLE III

G. FEINBERG AND M.G., PROC. NAT. ACAD. SCI., 45, 1301 (1958)
(UPDATED)

THE CHARGES OF THE KNOWN PARTICLES INTO WHICH A NUCLEON MIGHT DECAY ARE

$$Q_x = 0, \pm 1\, Q_e.$$

ELECTRIC CHARGES OF KNOWN PARTICLES CAN BE EXPRESSED IN TERMS OF A FEW INDEPENDENT CHARGES WHICH WE CAN CHOOSE TO BE

$$Q_e,\ Q_p,\ Q_n,\ Q_\mu \text{ (AND } Q_\tau),$$

OR WE CAN CHOOSE THE DIFFERENCES FROM UNIT CHARGE

$$\delta q_p,\ Q_n,\ \delta q_\mu\ (\delta q_\tau).$$

TABLE IV

L. J. FRASER, E. R. CARLSON AND V. W. HUGHES (1968)

(See ref. 7)

$$Q({}^{39}_{19}K_{20}) = 19\delta q_p + 20Q_n = (0.84 \pm 0.78) \times 10^{-18} Q_e;$$

$$Q({}^{133}_{55}Cs_{78}) = 55\delta q_p + 78Q_n = (1.62 \pm 0.70) \times 10^{-18} Q_e;$$

IT FOLLOWS THAT

$$\delta q_p = (0.9 \pm 2.0) \times 10^{-19} Q_e;$$

$$Q_n = (0.4 \pm 1.5) \times 10^{-19} Q_e.$$

FROM $n \to p + e^- + \bar{\nu}_e$ IT FOLLOWS THAT

$$Q_{\bar{\nu}_e} = -(0.5 \pm 3) \times 10^{-19} Q_e.$$

been used in the same setup).

If the charge differences were not exactly equal to zero, nucleon stability would trivially follow from charge conservation. But let us assum that the charge differences are exactly equal to zero (a widespread conjecture or tacit assumption, which would follow from Dirac's quantization condition if magnetic monopoles exist). Nucleons might then decay to some of the particles shown in Fig. 1. As mentioned above, there are two essentially different methods[4)] for measuring nucleon decay:

1. Detection of disappearance of a nucleon from a nucleus.
2. Detection of radiations emitted in the decay of a nucleon.

The first, the nuclear method, of which there are now two varieties, gives the results shown in Table V. These methods are fairly insensitive to the particular decay mode: If a nucleon disintegrates, or even "disappears without a trace" (when there was talk of "continuous creation" one might also have had to contemplate "continuous disappearance" of nucleons), it leaves a hole in the nuclear shell, which in ^{232}Th is, for most of its nucleons, equivalent to raising the energy above the fission threshold. From a limit on the spontaneous fission lifetime of ^{232}Th, one can deduce a limit on nucleon lifetime which is ~200 times as large. This method would also be valid if processes involving two or more nucleons played a role in the "disappearance" of nucleons.

The radiochemical variation of the nuclear method relies on the detection of a specific final nucleus resulting from the original one.

The nuclear methods give limits on total lifetimes for nucleons, whereas the counting methods give limits on partial lifetimes. Both methods ultimately suffer from neutrino-induced background due to charged or neutral currents. Perhaps such experiments will ultimately be carried out on the

TABLE V

NUCLEAR METHODS

1) SEARCH FOR FISSION INDUCED IN ^{232}TH (1954).
FLEROV ET AL (1958)[a]

$$\tau_N > 2 \times 10^{23} \text{ YR}$$

2) RADIOCHEMICAL METHODS.
GENERAL SUGGESTION BY P. ROSEN (1975)[b]

SPECIFIC EXAMPLES INVESTIGATED:

I. J. C. EVANS AND R. I. STEINBERG (1977)[c]

$$^{130}\text{TE} \rightarrow {}^{129}\text{XE} \qquad \tau_N > 1.6 \times 10^{25} \text{ YR}$$

II. FIREMAN ET AL (1977)[d]
STEINBERG ET AL (1977)[e]

$$^{39}\text{K} \rightarrow {}^{37}\text{AR} \qquad \tau_N > 2.2 \times 10^{26} \text{ YR}$$

a) See ref. 7.

b) Phys. Rev. Lett. 34, 774 (1975).

c) Science 197, 989 (1977).

d,e) Reported at International Conference on Neutrino Physics, Elbrus, USSR, June 1977.

moon to escape the effects of atmospheric neutrinos.

I have discussed the question of the history and of the empirical evidence concerning baryon conservation at such length because there has been no recent systematic review of this subject.

Theory teases the experimenters to bigger efforts. Within the general framework of unified gauge theories, the possibility of transitions from quarks to leptons has been discussed. Pati and Salam[9)] consider a nucleon to be made of three quarks of integer charge to which they assign a corresponding flavor. Other theories[10)] make use of the more "conventional" fractionally charged quarks. In each case the nucleon lifetime cannot be estimated without further assumptions on properties of heavy intermediate "lepto-quark" gauge bosons. However, it has also been demonstrated that in some unified gauge theories the proton can be absolutely stable[11)]. The theoretical situation has been reviewed by Gell-Mann, Ramond and Slansky[12)].

For conservation of electric charge the most sensitive tests are probably those that search for decay of an electron into neutral particles, e.g., $e \rightarrow 3\nu$ (detected by x-rays which would follow disappearance of a K-electron) or $e \rightarrow \nu + \gamma$. Some of the data are summarized in Table VI (see, e.g., ref. 7). The lifetime limits are good enough to indicate that none of the known interactions break charge conservation[6)], as mentioned above.

The search for $\mu \rightarrow e + \gamma$, which became a hot subject about two years ago, still continues. The present situation is still the same as that recently summarized at a Conference at SIN[13)]. Limits for $\dfrac{\mu^+ \rightarrow e^+ + \gamma}{\mu^+ \rightarrow e^+ + \nu_e + \bar{\nu}_\mu}$ have been reported from TRIUMF[14)] and SIN as $< 3.6 \times 10^{-9}$ and 1.6×10^{-9}, respectively. A LASL-Chicago-Stanford experiment now in progress expects to reach a limit of $\sim 10^{-10}$.

TABLE VI

SUMMARY OF EXPERIMENTAL LIMITS FOR ELECTRON HALF-LIVES

EXPERIMENTERS	ELECTRON LIFETIME (YR)	DETECTION METHOD	ELECTRON SOURCE
E. DER MATEOSIAN AND M. GOLDHABER, UNPUBLISHED (1958-1959)	$>10^{19}$	K X-RAYS OF IODINE	NaI CRYSTAL
	$>10^{20}$	γ RAYS FROM PROCESS $e^- \rightarrow \nu + \gamma$.	"
M. K. MOE AND F. REINES PHYS. REV. 140B, 992 (1965)	$>2 \times 10^{21}$	K X-RAYS OF IODINE	"
	$>4 \times 10^{22}$	γ RAYS FROM PROCESS $e^- \rightarrow \nu + \gamma$.	"
R. I. STEINBERG, K. KWIATKOWSKI, W. MAENHAUT, AND N.S. WALL PHYS. REV. D12, 2582 (1975)	$>5.3 \times 10^{21}$	K X-RAYS OF Ge	Ge(Li) DETECTOR

It is worth recognizing that the limit already found for the partial lifetime for $\mu \rightarrow e + \gamma$ exceeds the β-decay time of a neutron — a remarkable measurement!

At the SIN Conference[15] a limit was also reported for

$$\frac{\mu^- + S \rightarrow e^- + X}{\mu^- + S \rightarrow \nu + Y} < 4 \times 10^{-10}.$$

Some theories predict only slightly lower values. The theoretical situation was reviewed by S. Weinberg at the SIN Conference.

The recent experiments and theories connected with the possibility of nonconservation of muon number have revived interest in neutrino oscillations — another process which would not conserve L_X. This subject has just been reviewed by Bilensky and Pontecorvo[16].

REFERENCES

1) A. Pais, The Early History of the Theory of the Electron, 1897-1947, Chapter 5, in Aspects of Quantum Theory, ed. by A. Salam and E. P. Wigner, p. 79 (Cambridge University Press, 1972).

2) E. C. G. Stueckelberg, Helv. Phys. Acta 11, 299 (1939).

3) E. P.Wigner, Proc. Am. Philos. Soc. 93, 521 (1949) [see also Proc. Nat. Acad. Sci. 38, 449 (1952)].

4) F. Reines, C. L. Cowan, Jr., and M. Goldhaber, Phys. Rev. 96, 1157 (1954).

5) A. Pais, Proc. of Int. Conf. of Theor. Phys. (Kyoto), Published by Science Council of Japan, Tokyo, p. 157 (1954).

6) G. Feinberg and M. Goldhaber, Proc. Nat. Acad. Sci. 45, 1301 (1958).

7) M. Goldhaber, Proc. Am. Philos. Soc. 119, 24 (1975).

8) J. G. King, Phys. Rev. Lett. 5, 562 (1960); G. Gallinaro, M. Marinelli and G. Morpurgo, Phys. Rev. Lett. 38, 1255 (1977).

9) J. C. Pati and A. Salam, Phys. Rev. Lett. 31, 661 (1973).

10) See, e.g., H. Georgi and S. L. Glashow, Phys. Rev. Lett. 32, 438 (1974); H. Georgi, H. R. Quinn and S. Weinberg, ibid 33, 451 (1974).

11) M. Abud, F. Buccella, H. Ruegg and C. A. Savoy, Phys. Lett. 67B, 313 (1977); P. Langacker, G. Segrè and A. Weldon, U. of Penn. preprint.

12) M. Gell-Mann, P. Ramond and R. Slansky (to be published).

13) See Nature 269, 375 (1977).

14) See Depommier et al, Phys. Rev. Lett. 39, 1113 (1977).

15) See A. Badertscher et al., Phys. Rev. Lett. 39, 1385 (1977).

16) S. M. Bilensky and B. Pontecorvo, Comments on Nuclear and Particle Physics, 7, 149 (1977).

PRESENT EXPERIMENTAL STATUS ON THE SECOND CLASS CURRENTS (SCC) AND THE CONSERVED VECTOR CURRENT (CVC) THEORY

C.S. Wu
Columbia University

Introduction:

This International Conference was called to discuss the present status of experimental data in Particle Physics, Nuclear Physics, Atomic Physics and Astrophysics that may be related to fundamental gauge theories of the weak and electromagnetic interactions. As we are all aware that in the past few years, a heavy dark cloud had been hovering over our heads threatening a possible existence of the second class induced tensor current (SCC). The impact of the existence of the SCC on the elegant unified gauge theories would be enormous. Suddenly, in the last 2-3 months, the threatening storm of the existence of the SCC seemed to have been blown clearly away. What has happened? What had gone wrong before? How consistent is all the experimental evidence with the predictions of the CVC Theory in the absence of the Second Class Current? This comprehensive report is organized as follows:

The first section (I) is a short introduction on the origin of the name of the second-class induced tensor current. This is intended only to familiarize the readers with the notation used in this talk.

The second section (II) is on the ft-value asymmetry for mirror β-decays. The question of the possible existence of the SCC was reopened in 1970 from a study of the ft-value asymmetry for mirror β-decays.[1] Unfortunately, a substantial part of the observed asymmetry δ^{exp} ($\delta^{exp} = \delta^{nucl.} + \delta^{scc.}$) may come from the so-called "trivial nuclear effects" $\delta^{nucl.}$.[2] Furthermore, the relationship between the observed $\delta^{scc.}$ and the strength of the fundamental SCC is model dependent if meson-exchange effects due to $\omega \rightarrow \pi e \nu$ as well as off-energy shell effects are included in a model calculation.[3] Therefore, it is difficult to extract unambiguous information on the fundamental SCC coupling from the ft-value asymmetry for mirror β-decays.

The third section (III) is on the correlation type of β-experiments; the measurements of these experiments will give less ambiguous information on the second-class current. There are two types of correlations:

(i) The β-energy dependent angular distribution from oriented nuclei[4][5][6].

In 1975, the results of the β-energy dependent angular distribution $W(\theta)$ for the polarized nuclei of A=12 systems by Sugimoto, Tanihata and Goring[7], and for the A=19 (^{19}Ne) by Calaprice et al.[8] strongly suggested the existence of SCC of appreciable magntiude. However, it should be mentioned that a less well known contributed paper which was submitted to the Santa Fe Conference (1975) by Steels, Grenacs, Lehmann, Palffy and Possoz [9] of Louvain on "Search for SCC in the β-decay of A=12 Isobars" contained results only on aligned ^{12}B targets and found the correlation coefficient in ^{12}B to be very small. Because the correlation coefficient in the β^+ decay of the analogous $^{12}N \rightarrow {}^{12}C + \beta^+ + \nu_e$. transition had not been measured and since no theoretical calculations on the values of some relevant 2nd order forbidden matrix elements were available, no conclusions on the question of the existence of SCC could be made at that time.

In the summer of 1976, an explicit formula for the beta-ray angular distribution in the decay of oriented nuclei was given by Morita et al[5] in which the relevant nuclear matrix ratios were evaluated by impulse approximation with the nuclear wave functions of the Cohen-Kurath type. Later, the Louvain group improved their β-angular distribution measurements on the aligned ^{12}B and obtained results*[10] in agreement with that of their previous run.[9] By using the theoretical calculations of the ratios of relevant nuclear matrix elements by Morita et al [5], the Louvain group interpreted their results from ^{12}B as $g_{W.M.} \cong g_{I.T.}$ by (Eq. 11,12); in good agreement with the absence of the 2nd class current contribution.

In the summer of 1977, Sugimoto, Minamisono, Nojiri and Masuda [11] remeasured the β-energy dependent angular distributions from both aligned nuclei ^{12}B and ^{12}N with greatly improved experimental arrangements. The combined results $(\alpha_- - \alpha_+)$ and $(\alpha_- + \alpha_+)$ from the aligned ^{12}B and ^{12}N are consistent with absence of second-class induced-tensor currents and also with the theoretical calculations of the second-forbidden correstions [5] respectively. The ETH group also observed a near zero slope for the coefficients "α_-" in ^{12}B.[12]

(ii) The angular correlations between β-rays and subsequent radiations (α,γ,ν). (The results of angular correlations (a) of (β -α) in A=8 system [13]; (b) of (β -γ) in A=20 system [14] and (c) of (β -ν) in ^{6}He decay [15] all show no presence of SCC.)

The fourth section (IV) is on the recent experimental status of the spectral shape factors of the mirror nuclei ^{12}B and ^{12}N and of the ^{20}F and the validity of the CVC Theory.

The last section (V) is a summary on the overall experimental status and the theoretical predictions on the validity of the CVC and the non-existence of the SCC.

(I) The Origin of the Name of the Second-Class Induced Tensor Current

The second class current is originated from the G-parity operation which is defined as the product of the charge conjugation C and the charge symmetry, T_2; the latter is the rotation about the y axis through 180° in the charge space.

$$G \equiv C.T_2 \qquad \text{Eq (1)}$$

C = charge conjugation.

T_2 = charge symmetry $(e^{i\pi I_2})$

The G transformations for the relativistic covariants are expressed by the simultaneous replacements:

$$\psi_p \rightarrow \psi_{\bar{n}} \text{ and } \psi_n \rightarrow -\psi_{\bar{p}} \qquad \text{Eq (2)}$$

We restrict our discussion to the V-A Theory for Four-fermion interactions.

In the case of higher-momentum transfer,the vector and axial vector currents are made up of the Dirac matrices as well as the four-momentum transfer. The "V_α" and "A_α" weak currents are given by

$$V_\alpha = (\bar{\psi}_p \; [f_V \; \gamma_\alpha - i \; f_{W.M.} \; \sigma_{\alpha\beta} \; q_\beta + i \; f_{I.S.} q_\alpha] \; \psi_n)$$

and Eq (3)

$$A_\alpha = (\bar{\psi}_p \; [\; i \; f_A \; \gamma_\alpha \; \gamma_5 - f_{I.T.} \; \sigma_{\alpha\beta} \; \gamma_5 \; q_\beta - f_{I.P.} \; \gamma_5 \; q_\alpha] \psi_n)$$

with $\sigma_{\alpha\beta} = 1/2 \; (\gamma_\alpha \gamma_\beta - \gamma_\beta \gamma_\alpha)$ and $q = (p^{(n)} - p^{(p)})_\alpha$

We call f_V, f_A, $f_{W.M.}$, $f_{I.S.}$, $f_{I.T.}$ and $f_{I.P.}$ the nucleon form factors for the vector, axial vector, weak magnetism, induced scalar, induced tensor and induced pseudoscalar interaction respectively.

The G parities for six terms in Eg. 3 are

Currents	G= +1	G = -1
V_α	f_V, $f_{W.M.}$	$f_{I.S.}$
A_α	$f_{I.T.}$	f_A, $f_{I.P.}$

In "V_α" current, the G parity for f_V and $f_{W.M.}$ terms is different from that of $f_{I.S.}$. Similarly, in "A_α" current, f_A and $f_{I.P.}$ are different from $f_{I.T.}$. If we assume that each current of a definite G-parity, $f_{I.S.}$ and $f_{I.T.}$ should vanish since the existence of f_V, $f_{W.M.}$ and f_A, $f_{I.P.}$ is experimentally verified. Accepting the CVC theory ($\partial_\alpha V_\alpha = 0$), then we also have a vanishing $f_{I.S.}$ term. The same condition cannot be applied to $f_{I.T.}$, since $\partial_\alpha A_\alpha = 0$ is strictly forbidden. Weinberg [16] called $f_{I.T.}$ the second class interaction in contrast to the first class interaction of f_V, $f_{W.M.}$ f_A and $f_{I.P.}$. Actually $f_{I.T.}$ is simply the induced tensor term which happens also to be a G-parity or charge symmetry irregular.

II. The ft-Asymmetry for Mirror β-decays

In the early sixties, from the accurate determinations [17] of life times and branching ratios of ^{12}B and ^{12}N decays, it was noticed that the $(ft)^+$ value of the β^+ decay of ^{12}N was greater than the $(ft)^-$ of the β^- decay of ^{12}B by about ∿10%. The asymmetry parameter δ^{exp} is defined as

$$\delta^{exp} \equiv \frac{(ft)^+}{(ft)^-} - 1 \qquad \text{Eq (4)}$$

It was shown that this discrepancy can neither be attributed to the weak magnetism terms nor to other second forbidden corrections [2][18]. The two possible explanations [18][19][2] of this difference are $\delta^{nucl.}$, (the isospin violation due to the Coulomb interaction) and $\delta^{scc.}$ (the presence of the second-class induced tensor current (SCC) if G-parity is violated in nuclear weak interactions). Therefore

$$\delta^{exp} = \delta^{nucl.} + \delta^{scc.} \qquad \text{Eq (5)}$$

In the most interesting cases such as ^{12}B and ^{12}N, $\delta^{nucl.}$ comes mostly from the difference of binding energies of the decaying proton and neutron in the mirror nuclei. $\delta^{nucl.}$ has been called the "trivial" nuclear effect . Although the meaning of the $\delta^{nucl.}$ is not fundamental, its calculation is certainly far from being trivial. It has to be evaluated on the basis of full many-body nuclear wave functions. It has been shown [20][21] that even when the wave functions are agreed upon, the results are still very sensitive to the details of the computations and therefore, the results of many calculations of the $\delta^{nucl.}$ are not unique. Furthermore, since $\delta^{nucl.}$ is the substantial part of δ^{exp}., this makes the extraction of $\delta^{scc.}$ from $\delta^{exp.}$ just that much more uncertain. However, if the second class current exists and can be extracted from experimental measurements, then, from a naive impulse approximation of a single second class coupling constant $g_{I.T.}$, the $\delta^{scc.}$ should be proportional to $(w_0^+ + w_0^-)$[22].

$$\delta^{scc} = -\frac{4}{3}\frac{g_{I.T.}}{g_A}(w_0^+ + w_0^-) \qquad \text{Eq(6)}$$

In 1970, an extensive survey [1] of the exisiting $(ft)^{\pm}$ values of mirror nuclei was made. It was shown that the experimental ft-values plotted against the corresponding $(w_0^+ + w_0^-)$ exhibit a nearly linear plot with a slope of $(g_{I.T.}/g_A) = -2 \times 10^{-3}$ (Fig. 1). This exciting finding generated great interest and debates [23] and the possible existence of the second class currents was strongly rekindled. However, an ingenious experiment in investigating the asymmetry of the ft-value as a function of $(w_0^+ + w_0^-)$ in the beta decays of ^8Li and ^8B by Wilkinson and Alburger [24] in 1971 provided totally new insights on the SCC question. The A=8 case provides a rather exceptional case as it removes the ambiguity in $\delta^{nucl.}$. A unique feature here is that the final state in ^8Be (T_2 =0) fed from $e^+(e^-)$ -decay from $^8B(^8Li)$ is a broad

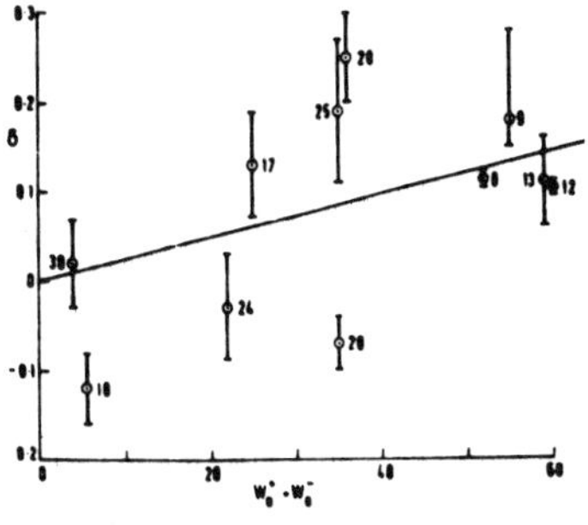

Fig. 1. Experimental value of $\delta = (ft)^+/(ft)^- - 1$ as a function of $W_o^+ + W_o^-$. The numbers beside the points are the A- values. The straight line is a least squares fit passing through the origin with all points given equal weight. It corresponds to $g_{I.T.} \simeq 2 \times 10^{-3}$ (Ref. 1).

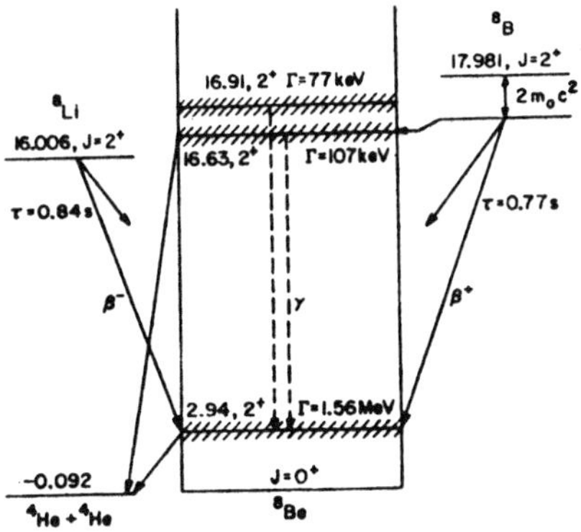

Fig. 2. Energy level diagram for A = 8 decays.

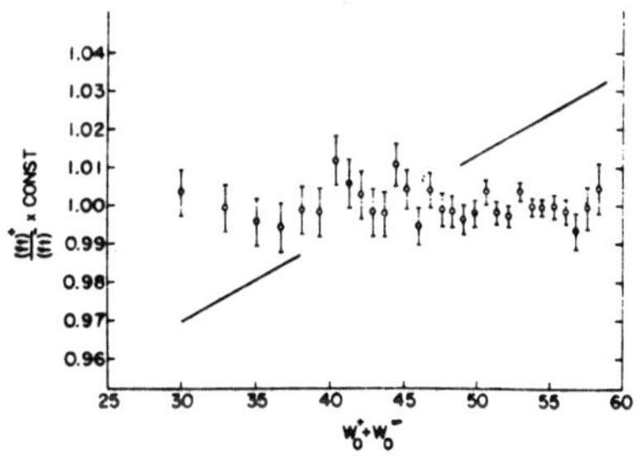

Fig. 3. $[N^-(E_x)/f(W_o)][N^+(E_x)/f(W_o^+)]-1 = \text{const}\ x(ft)^+/(ft)^-$ as a function of $W_o^+ + W_o^-$. The solid lines show the slope to which the points should conform for a second class current of $g_{I.T.} = 2 \times 10^{-3}$ (Ref. 24).

resonance, nominally at 2.90 MeV and with a level width of many MeV in excitation (Fig. 2). As long as the nucleons in the final state (T_2=0 in ^8Be) remained strongly bound (when n and p are bound by 16.00 and 14.35 MeV respectively), the above case provides a pair of mirror transitions where the wave functions would change little and so their overlap with $T_2 = \pm 1$ state would change little. If the observed asymmetry δ^{exp} were due to second class current; δ^{exp} should vary proportionally to $(w_0^+ + w_0^-)$. If, however, the δ^{exp} value changes little as the mass of the $T_2 = 0$ state changes, the observed ^{exp} is probably due to the differences in nuclear overlaps. The results of (ft^+/ft^-) X const vs $(w_0^+ + w_0^-)$ of A=8 system are shown in Fig.(3). $[(ft)^+/(ft)^-]$ shows no substantial $(w^+ + w^-)$ dependence and is almost constant at over a wide range of w. By a least squares fitting of the results, they quoted

$$|g_{I.T.}| < 7 \times 10^{-4}$$

at the 99% confidence level showing no evidence of the SCC on the naive impulse approximation formula (Eq. 6).

It was pointed out in several papers [(23)] that the naive impulse approximation is inadequate; off-mass-shell nucleon effects and explicitly mesonic effects play an esssential role and significantly complicate the interpretation of δ^{scc} even if that can be disentangled from $\delta^{nucl.}$. The naive impulse approximation (Eq.6) was then modified to the KDR formula [(13)].

$$\delta^{scc} = -\frac{4\lambda}{g_A} J + \frac{2}{3g_A}(\lambda L - 2\zeta)(w_0^+ + w_0^-) \qquad \text{Eq (7)}$$

where $\zeta \equiv g_{I.T.} + g'_{I.T.}$, and $g_{I.T.}$ and $g'_{I.T.}$ are the coupling constants of an on-mass-shell and an off-mass-shell nucleon in SCC respectively, λ is a parameter containing the $\omega \to \pi e \nu$ coupling constant; J and L are matrix elements of relevant exchange-current operators. It is to be noted that the first term of this formula is not explicitly proportional to W. The parameter λ varies from nucleus to nucleus. It is interesting to quote a comment in KDR's paper [3] that "one point which emerges from our study is that it will not be feasible to extract unambiguously fundamental second-class coupling constants from the mirror asymmetry δ^{scc}".

Meantime, Wilkinson [18] carried out an extensive evaluation of the $\delta^{nucl.}$ and thereby δ^{scc}. The δ^{scc} in even A nuclei are very small, although they are sizable in odd A nuclei which could be due to nuclear effects. A systematic analysis [25] of δ^{scc} from 8 even A nuclei (from A= 8 to 30) in extracting the parameters ζ and λ in KDR Formula gave

$$|\zeta| \leq 2.5 \times 10^{-3} \text{ /Mev}$$

and

$$|\lambda| \leq 10 \times 10^{-3}$$

The δ^{exp} and $\delta^{nucl.}$ for a few well studied cases in even A-nuclei are listed in Table (1) for reference. The differences between δ^{exp} and $\delta^{nucl.}$ are all within the quoted uncertainties, therefore $\zeta = \lambda = 0$ is not necessarily excluded. The study of the ft asymmetry although not fruitful, has indeed aroused an intense interest in the second-class current problem in nuclear β-decay. Now let us turn our attention to the correlation-type of β-experiments.

III. The Correlation-Types of β-Experiments

A less ambiguous way to search for the possible existence of the second class current in β-interaction is through the correlation-type experiments[4,5,6] where the nuclear structure effects largely cancel out in the lowest order and therefore the results are less susceptible to isospin symmetry breaking. The correlation experiments may be grouped into two distinctive types: (i) The β-ray energy dependence of angular distributions from oriented (polarized or aligned) nuclei and (ii) The angular correlations between β-rays and subsequent radiations. (No nuclear orientation is required). The expected effects in correlation type experiments are generally small, of the order of

Table 1

The $(ft)^{\mp}$ Values and "δ^{exp}" and "δ^{bind}"* of a Few Better Known Mirror Nuclei

A	$(ft)^{-}$sec	$(ft)^{+}$sec	"δ^{exp}"	"δ^{bind}"	Ref.
A = 8			0.106 ± 0.011	0.085 ± 0.045	(1)
A = 12 (G.S)			0.1155 ± 0.0085	0.135 ± 0.037	(1)
	$(1.182\pm0.005)\times10^{4}$	$(1.317\pm0.008)\times10^{4}$	0.114 ± 0.010	0.092 ± 0.161	(2)
	$(1.669\pm0.0037)\times10^{4}$	$(1.3178\pm0.0084)\times10^{4}$	0.129 ± 0.008	0.135 ± 0.037	(3)
A = 12^{*} (4.44MeV)			0.013 ± 0.066	0.091 ± 0.043	(1)
	$(1.287\pm0.040)\times10^{5}$	$(1.344\pm0.070)\times10^{5}$	0.043 ± 0.066	0.091 ± 0.043	(3)
A = 18			0.009 ± 0.013	0.004 ± 0.003	(1)
A = 20	$(9.498\pm0.025)\times10^{4}$	$(9.7400\pm0.12)10^{4}$	0.025 ± 0.013		(4)
			0.041 ± 0.011	0.029 ± 0.025	(1)
A = 24			-0.034 ± 0.063	0.024 ± 0.020	(1)

(1) D.H. Wilkinson, Phys. Lett. 48B, 169 (1973).
(2) H. Behrens and Szybisz, Z. Physik, A273, 177 (1975).
(3) D.E. Alburger, A.M. Nathan, Preprint, BNL-23141.
(4) H. Genz, A. Richter, B.M. Schmitz, H. Behrens, Nucl. Phys. A267, 13 (1976).

*Wilkinson set "δ^{Nucl}" = "δ^{bind}". "δ^{bind}" comes from the defacto difference of binding energies on the two sides of the mirror. The uncertainty in "δ^{bind}" exceeds the magnitude of these other sources in "δ^{Nucl}".

E_0/M or less, where E_0 is the maximum energy of the β-articles and M is the nucleon mass.

(i) The β-ray energy dependence of angular distributions from oriented nuclei.

The second class induced tensor current for an on-shell nucleon is characterized by the nucleon form factor $f_{I.T.}$ in the form of

$$f_{I.T.}\sigma_{\alpha\beta}\gamma_5 q_\beta$$

For complex nuclei, the elementary particle treatment has been used in formulating the SCC in correlation-type experiments in a nuclear model independent way. However, to relate the experimental results to the desired nucleon form factor ($f_{I.T.}$), all induced effects, high order corrections, etc. must be included in the correlation formula. Therefore, the energy gradient effects observed in the angular correlation type experiments depend not only on the induced nucleon form factors $f_{I.T.}$ and $f_{W.M.}$ but also on the first class second-forbidden terms, etc.(5). The first-class second-forbidden terms can be evaluated by using appropriate nuclear models and the impulse approximation method. How satisfactory this type of evaluation is, however, remains to be verified. Fortunately, on a pair of mirror transitions, and in a suitable combination of data from the correlation-type experiments, the ordinary second-forbidden terms are approximately canceled. Thus, the case of β-ray angular distribution from oriented nuclei ^{12}B and ^{12}N is particularly suited not only for the verification of the CVC theory, but also for the study of the question of second class current.

The β-ray angular distribution can be expressed as

$$W(\theta) = \text{const}\ pE(E_0-E)^2\ F(\pm Z,E)[1+F(E,E_0)][1\pm(8/3)aE]$$

$$X\{1\mp P(p/E)\ [1+\alpha_{\mp}E]\ P_1(\cos\theta) + A\alpha_{\mp}E\ P_2(\cos\theta)\} \qquad \text{Eq(8)}$$

for the $1^+ \to 0^+$ transition. Here $F(\pm Z,E)$ and $F(E,E_0)$ are the Fermi function and radiative correction, respectively. The double signs "$\mp$" refer to ^{12}B (β^-) and ^{12}N (β^+) decays. $P_n(\cos\theta)$ are Legendre polynomials. P and A are the nuclear polarization and alignment, respectively, and are defined from the magnetic substate populations a_m for I=1 as

$$P = a_{+1} - a_{-1}; \qquad A = 1-3a_0 \quad \text{and} \quad a_{+1} + a_{-1} + a_0 = 1 \qquad \text{Eq(9)}$$

Also, the energy dependent coefficients $\alpha_\mp$ are

$$\alpha_\mp = \pm \frac{2}{3} (a-b)$$

$$\text{with } a = Re(g_{W.M.}/g_A) \; , \qquad b = Re(g_{I.T.}/g_A) \qquad \text{Eq(10)}$$

Here a is the W.M. term, which also shows up in the spectral shape term, $[1\pm (8/3)aE]$. g_A, $g_{W.M.}$ and $g_{I.T.}$ are effective form factors which are real if T-reversal is invariant and can be related to the nucleon form factors in the impulse approximation by Morita et. al [5] as

$$g_A = f_A \mp E_0 f_{I.T.}$$

$$g_{W.M.} = x[f_{W.M.} - (f_V/2M)] \qquad \text{Eq(11)}$$

$$g_{I.T.} = f_{I.T.} \pm y(f_A/2M)$$

The $g_{I.T.}$ contains not only the nucleon form factor for the SCC $f_{I.T.}$ term, but also the first class second-forbidden term containing the parameter "y". The values of the parameters x and y are also given by Morita et al[5] by using the impulse approximation with Cohen-Kurath model.

$$x = 1+ (1/4.7)[\int \vec{r} \times \vec{p})/(\int \vec{\sigma})] \simeq 0.98$$

$$y = 1+ 2i[\int \vec{r}\, (\vec{\sigma} \cdot \vec{p})]/ (\int \vec{\sigma}) \simeq 3.55 \qquad \text{Eq(12)}$$

It should be noted that if both $\alpha_\mp$ are determined in ^{12}B and ^{12}N, the combined value of $\alpha_- - \alpha_+$ is essentially composed of terms of W.M. + I.T., as the contribution from the second term in $g_{I.T.}$ is canceled. Furthermore,

$\alpha_- + \alpha_+$ gives essentially the value of the second term in $q_{I.T.}$, and consequently, is a crucial test of the calculated value of the parameter "y".

From Eq.(8) it should also be noted that the measurement of the alignment coefficient ($\alpha_\pm E$) which contains no zeroth order effect is obviously more advantageous than that of the polarization coefficient ($1 + \alpha_\pm E$) since the alignment coefficient is less sensitive to any small β scattering effects.

a) The Polarized ^{12}B and ^{12}N Experiments(7):

The first of such polarized experiment is a measurement of angular distributions in the β-decays of polarized ^{12}B and ^{12}N by Sugimoto, Tanihata and Goring.(7) The energy levels of ^{12}B and ^{12}N are shown in Fig.(4).

Polarized ^{12}B and ^{12}N were produced through the reactions $^{11}B(d,p)^{12}B$ and $^{10}B(^3He,n)^{12}N$, respectively. The incident-particle energy and the product recoil angle (θ_R) were experimentally determined so as to optimize the polarization and the yield:

$$E_d = 1.5 \text{ MeV}, \ \theta_R \simeq 45^\circ \text{ and } P \simeq 13\% \text{ for } ^{12}B$$

$$E_{^3He} = 3.2 \text{ MeV}, \ \theta_R \simeq 24.5^\circ \text{ and } P \simeq 25\% \text{ for } ^{12}N$$

The degree of polarization was determined by measuring the $\beta^\mp$-asymmetry due to the parity-violation in $\beta^\mp$-decay. Notice that the polarization of ^{12}B is only half as large as that of ^{12}N. The recoil nuclei were implanted into an Al foil of 6 μm in thickness. The nuclear polarization was preserved by applying a static magnetic field H_0 = 2.2 kG. Beta rays emitted from the recoil stopper were detected by a pair of counter telescopes located above and below the reaction plane. (See Fig. 5). The product nuclear polarization was reversed by the use of the adiabatic-fast-passage method (26) in NMR. The decay asymmetry A is defined by

$$A = \frac{T(\theta=0,E_e) - T(\theta=\pi, E_e)}{T(\theta=0,E_e) + T(\theta=\pi, E_e)} = \mp P(1+\alpha_\pm E_e) \qquad \text{Eq(13)}$$

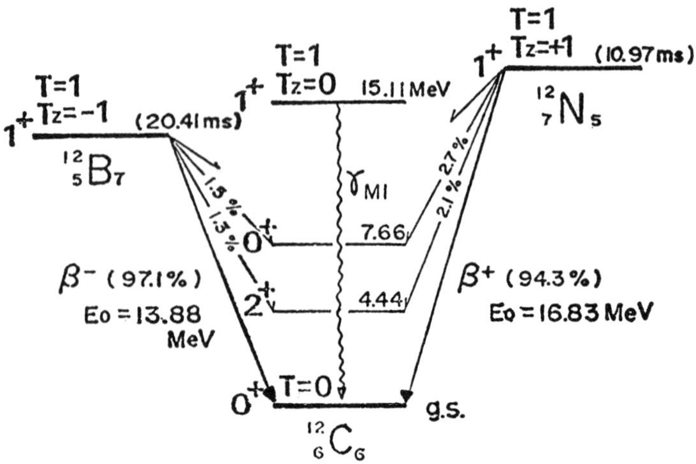

Fig. 4. Energy level diagram for A = 12 decays.

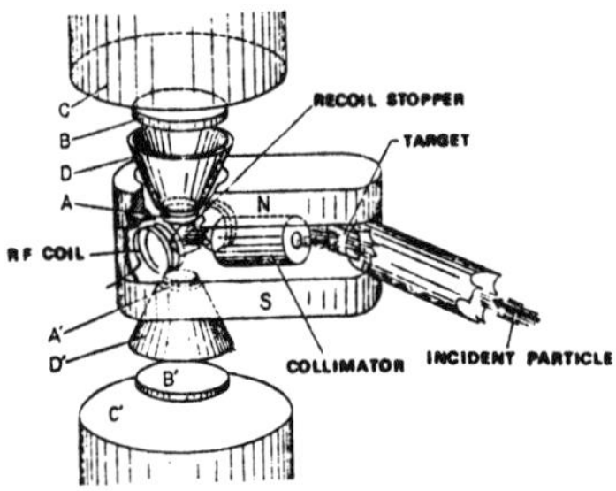

Fig. 5a. Experimental setup used in the polarized ^{12}B and ^{12}N studies by the Osaka group. Each counter telescope consisted of four plastic scintillators $ABC\bar{D}$ (Ref. 7).

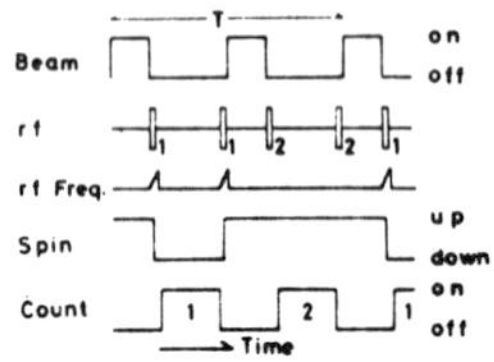

Fig. 5b. Time sequence program of producing polarized ^{12}B and ^{12}N and recording the decay data (Ref. 7).

The results are, see (Fig. 6):

$$\alpha_-\ (^{12}B) = (0.31 \pm 0.06)\%\ MeV^{-1}$$

$$\alpha_+\ (^{12}N) = -\ (0.21 \pm 0.07)\%\ MeV^{-1}$$

$$(\alpha_- - \alpha_+)_{exp} = (0.52 \pm 0.09)\%\ MeV^{-1}$$

These can be compared with the predictions of CVC with no SCC as given below:

$$(\alpha_- - \alpha_+)_{CVC,no\ scc.} \simeq 0.27\%\ MeV^{-1}$$

The experimental value is roughly twice as large as the CVC prediction due to the weak-magnetism term alone. The discrepancy seemed to be a serious one and was interpreted to support the existence of the second-class induced tensor current in nuclear β-decay with

$$f_{I.T.}/f_A = -3.5 \pm 1$$

b) <u>The Polarized ^{19}Ne Experiment</u>:

An experiment with the polarized ^{19}Ne undergoing a β^+ transition $^{19}Ne \rightarrow {}^{19}F + e^+ + \nu$. (^{19}Ne and ^{19}F are members of a $J^\pi = 1/2$ isospin doublet) was carried out by Calaprice et al.(18). The energy level diagram of ^{19}Ne β-decay is shown in Fig. (7). The experimental arrangement is shown in Fig. (8). and the results are shown in Fig. (9). The observed value of α is

$$\alpha_{exp} = -(0.55 \pm 0.15)\%\ MeV^{-1}\ ;$$

whereas CVC predicts with no SCC

$$\alpha_{CVC} = -(0.34 \pm 0.01)\%\ MeV^{-1}$$

The discrepancy between the above two values can be explained by the existence of a second class induced tensor current of

$$\frac{f_{I.T.}}{f_A} = (-8 \pm 3).$$

This discrepancy in this case is even more dramatic than that in A=12 system.

It should be noticed that the radioactive ^{19}Ne gas source in this experiment was contained in an extended hollowed disk (d = 10.2 cm, W = 0.64 cm).

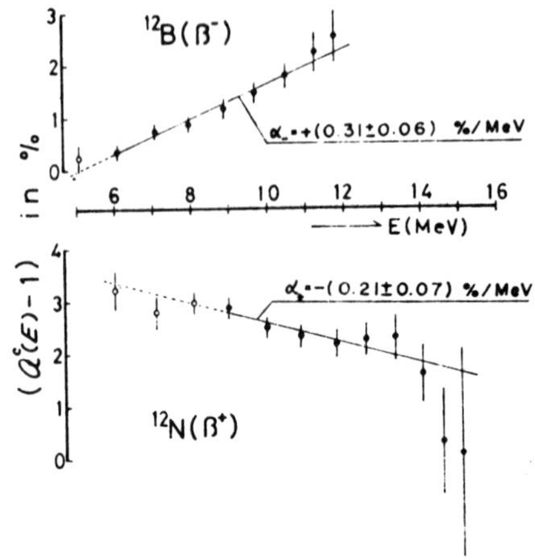

Fig. 6. Energy Dependence of the asymmetry coefficient in the correlation experiments of polarized ^{12}B and ^{12}N. The $A^c(E)$ stands for the normalized asymmetries after the corrections for branching transition and the factor (p/E) (Ref. 7).

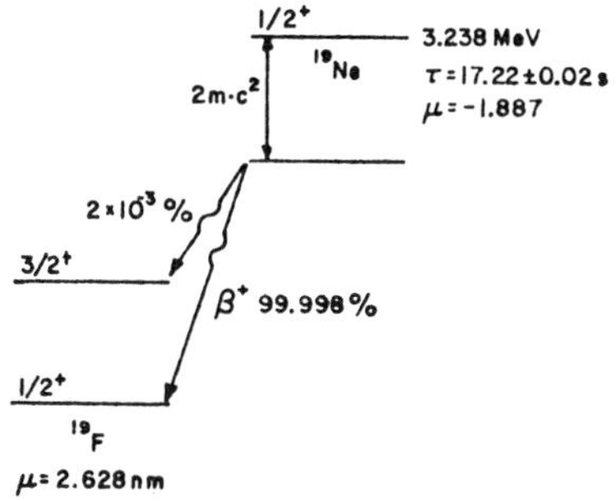

Fig. 7. Energy level diagram of ^{19}Ne β^+ decay.

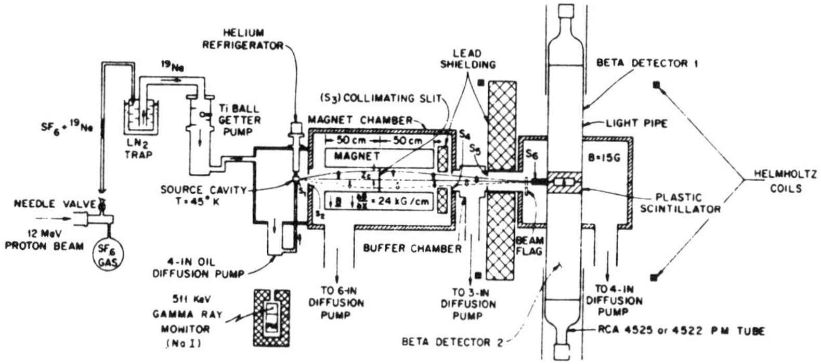

Fig. 8. Atomic beams apparatus used in the polarized ^{19}Ne β^+ decay experiment. The radioactive gas ^{19}Ne was stored in a hollowed disk of d = 10.2 cm and w = 0.64 cm. The β^+ spectrum was measured by the two plastic scintillator counters adjacent to each other and to the source volume (Ref. 8).

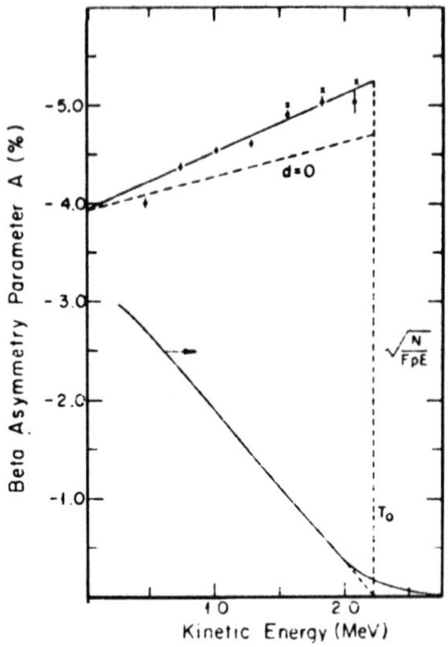

Fig. 9. Observed energy dependence of the asymmetry coefficient A(%) in the β^+ decay of ^{19}Ne as a function of β^+ energy. The dashed line labelled d = 0 is the expected energy dependence with no second class currents (Ref. 8).

Furthermore, the two large plastic β detectors were placed right against the source. The pulses due to multiple scattering of $e^{\pm}$ between the two $\beta^{\pm}$ detectors and the coincidences between e^{+} and its annihilation radiation which had escaped the coincidence discrimination would distort the true slope of the energy dependence plot of the asymmetry parameter A. It would be interesting to modify the experimental geometry to a smaller spherical or cubic gas volume and move the two β-detectors further apart in order to eliminate the multiple scattering and the annihilation radiation effects on each other. We wait for further improved results from the ^{19}Ne experiment.

c) The β-energy Dependent Angular Distribution from Aligned ^{12}B and ^{12}N.

In 1977, Sugimoto, Minamisono, Nojiri and Masuda measured the β-energy dependent angular distribution from the aligned ^{12}B and ^{12}N. As it has been pointed out earlier in this paper, the measurements from alignment "A" are less sensitive to any possible disturbing effects, such as β-scattering effects and the branching ratios, than those with polarization "P". The method of producing alignment is more involved than that for polarization; nevertheless, the process has been well-developed using the NMR techniques.[26]

The experimental setup is shown in Fig. (10). The basic method used in producing polarized nuclei was the same as that of the previous work[7]. A new rotating target arrangement was specially designed and employed in order to increase the production of polarized ^{12}B and ^{12}N. The pulsed- beam method was used to separate the production and counting periods. The polarization "P' was monitored by the up-down counting ratio of β-detectors.

$$W(\theta=0)/W(\theta=\pi) \simeq (1\mp P)/(1\pm P).$$

The conversion of the nuclear polarization to an alignment was accomplished by using NMR techniques. In order to preserve the polarization or alignment, the recoiled nuclei were implanted in a thin Mg single crystal

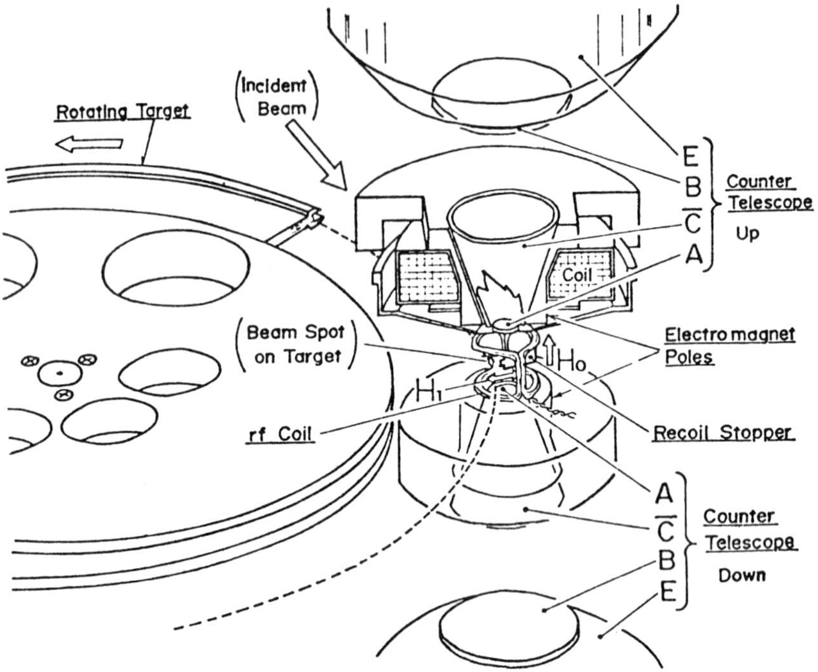

Fig. 10. The experimental setup. The ^{11}B or ^{10}B target evaporated on a tantalum ribbon was mounted in the groove of the target holder. The target system was rotated at a period of 60 ms. Each counter telescope consisted of four plastic scintillators, connected as $AB\bar{C}E$ (Ref. 11).

(150 μm in thickness), where the hexagonal C-axis of Mg was set parallel (or vertical) to the external magnetic field H_0(1-1.4 kG) which was applied normal to the reaction plane. Because of the superimposition of an electric quadrupole "eqQ" interaction on the magnetic interaction, "μH", the Zeeman splittings of ^{12}B and ^{12}N are unequal (Fig. 11). The unequal splittings were utilized to induce selective transitions between nuclear magnetic sub-levels m = +1↔0 (or 0↔ -1) by an equalization rf (HF or LF) in order to affect equalization processes $a'_{+1} \simeq a'_0$ ($a'_0 \simeq a'_{-1}$). In the next step, a sweeeping rf field for the adiabatic fast passage (AFP) between m = 0↔ -1 (+1↔ 0) was applied in order to reverse the populations a'_0 and a_{-1} (a_{+1} and a'_0), and a positive alignment A^+ (negative A^-) was produced with a small residual P_{II}. The A can be related ideally, to the initial P_0 and A_0 as

$$A^{\pm} = \pm \frac{3}{2} P_0 - \frac{1}{2} A_0 \qquad \text{Eq(14)}$$

assuming $P_{II} = 0$ and no spin-lattice relaxation. The alignment was reexamined by applying again AFP rf between m = o ↔-1 (+1↔0) and reproducing a polarization P_{III}. The average value of alignment obtained at the Section II was $A^+ - A^- \simeq 0.26$ with $|P^+_{II} - P^-_{II}| < 0.01$ for ^{12}B and $|A^+ - A^-| \simeq 0.55$ with $|P^+_{II} - P^-_{II}| < 0.01$ for ^{12}N. The original polarizations obtained for ^{12}B and ^{12}N were P ≃10% and ≃ 20% respectively.

The comparison of β-spectra with A^+ and A^- was made as follows: The β -spectrum was measured as a pulse height spectrum from the large plastic E counter of the counter telescope (Fig. 10). The ratio of the spectra with $A^{\pm}$ for the up-counter (θ=0) can be expressed

$$R_u\,(E) = W(0,A^+)/W(0,A^-)$$

$$\propto 1+(A^+ - A^-)\alpha_{\mp}E \mp (P^+_{II} - P^-_{II})\,[1 + \delta(E)]$$

and for the down-counter (θ = π),

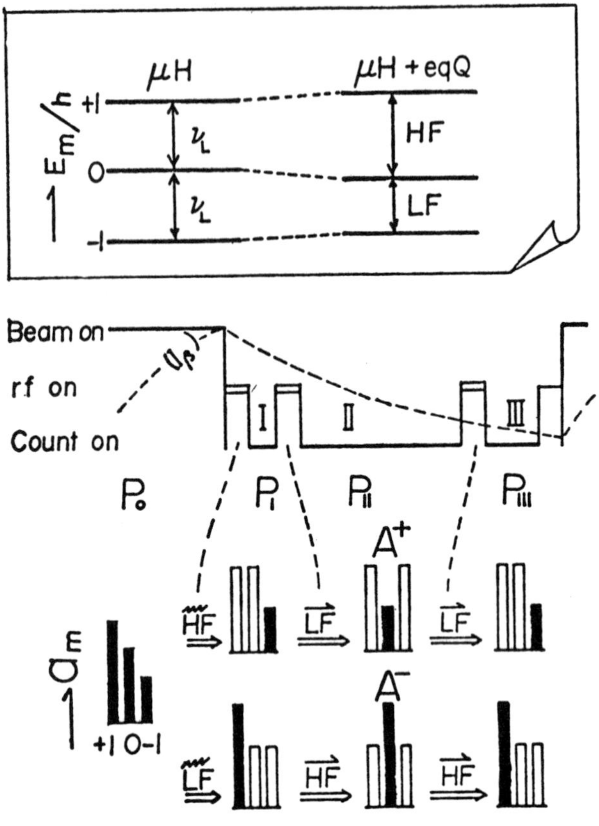

$$A^{\pm} = \pm\frac{3}{2}P_0 - \frac{1}{2}A_0$$

Fig. 11. Diagram to explain the method of conversion of the initial polarization P_0 into a positive A^+ or a negative A^- alignment by use of NMR technique. Equalization processes are denoted by HF and LF, and AFP by $\overrightarrow{HF}$ and $\overrightarrow{LF}$ (Ref. 11).

$$R_D(E) = W(\pi,A^+)/W(\pi,A^-)$$
$$\propto 1 + (A^+ - A^-)\alpha_{\mp}E \pm (P^+_{II} - P^-_{II})[1 + \delta(E)]$$

Here $\delta(E)$ includes effects due to branching transitions and β-ray scatterings besides $\alpha_{\pm}E$. The size of $\delta(E)$ can be estimated from a measurement of β -energy dependent asymmetry with a sizable polarization. In order to cancel the effect $\delta(E)$ as much as possible, data for R_u and R_D were averaged with the same weight for every series of runs. (Fig. 12).

$$\alpha_- \,(^{12}B) = +(0.02_5 \pm 0.03_4)\%\ MeV^{-1}$$
$$|\alpha_+ \,(^{12}N)\,| = (0.27_7 \pm 0.05_2)\%\ MeV^{-1}$$

The $\alpha_{\mp}$ values thus obtained are plotted in Fig.(13) and compared with theoretical predictions(5,21) without SCC and with the previous values from polarization measurements (7). The agreements of both α_- and α_+ from the alignment measurements with the values theoretically predicted in the absence of SCC are excellent. However, the previous polarization results showed large discrepencies with the theoretical predictions. The authors attributed these discrepencies mainly to electron scattering which resulted in an additional slope for the energy dependance of the asymmetry coefficients.

(ii) The angular correlations between β-rays and subsequent radiations
(a) in the (β-α) in A=8;(b) (β-γ)in A=20 and(c)(β-ν) in ^{6}He experiments.

(a) The (β-α) angular correlation in A=8 system.

During the early sixties, several beautiful experiments (27) on the (β-α) angular correlation in the ^{8}Li and ^{8}B beta-decays were carried out to test for the isotriplet CVC Theory. The objective then was to detect the effect of the weak magnetism term in comparison to that of the old Fermi Theory. The correlation results were within the range of the theoretical values predicted for the CVC theory and in disagreement with that of the old

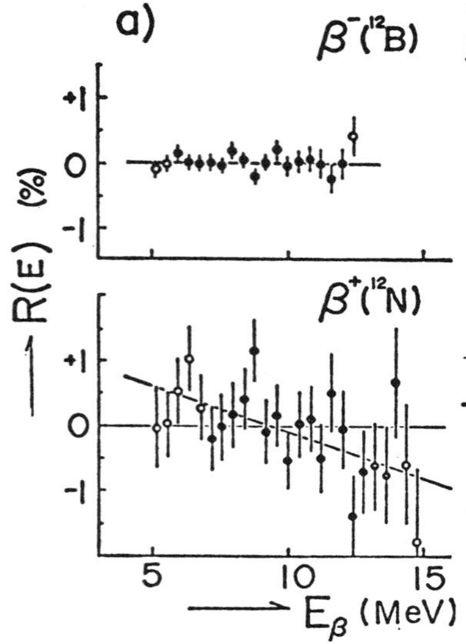

Fig. 12a. Typical examples of variations R(E) for individual runs; $R(E) \propto [1+(A^+-A^-)\alpha_\mp E]$. Data denoted by closed circles were used for the analysis.

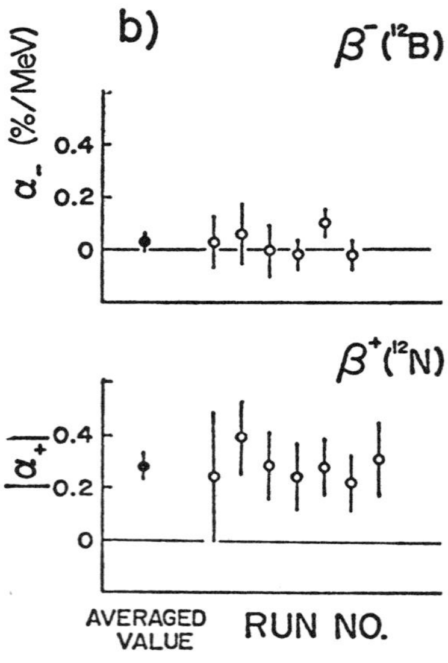

Fig. 12b. Values for $\alpha_-(^{12}B)$ and $|\alpha_+(^{12}N)|$ obtained from individual runs and the final averaged values by the Osaka group.[11]

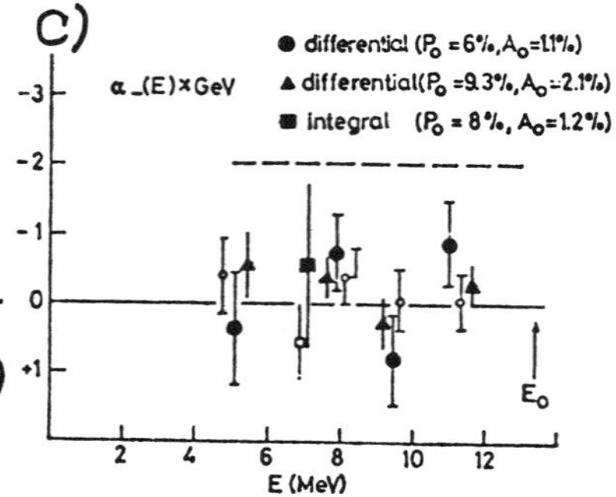

Fig. 12c. Values for $\alpha_-(^{12}B)$ obtained by the Louvain group.[10] Open circles and squares are tests with $P_o = A_o = 0$

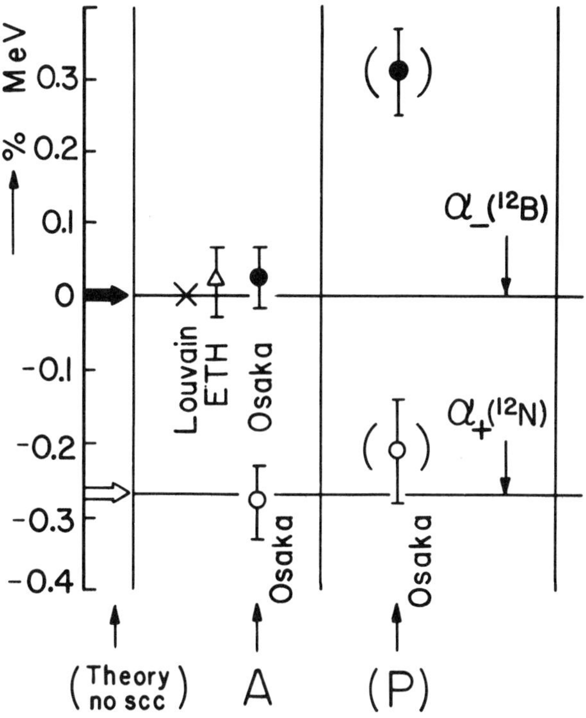

Fig. 13. Comparison of the $\alpha_{\mp}$ values obtained from the ^{12}B and ^{12}N alignment experiments by Osaka group[11] with those from theoretical predictions without SCC,[5,21] the α_- value from the Belgium group,[10] and the ETH group[12] and the previous polarization results.[7]

Fermi theory. When the question of possible existence of the second class induced tensor current was raised in the early seventies [1], Treiman and Holstein [4] made a thorough theoretical study of the usefulness of the correlation-type experiments in searching for the SCC. In the ($\beta^{\pm}$- α) angular correlation experiments in $^{8}Li \rightarrow {}^{8}Be^{*} \leftarrow {}^{8}B$, $^{8}Be^{*} \downarrow 2\alpha$, the distribution is expressed as

$$W^{\mp}(\theta) = 1 + \alpha^{\mp} \cos\theta + p^{\mp}\cos^{2}\theta \quad \text{(for } \beta^{\mp}\text{-decays)} \qquad \text{Eq(15)}$$

where θ is the angle between the $\beta^{\mp}$ and the α-particle. The coefficient α of the cos Θ term comes from the transformation from the $^{8}Be^{*}$ rest system to the laboratory system and therefore contains only kinematic terms. However, the coefficient $p^{\mp}$ of the $\cos^{2}\theta$ term contains the relevant information on the weak magnetism term and the SCC if it exists. Although the expression of the coefficients $p^{\mp}$ are rather complicated, the difference term ($p^{-} - p^{+}$) reduces to a simpler expression as all higher order axial vector form factors cancel. Let us denote the diference of the slopes as

$$\delta^{-} = p^{-} - p^{+} = \frac{E_{\beta}}{M_{N}}\left[\frac{b}{Ac}\left[1-\Delta_{\omega}\sqrt{\frac{5}{21}}\left(1 + 2\frac{E}{E_{0}}\right)\right] - \frac{d_{II}}{Ac}\right] \qquad \text{Eq(16)}$$

Where c,b, and d_{II} are the axial vector, the weak magnetism and the second class induced tensor form factors respectively. A is the atomic number. Δ_{ω} is the "weak" mixing ratio which in CVC theory should be equal to $\Delta_{E.M.}$. The value of "c" can be determined from the total ft values of β-decays, i.e., $c^{2} = \frac{6165}{ft.}$. To obtain the weak magnetism term "b" a better evaluation of the relevant electromagnetic form factor is required $b = 2.695A[\Gamma^{\Delta T=1}_{M1}/E^{3}_{\gamma}]^{1/2}$. Garvey and Tribble and their co-workers[13] have carried out, in the past several years, not only the precise measurements of the (β - α) angular correlations in $^{8}Li \xrightarrow{\beta^{-}} {}^{8}Be^{*} \xleftarrow{\beta^{+}} {}^{8}B$ ($^{8}Be^{*} \downarrow 2\alpha$) and obtained $(M_{N}/E_{\beta})\,\delta^{-} = (b-d_{II})/AC = 7.0 \pm 0.5$ but also have devised ways to determine the integrated radiative width at 90^{0} ($\Gamma_{90^{0}}$) for the two mixed states at 16.63 MeV (T=1) and 16.91 MeV (T=0) with α widths of Γ_α = 107 and 77 keV respectively[28]. See Fig.(2). Their measurements and analysis [13] yield the following results:

$$\Gamma_{90^0} = 7.87 \pm 0.63 \text{ eV}; \qquad \Gamma(M1, \Delta T=1) = 6.69 \pm 0.57 \text{ eV};$$

and the mixing ratio for the isovector amplitudes is

$$\bar{\Delta}_{E.M.} \equiv \frac{I^{\frac{1}{2}}(E2, \Delta T=1)}{I^{\frac{1}{2}}(M1, \Delta T=1)} = 0.19 \pm 0.03$$

The bar indicates the average mixing over the observed widths of the final state.

The experimental results for $\delta\frac{-M_N}{E_\beta}$ are shown as points with error flags in Fig.(14). The shaded area is the prediction of CVC using Eq.(16) with $d_{II}(scc) = 0$. The limit on the size of the SCC form factor is

$$d_{II}/Ac = 0.6 \pm 0.9$$

<u>It is in excellent agreement with the predictions of the CVC theory with no second-class current (Fig. 14)</u>.

(b) The (β- γ) angular correlations in the A=20 system.[14]

The (β -γ) angular correlations in a pair of mirror nuclei can also be used to test for the possible existence of the second class currents. The angular correlation of (β- γ) is of the form

$$W(\theta)^{\mp} \propto 1 + p_{\mp} \cos^2 \theta \qquad \text{Eq(17)}$$

where θ is the angle between the β and γ . The coefficient "$p_{\mp}$ " to first order momentum transfer is

$$p_{\mp} = \frac{E}{4M_N AC} (\pm b \pm d_{II} - d_I). \qquad \text{Eq(18)}$$

The c,b,d_{II} and d_I are the form factors of the axial vector, the weak magnetism, the induced tensor of the second class type and the induced tensor of the first class type respectively. The value ($\frac{b}{Ac}$) can be evaluated by calculating the "b" from the radiative width of the M1 analogue gamma ray and the "c" from the ft value of the β-decay.

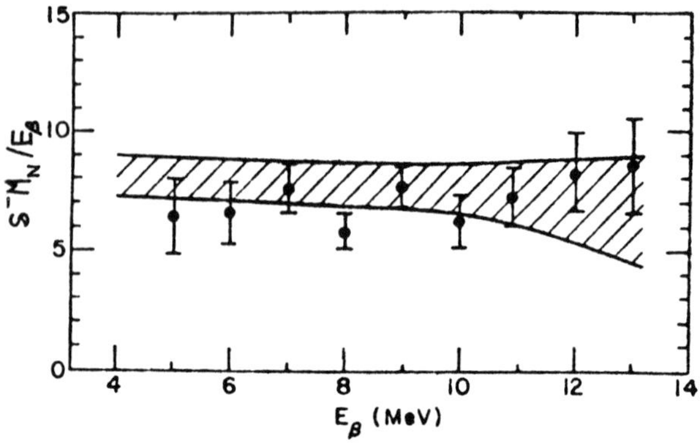

Fig. 14. The experimentally measured values of $\delta^- \equiv p^- - p^+$ vs the β energies E_β. The shaded area is the prediction of CVC with no SCC(d_{II}=0) (Ref. 13).

So far, the β decays of

$$^{20}F \rightarrow e^- + \bar{\nu} + {}^{20}Ne \rightarrow {}^{20}Ne + \gamma[E2, 1.63 \text{ MeV}]$$

and

$$^{20}Na \rightarrow e^+ + \nu + {}^{20}Ne \rightarrow {}^{20}Ne + \gamma\ [E2, 1.63 \text{ MeV}]$$

have been used in the (β - γ) correlation tests in two independent investigations. One is by Rolin, Deutsch et al.[14] of Louvain University; the other is by Tribble and May.[14] The latter group measured the correlation in ^{20}F only as reported in their contributed paper to the Tokyo Conference on Nuclear Structure (1977). The slopes of the (β-γ) correlation data in A=20 system are given in Table (2).

TABLE (2)

	Rolin, Deutsch et al.	Tribble and May
^{20}F	$P_- = (0.2 \pm 0.2) \times 10^{-3}E$	$P_- = (0.5 \pm 0.7) \times 10^{-3}E$
^{20}Na	$P_+ = (-4.0 \pm 0.7) \times 10^{-3}E$	————

The agreement of the data on ^{20}F between the two groups is satisfactory. The difference of the two measured slopes $P_- - P_+$ as that quoted in Tribble's paper.[14] was $\left(\frac{b}{Ac} - \frac{d_{II}}{Ac} \right) = 8.4 \pm 2.0$. For comparison, the first term b/Ac can be evaulated from "b" and "c" and it gives $\frac{b}{Ac} = 8.3 \pm 0.3$. Therefore, under the assumption of validity of the CVC Theory, the $\frac{d_{II}}{Ac}$ = 0.1 ± 2.0 (a vanishing second class current.)

(c) The (β-ν) angular correlations in ^6He β-decay.

Although all the correlation-type experiments are relatively free of Coulomb effects, the (e,ν) correlation is unique in that it's average over the energy spectrum depends only on the tensor term "d" rather than the combination of the weak magnetism "b" and the second class tensor form factor "d". The general expression for the (β-ν) angular correlation can be found in the paper of Holstein and Treiman[4]. Calaprice[15] and Sato et al.[30] analyzed the old results of (β,ν) experiments of ^6He by Johnson et al[29] and found no evidence of SCC. T. Sato et al [30] theoretically studied the various effects due to Coulomb interaction and nuclear structure on the (β,ν) correlation and found that all of them are smaller than 10^{-4} in affecting $a_{Ave} \simeq \frac{1}{3}$. They also concluded that the (β-ν) correlation results in ^{6}He did not indicate the presence of the SCC.

(IV) The Validity of the CVC Theory

In 1958, Feynman and Gell-Mann [31] proposed the Conserved Vector Current hypothesis [C.F. Gerstein and Zeldovitch [32] suggested this possibility earlier in 1955] which states that

(a) $$\frac{\partial V_\lambda}{\partial X_\lambda} = 0$$

where V_λ is the vector part of the strangeness-conserved current J_λ, and

(b) the iso-triplet vector current hypothesis which states that the three current operators

$$V_\lambda \ , \quad (g_V^\beta /e) \times (J_\lambda^{el})_{I=1} \ , \text{ and } V_\lambda^* \qquad \text{Eq(19)}$$

transform, respectively, like the I_Z = -1, 0 and +1 members of a single I=1 triplet, where $(J_\lambda^{el})_{I=1}$ is the iso-vector part of the nonleptonic electromagnetic current operator. It is important to note that hypothesis (b) implies hypothesis (a) since if $(J_\lambda^{el})_{I=1}$ is conserved, so must be V_λ. However, hypothesis (a) does not imply (b) as there exist infinite varieties of conserved vector currents [33], but only the hypothesis (b) which relates it to $(J_\lambda^{el})_{I=1} \times (g_V^\beta/e)$ through the isospin rotation. Therefore, only in the CVC theory which specifically implies also the iso-triplet vector current hypothesis are these unique properties predicted.

There are four different types of experiments designed to test for the validity of the isotriplet CVC theory. They are (1) The determination of the energy dependent spectral shape factors in the $^{12}B \overset{\beta^-}{\rightarrow} {}^{12}C \overset{\beta^+}{\leftarrow} {}^{12}N$ beta decays;

(2) The $\pi^+ \to \pi^0 + e^+ + \nu_e$ decay rates; (3) The (β-α) angular correlations in the ^8Li and ^8B beta decays and (4) The [β -α (circularly polarized)] correlations in the mirror transitions $^{24}Na \overset{\beta^-}{\to} {}^{24}Mg \overset{\beta^+}{\leftarrow} {}^{24}Al$.

All the experimental results were strongly in favor of the CVC theory.(34). However, starting around 1975, experimental evidence began to show the sizable existence of the second class current in beta decay of ^{19}Ne(8) and ^{12}B and ^{12}N(7). Furthermore, Calaprice and Holstein (35) suggested that the analysis of the ^{12}B and ^{12}N spectra in Lee-Mo-Wu's(36) work were in error because the published Fermi function for β^+-decay by Bhalla and Rose (37) which were used in Wu's analysis (36) in 1963 turned out to have some comutational errors as pointed out by Bühring(38). A crude estimate (35) by using the new Fermi functions of Behrens and Janeoke(39) prompted them to conclude that the shape factors of the β-spectra of ^{12}B and ^{12}N are in disagreement with the CVC prediction. Their hasty conclusion greatly undermined the confidence in the CVC theory. However, using improved and revised functions, Wu-Lee-Mo reanalyzed their 1962 experimental data in ^{12}B and ^{12}N in the early summer of 1976(40). It turned out that, while the replacement of the erroneous Fermi function of Bhalla and Rose by that of Behrens and Janecke does indeed alter the shape factors of ^{12}N and ^{12}B, the presently accepted values of the branching ratios and the ft-values, whose effects were considered negligible in Calaprice and Holstein article(35), actually alter the shape factors in the opposite direction to that of the Fermi function replacement ($F_{B-R} \to F_{B-J}$). The shape correction factors for ^{12}B and ^{12}N from the recent reanalysis(40) are shown in Fig.(15). The experimental values and theoretical predictions of a_-^0 and a_+^0 are summarized in table (3). The a_-^0 and a_+^0 are the shape factors based on F_0 (the Fermi function for a point charge nucleus). The agreements with the theoretical predictions of CVC are excellent. The uncertainty assigned to $(a_-^0 - a_+^0) = (0.36 \pm 0.24)\%\ MeV^{-1}$ includes a conservative estimate

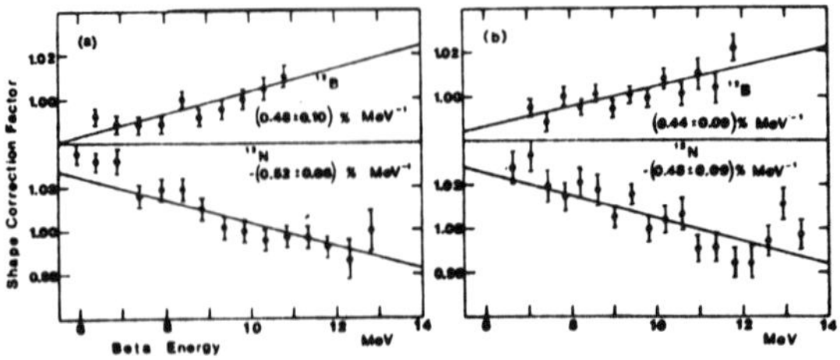

Fig. 15. Shape correction factors for ^{12}B and ^{12}N. $S_{exp}/S = 1+a_{\mp}E$ measured with (a) the narrow (3/16 in.) and (b) the wide (3/8 in.) annular slits. The open circles for ^{12}N are not used for fitting (Ref. 40).

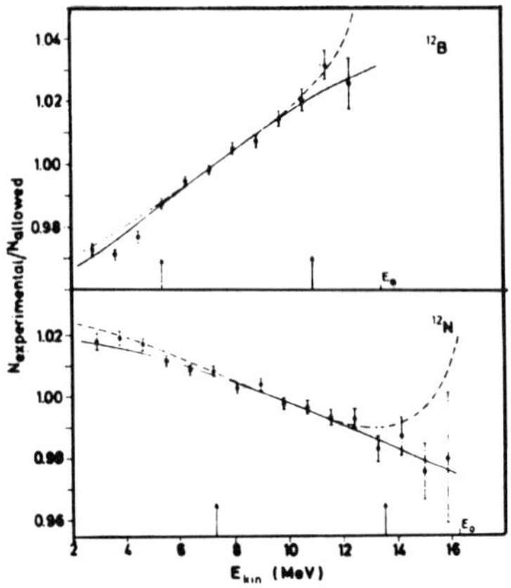

Fig. 16. Pulse height spectra of ^{12}B and ^{12}N sources normalized to those expected for an allowed shape. The solid curves give the spectra expected for a slope correction factor of $1+aW+bW^2$ with the coefficient b = b (theor.) and $a_- = 0.91 \pm 0.11$ and $a_+ = -0.07 \pm 0.09$ and $a_- - a_+ = 0.98 \pm 0.09$. The broken lines give the spectra expected for a calibration shift of 0.1% (Ref. 41).

Table 3

Experimental and Theoretical Values of the difference between a_-^o (^{12}B) and a_+^o (^{12}N).

Experimental 40)	0.41+ 0.45 = 0.86 0.24
Theoretical a)	
Gell-Mann and Berman[b)]	0.86 ± 0.14
Morita[c)]	0.86 ± 0.07
Huffaker and Laird[d)]	0.90 ± 0.07
Bohr and Mottelson[e)]	0.42+0.49 = 0.91
Calaprice and Holstein[15)]	0.37+ 0.47 = 0.84

a) $a_-^o - a_+^o$ in Ref b) was originally + 1.33% MeV^{-1} based on Γ_{M1} = 53ev. The Coloumb effect is -0.25% MeV^{-1}. After correction using Γ_{M1} = 37 ev, the total effect is 1.33 x $(37/53)^{1/2}$ - 0.25= 0.86% MeV^{-1}. Calaprice and Holstein's value is corrected for $\delta_{t.s.}$

b) M. Gell-Mann and S.M. Berman, Phys. Rev. Lett. 3. 99 (1959).

c) M. Morita, Phys. Rev. 113, 1584 (1959).

d) J.N. Huffaker and C.E. Laird, Nucl. Phys. A92, 584 (1967).

e) A. Bohr and B.R. Mottelson, Nuclear Structure (Benjamin, Reading,Mass., 1968) Vol. I, p. 419.

of systematic errors of 0.13% MeV^{-1} in the momentum calibration for each individual curves. The experimental value of ($a_{-}-a_{+}$) calculated with a finite size nucleus and higher order corrections is 0.96% MeV^{-1}.

In 1975, Behrens and Szybisz[21] made a theoretical study of the beta spectra of ^{12}B and ^{12}N by calculating the form factor coefficients using the Cohen and Kurath wave functions and also taking into account both the linear ($a^{\mp}$ E) and the quadratic energy ($b^{\mp}$ E^2) terms of the shape factors. The signs of the slopes of the β^- and β^+ shape factors were opposite but their magnitudes were unequal because of the higher order forbidden matrix element corrections. Unfortunately, there are no reliable methods to calculate these corrections. Nevertheless, the value of (a_{-} - a_{+}) is less sensitive to these high order corrections and was given as 0.98% Mev^{-1}.

In 1977, Kaina, Soergel, Thies and Trost[41] remeasured the β -spectra of ^{12}B and ^{12}N using a NaI crystal as spectrometer with a thin plastic scintillator counter in front of the NaI crystal to identify incoming electrons. The responses of the NaI crystal to e^- or e^+ in the energy ranges of the $\beta^{\mp}$ -spectra were calibrated by taking many pulse height spectra at small energy intervals using the magnetically analyzed e^- or e^+ beams. These calibrations are used to compose the pulse height spectra to be expected from the ^{12}B and ^{12}N decays. The observed pulse height spectra of ^{12}B and ^{12}N are then compared with that obtained from the spectral composition . Assuming the quadratic term b=0, Kaina's results (Fig. 16) give (a_{-}-a_{+}) = [+0.71 - (-0.38)] = (1.09 ± 0.09) x 10^{-2} MeV^{-1}. These are to be compared with Wu-Lee-Mo's reanalyzed results of (a_{-}- a_{+}) = [+0.46-(-0.50)] = (0.96 ± 0.24) x $10^{-2}MeV^{-1}$ and the theoretical value [21] of (a_{-} - a_{+}) = + 1.1 x 10^{-2} MeV^{-1}.

The predicted difference (a_{-} - a_{+}) is dominated by the weak magnetism term and is less effected by the quadratic terms and the high order terms since they cancel. Thus the overall agreements in the $(a_- - a_+)$ values as shown above[40,41,21] gives strong support to the CVC theory.

The ^{20}F spectrum:

The ^{20}F-spectrum was considered as a favorite case in studying the weak magnetism term (CVC theory). ^{20}F has a half-life of 11.2 sec. and decays mainly ($\simeq$ 99.98%) to the first excited state of ^{20}Ne at 1.63 MeV with a maximum electron energy of 5.38 MeV. From the measured radiactive width $\Gamma_\gamma(M1)$ of the isobaric analog γ-transition in ^{20}Ne [$\Gamma_\gamma = 4.08 \pm 0.4$ eV][42], a large shape factor of about 0.94×10^{-2} MeV^{-1} was expected. The failure to detect any observable slope in the shape factors of the ^{20}F spectrum due to excessive electron back scatterings from the high Z backing material was a real concern to many physicists for a short while. It was a great relief when a slope of the shape factors "a" = (+ 0.80 ± 0.44)% MeV^{-1} was reported by Genz, Richter and Schmitz[43] in 1976 (Fig. 17). Their theoretical calculation[43] gives a = +0.80% MeV^{-1} which is also in good agreement with that observed.

(V) Summary:

From a direct comparison between the experimentally determined and the theoretically predicted values of $(\alpha_- - \alpha_+)$ of the aligned ^{12}B and ^{12}N nuclei:

$$(\alpha_- - \alpha_+)_{exp} = 0.30_2 \pm 0.06\% \text{ MeV}^{-1} \quad (11,12)$$

$$= 0.28 \pm 0.06\% \text{ Mev}^{-1} \quad (10,11)$$

and

$$(\alpha_- - \alpha_+)_{\substack{CVC \\ no\ SCC}} = 0.270\% \text{ MeV}^{-1}$$

A limit of the SCC can be estimated to be $f_{I.T.} \leq -(0.45 \pm 0.9) f_A/2M$ which can be compared with the weak magnetism term $f_{I.T.}/f_{W.M.} \simeq 0.15 \pm 0.30$.

If no SCC exists, the asymmetry coefficients $(\alpha_- - \alpha_+)$ of the aligned ^{12}B and ^{12}N nuclei can be also directly compared to the slope of the shape factors "a" [in Eq.(8)] of the β-spectra of ^{12}B and ^{12}N by the relation $(\alpha_- - \alpha_+) = (4/3)a$

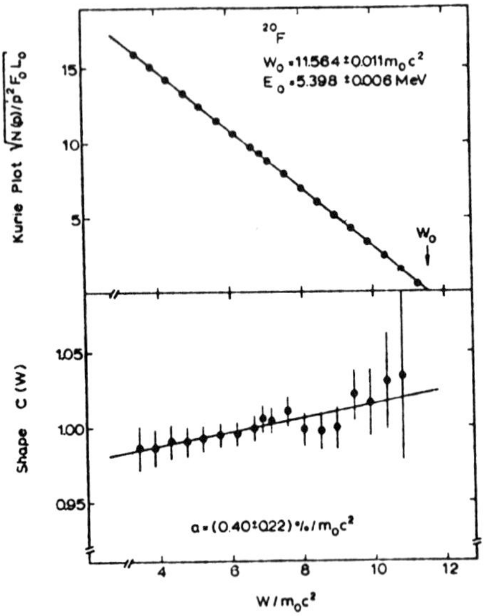

Fig. 17. The upper part shows the Fermi-Kurie plot obtained from the momentum spectrum of the β^- decay of ^{20}F. The bottom part exhibits the shape factors as a function of the beta energy W/m_oc^2. The slope 'a' is $(+0.40 \pm 0.22)\%(mc^2)^{-1}$ or $(+0.78 \pm 0.43)\%$ MeV^{-1} .

where a = 3/16 (a_- - a_+). Furthermore, according to the CVC theory, the value of "a" can be also calculated from the radiative width Γ_γ of the analogue γ-transition in ^{12}C (37.0 ±1.0 eV). Table (4) and Fig. (18) show the various "a" values thus obtained from both experiments and theoretical calculations. The consistencies among all "a" values obtained from these totally independent sources are truly satisfactory and very impressive.

The above are also in complete agreement with the results so far obtained from the analysis of(ft)symmetries [25], the (β -α) correlation in A=8 system[13] the (β - γ) correlation in A=20 system [14] and the (β-ν) correlation in ^{6}He [15,29,30] assuming that the CVC theory is valid and that SCC term is less than 10% of that of the weak magnetism.

What can one conclude on the validity of the CVC and the non-existence of the SCC from the results of muon capture in ^{12}C? Hwang and Primakoff[44] and Pascual[45] have shown by detailed analysis, that only under the assumption of the validity of the isotriplet CVC theory, the partially conserved axial vector current (PCAC) and no existence of the second class current could good overall predictions be made on the muon capture rate in ^{12}C [$\Gamma_{calc.}^{(44)}$ = (6.2 ±0.2) x $10^3 sec^{-1}$ and $\Gamma_{obs.}^{(46)}$ = (6.2 ± 0.3) x 10^{-3} sec^{-1}] and the average nuclear polarization of the recoil ^{12}B [$P_{Ave}^{calc.(44)}$ = 0.57 ±0.01 and $P_{Ave}^{exp\,(47)}$ = 0.53 ± 0.04] in (μ^- + ^{12}C $\rightarrow$ ^{12}B + ν_μ). Furthermore, they also predicted the variation of the asymmetry coefficients with energy in the oriented ^{12}B and ^{12}N as α_-= (0.008 ± 0.003) x 10^{-2} MeV^{-1} and α_+= - (0.275 ± 0.003) x 10^{-2} MeV^{-1} which are in excellent agreements with that determined in later experiments.[10,11,12]

What about the experimental evidences on the CVC and the SCC in elementary particle physics? In the hyperon $\Sigma^{\pm}$ β-decays, electromagnetic corrections should be much less important. If no SCC exists, the ratio of the (ft) value should be very close to 1. The experimental ratio[48] is:

Table 4

The Comparison of Present Results of $a = \frac{g_{W.M.}}{g_A}$ from Various Experiments and Calculations in A = 12 System.

1) From the shape factors of β-spectra "$a_- - a_+$".
2) From the radiative width of the analogue γ-transition.
3) From the energy dependence of the asymmetry coefficients "$\alpha_- - \alpha_+$".
4) From the theoretical calculations.

	"a"
1. The spectral shape factor "a"	
(i) Theoretical calculation for free nucleon (C.V.C.) $a \simeq \frac{f_V}{2M}\frac{\mu + 1}{f_A}$	$a \simeq \frac{4.7}{1.239 \times 2 \times 931} \simeq 0.2\%\ \mathrm{MeV}^{-1}$
(ii) Experimental Results $(1 \pm \frac{8}{3} a_{\mp}E)$ $\therefore (a_- - a_+) \simeq \frac{16}{3}a$	$a \simeq \frac{3}{16} \times 0.96 \simeq$ 0.18% MeV^{-1} Columbia (40) $a \simeq \frac{3}{16} \times 1.02 \simeq$ 0.19% MeV^{-1} $\times\ 0.97 \simeq$ 0.18% MeV^{-1} Heidelberg (41)
2. The Analogue γ-Transition $T_\gamma = 37.0 \pm 1.0$ ev $a = (\frac{1}{E_\gamma})\left[\frac{3}{4}\frac{T_\gamma(ft)}{\alpha E_\gamma (ft)_{0\to 0}}\right]^{\frac{1}{2}}$	$a \simeq 0.21\%$ MeV
3. The Energy Dependence of Asymmetry Coefficients	
(i) $a \simeq \frac{3}{4}(\alpha_- - \alpha_+)$ No scc.	$a \simeq (\frac{3}{4}) \times 0.30 \simeq 0.22_5\%\ \mathrm{MeV}^{-1}$ Osaka (11)
(ii) $a \simeq (\frac{3}{2})\alpha_- + b$ y = 3.55 No scc.	$a \simeq (\frac{3}{2}) \times {}^{-0.01}_{+0.02}{+0.02} + 3.55 \frac{f_A}{2Mf_A}$ $0.20_5\%\ \mathrm{MeV}^{-1}$ (Louvain) (10) $\simeq 0.22\ \%\ \mathrm{MeV}^{-1}$ (Osaka) (11) $0.22\ \%\ \mathrm{MeV}^{-1}$ (E.T.H.) (12)
(iii) $a \simeq -(\frac{3}{2})\alpha_+ - b$ y = 3.55 No scc.	$a \simeq +(\frac{3}{2}) \times 0.27 - 3.55\frac{f_A}{2Mf_A}$ $\simeq 0.22\%\ \mathrm{MeV}^{-1}$ (Osaka) (11)
4. Theoretical Calculations	
(i) Morita et al (5)	$a \simeq 0.21\%\ \mathrm{MeV}^{-1}$
(ii) Behrens and Szyb sz (21)	$a \simeq 0.18\%\ \mathrm{MeV}^{-1}$

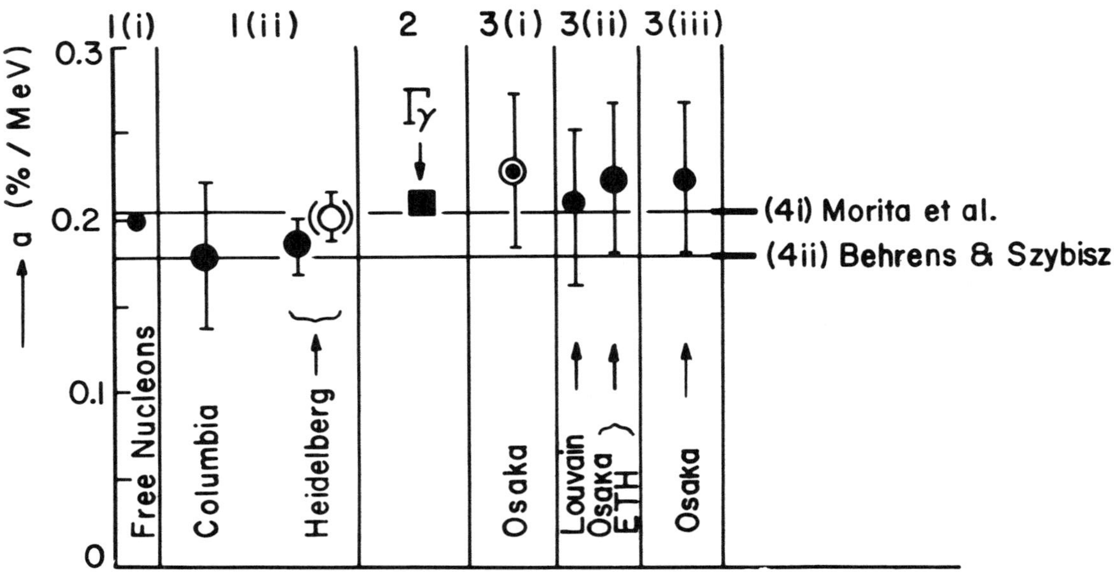

Fig. 18. A CVC test in A=12 system by plotting the determined or calculated quantity $a = g_{W.M.}/g_A$ from (1) the spectral shape factor $(a_- - a_+)$ (2) the radiative width of the analogue γ-transition; (3) the symmetry coefficient in the aligned experiment $(\alpha_- - \alpha_+)$; and (4) the theoretical predictions. Detailed explanation, see Table 4.

$$\left[\frac{ft(\Sigma^+ \to \Lambda + e^+ + \nu_e)}{ft(\Sigma^- \to \Lambda + e^- + \bar{\nu}_e)}\right]_{exp} = 0.96 \pm 0.24$$

The ratio is close to 1, but its experimental error is too large to give a decisive conclusion. Several neutrino reactions which would be sensitive to the existence of the SCC, have been discussed in detail by Walecka[(49)] and Holstein and Treiman[(50)]. However, we may have a long time to wait for the outcomes.

At the recent International Conference on Nuclear Structure held at Tokyo on September 5-10, 1977, Dr. J.P. Deutsch wittingly and aptly summarized the episode on the search for the possible existence of the SCC by quoting the title of a Shakespearean play Much Ado About Nothing . To those who are happy for the final outcome, the title of another play, All's Well That Ends Well, is equally appropriate.

REFERENCES:

1. D. H. Wilkinson, Phys. Lett. 31B,447 (1970);
D. H. Wilkinson and D. E. Alburger, Phys. Rev. Lett. 24,1134 (1970);
D. H. Albierger and D. H. Wilkinson, Phys. Lett. 32B, 190 (1970)

2. R. J. Blin-Stoyle and M. Rosina, Nuclear Phys. 70,321 (1965)

3. K. Kubodera, J. Delorme, and M. Rho, Nucl. Phys. B66,253 (1973)

4. B. R. Holstein and S. B. Treiman, Phys. Rev. C3, 1921 (1971)
B. R. Holstein, Rev. Mod. Phys. 46, 789 (1974)

5. M. Morita, M. Nishimura, A. Shimizu, H. Ohtsubo and K. Kobodera. Supp. of the Prog. of Theor. Phys. No.60, 1 (1976)

6. J. Delorme and M. Rho, Nucl. Phys. B34, 317, (1971); Phys. Letters 34B, 238 (1971)

7. K. Sugimoto, I. Tanihata, and J. Goring, Phys. Rev Lett. 34,1533 (1975)

8. F. P. Calaprice, S. J. Freeman, W.C. Freeman, W.C. Mead and H.C. Vantine, Phys. Rev. Lett. 35, 1566 (1975)

9. M. Steels, L. Grenacs, J. Lehman, L. Palffy and A. Possoz, A contributed paper to the VIth Inst. Conf. on High Energy Physics and Nuclear Structure at Santa-Fe and Los Alamos (1975)

10. P. Lebrun, Ph. Deschepper, L. Grenaces, J. Lehmann, C. Leroy, L. Palffy, A. Possoz, and A. Maio. A Contribution to the VIIth Int. Conf. on High-Energy Physics and Nuclear Structure, Zurich, Switzerland, Aug. 28-Sept. 2 (1977); $\alpha^- = (0.007 \pm 0.020)\%\ MeV^{-1}$.
*A Mg single crystal was used as the implantation foil for the oriented ^{12}B nuclei and an adequate external magnetic decoupling field was maintained.

11. K. Sugimoto, T. Monamisono, Y. Nojiri, and Y. Masuda presented by Sugimoto at the Int. Conf. on Nuclear Structure, Tokyo, Japan, Sept. 5-10, 1977.

12. H. Brandle, L. Grenaces, J. Lang, L. Ph. Roesch, V.L. Telegdi; P. Truttmann, A. Weis, and A. Zehnder, (preprint) $\alpha^- = (0.024 \pm 0.044)\%\ MeV^{-1}$.

13. A. M. Nathan, G.T. Garvey, P. Paul, and E.K. Warburton Phys. Rev. Letters 35, 1137 (1975); Also presented by G. T. Garvey at the Int. Conf. on Nucl. Phys. at Tokyo (1977); J. M. Long and D.A. Bromley, Contributed papers to the above conference, P. Paul, M. Suffert and Ph. Gorodetsky, Contributed papers to the above conference.

14. N. Rolin, J.P.Deutsch, O. Favart, M. Lebrun and R. Prierls. (Presented by J.P. Deutsch) at the Int. Conf. on Nuclear Physics at Tokyo (1977); R. E. Tribble and D. P. May. A contributed paper to the above conf.

15. F. P. Calaprice, Phys. Rev. C. 12,2016 (1975)

16. S. Weinberg, Phys. Rev. 112, 1375 (1958)
T. D. Lee and C. N. Yang Nuovo Cim. 3, 749 (1956)

17. R. W. Peterson and N.W. Glass. Phys. Rev. 130, 292 (1963)
T. R. Fisher, Phys. Rev. 130, 2388 (1963)
T. R. Fisher and W. Whaling, Phys. Rev. 133, B1502 (1964)
R. W. Kavanagh, Phys. Rev. 133B, 1504 (1964)

18. D. H. Wilkinson, Nucl. Phys. A178, 65 (1971)

19. J. Eichler, T. A. Tombrello, J. N. Bahcall, Phys. Lett. 13, 146 (1964); M. E. Mafethe, P. E. Hodgson, Proc. Phys. Soc., London 87, 429 (1966).

20. D. H. Wilkinson, Phys. Rev. Lett. 27, 1018 (1971)
J. Blomquist, Phys. Lett. 35B, 375 (1971)

21. H. Behrens and L. Szybisz, Z. Physik A273, 177 (1975).

22. J. N. Huffaker and E. Greuling, Phys. Rev. 132, 738 (1963)

23. Question Period after Wilkinson's paper recorded in the Proceedings of the 15th Solvay Conference on Symmetry Properties of Nuclei,held on Sept. 28-October 3, 1970, published by Gordon and Breach Science Publishers; p.178-185.

24. D. H. Wilkinson and D. E. Alburger, Phys. Rev. Lett., 26, 1127 (1971)

25. D. H. Wilkinson, Phys. Lett. 48B, 169 (1975)

26. T. Minamisono, J. Phys. Soc. Japan, 34, Suppl. 324, (1973). It was contained in the proceedings of the Int. Conf. on Nuclear moments and Nuclear Structure, Osaka, Japan, (1972) p.324.

27. M. E. Nordburg, F. B. Morinigo, and C.A. Barnes, Phys. Rev. 125, 321 (1962).
W. Gruhle, K. H. Lauterjung and B. Schimmer, Nucl. Phys. 42, 321 (1963).

28. F. Ajzenberg-Selove and[T. Lauritsen], Nucl. Phys. A227, 1 (1974).
F. Ajzenberg-Selove, Nucl. Phys. A248, 1 (1975)

29. C. H. Johnson, F. Pleasonton and T. A. Carlson, Phys. Rev. 132, 1149 (1963)

30. T. Sato, J. Yamane, and M. Norita, Proceedings of the Tokyo Conf. on Nuclear Structure, Sept. 5-10, 1977 p.790.

31. R. Feynman and M. Gell-Mann, Phys. Rev. 109, 193 (1958)

32. S. S. Gershtein and J. B. Zeldovich, Zh. Eksperim. i. Theor. Fiz 29, 698 (1955) (English transl: Soviet Phys-JETP2, 576,(1955)

33. T. D. Lee and C. S. Wu. Ann Rev. Nucl Phys. V15, 381, (1965)

34. C. S. Wu, Rev. Mod. Phys. 36, 618 (1964)

35. F. P. Calaprice and B. R. Holstein, Nucl. Phys. A273, 301 (1976)

36. Y. K. Lee, L. W. Mo and C. S. Wu, Phys. Rev. Lett. 10, 253(1963); C. S. Wu, Rev. Mod. Phys. 36, 618 (1964) Fermi function F_{B-R} was used in this paper.

37. C. P. Bhalla, M. E. Rose, Oak Ridge National Laboratory Report No.ORNL-3207 (Unpublished) - C. P. Bhalla, Tables of Electron Radical Functions and Tangents of phase shifts for Light Nuclei (z = 1 through 10), National Bureau of standards Monograph No.81 (U.S. GPO, Washington, D.C. 1964).

38. W. Bühring, Kernforschungszentrum Karlsruhe report KFK-559 (1967)

39. H. Behrens; J. Janecke, Numerical tables for beta decay and electron capture, Lendolt-Bornstein, New series, Vol 1/4 Berlin-Heidelberg-New York: Springer 1969.

40. C.S. Wu, Y.K. Lee, and L.W. Mo, Phys. Rev. Lett. 39, 72 (1977). J.P. Deutsch's *Comments on the Annihilation Correction was taken care of in the original Wu-Lee-Mo paper (36). Therefore the comments of Deutsch et al. are irrelevent to the results of Ref. (34) or (36).

* J.P. Deutsch, P.C. Macq, and L. Van Elmst, Phys. Rev. C 15, 1587 (1977).

41. W. Kaina, V. Soergel, H. Thies and W. Trost. Phys. Lett. 70B, 411 (1977).

42. Γ_γ = (4.08 ± 0.33) ev. (Ingalls); Γ_γ = (4.65 ± 0.5) ev. Pearson and Spear)

43. H. Genz, A. Richter, B. M. Schmitz and H. Behrens, Nucl Phys. A267, 13, (1976)

44. P. Huang and H. Primakoff, Phys. Rev. C 16, 397 (1977)

45. P. Pascual, preprint, TH 2383-CERN

46. G. H. Miller, M. Eckhause, F. R. Kane, P. Martin and R. E. Welsh, Phys. Lett. 41B, 50 (1972)

47. A. Possoz, D. Favart, L. Grenacs, J. Lehmann, P. Macq, D. Meda, L. Paeffy, J. Julien and C. Samour, Phys. Lett. 50B, 438 (1974)

48. Review of Particle Properties, Rev. Mod. Phys. 45 (1973)

49. T. W. Donnelly and J. D. Walecka, Phys. Lett. 41B, 275, (1972)

50. B. R. Holstein and S. B. Treiman (Princeton University Preprint 1975).

QUANTUM CHROMODYNAMICS AT HIGH ENERGIES†

H. David Politzer
California Institute of Technology, Pasadena, California 91125

I outline the method for computing any parton-semi-inclusive process in QCD. The salient features are logarithmic scaling violations in effective parton distribution and decay functions and $O(g^2)$ transverse momenta that grow linearly with the large invariants. Phenomenological topics discussed are: heavy hadron production, the $\langle \vec{p}_\perp^{\,2} \rangle$ in $p + p \to \mu^+\mu^- + X$, and details of $\mu + p \to \mu + \pi + X$.

† Work supported in part by U.S. Energy Research and Development Administration under Contract EY76-C-03-0068.

Can we understand hadrons in terms of a gauge theory, in particular the gauge theory of colored quarks and gluons commonly called QCD? The answer to this question is not yet clear. But why then do we even suspect that QCD may be right? Aside from aesthetic or qualitative considerations, there are two problems where the quantitative predictions of QCD have been spectacularly successful.

In inclusive lepton-hadron scattering ($\ell + h \rightarrow \ell + X$), QCD offers the only known self-consistent explanation of scaling[1]. But with that explanation come specific predictions of scaling violation which have since been confirmed[2]. In inclusive electron-positron annihilation ($e^+ + e^- \rightarrow X$), charmonium had to be invented to reconcile the apparent leveling off of the cross section with the predictions of QCD[3]. While there remain several puzzles connected with charmonium spectroscopy, those predictions that follow unequivocally from QCD have been verified. In fact, all tests of unequivocal predictions of QCD have been successful. Unfortunately, they are precious few. So an important question is: are there any further possible tests of QCD?

Attempts to compute the hadron spectrum from QCD have been unsuccessful at providing a quantitative, comprehensive scheme. This constitutes a serious flaw in the gauge theory picture. But it is worth remembering that there is nothing holy about the spectrum. There may well be other problems of experimental interest which can serve as tests and applications of QCD.

I would like to describe some recent advances in extending the predictions of QCD[4]. These new predictions can be summarized as follows: all semi-inclusive problems that have been addressed by the parton model can be computed in QCD. The results are characterized by logarithmic scaling violations in the distribution and decay functions. These violations are the same for all processes. Furthermore, the scaling violations in the decay functions are determined by the same scale-differential equations (anomalous dimensions) as those in the distribution functions.

I would like to emphasize here aspects of the QCD predictions which differ radically from the naive parton picture, i.e., not just logs instead of constants but powers instead of constants. In particular, QCD predicts there will be large and growing transverse momenta, $\vec{p}_\perp$'s, where $\vec{p}_\perp$ is the transverse momentum of a quark within a hadron or a hadron within a jet. Crudely speaking, QCD predicts $\vec{p}_\perp^{\,2} \sim g^2(Q^2)Q^2$ where Q^2 is some large invariant characterizing the scattering.

I wish to emphasize that I now have a method of computing details of differential cross sections and not simply estimates or analogies. In the following I will outline the method and emphasize the two remaining loop-holes or open questions. I will then discuss the most striking predictions.

Electroproduction serves as a simple example of the method. In this problem I guarantee the new method to be 100% equivalent to the orthodox (operator product) analysis. Begin with a parton description of the e-p cross section dσ(see Fig. 1):

$$d\sigma(x,Q^2) = \int_0^1 d\xi \; f(\xi)\, d\sigma_{parton}(x_{parton}, Q^2)$$

where $\xi = p^+/P^+$, $x_{parton} = x/\xi$ and $\vec{p}_\perp^{\,2} \sim 0(300 \text{ MeV})^2$. A convolution, like the above, factorizes in terms of moments:

$$\int_0^1 x^n \, d\sigma \, dx = \int_0^1 d\xi \; \xi^{n+1} f(\xi) \int_0^1 dy \; y^n \; d\sigma_{parton}(y,Q^2) \quad .$$

The algorithm is to compute $d\sigma_{parton}$ in quark-gluon perturbation theory. In general

$$\int_0^1 dy \; y^n \; d\sigma_{parton} \propto 1 + g^2 a_n \log \frac{p^2}{Q^2} + g^2 b_n + \ldots$$

for massless partons. While p^2 may not be exactly zero for bound partons, it is small and $\log p^2/Q^2$ is arbitrarily large, rendering the above perturbative

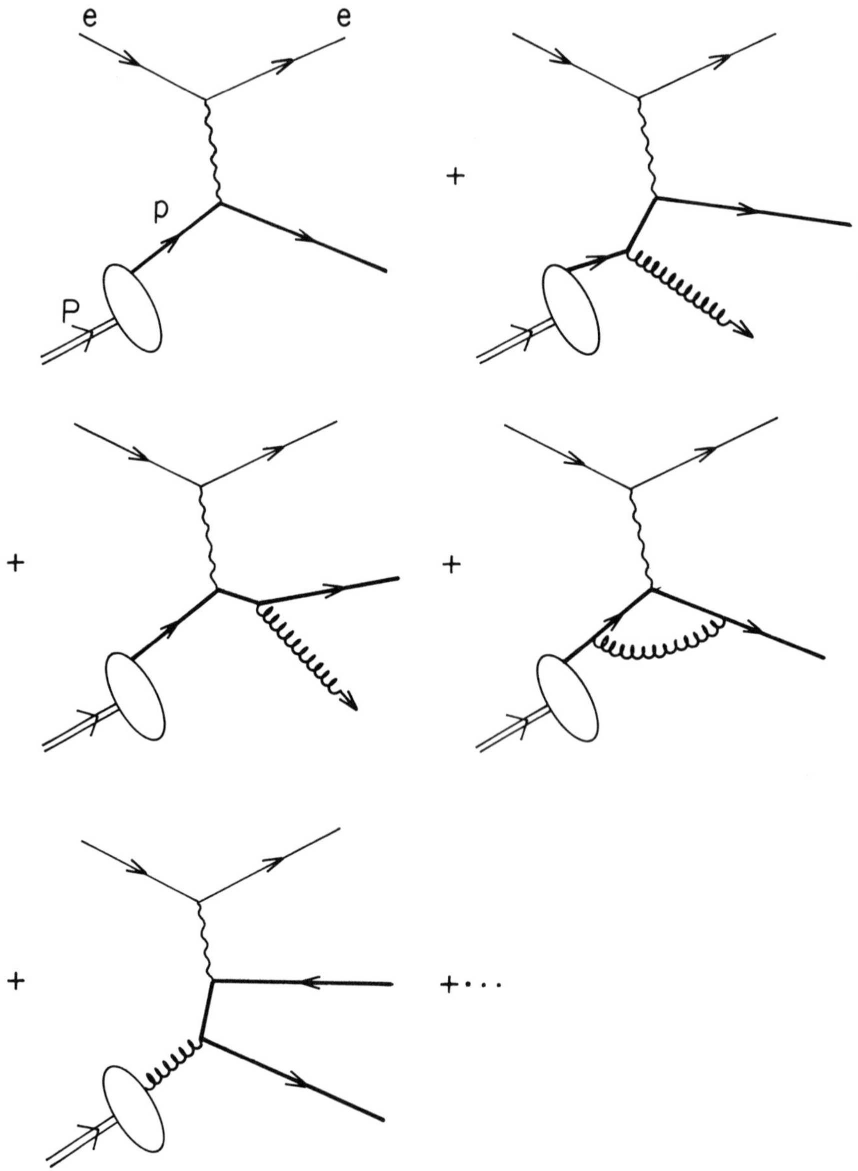

Fig. 1: e+p → e+X including the first radiative corrections

expansion relatively useless. However we can factorize it,

$$(1+g^2a_n \log \frac{p^2}{Q^2}+g^2b_n+\ldots) = (1+g^2a_n \log \frac{p^2}{Q_0^{\,2}}+\ldots)(1+g^2a_n \log \frac{Q_0^{\,2}}{Q^2}+g^2b_n+\ldots) ,$$

at the expense of introducing an arbitrary $Q_0^{\,2}$. All the infrared difficulties reside in the first factor (if we choose $Q_0^{\,2} \sim 0(Q^2)$), which is independent of Q^2. The second factor admits a well-behaved power series expansion in g^2 and allows us to compute the Q^2 dependence reliably. Now absorb the infrared sensitive factor into $f(\xi)$, where it really belongs, i.e., define $\tilde{\tilde{f}}$:

$$\int_0^1 d\xi\ \xi^{n+1}\ \tilde{f}(\xi,Q_0^{\,2}) = \int_0^1 d\xi\ \xi^n f(\xi)(1+a_n g^2 \log \frac{p^2}{Q_0^{\,2}}+\ldots) .$$

We are left with

$$d\sigma(x,Q^2) = \int_0^1 d\xi\ \tilde{f}(\xi,Q_0^{\,2})d\tilde{\sigma}_{parton}(x_{parton},Q^2,Q_0^{\,2}) ,$$

where $d\tilde{\sigma}_{parton}$ is calculable and $\tilde{f}$ must be measured experimentally.

The operator product, light cone, or short distance analyses serve only to prove that the above discussed factorization of p^2 dependence from Q^2 dependence persists to all orders in perturbation theory.

We can apply this method to any parton process. I have checked the factorization to lowest non-trivial order in the following processes

$e^+ + e^- \to \pi + X$[5,6,7]

$\mu + p \to \mu + \pi + X$[7,8]

$p + \overset{(-)}{p} \to \mu^+\mu^- + X$[9] .

Not only did factorization occur, but the same factors appeared in each calculation. (This is a necessary condition that the parton functions be the same functions in each process.) Some of the graphs relevant to the last process are shown in Fig. 2.

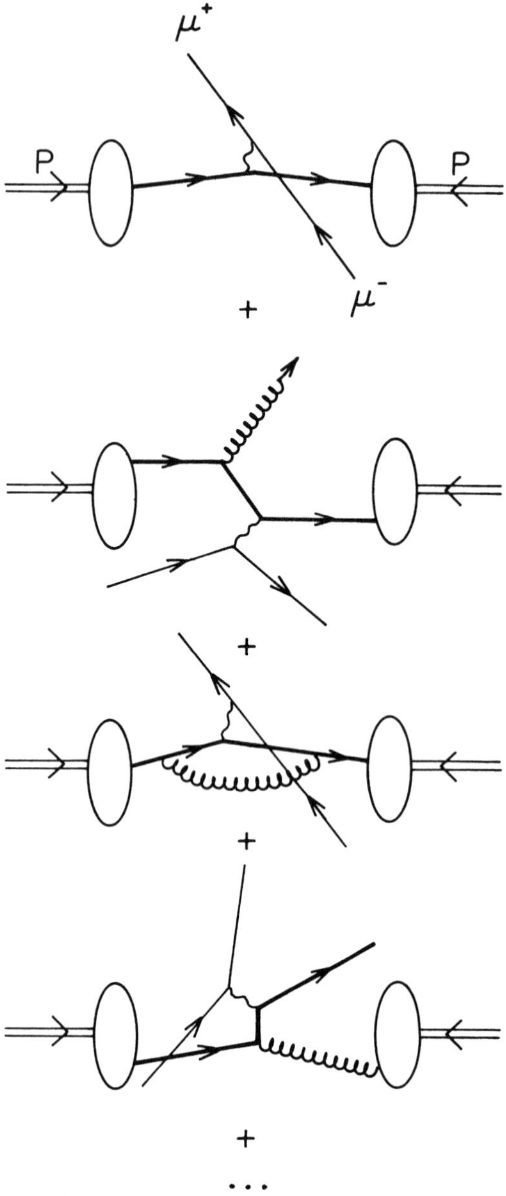

Fig. 2: p+p → $\mu^+ + \mu^- + X$ including the first radiative corrections.

The first open question is whether factorization with the same factors occurs in processes where color is exchanged in the Born graph. An important example is

$$p + \overset{(-)}{p} \to \pi + X$$

at large angles.

(A brief comment regarding angles and dimensionless variables: the $d\tilde{\sigma}_{parton}$ does not contain any factors of $\log p^2$, but it may contain $\log x$ or $1/(1-x)$ where x is some ratio of large invariants $0 < x < 1$. These factors signal the breakdown of perturbation theory in certain limits - usually of obvious physical significance.)

The parton cross section relevant to $p + \overset{(-)}{p} \to \pi + X$ involves color exchange and new graphs (see Fig. 3) and potentially new physics. Does it factorize? If so are the factors the same as encountered in $\mu + p \to \mu + \pi + X$? I don't yet know.

The second open question is whether the factorization persists to all orders. For electroproduction, a proof of the light-cone operator product expansion (OPE) sufficed to prove factorization. For these other processes the OPE is just not enough. That of course does not mean that the factorization does not occur, but rather more powerful tools are needed to prove it.

In fact, I am certain that the desired factorization is a corollary of the great theorem of Lee and Nauenberg[10] and Kinoshita[11] regarding the cancellation of infrared divergences in theories with massless particles. This theorem has not been proven (at least diagrammatically) for QCD, but it has been tested, to several non-trivial orders[12]. I have discussed the relation of this theorem to factorization elsewhere[4,9].

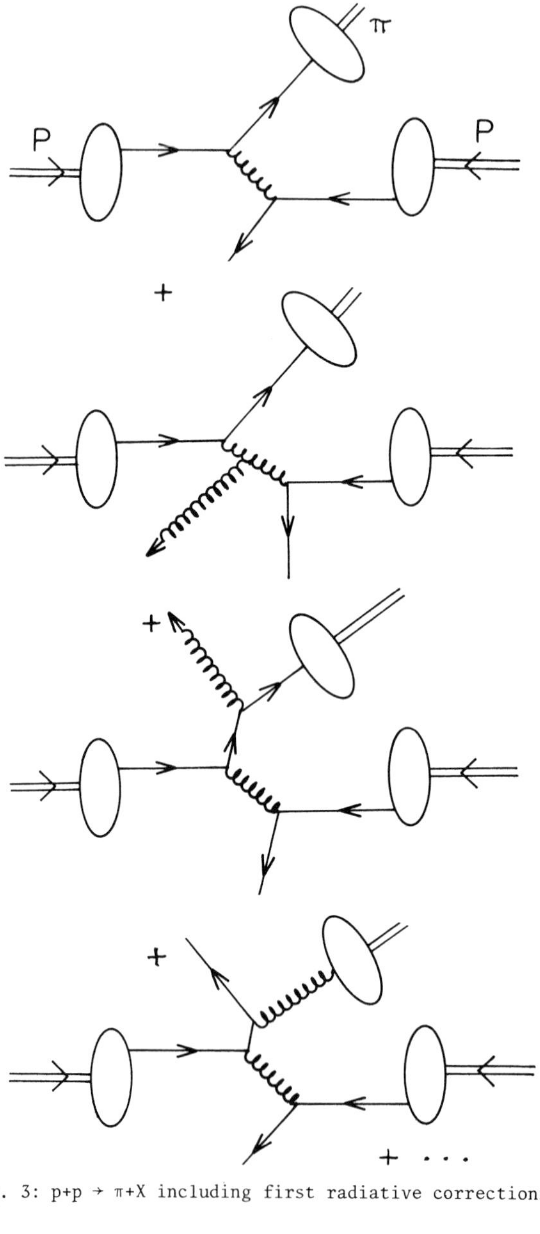

Fig. 3: p+p → π+X including first radiative corrections.

I will discuss the following outstanding consequences of these QCD calculations as examples: heavy quarks and heavy hadrons, $\langle \vec{p}_\perp^{\,2} \rangle$ of the $\mu^+\mu^-$ pair in $p + p \to \mu^+\mu^- + X$, and $\vec{p}_\perp$ of the π in $\mu + p \to \mu + \pi + X$.

The charm or heavy quark content of the proton has been calculated in terms of the gluon distribution[13,14]. (The model essentially takes account of the possibility of a gluon dissociating into a quark pair.) Similar considerations now applied to decay functions allow the calculation of the decay function describing how a heavy (charmed) hadron is produced by light partons[7]. The only input function needed is one which we know a priori: The "decay" of a heavy quark h into a generic heavy hadron H (containing one h) - summed over possible H's - is $\delta(1-z)$, where z is the fraction of h's momentum carried by H, for $Q^2 \approx m_h^2$; the "decay" of a light parton into H is virtually zero for $Q^2 \approx m_h^2$. This latter function remains zero for $Q^2 \lesssim m_h^2$, but grows for $Q^2 > m_h^2$ as determined by perturbation theory (operator mixing[14]). The model is suggested by Fig. 4.

Since gluons are much more effective at producing charmed hadrons than light quarks, events with large $\vec{p}_\perp$'s will have a relatively higher fraction of charm. It is also a straightforward excercise to work out the quantitative predictions for charm production in terms of the four structure functions for $\mu + p \to \mu + H + X$.

In any process, the radiative corrections will introduce $\vec{p}_\perp^{\,2}$'s of order $g^2(Q^2)Q^2$. A little more precisely, one finds from $\mu + p \to \mu + X$

$$\langle \vec{p}_{\perp\,quark}^{\,2} \rangle \approx \frac{Q^2(1-x)}{\text{"16"} \log Q^2/\Lambda^2} + 0(0.300 \text{ GeV})^2 \qquad ,$$

where $\Lambda \approx 0.5$ GeV as determined[2] from the measured magnitude of the quark=gluon coupling g; and "16" is a crude approximation to a complex convolution over distribution functions. The $\vec{p}_{\perp quark}$ is the mean struck quark's momentum transverse to the proton-photon axis. The first, perturbative term comes from the

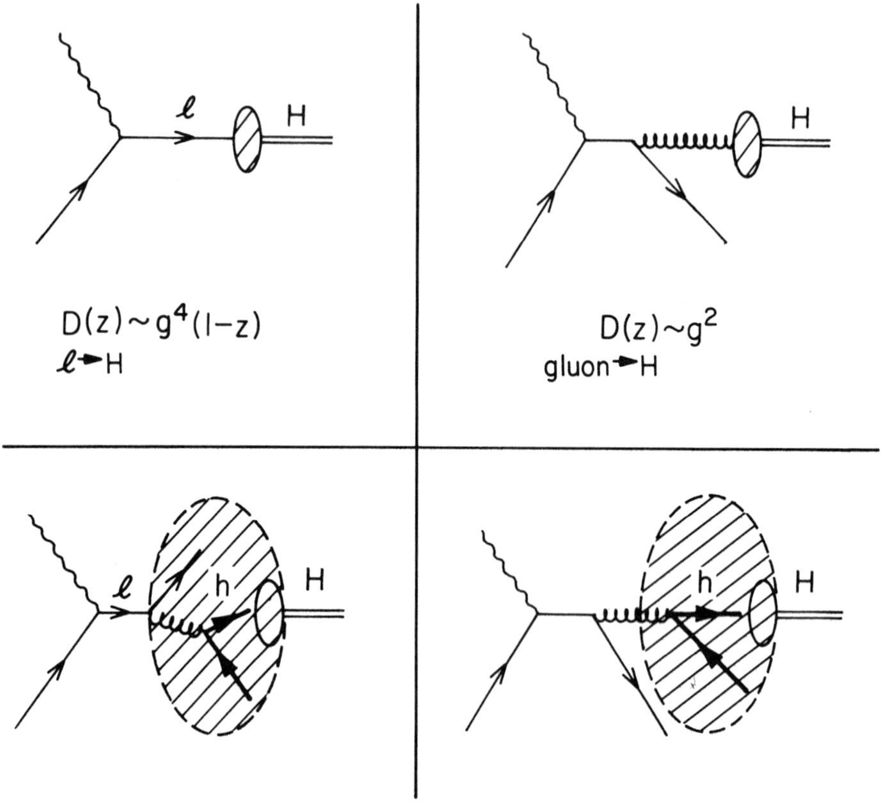

Fig. 4: Schematic description of the calculation of the decay of light quarks and gluons into heavy hadrons.

possibility of radiating a gluon just before the quark-photon collision. (The $\vec{p}_\perp$ refers to the quark after radiating the gluon but before - or after - absorbing the photon.) The $(0.300\ \text{GeV})^2$ is an estimate of non-perturbative effects of confinement. In a proton beam, quarks are confined in the transverse direction to within a proton radius; from the uncertainty principle, they must have a transverse momentum of order $(\text{radius})^{-1}$.

In $p + p \to \mu^+\mu^- + X$, each of the annihilating quarks can radiate before annihilating[9]. The net effect is that the μ-pair will have a $\langle \vec{p}^{\,2}_{\perp\mu^+\mu^-} \rangle$ of

$$\langle \vec{p}^{\,2}_{\perp\mu^+\mu^-} \rangle \approx 2\, \frac{M^2(1-x)}{16 \log 4M^2} + 2(0.3\ \text{GeV})^2 \quad ,$$

where $x = M/\sqrt{s}$.

I plot $\langle p^2_{\perp\mu^+\mu^-} \rangle^{1/2}$ for $S = 400\ \text{GeV}^2$ in Fig. 5, along with some recent data. (Note that while the CPII data is for $S = 225\ \text{GeV}^2$, the correction of the theoretical curve is only ~ 0.8% for those particular x's.)

In $\mu + p \to \mu + \pi + X$ we can estimate the expected $\vec{p}^{\,2}_{\perp\pi}$. The paradigm decay function calculation $(e^+e^- \to \pi + X)$ gives a $\langle \vec{p}^{\,2}_{\perp\pi} \rangle$ relative to the jet axis of

$$\langle \vec{p}^{\,2}_{\perp\pi} \rangle_{e^+e^- \to \pi + x} \approx \frac{S(1-z)}{\text{"16"} \log 4S} + O(0.3\ \text{GeV})^2 \quad .$$

Putting this together with the expected $\vec{p}^{\,2}_\perp$ of the parent quark, we expect

$$\langle \vec{p}^{\,2}_\perp \rangle_{\mu+p \to \mu + \pi + X} = z \left(\frac{Q^2(1-x)}{16 \log 4Q^2} + (0.3\ \text{GeV})^2 \right)$$

$$+ \frac{Q^2(1-z)}{16 \log 4Q^2} + (0.3\ \text{GeV})^2$$

where $x = Q^2/2p\cdot q$, $z = p'\cdot p/p\cdot q$ and p is the proton momentum and p' is the π momentum.

We can do far better and compute the $\langle \vec{p}_\perp \rangle$, in magnitude and direction[8]. The details are rich and complicated, but a simple sample consequence was

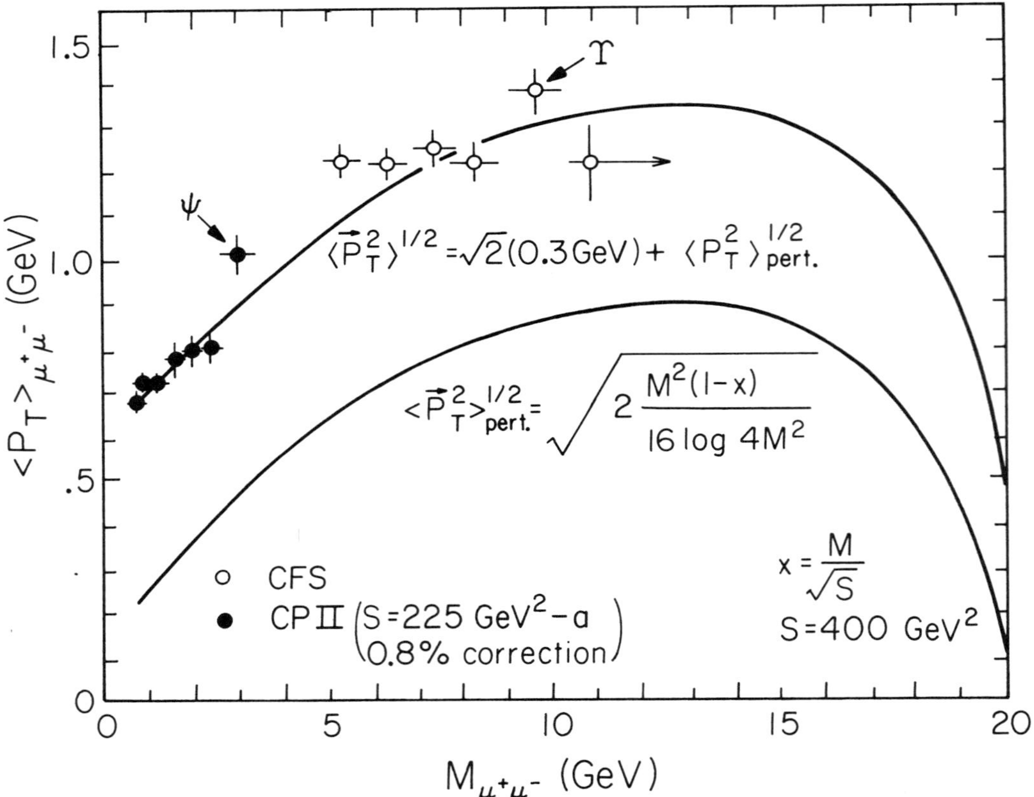

Fig. 5: $\langle p_\perp \rangle_{\mu^+\mu^-}$ versus $M_{\mu^+\mu^-}$ in $p+p \to \mu^+\mu^- + X$.

pointed out by Howard Georgi. If $\vec{k}_\perp$ if the incoming μ 3-momentum transverse to $\vec{q}$ and ϕ is the π azimuthal angle about $\vec{q}$ measured from $\vec{k}_\perp$, then $\langle \cos\phi \rangle < 0$ and explicitly calculable in terms of known parton functions. Qualitatively $|\langle \cos\phi \rangle| \sim 0\left(\frac{g^2(Q^2)}{4\pi^2}\right)$ which is something like 0.15 at present Q^2's. This angular correlation depends precisely on the spins of the relevant partons and should be absent from the non-perturbative confinement effects, which are expected to be roughly spin independent.

There are certainly other sensitive tests of the details of the quark-gluon couplings which can be made in the near future.

References

1. H. D. Politzer, Phys. Reports 14C, 129 (1974).

2. e.g. A. D. Rújula, H. Georgi, and H. D. Politzer, Ann. Phys. 103, 315 (1977).

3. T. Appelquist and H. D. Politzer, Phys. Rev. Lett. 34, 43 (1975); and Phys. Rev. D12, 1404 (1975).

4. H. D. Politzer, "QCD Off the Lightcone and the Demise of the Transverse Momentum Cut-Off," to be published in Phys. Lett.

5. A. Mueller, Phys. Rev. D9, 963 (1974).

6. C. Callan and M. Goldberger, Phys. Rev. D11, 1553 (1975).

7. H. Georgi and H. D. Politzer, to be published.

8. H. Georgi and H. D. Politzer, "Clean Tests of QCD in μ-P Scattering," Harvard Preprint HUTP-77/A063.

9. H. D. Politzer, "Gluon Corrections to Drell-Yan Processes," to be published in Nucl. Phys.

10. T. D. Lee and M. Nauenberg, Phys. Rev. 133, 1549 (1964).

11. T. Kinoshita, J. Math. Phys. 3, 650 (1962).

12. T. Appelquist et al., Phys. Rev. Lett. 36, 768 (1976).

13. E. Witten, Nucl. Phys. B104, 445 (1976).

14. H. Georgi and H. D. Politzer, Phys. Rev. D14, 1829 (1976).

Some Problems in Gauge Field Theories

F. Wilczek†

The Institute for Advanced Studies

Princeton, New Jersey 08540

Some problems relating to symmetry breaking, properties of Higgs particles, the cosmological constant, and strong and gravitational P and T invariance are discussed.

†A.P. Sloan Foundation Fellow, on leave from Princeton University.

In this talk we shall discuss some problems in gauge thoeries connected with symmetry breakdown and with the discrete space-time symmetries P and T.

I. Higgs boson and symmetry breakdown

1. Gravitation, electromagnetism, and very likely the strong interaction are governed by unbroken local gauge symmetries. Gauge theories of weak interactions require that the gauge invariance be spontaneously broken - that is, that the gauge symmetry of the Lagrangian is not a symmetry of the physical vacuum. The nature of this symmetry breakdown is, from a fundamental point of view, the major feature of gauge theories most in need of both experimental and theoretical clarification.

The form of symmetry breaking discussed almost exclusively in the literature involves the introduction of scalar fields into the theory, which acquire vacuum expectation values spoiling the gauge invariance. This procedure is neither logically compelling nor esthetically pleasing, but it has the practical virtues of leading to a weak-coupling theory and, most important, of giving definite predictions which may be tested against experiment.

The most characteristic feature of this type of symmetry breakdown is the existence of scalar particles (Higgs bosons) with characteristic properties.[1]

2. In the Weinberg-Salam model[2] the coupling of the physical Higgs H particle to quarks and leptons takes the form

$$\mathcal{L}_{int.} = 2^{1/4} G_f^{1/2} m H(\bar{q}q \text{ or } \bar{\ell}\ell) \qquad (1)$$

where m is the mass of the quark or lepton. From this coupling several facts are evident:

a) The Higgs scalar does not mediate any flavor-changing interactions, nor do its interactions violate the discrete symmetries P,C: we assign $J^{PC} = 0^{++}$ (vacuum quantum numbers) to H.

b) The Higgs scalar couples with a characteristic "semi-weak" strength and preferentially to heavy particles.

These properties make the Higgs scalar very difficult to produce, and especially to detect, unless it is very light.

3. In more complicated theories often several physical Higgs particles are generated, some of which may be charged. These theories generally have many undetermined parameters characterizing the Higgs particle couplings and masses. The possible phenomenology of these particles, should they exist, has been studied by D. Toussaint.[3]

In a very interesting paper M. Veltman[4] has explored the possibility that Higgs particles are very heavy and (as a consequence) strongly interacting. The consequences of this are hard to extract in a convincing way, but according to Veltman it might lead to violations of μ-e universality and/or light mass composite particles formed from several Higgs particles.

In §4-7 we shall, mainly for simplicity, focus on the original Weinberg-Salam model with a single Higgs doublet and with quarks and leptons added sequentially.

4. The form of Eqn. (1), which shows that the Higgs particle couples preferentially to heavy fermions, suggests that it will be profitable to exploit the newly discovered heavy vector mesons to produce H. Indeed,

the most promising way to produce and detect Higgs particles which are not very light seems to be through the radiative vector meson decay $V \to H + \gamma$.[5] Assuming that V is a non-relativistic bound state formed of quarks of mass m_q, we can compute

$$\frac{\Gamma(V\to H\gamma)}{\Gamma(V\to\mu\mu)} = \frac{G_F m_q^2}{\sqrt{2}\pi\alpha}\left(1-\frac{m_H^2}{m_V^2}\right)^{1/2} \qquad (2)$$

which is .007 times the phase-space factor when $m_q = 4.5$ GeV. In this process the difficulty of detecting H is reduced because of the accompanying monochromatic photon.

5. According to Eqn. (1), the Higgs particle will decay preferentially into the heaviest kinematically allowed quarks and leptons. There is, however, an important and very interesting qualification to this. That is, light hadrons through the "loop-annihilation" diagram, Fig. 1. In this process, H couples to a virtual heavy quark which then annihilates into color gluons. According to standard arguments, this represents a contribution to the total decay rate. It may be calculated to give[5]

$$\frac{\Gamma(H\to\text{light hadrons})}{\Gamma(H\to\tau\tau)} = \left(\frac{\bar{g}^2(m_H)}{8\pi^2}\right)^2 \frac{16N^2}{9\left(1-\frac{4m_\tau^2}{m_H^2}\right)^{3/2}} \qquad (3)$$

where $\bar{g}$ is the running strong-interaction coupling constant, τ is the heavy lepton, and N is the number of quark flavors comparable in mass or heavier than H. This result is remarkable in that it allows a direct measurement of the number of (arbitrarily) heavy quarks.

It goes without saying that Higgs particles cannot be produced in radiative decays if they are too heavy. I would like to show you two speculative and mutually contradictory ideas which suggest that the mass range ∿4-10 GeV is a very interesting one for a "Higgs hunt".

6. In the Weinberg-Salam model, there is one coupling constant with the dimensions of mass, that is the bare mass of the scalar Higgs field. Following the idea of S. Coleman and E. Weinberg[6], E. Gildener and S. Weinberg[7] explored the possibility that this bare mass might be zero. We would then have a theory with no explicit scale of mass at all! (A mass scale is brought in implicitly because of the necessity to normalize the couplings off-shell .) Following this idea leads to a Higgs mass $m_H \approx 7$ GeV.[7]

7. Dreitlein[8], and Veltman[9] have emphasized that symmetry breaking through the Higgs mechanism is associated with changes in the energy of the vacuum. We are accustomed to sweeping vacuum energies under the rug (there is generally an infinity connected with the zero-point motion of the quantum fields in any case), but gravitational interactions are sensitive to this energy. Leaving aside the divergent common zero-point energy, there is a finite difference in energy densities between the symmetric and asymmetric vacuua. Suppose, as seems reasonable, that we normalize the cosmological constant to zero in the symmetric vacuum. Then these differences would generate a cosmological constant far in excess of what is allowed by observations, unless the symmetric and asymmetric vacuua have very nearly the same energy density. Requiring that these energeis be equal leads to a definite value of the Higgs mass, $m_H \approx 5$ Gev.[10]

Discovery of a Higgs particle in the mass range 5-10 GeV, which might be possible through radiative decays of the Υ, would have, as we have seen, not only a profound meaning for weak interaction theory but also implications concerning the possibility that all scale invariance breaking is dynamical or the smallness of the cosmological constant.

8. Along these lines, there have been several interesting papers dealing with the transitions between metastable vacuua.[11] To my knowledge the gravitational aspect of this problem has not been treated, although the change in vacuum energy densities would seem to have a profound effect on the characteristics of such transitions in space and time.[12]

II. Problem of Strong P and T Invariance

1. One main advantage of the color gauge theory of strong interactions is that so many of the observed symmetries of strong interactions seem to follow automatically from the principles of gauge invariance and renormalizability - P,C,T, flavor conservation, the algebra of currents, $(3\bar{3})$ chiral symmetry breaking, asymptotic scale invariance. It is not only satisfying to see these observed symmetries arise from more abstract and possibly deeper ideas - it also guarantees that when we adjoin the weak interactions to the strong, these symmetries are only violated by small amounts at low energies (potentially dangerous renormalization effects are under control).[13]

A cloud has recently appeared which threatens to darken this sunny picture. That is, through the work of Polyakov[14]; 't Hooft[15]; Callan, Dashen, and Gross[16]; Jackiw and Rebbi[17]; and others, it has become established that an interaction of the form

$$\mathcal{L}_{\text{int.}} = \frac{\theta}{32\pi^2} \text{ tr } G_{\mu\nu}\tilde{G}_{\mu\nu} \tag{4}$$

(where $G_{\mu\nu}$ is the field strength, $\tilde{G}_{\mu\nu}$ its dual, written in matrix form) although it is formally a total divergence, cannot be ignored.

This is most easily seem in a path integral formulation, where there are field configurations with finite action (e.g., instantons) whose potential only decreases as $A \sim 1/r$ for large r.[14] Such configurations have a finite measure in the path integral and for them partial integration is forbidden.

2. The interaction in Eqn. (4) violates P and T invariance while conserving C. It can, therefore, contribute directly to the electric dipole moment of the neutron. From the very good experimental limits on this quantity, we conclude conservatively[18]

$$\frac{\theta}{32\pi^2} \lesssim 10^{-7} \tag{5}$$

3. The problem facing us is, then, to explain in a credible way how such a small number can arise. It is useful to introduce the following classification of gauge theories of strong, weak, and electromagnetic interactions[19]:

i) We say strong P and T invariance is <u>unnatural</u> if θ is infinitely renormalized by the interactions which break P and T .

ii) We say strong P and T invariance is <u>natural</u> if θ is only finitely renormalized by the interactions which break P and T. In this case, if $\theta = 0$ is imposed as a symmetry condition in the bare theory, a finite value in the full theory emerges and may give a satisfactory theory with no further adjustment.

iii) Because of the deep connection between instanton effects and chiral invariance discovered by 't Hooft[15], in theories with chiral invariances θ can be eliminated as a physical parameter[16,17,20]. If such theories give strong P and T conservation, we shall say the conservation is <u>automatic</u>.

4. Naively, the interaction in Eqn. (4) has dimension 4 and we would expect in general logarithmically infinite renormalization of θ. This expectation has been verified in a simple model[21,22]. We therefore expect that in a theory in which P and T are both broken by "hard" (i.e., dimension four) interactions strong P and T invariance is <u>unnatural</u>. This includes most models of T violation in the literature.

5. If the interactions which break P or T are soft (in particular, if either P or T is spontaneously broken) then θ is only finitely renormalized, and may be acceptably small in suitable models.

Spontaneous breaking of P has been advocated, mainly on esthetic grounds, by many authors in the literature.[23] To my knowledge the fundamental problem of understanding the two-component neutrino in these models has never been resolved in a convincing way.

The possibility of spontaneous T violation has been explored by T. D. Lee[24]. It the axion (see below) does not exist, this may be the most attractive way of trying to understand (approximate) strong T invariance.

6. If the mass of some quark vanishes, then θ can be removed from the theory[16,17]. The most plausible candidate for this is the up-quark. Standard soft-meson techniques[25] yield in this case the relation

$$m_{K^0}^2 - m_{K^+}^2 = m_\pi^2 + \text{(electromagnetic corrections)} \quad (5)$$

This can only be consistent with experiment if the electromagnetic contribution to $m_{K^0} - m_{K^+}$ is -16 MeV, which seems rather large.

7. Peccei and Quinn[20] have proposed a different type of chiral symmetry, involving Higgs bosons as well as quarks, which also gives automatic strong P and T invariance. As pointed out by S. Weinberg and the author independently[26,19] this proposal requires the existence of a new particle, an <u>axion</u>, with remarkable properties. We cannot begin to give a full analysis here,[19,20,26] but the following should be noted:

i) The axion a has $J^{PC} = 0^{-+}$ and its mass is estimated as 100 keV $\times 10^{\pm 1}$.

ii) The coupling of a to leptons and quarks (in the simplest model) is given by

$$\mathcal{L}_{\text{int.}} = 2^{\frac{1}{4}} G_F^{\frac{1}{2}} a \left(\tan\lambda \sum_{\substack{Q=2/3 \\ \text{Quarks}}} m_i \bar{q}_i \gamma_5 q_i + \cot\lambda \sum_{\substack{Q=-1/3 \\ \text{Quarks}}} m_j \bar{q}_j \gamma_5 q_j \right. \quad (6)$$

$$\left. + \cot\lambda \sum_{\substack{Q=-1 \\ \text{leptons}}} m_K \bar{\ell}_K \gamma_5 \ell_K \right)$$

where λ is an undetermined angle.

iii) The main decay mode of a is into $\gamma\gamma$; its lifetime is estimated as

$$\tau_a \sim 2\times10^{-2} \left(\frac{100 \text{ keV}}{m_a} \right)^3 \quad \text{sec.} \qquad (7)$$

for $\lambda = \pi/4$, 6 quarks and 6 leptons.

iv) Analogously to the H particle described in Part I, we have the heavy vector meson decay

$$\frac{\Gamma(V\to a\gamma)}{\Gamma(V\to\mu\mu)} = \frac{G_F m_q^2}{\sqrt{2}\ \pi\alpha} \qquad (8)$$

for $\lambda = \pi/4$.

v) The axion is only marginally consistent with limits from experiments which could have detected $K^+\to\pi^+ a$, production of a's in nuclear reactors, and beam dump experiments. None of these limits seems conclusive, however.[26]

If the axion were discovered, it would be a striking confirmation of gauge theories of particle interactions.

8. In the theory of gravity, there are two interaction terms which are formally total divergences, associated with the Euler and Pontryagin indices, viz:

$$\mathcal{L}_{\text{int.}} = \frac{\chi}{32\pi^2} \int \sqrt{g}\ R_{\alpha\beta}{}^{\gamma\delta} R_{\rho\sigma}{}^{\tau\nu} \varepsilon^{\alpha\beta\rho\sigma} \varepsilon_{\gamma\delta\tau\nu} \qquad (9)$$

$$\mathcal{L}_{\text{int.}} = \frac{\theta}{32\pi^2} \int \sqrt{g} \; R_{\mu\nu\alpha\beta} R^{\mu\nu}{}_{\gamma\delta} \varepsilon^{\alpha\beta\gamma\delta} \tag{10}$$

The first of these interactions, Eqn. (9) does not violate any quantum numbers, and might be expected to arise in pure gravity (if, as we have argued[19,22], these kinds of interactions get renormalized). This kind of interaction is amenable to dimensional regularization, and has indeed been found[27] with a divergent coefficient.

The other interaction, Eqn. (10), is directly analogous to other gauge theories and would be induced by P and T violating interactions. Indeed, from a point of view we have discussed earlier[28] the gravitational interaction is nothing but an SO(3,1) gauge theory, and the discussion given for the strong interaction carries over with only group-theoretic changes. There is, however, one relevant important difference between the Einstein and the Yang-Mills action. That is, in the Yang-Mills case, field configurations with non-trivial topological properties (i.e. sensitive to the interactions (9), (10)) have a certain minimal action and are thereby suppressed for small coupling. In the gravitational case, configurations of non-trivial topological character can be found with arbitrarily small action, so that the P and T violating effects could conceivably be substantial in some circumstances.

References

1. P. W. Higgs, Phys. Rev. 145, 1156 (1966).

2. S. Weinberg, Phys. Rev. Lett. 19, 1264 (1967);
A. Salam, Proc. 8th Nobel Symposium, New York, Wiley (1968).

3. D. Toussaint (private communication and to be published).

4. M. Veltman, Acta Physica Polonica B8, 475 (1977).

5. F. Wilczek, Phys. Rev. Lett. 39, 1304 (1977).

6. S. Coleman, E. Weinberg, Phys. Rev. D7, 2887 (1973).

7. E. Gildener, S. Weinberg, Phys. Rev. D13, 3333 (1976).

8. M. Dreitlein, Phys. Rev. Lett. 33, 1243 (1974).

9. M. Veltman, Phys. Rev. Lett. 34, 772 (1975).

10. This is the condition that the Linde-Weinberg bound on the Higgs mass be saturated, see A. Linde, JEPT Lett. 23, 64 (1976); S. Weinberg, Phys. Rev. Lett. 36, 294 (1976). The value of the mass quoted here is based on a value $\theta = 35°$ for the weak mixing angle and on neglect of presumably sub-dominant scalar and fermion contributions. For the exact form, see the above papers.

11. M. Voloshen, L. Kobrazev, L. Okun, Sov. J. Nucl. Phys. 20, 644 (1975). Further work along this line has been done by Coleman, Frampton, Linde, and others. For latest references see e.g. P. Frampton, Phys. Rev. D15, 2922 (1977), A. Linde, Phys. Lett. 70B, 306 (1977).

12. The precise connection between the smallness of the cosmological constant and the Higgs mass $\sim$5 GeV. I believe is presented here for the first time, although it requires nothing more than bodily joining the ideas of Ref. 8-9 and Ref. 10. The role of gravitational interactions in connection with metastable vacuua is being further investigated.

13. S. Weinberg, Phys. Rev. Lett. 31, 494 (1973).

14. A. Polyakov, Phys. Lett. 59B, 82 (1975), Nucl. Phys. B120, 82 (1975); A. Belavin, et al., Phys. Lett. 59B, 85 (1975).

15. G. 't Hooft, Phys. Rev. Lett. 37, 8 (1976), Phys. Rev. D14, 3432 (1976).

16. C. Callan, R. Dashen, D. Gross, Phys. Lett. 63B, 334 (1976).

17. R. Jackiw, C. Rebbi, Phys. Rev. Lett. 37, 172 (1976).

18. This bound could be sharpened by computing the interaction of a photon with a quark propagating in an instanton field weighted by $e^{i\theta}$ plus an anti-instanton field with weight $e^{-i\theta}$, e.g. using the results of L. Brown, R. Carlitz, D. Creamer, C. Lee, Phys. Lett. 71B, 103 (1977).

19. F. Wilczek "Problem of Strong P and T Invariance in the Presence of Instántons" Institute for Advanced Studies preprint (1977).

20. R. Peccei, H. Quinn, Phys. Rev. D16, 1791.

21. We are sloughing over an important technical point here, that renormalization of θ must be considered in cinjunction with renormalization of the quark mass matrix. This is discussed carefully in Ref. 22.

22. F. Wilczek "Example of Induced Strong P and T Violation" Institute for Advanced Studies preprint (in preparation).

23. For example:
J. Pati, A. Salam, Phys. Rev. D10, 275 (1974);
R. Mohapatra, J. Pati, Phys. Rev. D11, 2558 I1975);
A. DeRujula, H. Georgi, S. Glashow, Phys. Rev. D12, 3589 (1975).

24. T. D. Lee, Phys. Reports 9C, #2 (1974).

25. For an excellent review, see S. Weinberg "The Problem of Mass" to

appear in I. I. Rabi Festschift, published by N.Y. Academy of Science.

26. S. Weinberg "A New Light Boson?" (Harvard Preprint).

27. G. 't Hooft, M. Veltman, Ann. Inst. Henri Poincare XX, 69 (1974).

28. F. Wilczek, in Quark Confinement and Field Theory, p. 211 (Wiley, 1977).

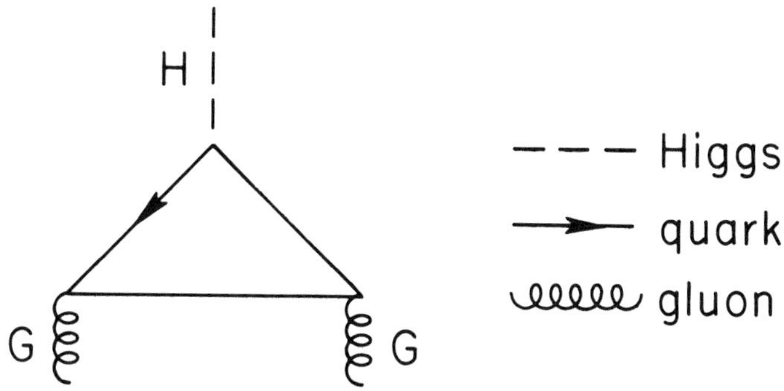

Fig. 1. Loop annihilation contribution to H→light hadrons.

PHENOMENOLOGY OF GAUGE THEORIES

S. L. Glashow
Harvard University, Cambridge, Mass.

Electrons and protons were the only elementary particles in the First Age of lepton-hadron symmetry. Atomic structure was explained, but atomic nuclei posed new mysteries. The Second Age began as I was born with the discovery of the neutron and the conjecture of the neutrino. For a time, it seemed as if the ancient Greeks had the number right: just four fundamental Fermion fields were needed. War-time discoveries of muons and of strange particles led to a Third Age of lepton-quark symmetry which lasted until 1962. The next dozen years, formative for many of us, constitute a unique interregnum of asymmetry wherein there were four observed leptons but only three "observed" quarks. The discovery of J/ψ, and the experimental demonstration that it was what it had to be, led to the restoration of lepton-quark symmetry and began the Fourth Age.

Is lepton-quark symmetry an essential aspect of Nature, and if so, what does it mean? To answer this, we need a unified description of quarks and leptons. What we have is just a formal argument that there must be no triangle anomalies in the gauge theory, a condition which is most easily and elegantly achieved with lepton-quark symmetry.

The short-lived Fourth Age marks a point of parity between the relevant building blocks -- those we know we need to build the universe -- and the seemingly irrelevant. There is even more to heaven and earth than is dreamt of in our cosmology. Of all the coequal quarks and leptons, we are made of those which by accident of birth are lightest, the runts of the litter.

Now, it seems that the relevant particles are a mere minority. A *fifth* lepton has been found, named, and confirmed. A newly-seen bump in the invariant mass of lepton pairs is generally believed to indicate the existence of a *fifth* quark. How prejudices change in three years, and how fitting it would be if *upsilon* were due to something other than just another quark. It is the *conservative* view that we have entered the *fifth* age, and that the *sixth* age cannot be far away. No one has predicted the masses of quarks and leptons yet unseen, nor even how many there must be. Bear in mind, though, that Mendeleev needed sixty-three hints to predict his new elements, and even then was half a century away from understanding what was going on.

Thus it is that we must address ourselves today to a more modest question. We are sold on a gauge theory of strong interactions and another gauge theory of weak and electromagnetic interactions. We dream of a grand unification of these two systems, but first we must know exactly what it is we are unifying. We do not know whether chromodynamics is correct, but we do know that it is *unique:* the group is SU(3), quarks are triplets, and leptons are singlets. The same may not be said for our theory of weak and electromagnetic interactions. In the "standard model" the group is SU(2) × U(1), left-handed fermions are doublets, and right-handed fermions are singlets. The standard model is surely the simplest model, but this does not mean that it is the correct theory: experiment must have its say.

Neutrino-induced neutral currents were predicted, and they were found. Four numbers characterize the neutral-current couplings of neutrinos to valence quarks. In the standard model, they are given in terms of a single parameter $\sin^2\theta$. All recent analyses of the data indicate perfect accord between experiment and theory with a value of $\sin^2\theta \simeq \frac{1}{4}$.

The standard model also requires the existence of parity-violating couplings between electrons and quarks, thereby predicting measurable parity violation in atomic physics. No effect has yet been seen, and there is a serious apparent conflict with the standard model. De Rújula, Georgi and I consider the implications of two simple alternatives to the standard model which have the following properties:

(1) The neutral-current couplings of neutrinos to hadrons coincide with those of the standard model.

(2) There is no coupling of the form

$$(\bar{e}\gamma_\mu\gamma_5 e)(\bar{q}\gamma^\mu q).$$

Thus, in our alternative models, there is no parity-violating electron-hadron coupling of the type to which the Bismuth experiments are most sensitive.

The hybrid model is a simple modification of the standard model with the right-handed electron put into a weak doublet. The ambi model is based on the $SU(2)\times SU(2)\times U(1)$ gauge group, and is naturally rigged to satisfy the above conditions. We shall show how experiments soon to be done may distinguish among the three models.

Atomic Parity Violation

The relevant couplings are:

$$\frac{G}{\sqrt{2}}(\bar{e}\gamma_\mu\gamma_5 e)(f_u\bar{u}\gamma_\mu u + f_d\bar{d}\gamma_\mu d)$$

$$\frac{G}{\sqrt{2}}f_e(\bar{e}\gamma_\mu e)(\bar{u}\gamma_\mu\gamma_5 u - \bar{d}\gamma_\mu\gamma_5 d)$$

$$\frac{G}{\sqrt{2}}f_e'(\bar{e}\gamma_\mu e)(\bar{e}\gamma_\mu\gamma_5 e)\ .$$

Using the notation $\sin^2\theta = \frac{1}{4} + \delta$, we obtain the following results:

	f_u	f_d	f_e	f_e'
Standard	$-\frac{1}{6}+\frac{4}{3}\delta$	$\frac{1}{3}-\frac{2}{3}\delta$	2δ	2δ
Ambi	0	0	0	0
Hybrid	0	0	$-\frac{1}{2}+2\delta$	0

Evidently, atomic physics experiments may in principle tell us which (if any) of the three models is viable.

Nuclear Parity Violation

The standard and hybrid models involve identical neutral-current contributions to nuclear parity violation, as well as charged current contributions. In the ambidextrous model, there are only charged current contributions. If neutral current contributions can be shown to be necessary, the ambi theory is excluded.

Neutrino-Electron Scattering

The relevant interactions are:

$$\frac{G}{\sqrt{2}}[\bar{\nu}_\mu \gamma_\lambda (1+\gamma_5)\nu_\mu][g_V \bar{e}\gamma_\lambda e + g_A \bar{e}\gamma_\lambda \gamma_5 e]$$

$$+\frac{G}{\sqrt{2}}[\bar{\nu}_e \gamma_\lambda (1+\gamma_5)\nu_e][g_V' \bar{e}\gamma_\lambda e + g_A' \bar{e}\gamma_\lambda \gamma_5 e] \ ,$$

and for all three models

$$g_V' = g_V + 1$$

$$g_A' = g_A + 1 \ .$$

In the three models, we obtain:

	g_V	g_A
Standard and Ambi	2δ	$-\frac{1}{2}$
Hybrid	$-\frac{1}{2}+2\delta$	0

In particular, the ratio of cross sections $\sigma(\nu)/\sigma(\bar{\nu})$ must be unity in hybrid, and must be $\sim(1+8\delta)$ in the other models. None of the models is as yet excluded by data.

Asymmetry in $e^+e^- \to \mu^+\mu^-$

This does not signify parity violation, but requires an axial-vector contribution due to virtual Z production. The overall asymmetry $(B-F)/(B+F)$ is given by

$$(6.6\times10^{-5}\ \mathrm{GeV}^{-2})fs$$

where s is the square of the total energy and f is given by

	f
Standard	1
Ambi	2
Hybrid	0

With 30 GeV in the center-of-mass, the three models give readily distinguishable effects: 6%, 12%, or none.

IVB Spectroscopy

In the more distant future, we may study weak interactions more directly through the study of the intermediary mesons themselves. For the models at hand, the predicted states are as follows:

	W's	Z's
Standard and Hybrid	75 GeV	86 GeV
Ambi	75 GeV; 75 GeV	75 GeV; 106 GeV

Think of an e^+e^- machine tuned in to the Z resonance, where R is more than a thousand! Such a particle will decay democratically into all quark species, leptons, and fundamental mesons with weak couplings. Then, and perhaps only then, can we discover *all* the fundamental entities with masses less than $\frac{1}{2}M_Z$.

Meanwhile, attempts to construct more grandiose theories are stymied. Perhaps we simply lack the courage of our convictions, but first we must beg the experimenter for guidance. He must eliminate at least two of the models we have discussed before we dare begin.

UNIFICATION AT LOW AND HIGH ENERGIES

by

Jogesh C. Pati*
Department of Physics and Astronomy
University of Maryland
College Park, Maryland 20742

I. INTRODUCTION

"Unification" is still in its infancy. It is the postulate that the observed asymmetries between basic particles (such as electrically charged electrons versus neutral neutrinos[1] or quarks versus leptons[2]), and the distinctions between their forces[2,3] (weak, electromagnetic and strong) are not intrinsic to the basic equations of motion[4]. These seemingly different particles are members of one set intimately linked to each other through the basic lagrangian; so also are their forces. The observed asymmetries between these particles and their forces arise through spontaneously induced asymmetry in the ground state. Such asymmetries should prevail only at low energies and must disappear at appropriately high energies.

Such a postulate raises a number of intriguing questions. In line with the emphasis of this conference being on the topics: parity non-conservation and the existence of observable new particles beyond charm, I shall discuss here primarily two such questions, pertaining to the origin of parity nonconservation on the one hand and the confinement or liberation of the basic constituents (quarks and gluons) on the other. The first would have its implications on the much discussed issue of this conference: namely the extent to which parity should be expected to be conserved in

* Supported in part by National Science Foundation under Grant No. GP43362X.

atoms[5], and the second on the other topical issue concerning the interpretation of the so-called heavy lepton[6].

I. The Question of Parity NonConservation: The relevant issues here are:

(a) Is parity nonconservation an intrinsic property of the basic equations of motion; or is it induced spontaneously in the same manner as the asymmetry between basic particles and their forces, and consequently should disappear at appropriately high energies?

(b) A related question: are neutrinos truely massless two-component objects, or are they four component entities, like other fermions, having a mass[7], though tiny?

(c) A more practical question: even if the basic equations of motion are left-right symmetric, is the symmetry retained up to relatively low energies in the hierarchical scheme of spontaneous symmetry breaking, or is it lost spontaneously at the primary stage of symmetry-breakdown by a mass-scale $M \gg m_W \sim 100$ GeV? (m_W denotes the mass of the "ordinary" weak gauge bosons).

(d) A still more practical question: even if the basic equations are left-right symmetric, is the symmetry retained in the neutral current-sector to a high degree of accuracy despite spontaneous symmetry breaking, or is it lost spontaneously in the charged as well as the neutral current-sectors to nearly the same extent? (The answer to this question is of course relevant to the extent to which parity violation would show in atoms[5]).

II. Quark-Confinement or Liberation? Independently of the answers to these questions, there is the important issue: are quarks and gluons truely confined permanently inside normal hadrons, or are they observable? As stressed elsewhere[2,8], equivalently are quarks fractionally charged, the octet of gluons neutral and massless; or are they integer charged, the gluons acquiring a mass and possessing charged members?

The first part of my talk would be devoted to showing the origin of these twin alternatives indicated by each of these questions within the unification-postulate, and the second half to their observational implications.

To begin with I should emphasize that the set of questions I (a), (b), (c) and (d) related to parity nonconservation on the one hand, and II related to quark-confinement (or liberation) on the other are logically independent of each other. For example, starting with a left-right symmetric basic lagrangian[9], it is possible to realize through spontaneous symmetry breaking a low energy manifestation of the basic symmetry, which is either "left-handed" or left-right symmetric; and in either case, quark-charges may be either integral or fractional.[2] As to which of the two low-energy-manifestations (left handed or left-right symmetric) and which of the two quark-charges (integral or fractional) are actually realized in nature, depend very much on the nature of spontaneous symmetry-breaking, as much as they do on the choice of the basic lagrangian.

This feature of the unification-hypothesis, exhibiting fundamental alternatives, which I refer to as the "many facets of unification", is outlined schematically in Figure 1. To be more specific, assume, following the unification-postulate, that the basic lagrangian L is generated by a unifying local symmetry G. It generates the so-called weak, electromagnetic and strong forces of quarks and leptons, and is governed by one universal gauge coupling constant g. Assume furthermore that the lagrangian is invariant under the discrete symmetry, which interchanges left and right, so that it conserves parity. [Examples of such unifying local symmetries, incorporating left $\leftrightarrow$ right symmetry as well, are[2]

$[SU(4)]^4$ and[10] SO(10)].

Now, at the primary stage of spontaneous symmetry breaking (stage-1), quark-lepton unification must be lost by a mass-scale $M >> m_W \sim 100$ GeV in order that we may account for the observed distinctions between quarks and leptons on the one hand and simultaneously between the effective low energy weak/electromagnetic versus strong coupling constants on the other. There is the choice at this stage, however, that either left $\leftrightarrow$ right symmetry of the basic lagrangian is still preserved (1a), or it is lost spontaneously to the same extent as quark-lepton-unification (1b). An example of the residual symmetry retaining left $\leftrightarrow$ right-symmetry[11] would be $SU(2)_L \times SU(2)_R \times U(1)_{L+R} \times SU(3)'_{Color}$, and that for which left $\leftrightarrow$ right symmetry is lost would be[1] $SU(2)_L \times U(1) \times SU(3)'_{Color}$. For either Case (1a or 1b), low energy effective strong interactions of quarks are generated by the vectorial SU(3)'-color-symmetry. Note, even if the low-energy-manifestation of weak/electromagnetic symmetry is "weakly" left handed (e.g. $SU(2)_L \times U(1)$) the basic lagrangian may still be left-right symmetric.

At the subsequent stage (stage-2) of spontaneous symmetry breaking, the charged and neutral gauge particles generating observed weak interactions acquire masses of order 50-100 GeV. The residual symmetry of stage-1 (1a or 1b) now reduces further spontaneously by this mass-scale of order 100 GeV to

$$U(1)^{flav}_{EM} \times SU(3)'_{Color}$$

where the abelian $U(1)^{flav}_{EM}$ denotes the color-singlet part of the electromagnetic charge-symmetry. At this stage, left-right symmetry, even if it was preserved through stage 1, must be broken at least in the charged current sector to account for the observed predominance of V-A over V+A β-decay and μ-decay interactions. This could come about through the charged gauge particles $W_R^{\pm}$ coupled to V+A β-decay and μ-decay-currents acquiring a mass at least three times heavier[2] than that of the charged $W_R^{\pm}$, coupled to parallel V-A-currents. But there is the allowed choice[11,12] still that left-right symmetry may be preserved in the neutral-current sector to an extent much better than in the charged one; or that it may be broken in the charged as well as neutral-sectors to nearly the same extent. Correspondingly atomic parity may be conserved to much better than $O(G_{Fermi})$, or it may be violated by an effective interaction of strength of order G_{Fermi}.

Following stage-2, there are two important alternatives:

(i) Quark-Liberation: Either the residual symmetry $U(1)^{flav}_{EM} \times SU(3)'_{Color}$ breaks spontaneously[2] softly by a mass-scale ~ 1 GeV; the octet of color gluons acquire a mass of this order; local SU(3)'color-symmetry is broken, yet a good effective global color-symmetry is preserved.[8] The residual local symmetry, that emerges in this case, is just electromagnetism $U(1)_{EM}$, which is a linear combination of flavor and SU(3)'- color generators.

$$Q_{em} = Q_{flav} + Q_{Col} \tag{1.1}$$

Quarks in this case acquire integer charges like the leptons; the octet of gluons possess four charged and four neutral members. Quarks as well as gluons become unstable against decay into leptons[2] possessing lifetimes $\leq 10^{-11}$-10^{-13} sec. Emission of such short lived quarks and gluons in hadronic and leptonic collisions do not contradict observation. This opens the possibility that the $\bar{\mu}e$-events[6] observed at SPEAR and DESY may have their origin in the pair-production and decay of integer-charge quarks[13] (rather than heavy leptons). In short quarks and gluons are liberated. Their confinement is partial[14] in the sense that they still have to tunnel through a barrier in order that they may be emitted; but the barrier is finite.

(ii) Quark-Confinement: Alternatively, the residual symmetry of stage-2 $U(1)^{flav}_{EM} \times U(1)^{flav}_{EM} \times SU(3)'_{Color}$ breaks no further; electromagnetism is composed of flavor-nonsinglet but color-singlet generators only[15]:

$$Q_{em} = Q_{flav} \tag{1.5}$$

SU(3)'-Color as a local symmetry remains preserved. Quarks in this case acquire fractional charges, even though leptons are integer-charged. The octet of color-gluons remain neutral and massless. Nonobservations of such massless gluons[16] (which could be emitted in pairs for example in decays such as $\pi^+ \to \mu^+\nu + V° + \bar{V}°$), and fractionally charged (therefore stable) quarks force one to assume that exactness of the nonabelian SU(3)'-color local symmetry help confine[17] (nonperturbatively) quarks and gluons permanently inside normal hadrons. This remains to be shown.

COMMON FEATURES AND ADVANTAGES

The two alternatives - softly broken or exactly conserved local SU(3)'-color-symmetry (equivalently integer or fractionally charged quarks)- share a number of common features and advantages:

(i) Asymptotic Freedom and Scaling: In the context of a unified gauge theory of weak electromagnetic and strong interactions, they are both (at least temporarily) asymptotically free[18]. Both solutions generate basic strong interactions through nonabelian SU(3)'-color-symmetry. Thus the strong gauge interactions in either case by themselves are asymptotically free. Asymptotic freedom is lost only due to the quartic couplings of the scalar fields, which need be introduced in either case at least to break the local symmetry that generates weak interactions. Since effective renormalised values of such quartic couplings (for either the field ϕ that breaks "weak" local symmetry, or the field σ that breaks the "strong" symmetry[2]) are typically $\lesssim e$ at presently attainable momentum ranges, loss of asymptotic freedom due to such couplings is not serious for present considerations. Both alternatives thus provide a good description of observed scaling and its violations and simultaneously offer several testable predictions[19] for experiments in the immediate future.

(ii) The Archimedes-Effect: In both cases, one must essentially assume (one way or another) that the effective masses of the valence-quarks and gluons inside the nucleonic environment are small ($\lesssim$ 300 MeV getting even smaller for high momentum probes) compared to 1 GeV in order to understand precocious scaling. Their masses outside of the environment are, of course, considerably larger being infinite, if confinement should materialise, and finite, but rather large ($\approx$ 1 to few GeV), if quarks are to be liberated. A true understanding of this "Archimedes-Effect"[14] i.e. the realisation of very light masses inside the nucleonic-bath and heavier masses outside is something, which has yet to emerge compulsively from the basic theory.

(iii) Spectroscopy and Regge Trajectories: Related to the above remark, since the effective mass of the gluons is very light or nearly zero for either case inside the nucleonic environment ($\lesssim$ 100 MeV), we would expect that the quark-quark and quark-antiquark potentials are the same for both cases for short (<< 1 fermi) as well as medium-short distances ($\lesssim$ 1 fermi ~ 1/100 MeV) perhaps rising linearly up to the point to accord with Regge-trajectories rising linearly up to masses ~ 3 GeV. The differences between them should begin to be apparent only in the long-distance (infra-red regime ($r > 1$ fermi), beyond which the potential for one case (liberated quarks) should turn over and begin to fall, while that for the alternative (confined quarks) should continue rising to infinity, if confinement has to materialise.

In other words, the low-energy spectroscopy of color-singlet composite hadrons with masses up to 3 or 4 GeV (including charm composites) are expected to be the same in both cases, since the potentials for the two cases coincide in the region where the bulks of the wave functions of these composites reside.

It needs to be stressed, that the approximate linear rise of the $(q\bar{q})$-potential up to distances of order 1 fermi (as opposed to some other form of rise) is at present purely a dynamical assumption for either alternative, suggested by the observed linear rise of Regge-trajectories, but not proven from the field theory of quarks and gluons. Perhaps multigluon-exchange, which are important beyond the short distance regime, will provide the answer.

(iv) Naturalness of Symmetries of Strong Interactions: Parity, Strangeness, etc. Observed symmetries such as parity and strangeness within either alternative turn out to be "natural"[20] in the sense that their violations are $O(\alpha/m_W^2) \sim O(G_F)$ rather than being of order α. This becomes simply a consequence of gauge principle and renormalisability: for weak, electromagnetic as well as strong interactions. [The two alternatives may provide different solutions, however, to the famous U(1)-problem , in as much as for one SU(3)'-color is an exact local symmetry , while for the other it is softly broken by the Higgs-mechanism.]

(v) $\pi^\circ \to 2\gamma$ -Decay: Both alternatives lead to the same rate for the $\pi^\circ \to 2\gamma$-decay and one which is compatible with the data.

(vi) Octet-Enhancement: Since the short distance behaviour of the two theories are the same, the enhancement[21] of the effective octet operators (relative to 27-plet) for nonleptonic weak transitions due to short-distance contributions are the same in both cases.

(vii) Quark-charges as measured in deep-inelastic lepton-hadron scattering: Within the unified gauge theory approach to integer-charge quarks, as opposed to a phenomenological theory, it turns out that electroproduction experiments measure only effective quark-charges (rather than their true charges) given by the formula[22]

$$\left(Q^{\text{quark}}_{\text{eff}}\right)^2 = \left(Q^{\text{quark}}_{\text{flavor}}\right)^2 + \left(\frac{m_U^2}{|q^2| + m_U^2}\right)^2 \left(Q^{\text{quark}}_{\text{color}}\right)^2 \tag{1.3}$$

where m_U denotes the mass of the gluon. From the above relation, we see that asymptotically (i.e. for $|q^2| >> m_U^2$), $Q_{\text{eff}}^2 \approx Q_{\text{flavor}}^2$; this implies that even integer-charge quarks (within unified theories) would asymptotically behave just like fractionally charged quarks in deep inelastic lepton-hadron scattering. Now coming to the experimental situation, deep inelastic lepton-hadron scatterings indeed appear to reveal fractional character for quark charges[23]. But alas! precisely for the reason mentioned above, this does not help decide whether the true quark-charges are fractional or integral, contrary to familiar claims.

This list of common features and advantages is not to say that Nature has conspired all along so as not to let us distinguish between the two solutions; quite to the contrary, there are indeed several crucial distinctions between the two, which I enumerate at the end; but just to stress that at least up until now, there appears to be no experimental or theoretical basis to help choose between these two important alternatives.

In the next three chapters, I discuss the three questions, raised in this introduction:

I. How to test basic left $\leftrightarrow$ right symmetry?

II. How to test quark-liberation?

III. What is the super unifying mass-scale M?

II. LEFT-RIGHT SYMMETRY

2.1. The Hypothesis

"Left-right symmetry" is the hypothesis that for every left-handed (V-A) current coupled to a set of gauge particles $\vec{W}_L$, there exists a parallel right-handed (V+A)-current coupled to a distinct set of gauge particles $\vec{W}_R$ with equal strength. Symbolically, the gauge-lagrangian has the form[11]

$$L = g[W_L\ (V\text{-}A) + W_R\ (V + A)] \tag{2.1}$$

A prerequisite for this hypothesis is that for every left matter, there must exist corresponding right matter. In particular, corresponding to left-handed neutrinos $\nu_L^{e,\mu}$ there must exist right handed neutrinos $\nu_R^{e,\mu}$ distinct from right handed antineutrinos $\bar{\nu}_R^{e,\mu}$.

Note that if the SU(3)'-color-gauges are vectorial, they generate parity conserving interactions, but they are not left-right symmetric in the sense defined above, since (V-A), and the parallel (V + A) currents generated by them couple to the same gauge particle. In order for them to be strictly left-right symmetric, they would have to be chiral[24] rather than vectorial in nature. For weak transitions, it is, however, immaterial whether the color-sector is vectorial or chiral; both conserve parity. Furthermore, it is also immaterial for most considerations (i.e. those involving for example neutrino-reactions, atomic parity and CP-nonconservation), whether the abelian gauges (if any) are vectorial, or chiral (left-right symmetric). Thus for practical considerations, I shall call a theory "left-right symmetric" as long as its abelian and color gauge-interactions are either vectorial or chiral (left-right symmetric) in nature, but its nonabelian flavor-gauge-interactions are left-right symmetric in the sense defined above (see (2.1)).

The standard so-called vector-like models of weak and electromagnetic interactions based for example on $SU(2)_{L+R} \times U(1)_{L+R}$-group[25] are not left-right symmetric even in this limited sense, since they lead to couplings of flavor charged (V - A) and corresponding (V+A)-currents to the same gauge-particle. [The experimental distinctions between left-right symmetric and standard vector-like theories are many and noted later].

The simplest example of the (V-A) $\leftrightarrow$ (V+A)-symmetric gauge-structure as mentioned above incorporating weak and electromagnetic interactions is provided by the local symmetry[11]

$$\bar{G}_{LR} = SU(2)_L \times SU(2)_R \times U(1)_{L+R} \qquad (2.2)$$

The components $SU(2)_L$ and $SU(2)_R$ generate parallel (V-A) and (V+A)-currents, while $U(1)_{L+R}$ generates an abelian vector current[26]. The abelian component and for that matter the entire symmetry $\bar{G}_{LR}$ are to be viewed as part of a unifying nonabelian symmetry[2,10]. With only four flavors (p,n,λ,c) and four leptons $(\nu_e, e^-, \mu^-, \nu_\mu)$, the $SU(2)_L \times SU(2)_R$-gauge interactions may symbolically be written as:

$$g\, W_L \left[\begin{pmatrix} \nu_e \\ e^- \end{pmatrix}_L + \begin{pmatrix} p \\ n_c \end{pmatrix}_L + \begin{pmatrix} \nu_\mu \\ \mu^- \end{pmatrix}_L + \begin{pmatrix} c \\ \lambda_c \end{pmatrix}_L \right]$$

$$+ g\, W_R \left[\begin{pmatrix} \nu_e \\ e^- \end{pmatrix}_R + \begin{pmatrix} p \\ n_c \end{pmatrix}_R + \begin{pmatrix} \nu_\mu \\ \mu^- \end{pmatrix}_R + \begin{pmatrix} c \\ \lambda_c \end{pmatrix}_R \right] \qquad (2.3)$$

where n_c and λ_c are orthogonal Cabibbo-mixtures of physical (n,λ)-fields[27]: $(n_c)_L = n_L \cos\theta_L + \lambda_L \sin\theta_L$, similarly $(\lambda_c)_L$. The remaining fields (p,c, etc.) are physical. The symmetry $SU(2)_L \times SU(2)_R$ generates two sets of charged gauge particles $W_L^{\pm}$ and $W_R^{\pm}$ and two neutral ones W_L^3 and W_R^3, while the symmetry $U(1)_{L+R}$ generates a vectorial gauge particle S°, whose coupling is determined by the nature of the electromagnetic charge.

Note that the theory would conserve parity in the charged as well as neutral-current sectors if $\vec{W}_L$ and $\vec{W}_R$ acquired symmetrical masses.

In this case $\frac{\vec{W}_L \pm \vec{W}_R}{\sqrt{2}}$ are mass-eigenstates as much as $\vec{W}_L$ and $\vec{W}_R$, and the lagrangian (2.1) may be written as:

$$L = \frac{g}{\sqrt{2}} \cdot \left[\frac{W_L + W_R}{\sqrt{2}} \ (V) + \frac{W_L - W_R}{\sqrt{2}} \ (-A)\right] \tag{2.4}$$

This generates in second order an effective interaction of form

$$L_{eff} \propto (VV + AA) \tag{2.5}$$

Departures from parity conservation may in general arise within such a theory in the charged as well as the neutral current-sectors to the extent of the spontaneously induced mass-asymmetry between $\vec{W}_L$ and $\vec{W}_R$. Parity violation may in addition arise in the charged current-sector due to possible differences between left and right handed Cabibbo-like angles[28] defined by the Fermi-mass-matrix.

Charged current interactions encountered in β-decay, μ-decay and μ-capture exhibit dominance of (V-A) over (V+A). From the longitudinal polarization of electrons in β-decay, which is empirically known to be (-v/c) up to about 1%, one may deduce that the ratio of (V+A) to (V-A) amplitudes is less than or of order 10%. In turn this implies[2]

$$\left(m_{W_L^+} / m_{W_R^+}\right)^2 \left(\cos\theta_R / \cos\theta_L\right) \lesssim 10\% \tag{2.6}$$

Assuming[28] $\cos\theta_R \approx \cos\theta_L$, one thus obtains

$$m_{W_R^+} \gtrsim 3\, m_{W_L^+} \tag{2.7}$$

Despite this rather large asymmetry between the masses of the charged gauge particles W_L^+ and W_R^+ the neutral $(W_L^3 - W_R^3)$-mass-matrix may still be nearly or exactly left-right symmetric (in the tree-approximation); correspondingly atomic interactions may be nearly or exactly parity conserving to order $G_F\alpha$. Such a <u>dichotomy</u> between the mass-matrices of the charged versus neutral gauge-mesons may in general arise through the following pattern of spontaneous symmetry breaking.

2.2. Spontaneous Breaking of the Symmetry $\bar{G}_{LR}$:

Introduce[11] Higgs-Kibble-scalar fields transforming as doublets as well as triplets under $SU(2)_{L,R}$

$$B = (1,2,Y = +1); \quad C = (2, 1, Y = +1)$$

$$E_R = (1,3, \; Y = 0); \quad E_L = (3, 1, Y = 0) \tag{2.8}$$

Note the correspondence $B \leftrightarrow C$ and $E_R \leftrightarrow E_L$ under the symmetry operation left $\leftrightarrow$ right. Y denotes the U(1)-quantum number for the group $\bar{G}_{LR}$. Consistent with the minimum of the potential as well as "natural" left-right symmetry[29], one may assume the following pattern of vacuum expectation values:

VEV	Mass To
$\langle B\rangle = \binom{o}{b}$	$(W_R^{\pm}, W_R^3, S^\circ)$
$\langle C\rangle = \binom{o}{c}$	$(W_L^{\pm}, W_L^3, S^\circ)$
$\langle E_R\rangle = \begin{pmatrix} o \\ \varepsilon_R \\ o \end{pmatrix}$	$W_R^{\pm}$
$\langle E_L\rangle = \begin{pmatrix} o \\ \varepsilon_L \\ o \end{pmatrix} \simeq 0$	$W_L^{\pm}$

(2.9)

Thus we see that with[11,12]

$$\langle E_R\rangle \gg \langle E_L\rangle, \quad \langle B\rangle, \langle C\rangle; \text{ but } \langle B\rangle \sim \langle C\rangle \tag{2.10}$$

one can obtain large mass-asymmetry between the charged W's ($W_L^{\pm}$ and $W_R^{\pm}$), while that between the neutral ones (W_L^3 and W_R^3) may be small or "zero". Correspondingly parity violation in the charged current-sector would be large $O(g_L^2/8m_{W_L^+}^2) \equiv O(G_F/\sqrt{2})$; while that in the neutral-current-sector could be vanishingly small depending upon the value of the neutral-current left-right symmetry violating parameter

$$x \equiv (b^2 - c^2)/c^2 \tag{2.11}$$

Since the parameter c is the major-contributor to W_L^{+}-mass and therefore not zero, x may in general vary between the limits -1 and + ∝ as b^2 varies between 0 and ∞.

In the extreme limit $x \to \infty$, the symmetry would reduce to $SU(2)_L \times U(1)$, while in the limit[30] $x = 0$ $(b = c)$, and with $g_R = g_L$, neutral current-interactions would acquire the effective parity conserving form[12]

$$L_{NC} = \alpha\, VV + \beta\, AA \tag{2.12}$$

This is barring finite $O(\alpha)$ radiative corrections[2,29] to $(g_R - g_L)/g_L \sim O(\alpha\, \ell n\, (m_{W_R^+}/m_{W_L^+})$, which arise from $m_{W_R^+} \neq m_{W_L^+}$; where $g_{L,R}$ denote effective renormalized gauge coupling constants for $SU(2)_{L,R}$. Such corrections would induce atomic parity violation[31] in order $G_F^{(N)}\alpha$, where $G_F^{(N)}/\sqrt{2} \equiv g_L^2/(8m_{N_1}^2)$ and m_{N_1} is the mass of the lightest neutral weak gauge particle.

At the present level of accuracy of theory and experiments[5], this extreme limit <B> = <C> (i.e. x = 0) as a zeroth order solution rather than <B> ~ <C>, is of course not warranted. It is worth noting however that even for this extreme limit, i.e. even for the parity conserving effective interaction $\alpha\, VV + \beta\, AA$ (Eq. (2.11)), neutral current interactions of left handed ν_L would be different from that of right handed $\bar{\nu}_R$:

$$\sigma(\nu_L p)_{NC} \neq \sigma(\bar{\nu}_R p)_{NC} \tag{2.13}$$

This is because of the simultaneous presence of VV as well as AA-pieces $(\alpha \neq 0)$, $\beta \neq 0)$. Such a distinction between ν_L and $\bar{\nu}_R$ is indeed observed experimentally[32].

This should be contrasted from the so-called vector-like theories[25] based on the group-structure $SU(2)_{L+R} \times U(1)_{L+R}$. In these theories, as mentioned before, (V-A) and (V+A)-currents co-exist, but they couple to the same set of gauge particles ($W^{\pm}$, W^3). Thus these vector-like theories immediately account for the null results of present atomic parity experiments; but with only vectorial neutral quark-current (i.e. having no AA-effective interaction, i.e. $\beta = 0$), they also predict that $\sigma(\nu_L p)_{NC} = \sigma(\bar{\nu}_R p)_{NC}$ in clear contradiction with the data[33].

Some general properties and advantages of the left-right symmetric theories are listed below.

2.3. General Properties and Advantages of the Left Right Symmetric Theories:

(i) The Equivalence Theorem:[33] It is a remarkable property of the left-right symmetric substructures such as $\bar{G}_{LR} = SU(2)_L \times SU(2)_R \times U(1)_{L+R}$ or the extended symmetry[34,35] $G_{LR} = SU(2)_L \times SU(2)_R \times U(1)_L \times U(1)_R$ that at zero-momentum transfers and in the limit of vanishing fermion masses as well as vanishing W_L-W_R-mixing[36], their predictions for neutral current-neutrino-processes coincide with those of the single parameter standard left-handed theory[1] based on the group-structure $SU(2)_L \times U(1)$. This equivalance holds regardless of the value of the neutral-current parity violating parameter $x = (b^2-c^2)/c^2$ (See Eq. (2.11)).

The general proof of this equivalence is given in the last two papers of Ref. 33. It is however easy to comprehend qualitatively the reason for this equivalence: Laboratory neutrinos produced by charged current interactions are dominantly (at least 99%) left-handed. In the chiral symmetric world (i.e. no W_L-W_R mixing and zero fermion-masses), left-handed neutrinos (or right-handed antineutrinos) do not couple to W_R's; therefore they are simply blind to right-handed (V+A)-gauge interactions. Thus even a left-right-symmetric substructure would appear to such neutrinos as though it possessed only the left-handed $SU(2)_L \times U(1)$-piece. The electrical neutrality of neutrinos also turns out to play a relevant role in the argument. By contrast, electrons in atoms (possessing all four components) respond to V-A as well as V+A-interactions with equal affinity. Thus atomic interactions, as also high energy electron-positron annihilation can help distinguish between the left-right symmetric versus left-handed solutions.

In summary, considering that typical momentum transfers in neutral current experiments, as well as W_L-W_R mixing are negligible compared to the masses of neutral Z°'s and $W_L^{\pm}$ respectively, we see that the recent experimental successes[37] of the standard $SU(2)_L \times U(1)$-symmetry for neutrino-scattering experiments are retained entirely by the left-right symmetric theories $\bar{G}_{LR} = SU(2)_L \times SU(2)_R \times U(1)_{L+R}$ and $G_{LR} = SU(2)_L \times SU(2)_R \times U(1)_L \times U(1)_R$.

(ii) Left-Right Symmetry and CP

A further distinct advantage of the left-right symmetric theory $SU(2)_L \times SU(2)_R \times U(1)_{L+R}$ as well as all its quark-lepton-unifying extensions (for example[2] $SU(2)_L \times SU(2)_R \times SU(4)'_{L+R}$ or $[SU(4)]^4$) is that it allows one to generate a desirable milliweak theory of CP violation[38] through phase angles in the Cabibbo-mass matrix with only four quark-flavors[39]. The attractive properties of this theory are listed below.

First one finds that within such a theory, one can link the smallness of CP violation to the observed smallness of (V+A) compared to (V-A)-interactions, one obtains:

$$|\eta_{+-}| \approx |(m_{W_L^+}/m_{W_R^+})^2 \ (\sin 2\theta_R/\sin 2\theta_L)\sin(\delta_R-\delta_L)| \tag{2.14}$$

where $\theta_{L,R}$ are the Cabibbo-angles and $\delta_{L,R}$ the phase angles defined in the space of (left, right) fermions via their mass-matrix. Thus even if $|\delta_R-\delta_L| \sim \pi/2$

and $\theta_R \simeq \theta_L$, CP-violation is naturally suppressed because $m_{W_R^+} >> m_{W_L^+}$. Second, the effective nonleptonic $|\Delta S| = 1$ CP conserving (P^+) and CP-violating (P^-)-interactions arising via the gauge interactions of the model, are found to satisfy the isoconjugate relation $[I_3,P^-] = (1/2)\ i(\tan\xi)P^+$, where I_3 is the third component of isospin and $\tan\xi$ is related to the phase angles. Such a relation automatically yields (regardless of the choice of ξ)

$$\eta_{+-} = \eta_{oo},\ \phi_{+-} = \tan^{-1}(2\Delta m/\Gamma_s) \qquad (2.15)$$

where $\eta_{ij} = \mathrm{Amp}(K_L \to \pi^i\pi^j)/\mathrm{Amp}(K_S \to \pi^i\pi^j)$ and ϕ_{+-} is the phase of η_{+-}. Both these relations, which incidentally coincide with the predictions of the superweak theory of CP-violation[40], are in full accord with the data.

Distinctions of the isoconjugate milliweak model of CP-violation lie in that on the one hand, it predicts milliweak CP-violation in $\Delta S = 1$ -processes such as $\Lambda \to N + \pi$-decays, like most milliweak models. However, its $\Delta S = 0$ gauge interactions are CP conserving. Hence in such a model, electric dipole moment of the neutrino arises (a) through fourth order gauge interactions (i.e. in order $G_F^2\varepsilon$, where $\varepsilon \approx 10^{-3}$), (b) through W_L-W_R mixing[36] and (c) through Yukawa interactions between fermions and left over Higgs bosons. The fourth order gauge interactions would make a contribution to the e.d.m. of the neutron (d_n) in the range of $\simeq 10^{-29}$ ecm. The W_L-W_R-mixing (given that the left-right symmetric theories need to agree with $SU(2)_L \times U(1)$-predictions for neutrino-reactions reasonably well) make a contribtuion[41] to $d_n \lesssim 10^{-24}$ e cm. Finally the Higgs-boson-contribution (given that[42] the relevant left over Higgs bosons are relatively heavy ($m_\sigma \gtrsim 30$ GeV), but that they are not excessively heavy ($m_\sigma \lesssim 1000$ GeV), or else the Higgs system would become strongly interacting) lead to a dipole moment in the r ange of 10^{-24}-10^{-27} e cm. Combining these contributions, one is led to predict

$$d_n \approx 10^{-24}\text{-}10^{-27} \text{ ecm} \qquad (2.16)$$

for the left-right symmetric theories. This is on the one hand lower than the prediction of a number of milliweak models, which typically predict $d_n \gtrsim 10^{-23}$ ecm, on the other hand it is orders of magnitude higher than the prediction of genuine superweak models; the latter predict $d_n \lesssim 10^{-29}$ ecm. The present experimental value[43] $d_n = (0.4 \pm 1.1) \times 10^{-24}$ ecm is compatible with the predictions of either model (isoconjugage milliweak as well as superweak). Improvement in this measurement in planning should be able to distinguish between these two classes of theories.

The point worth noting is that the isoconjugate milliweak CP-violation with the desirable features as noted above is special to the left-right symmetric theory $SU(2)_L \times SU(2)_R \times U(1)_{L+R}$ and its unifying extensions. Such a theory of CP-violation can not be generated within the $SU(2) \times U(1)$-group structure regardless of the choice of quark-flavors and the chirality of different couplings.

(iii) Left-Right Symmetry and Atomic Parity:

The left-right symmetric theory provides the scope for accomodating small atomic parity violation[12], should this be desired experimentally . Depending upon the value of the neutral-current parity violation-parameter $x = 0$ or $|x| << 1$ (see Eq. (2.11)), atomic parity violation would be of order $G_F\, \alpha\, \ell n\, (m_{W_L^\mp}/m_{W_L^+})$, or of order $G_F \kappa$ ($\kappa << 1$). A small atomic parity violation of order $<< G_F$ can not however be accomodated within the left-handed $SU(2)_L \times U(1)$-theory . Correlated with the degree of atomic parity violation, the left-right symmetric theory offers several testable predictions involving high energy e^-e^+ collisions to be available at PETRA and PEP and pp as well as $\bar{p}p$ collisions .

(iv) Light Z°'s:

The left-right symmetric theories in general offer the possibility that one or more neutral weak gauge particles (Z°'s) can be much lighter than 80 GeV, the lightest among them being perhaps as light as nearly 40 GeV[35], consistent with all present experiments (see discussions later). This is to be contrasted from the mass of ≃ 83 GeV for the Z° of the left-handed $SU(2)_L \times U(1)$-theory. A light Z° of mass ≲ 60 GeV offers the practical advantage of easier detection.

(v) "Naturalness" of Isospin Symmetry:

The left-right symmetric gauge-structure renders a number of important parameters to be finite and calculable: notably the mass-difference between n and p-quarks, relevant to isospin-conservation[44]. The same symmetry-structure has also recently been used[45] to calculate Cabibbo angle in terms of quark-mass-parameters.

I now turn towards a discussion of the experimental tests of left-right symmetry. The mass-matrix, eigenstates and the interactions of the neutral gauge particles are presented to facilitate this discussion.

2.4. Mass-Matrix and Eigen States:

The "general" mass matrix of the neutral gauge particles W_L^3, W_R^3 and S° generated by the symmetry $\bar{G}_{LR}$, is given by[3]:

$$m^2_{\text{neutral}} = \frac{1}{4} \begin{bmatrix} g^2(a^2+c^2) & -g^2 a^2 & \sqrt{3/2}\, f_{15}\, gc^2 \\ -g^2 a^2 & g^2(a^2+b^2) & \sqrt{3/2}\, f_{15}\, gb^2 \\ \sqrt{3/2}\, f_{15} gc^2 & \sqrt{3/2}\, f_{15} gb^2 & 3/2\, f_{15}^2(b^2+c^2) \end{bmatrix} \quad (2.17)$$

(columns: W_L^3, W_R^3, S°)

The above mass-matrix includes contributions from the VEV of the multiplets B and C, introduced before (Eq. (2.8)), as well as the multiplet A = (2,2 Y = 0) with VEV $\begin{pmatrix} a_1 & o \\ o & a_2 \end{pmatrix}$; $a^2 \equiv a_1^2 + a_2^2$. The multiplet A must be introduced in order that fermions may acquire masses. In writing the above, it is assumed that $g_L^2 = g_R^2 \equiv g^2$. The constant f_{15} denotes the effective coupling constant associated with S°. [Once the subsymmetry $\bar{G}_{LR}$ is viewed as a low energy-descent from a higher unifying symmetry G, as discussed in the introduction, the coupling constants f_{15} and g would be related via the unifying mass-scale M. These relations are given in Chapter IV].

The eigenstates of the mass-matrix including the massless photon A_μ are given by[46]:

$$A_\mu = \frac{W_L^3 + W_R^3}{\sqrt{2}} \cos\alpha - S^o \sin\alpha \qquad \text{(Frac. ch. quarks}^{47}\text{)}$$

$$N_1 = Z_A \cos\beta + Z_V \sin\beta$$

$$N_2 = -Z_A \sin\beta + Z_V \cos\beta \tag{2.18}$$

where,

$$Z_A \equiv (W_L^3 - W_R^3)/\sqrt{2}$$

$$Z_V \equiv (W_L^3 + W_R^3)/\sqrt{2} \sin\alpha + S^o \cos\alpha$$

$$\cos\alpha = \sqrt{2}\,(e/g);\ \sin\alpha = \sqrt{2/3}\,(e/f_{15}) \tag{2.19}$$

Thus,

$$1/e^2 = 2/g^2 + (2/3)\,(1/f_{15}^2) \tag{2.20}$$

$$\tan 2\beta = \frac{(3r+1)^{1/2}(\eta-\zeta)}{(3r/2)(\eta+\zeta)-2(1-\zeta)} \tag{2.21}$$

$$\begin{pmatrix} m_{N_1}^2 \\ m_{N_2}^2 \end{pmatrix} = \frac{m_{W_L^+}^2}{4}\left[(3r+2)(\eta+\zeta)+4(1-\zeta)\mp\{3r(\eta+\zeta)-4(1-\zeta)\}\sec 2\beta\right] \tag{2.22}$$

$$m_{W_L^+}^2 = [(3r+1)/3r](e^2/\sqrt{8})G_F^{-1} \simeq \left[(3r+1)/3r\right](53\text{ GeV})^2$$

$$\simeq (g^2/4)(c^2+a^2) \tag{2.23}$$

where,

$$r \equiv (f_{15}^2/g^2);\ \zeta \equiv c^2/(c^2+a^2);\ \eta \equiv b^2/(c^2+a^2) \tag{2.24}$$

$$0 \lesssim \zeta \lesssim 1;\quad 0 \lesssim \eta < \infty \tag{2.25}$$

The fields Z_A and Z_V - given by (2.19) - couple to purely axial and purely vector currents respectively. The eigenstates N_1 and N_2 are mixtures of the parity eigenstates Z_A and Z_V, the mixing being determined by tan2β. Note for b=c, $\tan 2\beta \to 0$, $N_1 \to Z_A$, $N_2 \to Z_V$. Thus, as expected, in this limit (b=c), neutral current-interactions of the theory would conserve parity (to $\mathcal{O}(G_F\alpha)$); the effective interaction would have the parity conserving form VV + AA. In general the theory would permit parity violation in the neutral-current sector to the extent that $\tan 2\beta \neq 0$.

The neutral current interactions of the theory depend in general upon three independent parameters: (i) the coupling constant-ratio $r = f_{15}^2/g^2$; (ii) $\eta \equiv b^2/(c^2+a^2)$ and (iii) $\zeta \equiv c^2/(c^2 + a^2)$. The coupling constant ratio r relates to the familiar weak angle $\sin^2\theta_W$ of the SU(2) x U(1)-theory (see later). As we will see, in practice, the theory would depend upon only two parameters r and η. For neutrino-physics a single parameter (r) will suffice as in SU(2) x U(1)-theory.

2.5 Interaction:[46]

The neutral current interaction induced by N_1 and N_2 is given by

$$L_{Nc} = (J_\mu^A \cos\beta + J_\mu^V \sin\beta)N_{1\mu} + (-J_\mu^A \sin\beta + J_\mu^V \cos\beta)N_{2\mu} \tag{2.26}$$

$J_\mu^{A,V}$ are the source currents of $Z_\mu^{A,V}$. In the four-flavor four-color-model[2], their fermionic contents are:

$$J_\mu^A = (g/2\sqrt{2})\,[(\bar{\nu}_e\gamma_\mu\gamma_5\nu_e - \bar{e}\gamma_\mu\gamma_5 e) + (e \to \mu) + (\bar{p}\gamma_\mu\gamma_5 p + p \to c) - (\bar{n}\gamma_\mu\gamma_5 n + n \to \lambda)] \tag{2.27}$$

$$J_\mu^V = g/(2\sqrt{2}\,\sqrt{3r+1})\;[(3r+1)\bar{\nu}_e\gamma_\mu\nu_e + (3r-1)\bar{e}\gamma_\mu e + (e \to \mu) + (1-r)(\bar{p}\gamma_\mu p + p \to c) - (1+r)(\bar{n}\gamma_\mu n + n \to \lambda)] \tag{2.28}$$

Note that neutrinos are in general four component fields with arbitrary mass in the theory (alternatively one may substitute $\nu \equiv \nu_L + \nu_R$).

(a) Amplitudes for Neutrino-Processes

Define the relevant amplitudes for processes involving left-handed neutrinos and right-handed antineutrinos by the effective lagrangian as follows:

$$\begin{aligned} L^{eff}_{(\nu_L,\bar{\nu}_R)} = \frac{G_F}{\sqrt{2}} (\bar{\nu}_\mu \gamma_\mu (1+\gamma_5)\nu_\mu)[\bar{p}\gamma_\mu (g^p_V + g^p_A \gamma_5)p \\ + \bar{n}\gamma_\mu (g^n_V + g^n_A\gamma_5)n + \bar{e}\gamma_\mu (M^e_V + M^e_A\gamma_5)e + \ldots] \\ + \frac{G_F}{\sqrt{2}} \bar{\nu}_e\gamma_\mu (1+\gamma_5)\nu_e [\bar{e}\gamma_\mu (E^e_V + E^e_A\gamma_5)e + \ldots] \end{aligned} \quad (2.29)$$

Here p and n denote quark-fields. The effective amplitudes $g^{p,n}_{V,A}$ etc. may be obtained straightforwardly from Eqs. (2.21)-(2.28). They are:

$$g^p_V = (r-1)G_V \quad ; \quad g^p_A = -G_A$$

$$g^n_V = (r+1)G_V \quad ; \quad g^n_A = G_A$$

$$M^e_V = -(3r-1)G_V \quad ; \quad M^e_A = G_A$$

$$E^e_V = -1-(3r-1)G_V ; \quad E^e_A = -1 + G_A$$

All neutrino-amplitudes are determined in terms of just two master-functions G_V and G_A; they are:

$$G_V = \frac{\eta + 2(1-\zeta)}{(3r+1)(\zeta+\eta-\zeta^2)} \quad (2.31)$$

$$G_A = \frac{\eta}{(\zeta+\eta-\zeta^2)} \quad (2.32)$$

(b) Amplitudes for Atomic Parity Violation

Define $L_{Atomic} = (G_F/\sqrt{2})\ (\bar{e}\gamma_\mu\gamma_5 e)\ (C^p_V\bar{p}\gamma_\mu p + C^n_V\bar{n}\gamma_\mu n)$ (2.37)

Using Eqs. (2.21)-(2.28), we obtain:

$$C_V^{\binom{p}{n}} = - \frac{(r\mp 1)m_{W_L^+}^2 \; (1/m_{N_1}^2 - 1/m_{N_2}^2)\sin 2\beta}{2(3r+1)^{1/2}} \tag{2.34}$$

The effective charge Q_W characterising atomic parity violation for an atom (A,Z) is given by

$$\begin{aligned} Q_W &= -2\{(2C_V^p + C_V^n)Z + (2C_V^n + C_V^p)\;(A-Z)\} \\ &= \frac{2Z - (3r+1)A}{(3r+1)} \; \frac{(\eta-\zeta)}{\zeta+\eta-\zeta^2} \end{aligned} \tag{2.35}$$

(c) Amplitudes for e^-e^+-Processes

Define the four-fermion amplitude for $e^-e^+ \to \mu^-\mu^+$ induced by one photon plus N_1 and N_2-exchanges by:

$$\begin{aligned} A(e^-e^+ \to \mu^-\mu^+) = {} & (\frac{e^2}{s} + G_{VV}^\mu)\;(\bar{e}\gamma_\mu e)(\bar{\mu}\gamma_\mu\mu) + G_{VA}^\mu(\bar{e}\gamma_\mu e)(\bar{\mu}\gamma_\mu\gamma_5\mu) \\ & + G_{AV}^\mu(\bar{e}\gamma_\mu\gamma_5 e)(\bar{\mu}\gamma_\mu\mu) + G_{AA}^\mu(\bar{e}\gamma_\mu\gamma_5 e)(\bar{\mu}\gamma_\mu\gamma_5\mu) \end{aligned} \tag{2.36}$$

Using (2.21-2.28), we obtain:

$$\begin{aligned} G_{VV}^\mu &= (G_F/\sqrt{2})(3r-1)^2/(3r+1)\;[D_1(s)\;\sin^2\beta + D_2(s)\cos^2\beta] \\ G_{VA}^\mu &= G_{AV}^\mu = (G_F/2\sqrt{2})\;\frac{(3r-1)}{\sqrt{3r+1}}\;\sin^2\beta\;(D_2(s) - D_1(s)) \\ G_{AA}^\mu &= (G_F/\sqrt{2})\;(D_1(s)\;\mathrm{Cos}^2\beta + D_2(s)\;\mathrm{Sin}^2\beta) \end{aligned} \tag{2.37}$$

where s = Total Center of Mass $(\text{energy})^2$, $D_{1,2}(s) = m_{W_L^+}^2/(s-m_{N_{1,2}}^2)$. For s in the vicinity of $m_{N_1}^2$ of $m_{N_2}^2$, the N_1 or N_2 widths (~1 GeV) needs to be inserted into $D_{1,2}$.

Likewise one may write the amplitudes for $e^-e^+ \to p\bar{p}(=c\bar{c})$ and $e^-e^+ \to n\bar{n}(=\lambda\bar{\lambda})$ due to N_1 and N_2-exchanges (without photon) as:

$$G_{VV}^{(p)}_{(n)} = \begin{bmatrix} (1-r) \\ -(1+r) \end{bmatrix} G_{VV}^{\mu}; \qquad G_{AV}^{(p)}_{(n)} = \begin{bmatrix} (1-r) \\ -(1+r) \end{bmatrix} (G_{AV}^{\mu})/(3r-1) \tag{2.38}$$

$$G_{VA}^{p} = -G_{VA}^{n} = -G_{VA}^{\mu}; \qquad G_{AA}^{p} = -G_{AA}^{n} = -\,G_{AA}^{\mu}$$

The pure vector amplitudes should be supplemented by photon-contributions (for example $-(2/3)e^2/s$ for p and $+(1/3)(e^2/s)$ for n-quarks in the fractionally charged quark model. For the integer-charge quarks photon as well as color-gluons will contribute above threshold for color-production . This case needs to be worked out separately.

Two Limits:

Before presenting the experimental tests, I first draw attention to two interesting limits of the left-right symmetric theory.

(I) The $SU(2)_L \times U(1)$-Limit ($b^2 >> c^2 + a^2$; or $\eta >> 1$):

The vacuum-expectation value b of the "right-handed" doublet scalar field B = (1,2 Y = +1) reduces the $SU(2)_R \times U(1)_{L+R}$ - component of the left-right symmetric theory $\bar{G}_{LR} = SU(2)_L \times SU(2)_R \times U(1)_{L+R}$ into just U(1), which is the diagonal sum of the neutral generators within $SU(2)_R$ and the $U(1)_{L+R}$. Thus when b tends to be very large $\to \infty$ ($\therefore\ \eta \to \infty$), the left-right symmetric theory $\bar{G}_{LR}$ reduces to its left-handed limit the familiar $SU(2)_L \times U(1)$. In this limit ($\eta >> 1$ with $r \neq 0$), we obtain (using Eqs. (2.18)-(2.22)):

$$\begin{aligned} A &= W_L^3 \sin\theta_W + B^\circ \cos\theta_W \\ N_1 &= W_L^3 \cos\theta_W - B^\circ \sin\theta_W \\ N_2 &= (\tfrac{3}{2} r + 1)^{-1/2} [\sqrt{3/2r}\, V_{15} + W_R^3] \quad (V_{15} \equiv S^\circ) \end{aligned} \tag{2.39}$$

where

$$B^\circ \equiv (f_{15}^2 + \frac{2}{3} g^2)^{-1/2} [f_{15} W_{3R} - \sqrt{2/3}\, g\, V_{15}] \tag{2.40}$$

$$\sin^2\theta_W = 3r/(6r + 2) \tag{2.41}$$

$$m_{N_1}^2 \simeq m_{W_L^+}^2 \quad (6r + 2)/(3r + 2) = m_{W_L^+}^2/\mathrm{Cos}^2\theta_W \tag{2.42}$$

$$m_{N_2}^2 = (m_{W_L^+}^2/2)(3r + 2)\eta \to \infty \tag{2.43}$$

Note that the composition of B° (Eq. (2.40)) is the same as that of the U(1)-gauge field in the $SU(2)_L \times U(1)$-theory. Thus N_1 (in the limit $\eta \to \infty$) reduces to the Z_o of the $SU(2)_L \times U(1)$-theory with the identification $\sin^2\theta_W = 3r/(6r + 2)$; N_2 in the same limit becomes infinitely heavy. Note also the familiar relationship (Eq. (2.42)) between the masses of N_1 (i.e. Z_o) and $W_L^{\pm}$.

The master functions G_V, G_A determining all neutrino-amplitudes (see Eq. (2.31)-(2.32) and the atomic parity violation-parameter Q_W(Eq. (2.35)) reduce in the limit $\eta \to \propto$ to:

$$G_V \xrightarrow[0\le\zeta\le1]{\eta \to \propto} 1/(3r + 1); \qquad G_A \xrightarrow[0\le\zeta\le1]{\eta \to \propto} 1 \tag{2.44}$$

$$Q_W \xrightarrow[0\le\zeta\le1]{\eta \to \propto} [2Z-(3r+1)A]/(3r+1) \tag{2.45}$$

Note that G_V, G_A and Q_W are <u>independent</u> of the parameter $\zeta \equiv c^2/(c^2+a^2)$ in the $\eta \to \propto$ limit. <u>All amplitudes are determined in this case in terms of the single parameter $r = (f_{15}^2/g^2)$ or equivalently $\sin^2\theta_W = 3r/(6r+2)$.</u> This is the familiar result of $SU(2)_L \times U(1)$.

(II) <u>The Chiral Symmetric Limit (The Equivalence Theorem)</u>:

In the limit $a^2 \to o$ (i.e. $\zeta \to 1$), W_L-W_R mixing mass as well as all fermion-masses vanish. Appropriately this limit will be referred to as the global (γ_5-invariant) chiral symmetric limit. It is important to note one remarkable general property[46] of this limit.

In the limit of exact global chiral symmetry ($a^2 \rightarrow 0$, i.e. $\zeta \rightarrow 1$), even though the masses of the two neutral eigenstates N_1 and N_2, the mixing angle β, as well as the atomic parity violation-parameter Q_W vary with b^2 (i.e. with $\eta = b^2/(c^2 + a^2)$; the amplitudes of all left-handed neutrino (or right-handed antineutrino)-induced processes at Zero momentum-transfers are independent of the vacuum-expectation-values of "right-handed" scalar-fields such as B. In the present case, they are independent of b^2 and therefore of $\eta = b^2/(c^2 + a^2)$. They depend just on the single parameter $r = f_{15}^2/g^2$. As they are independent of η, they remain unaltered even if we pass to the limit $\eta \rightarrow \infty$, i.e. even when the left-right symmetric theory approaches its $SU(2)_L \times U(1)$ limit. (See discussion under I). As a consequence, in the limit of chiral symmetry ($a^2 \rightarrow 0$), the predictions of the left-right symmetric theory with two light neutral gauge particles (N_1 and N_2) coincide with those of the left-handed $SU(2)_L \times U(1)$-theory with only one light gauge particle Z° for all low-momentum transfer processes induced by left-handed neutrinos and right-handed antineutrinos.

The qualitative reason for this result has been mentioned before. The result follows immediately by taking the chiral symmetric limit ($\zeta \rightarrow 1$) of the neutrino-amplitudes, i.e. of the master functions G_V and G_A (Eqs. (2.31) and (2.32)) for the left-right symmetric theory.

$$G_V(\zeta = 1) = 1/(3r + 1)$$
$$G_A(\zeta = 1) = 1 \qquad (2.46)$$

Note that they are independent of η, and are identical to the G_V and G_A of the left-handed $SU(2)_L \times U(1)$-theory as claimed. This is the equivalence theorem stated before[33].

By contrast in the same limit ($\zeta \to 1$), we obtain

$$Q_W(\zeta = 1) = \frac{\{2Z - (3r+1)A\}}{(3r + 1)} \quad \frac{(\eta-1)}{\eta} \tag{2.47}$$

Thus the atomic parity violation parameter Q_W for the left-right symmetric theory depends upon η (even in the limit of chiral symmetry); it differs from the $SU(2)_L \times U(1)$-prediction for Q_W except of course in the limit $\eta \to \infty$, when the two theories become identical to each other in all respects.

In short the two theories in general would differ from each other (even in the chiral symmetric limit) in respect of (a) atomic parity experiments, (ii) rates and asymmetry-parameters in $e^-e^+ \to \mu^-\mu^+$ and (iii) rates and asymmetry-parameters in $pp \to \mu\bar{\mu} + x$, i.e. for processes, in which all four components of fermions get a chance to play their roles. In addition they will differ from each other in respect of the masses of the neutral gauge particles.

I now turn into more detailed experimental tests of left-right symmetry.

2.6. Tests of Left-Right Symmetry:

(a) Neutrino-Scattering:

As noted above, predictions of the left-right symmetric theory for left-handed neutrino-scattering differs from those of the left-handed $SU(2)_L \times U(1)$-theory only to the extent that chiral symmetry is not exact in nature ($a^2 \neq 0$). Since the smallness of fermion-masses (i.e., m_q and $m_{lepton} << m_{W_{L+}}$) depends on the parameter a^2 as well as on the relevant Yukawa Coupling Constants, it is not possible to put apriori a stringent upper limit on the chiral symmetry breaking parameter $a^2/(c^2+a^2) = 1 - \zeta$ on the basis of fermion-masses only.

Examining the neutrino-data as well as the atomic parity violation-parameter Q_W, however, one may draw the following important conclusion[46] for the chiral symmetry-breaking-parameter:

If Q_W is indeed small (say at least a factor of 3 lower in magnitude compared to the $SU(2)_L \times U(1)$-value), and if the neutrino-parameters retain their present values (especially R_ν, $R_{\bar{\nu}}$ and $R = \sigma^{\nu N}_{Nc}/\sigma^{\nu}_{Nc}$); both these features can be well accounted for within the left-right symmetric theory provided the chiral symmetry-breaking parameter $(1-\zeta)$ is small compared to unity. To be precise, one needs :

$$1 - \zeta = a^2/(c^2 + a^2) < .2$$

For such small values of $(1-\zeta)$, predictions of the theory for neutrino-parameters remain essentially the same as those for $(1-\zeta) \to 0$.

For simplicity, the precictions of the theory pertaining to neutrino-experiments are presented in Table I for the case of exact chiral symmetry ($a^2 \to 0$, i.e. $1 - \zeta = 0$) with two different values of the coupling constant ratio $r \equiv f_{15}^2/g^2 = 1/3$ and $4/9$. (Cōrrespondingly $\sin^2\theta_W = (3r/6r+2) = .25$ and .29). The predictions are based on the use of the quark-model with the neglect of sea-partons.

We find that the overall agreement of the left-right symmetric theory (as also of the $SU(2)_L \times U(1)$-theory) with neutrino-data is rather impressive for r lying in the range (see Fig. 2):

$$0.3 \lesssim r = f_{15}^2/g^2 < 0.66 \qquad (2.48)$$

Correspondingly,

$$0.23 \lesssim \sin^2\theta_W = \frac{3r}{6r+2} \lesssim 0.33 \qquad (2.49)$$

The lower value of $\sin^2\theta_W \approx 1/4$ (i.e. $r \approx 1/3$) would be preferred if the data settles near the recent results of the CDHS-measurements[37] $R_\nu = 0.30 \pm .015$.

It is important to observe that should the elastic scattering parameters for example R_ν^{el} settle near the upper edge of the error bar (.17 + .05) rather than near the lower edge (see Table I), it would be difficult to accomodate such a finding within either the left-right symmetric or the

TABLE I†

Predictions of Neutral-Current Neutrino Parameters in the Chiral Limit ($a^2 = 0$, or $\zeta = 1$)

PROCESS	PARAMETERS	PREDICTIONS $r \equiv f_{15}^2/g^2$ 0.33	0.44	EXPERIMENT GGM	HPWF	CITF	CDHS
$\frac{(\nu_\mu N)_{NC}}{(\nu_\mu N)_{CC}}$	R_ν	0.30	0.27	0.25 ±0.04	0.29 ±0.04	0.24 ±0.02	0.29 ±.01
$\frac{(\bar{\nu}_\mu N)_{NC}}{(\bar{\nu}_\mu N)_{CC}}$	$R_{\bar{\nu}}$	0.39	0.40	0.39 ±0.06	0.39 ±0.10	0.34 ±0.09	0.34 ±.03
$\left(\frac{\bar{\nu}N}{\nu N}\right)_{NC}$	R	0.44	0.48	0.59 ±0.14	≤ 0.61 ±0.25		
$\bar{\nu}_e e \rightarrow \bar{\nu}_e e$	$\frac{C_{\bar{\nu}_e e}}{0.57}$	1.00	1.15	0.87 ± 0.25 1.70 ± 0.44		1.5 ~ 3 MeV 3 ~ 4 MeV	Reines *et al.*
$\bar{\nu}_\mu e \rightarrow \bar{\nu}_\mu e$	$C_{\bar{\nu}_\mu e}$	0.14	0.17	$0.11^{+0.21}_{-0.09}$ GGM 0.54 ± 0.17 Aachen-Padova			
$\nu_\mu e \rightarrow \nu_\mu e$	$C_{\nu_\mu e}$	0.14	0.13	0.24 ± 0.12 Aachen-Padova			
$\nu_\mu P \rightarrow \nu_\mu P$	R_ν^{el}	0.13	0.12		0.17 ± 0.05 (HPWF) 0.23 ± 0.09 (CIR)		
$\bar{\nu}_\mu P \rightarrow \bar{\nu}_\mu P$	$R_{\bar{\nu}}^{el}$	0.18	0.19		0.2 ±0.10 (HPWF)		
$\left(\frac{\bar{\nu}_\mu P}{\nu_\mu P}\right)$ Elastic	R^{el}	0.46	0.53		0.4 ±0.2 (HPWF)		

†The predictions of elastic νP and $\bar{\nu}P$-scattering parameters are for a choice of axial mass $m_A = 1.08$ GeV. The values of $r = .33$ and $.44$ correspond to $\sin^2\theta_W = 3r/(6r+2) = 0.25$ and 0.28 respectively.

left-handed $SU(2)_L \times U(1)$ theory. Thus these theories will find their more stringent tests through improved elastic semileptonic as well as pure leptonic neutral current neutrino-scattering experiments. Eventually, one would need high momentum transfer high precision experiments for a distinction between left-right symmetric and left-handed theories on the basis of neutrino-experiments.

(b) Atomic Parity Violation Parameter Q_W and the Masses of the Two Neutral Gauge Particles N_1 and N_2:

The variation of Q_W for atomic bismuth as a function of $x \equiv (\eta-\zeta)/\zeta = (b^2-c^2)/c^2$ is exhibited in Fig. 3, while the variations of the masses of N_1 and N_2 as a function of $[Q_W/Q_o]$ are exhibited in Fig. 4; Q_o denotes the $SU(2)_L \times U(1)$-prediction for Q_W. Both these figures are drawn for a typical value of $r = f_{15}^2/g^2 = 0.44$. (This is equivalent to setting $\sin^2\theta_W \simeq 0.29$).

As we see, Q_W passes through zero[48] at $x = 0$ (i.e. $b=c$) approaching its $SU(2)_L \times U(1)$-value in the limit $x \to \infty$ (i.e. $b^2 \to \infty$ (with $\zeta = 1$). While for negative x ($b^2 < c^2$), Q_W rises rapidly with $|x|$ acquiring large positive values, (see Fig. 3). This is opposite in sign compared to Q_o, the $SU(2)_L \times U(1)$ value for Q_W. Thus $[Q_W/Q_o]$ acquires large negative values as x decreases below zero (i.e. as b^2 becomes smaller than c^2). Such negative signs for (Q_W/Q_o) has the potential interest in that it implies a particularly light neutral gauge particle N_1 ($m_{N_1} < m_{W_L^+}$), see Fig. 4.

From the recent experimental values of the optical rotation parameters for atomic Bismuth, together with the existing theoretical calculations[5], one infers that (Q_W/Q_o) lies in the range:

$$-\frac{1}{2} \lesssim (Q_W/Q_o) \lesssim +1/3 \tag{2.50}$$

Either sign for (Q_W/Q_o) appears to be consistent with the data at present.

For illustration, we list the predicted masses of N_1 and N_2 for a range of (Q_W/Q_o) with r = 0.44 (i.e. $\sin^2\theta_W \simeq 0.29$):

(Q_W/Q_o)	m_{N_1} (GeV)	m_{N_2} (GeV)	$m_{W_L^+}$ (GeV)
+1	83	∞	70
0.33	76	120	70
0	70	107	70
-0.5	58	100	70
-1	50	99.5	70

(c) Asymmetries in High Energy e^-e^+ Collisions at PETRA and PEP Regions

Assuming time reversal invariance, the differential cross section for $e^-e^+ \to \mu^+\mu^-$ involving in general a polarized incident beam (but no final state polarization measured) is given by five invariant quantities[49] $(F_1, F_2, \cdots, F_5)$, which in turn are given by the effective amplitudes G^μ_{VV}, G^μ_{VA}, G^μ_{AV}, G^μ_{AA} defined before (Eqs. (40)-(41)). The F_i's are related[50] to the G's by:

$$
\begin{aligned}
F_1 &= (1 + (s/e^2)G^\mu_{VV})^2 + (s^2/e^4)](G^\mu_{AA})^2 + 2\,(G^\mu_{VA})^2] \\
F_2 &= (1 + (s/e^2)G^\mu_{VV})^2 - ((s/e^2)G^\mu_{AA})^2 \\
F_3 &= 2(s/e^2)G^\mu_{AA} + 2(s/e^2)^2\,[G^\mu_{VA}G^\mu_{AV} + G^\mu_{VV}G^\mu_{AA}] \\
F_4 &= F_5 = -2(s/e^2)G^\mu_{VA}\,[1 + (s/e^2)(G^\mu_{VA} + G^\mu_{AA})]
\end{aligned}
\qquad (2.51)
$$

The relevant asymmetry parameters are:

(A) Longitudinal Asymmetry in $e^-e^+ \to \mu^-\mu^+$ with longitudinally polarized incident beam: suppressing kinematic polarization factor;

$$L_p^{\mu\mu}(\theta) = [2\cos\theta\, F_5 + (1 + \cos^2\theta)F_4]/[(1 + \cos^2\theta)F_1 + 2\cos\theta F_3]$$

This is a parity violating parameter. A measure of this ratio is given by (F_4/F_1).

(B) The mean longitudinal polarization of μ^+ with unpolarized incident beam: $\bar{P}_L(\mu) = -F_4/F_1$.

(C) The forward/backward-asymmetry with unpolarized incident beam (no final state polarization measured): $A^{\mu\mu} = (N_f - N_b)/(N_f + N_b) = (3/4)\ (F_3/F_1)$. Here N_f and N_b are the number of μ^+'s in the forward and backward hemispheres relevant to incident e^+. This is a parity conserving parameter.

(D) Longitudinal asymmetry for $e^+e^- \to \pi^+\pi^-$: suppressing kinematic polarization factor; $\rho_L^{\pi\pi} = -2\sigma/(1+\sigma^2)$, where $\sigma = (G^p_{AV} - G^n_{AV})/(e^2/s + G^p_{VV} - G^n_{VV})$.

The asymmetry-parameters for two centre of mass energies $\sqrt{s}$ = 28 GeV and 38 GeV covering the PETRA and PEP energy regions for a range of values of the atomic parity violation parameter Q_W are given in Table II (with r = 0.55, i.e. $\sin^2\theta_W \simeq 0.32$ and $\zeta = 1$). These predictions are compared with those of $SU(2)_L \times U(1)$.

<u>TABLE II</u> **(Ref. 50)**

e^-e^+ ASYMMETRIES AS A FUNCTION OF Q_W†

		$SU(2)_L \times SU(2)_R \times U(1)$ THEORY					$SU(2) \times U(1)$ LIMIT
Q_W INPUT		245	147	88	0	-49	-146
m_{N_1} (GeV)		47	53	58	67	73	81
m_{N_2} (GeV)		95	98	100	110	124	∞
$L_p^{\mu\mu}$	$\sqrt{s}$=28	5%	3%	1.8%	0	-1%	-2.5%
	$\sqrt{s}$=38	~8%	3.1%	2.9%	0	-1.8%	-4.9%
$A^{\mu\mu}$	$\sqrt{s}$=28	-42%	-31%	-25%	-17%	-14%	-8%
	$\sqrt{s}$=38	-65%	-71%	-58%	-38%	-30%	-16.5%
$\rho_L^{\pi\pi}$	$\sqrt{s}$=28	-22%	-11%	-6%	0	+2.9%	7.9%
	$\sqrt{s}$=38	-86%	-37%	-18%	0	+6.7%	16%

†$\sqrt{s}$ is given in GeV (r = 0.55, ζ= 1). m_{N_1} and m_{N_2} are predictions of the theory, as also the asymmetry-parameters for a given Q_W. $L_p^{\mu\mu}$ denotes F_4/F_1.

DISCUSSION

(i) Note that as long as Q_W is either small[5] in magnitude (compared to its SU(2) x U(1)-value), or opposite in sign, the left-right symmetric theory would inevitably predict that the parity conserving forward-backward asymmetry parameter $|A^{\mu\mu}|$ must be strongly enhanced by a factor (≈ 2 to 5) compared to its SU(2) x U(1)-value even at PETRA and PEP energies.

(ii) Second, in general there is a correlation between the signs of the parity violating asymmetry-parameters for $e^-e^+ \to \mu^-\mu^+$ and $\pi^-\pi^+$ on the one hand and the sign of Q_W on the other.

Thus e^-e^+ asymmetry parameters (as also atomic parity violation-parameter Q_W) unlike neutrino data, provide clear distinctions between the left-right symmetric theory with a light neutral mass ($\lesssim m_{W_L^+}$) on the one hand and the SU(2) x U(1) theory on the other.

Extended Left-Right Symmetry: Asymmetry Measurements at SPEAR and DORIS Energies

It should be recalled that unifying symmetries such as $[SU(4)]^4$ permit an extended low energy weak/electromagnetic symmetry of the form[34,35]

$$G_{LR} = SU(2)_L \times SU(2)_R \times U(1)_L \times U(1)_R$$

which is left-right symmetric like $\bar{G}_{LR} = SU(2)_L \times U(1)_{L+R}$. G_{LR} in fact contains $\bar{G}_{LR}$ as a subgroup. Following the results of Ref. 33, it may be argued that G_{LR} (like $\bar{G}_{LR}$) retains all the successful features of the "left-handed" $SU(2)_L$ x U(1)-symmetry for neutrino-experiments; simultaneously (like $\bar{G}_{LR}$) it can accommodate the null or near null results of atomic parity experiments. In these two respects, the symmetry G_{LR} resembles its subgroup $\bar{G}_{LR}$. The major difference between the two symmetries is this: $\bar{G}_{LR}$ permits in general two relatively light neutral weak bosons (N_1 and N_2); the mass of the lightest one (N_1) is constrained by present atomic parity experiments to lie above about 58 GeV (see Eq. (2.50) and the table following the same equation). By contrast the extended symmetry G_{LR} permits in general three relatively light neutral weak bosons (Z_1, Z_2 and Z_3); atomic parity experiments do not set any constraint on how light the lightest among them (Z_1) can be.

It turns out[35] that Z_1 (especially for $m_{Z_1} \lesssim 50$ GeV) is mostly like a massive axial photon. Such a gauge-particle contributes significantly to hyperfine splitting in hydrogen, (g-2) of muon, as well as to forward-backward $e^-e^+ \to \mu^-\mu^+$-asymmetry parameter even at existing SPEAR and DORIS energies. Considerations based on hyperfine splitting and (g-2) allow one to set the following lower limits on the mass of Z_1

$$m_{Z_1} \gtrsim 15 \text{ GeV (hfs)} \qquad \text{(Ref. 35)}$$

$$m_{Z_1} \gtrsim 35 \text{ GeV } ((g-2)_\mu) \qquad \text{(Ref. 51)} \tag{2.52}$$

This permits the possibility that Z_1 may even be discovered at PETRA and PEP. At any rate, even if it lies in the mass-range 40 to 60 GeV range, its virtual effects at PETRA and PEP would be dramatic. The presence of a light Z_1 ($m_{Z_1} \lesssim 60$ GeV) can be felt sensitively even at SPEAR and DORIS energies. This may be seen from table III, which gives the values of the forward-backward asymmetry parameter $A^{\mu\mu}$ as a function of m_{Z_1} at SPEAR-DORIS as well as PETRA-PEP energies (for $\sin^2\theta_W = 0.3$).

TABLE III (Ref. 35)

m_{Z_1}	$A_{\mu\mu}$ (for $SU(2)_L \times SU(2)_R \times U(1)_L \times U(1)_R$)		
(GeV)	$\sqrt{s}$ = 7 GeV	28 GeV	40 GeV
21	-0.18	.56	.73
35	-0.06	-.63	.41
45	-0.03	-.64	-.36
49	-0.03	-.52	-.64
64	-0.01	-.20	-.48

We see that accurate measurement[52] of $A^{\mu\mu}$ at SPEAR and DORIS energies can decide on the mass or set an improved lower limit on the mass of the lightest Z_1. In turn, this could help determine whether a real or a virtual Z_1

will be seen prominently at PETRA and PEP.

As a general remark, it is worth noting that the symmetry $G_{LR} = SU(2)_L \times SU(2)_R \times U(1)_L \times U(1)_R$, the left-right symmetric subgroup $\bar{G}_{LR} = SU(2)_L \times SU(2)_R \times U(1)_{L+R}$ and the standard left-handed symmetry $SU(2)_L \times U(1)$ are three alternative low energy manifestations of the <u>same parent symmetry</u> $[SU(4)]^4$. As to which of these manifestations may correspond to nature, depends largely on the mass of the lightest Z°. Neither neutrino, nor atomic parity experiments can help resolve a choice between G_{LR} and $\bar{G}_{LR}$, but forward-backward asymmetry measurements at SPEAR and DORIS and at PETRA and PEP can help choose unambiguously between the three.

(d) <u>Asymmetry Parameters in ep, pp and $\bar{p}$p-experiments</u>:

The left-right symmetric theory with a light neutral mass-solution ($m_{N_1} \lesssim m_{W_L^+}$) may be distinguished from the $SU(2)_L \times U(1)$-theory by other means, notably via measurements of parity violating effects in polarized deep inelastic eN-scattering on the one hand and via measurements of rates as well as asymmetry parameters of dilepton production in high energy pp and $\bar{p}$p-collisions on the other.

A calculation of the asymmetry-parameter for deep inelastic eN-scattering has recently been done by Cahn and Gilman[53] for $SU(2)_L \times U(1)$-theory and by Janah[54] for the $SU(2)_L \times SU(2)_R \times U(1)_{L+R}$-theory permitting the general possibility[46] of Q_W (Atomic) $\neq 0$.

For $\bar{p}p \to \mu^-\mu^+ + X$, a recent calculation by Mani, Rajpoot, Salam and myself[55] shows that the parity conserving forward-backward asymmetry turns out to be a particularly sensitive parameter (with dilepton-mass $\gtrsim$ 30 GeV), to help distinguish between the two theories. (This is similar to the situation encountered for $e^-e^+ \to \mu^-\mu^+$). An optimum center of mass-energy, where the dilepton-production-rate is descent (i.e. $(d\sigma/dm_{\mu\mu}) \gtrsim 10^{-37}$ cm^2/GeV for $m_{\mu\mu} \simeq$ 40~70 GeV) and the forward-backward asymmetry-parameter $A_{\mu\mu}$ is also generally large (though varying with $m_{\mu\mu}$), appears to be around 200 GeV. Plots of $(d\sigma/dm_{\mu\mu})$ and $A_{\mu\mu}$ varying with $m_{\mu\mu}$ are given in Figs. 5 and 6 for 3 different values of Q_W for the center of mass-energy $E_{Cm} = \sqrt{s} =$ 200 GeV.

It appears that (a) $e^-e^+ \to \mu^-\mu^+$-measurements at PETRA and PEP, (b) deep inelastic ep- and μp-scatterings at SLAC and Fermilab, (c) measurements of pp or $\bar{p}p$ dilepton-forward backward asymmetry-parameter as well as the dilepton-rate, and last but not least (d) accurate measurements of atomic parity violation not only in Bismuth, but also in other atoms such as Cesium, thallium, and in particular in hydrogen should help us decide which of the two (if any) - the left-right symmetric theory with the light-neutral mass solution and the left-handed $SU(2)_L \times U(1)$-theory - may correspond to nature. They should also help fix the mass of the light neutral weak gauge particle; some of them may even produce it.

REMARKS

The advantages of the postulate of left-right symmetry[2] are many:

(1) It helps us generate through spontaneous symmetry breaking a desirable milliweak theory of CP-violation[38], the strength of which is directly related to the observed left-right-asymmetry (dominance of V-A over V+A) in nature.

(2) It retains the successes of the left-handed $SU(2)_L \times U(1)$-theory for neutrino-processes[33] and simultaneously provides the scope for accommodating small atomic parity violation (should this be desired experimentally)[12].

(3) The left-right symmetric gauge-structure offers the possibility of accounting for isospin as a "natural" symmetry[44].

(4) Last but not least, it meets one's aesthetic requirement that Nature is intrinsically parity conserving; the observed left-right asymmetry being a

"low"-energy phenomenon to disappear at high energies.

Looking ahead, the left-right symmetric theory raises two important questions: (1) Where in the mass-scale should the charged right handed gauge-particle $W_R^{\pm}$ lie? (2) How and at what level should one expect to see right handed neutrinos? The answer to the second question is, of course, related to that of the first. Unfortunately neutral current-phenomena do not help put any experimental constraint on $m_{W_R^{\pm}}$. I end with two remarks pertaining to this question. Both signify extreme possibilities. On the one hand, the fact that the left-right symmetric theory naturally generates a milliweak theory of CP-violation[38] with

$$|\eta_{+-}| \simeq |(m_{W_L^{\pm}}/m_{W_R^{\pm}})^2 \ (\sin 2\theta_R/\sin 2\theta_L)\sin(\delta_R-\delta_L)| \approx 10^{-3}$$

suggests (assuming that the phase angle is not tiny and that $\theta_R \simeq \theta_L$), that perhaps $W_R^{\pm}$ is a factor of 10 to 30 heavier than $W_L^{\pm}$. This would be in line with the attractive possibility[56] that spontaneous symmetry breaking is triggered at one place through the vacuum-expectation value of the "right-handed" vectorial scalar field E_R contributing to the mass of $W_R^{\pm}$ being non zero, which then feeds through finite order α-radiative corrections into the neutral and the charged $W_L^{\pm}$-sectors $\left[\text{in this picture } (m^2_{W_L^{\pm}}/m^2_{W_R^{\pm}}) \text{ is naturally order } \alpha\right]$. This however makes the possibility of observing right handed neutrinos in β-decay or $\pi^{\pm}$-decay remote:

$$\text{Rate}(\nu_R)/\text{Rate}(\nu_L) = (m_{W_L^{\pm}}/m_{W_R^{\pm}})^4.$$

On the optimistic side, there is the possibility that CP-violation within the left-right symmetric theory may itself be a radiative phenomenon, i.e. it vanishes in the tree-approximation, but arises at the one-loop level. In this case, the phase $(\delta_R-\delta_L)$ might be naturally order α permitting the ratio $(m_{W_L^{\pm}}/m_{W_R^{\pm}})$ to be $\gtrsim$ (1/10). This might suggest that right-handed neutrino-emission (i.e. the rate of (V+A)-interaction) in β-decay and π^+-decay may actually be occurring at about 1% level. In this connection, improvement in the measurement of longitudinal polarization of β-decay electrons can not be overemphasized.

I now turn briefly to the second question: Testing Quark-Liberation.

III. QUARK LIBERATION

3.1 General Remarks

The question I discuss now is whether quarks acquire integer-charges, gluons acquire a light mass in the process of spontaneous symmetry breaking, both quarks and gluons are "liberated" out of normal hadrons turnelling through a barrier, they have so far escaped detection mainly because they die out rapidly decaying into leptons; or whether quarks acquire fractional charges, gluons remain massless, quarks and gluons are not observed because they are "confined".

I have indicated in the introduction the origin of these two alternatives within unified theories. I should add now that consistent with the physical requirement that an effective SU(3)'-Color symmetry must emerge at least as a good global symmetry, it turns out[8] that both of these (integer and fractional charges for quarks) are allowed solutions only for such unifying symmetries, which are semisimple of the form $G_{flavor} \times G_{color}$. Example of such an unifying symmetry[2] is $[SU(4)]^4 = [SU(4)_A \times SU(4)_B]_{flavor} \times [SU(4)'_C \times SU(4)'_D]_{Color}$.
For (alternative) simple symmetries[3], such as SU(5), SO(10) or E_7, it turns out[8] that SU(3)'-color must remain unbroken as a local symmetry (therefore quarks should acquire fractional charges and gluons should remain massless), if SU(3)'-color must emerge as a good global symmetry.

As a corollary to this result, should the octet of color gluons have a finite mass, it appears that within the premises of a quark-lepton unified theory, quarks must be integer charged and the unifying symmetry must be of the form $G_{flavor} \times G_{color}$.

There are distinct advantages of unifying symmetries of the form $G_{flavor} \times G_{color}$, such as $[SU(4)]^4$: (a) They are left-right symmetric in the sense discussed before. (b) They lead to desirable values for the weak angle $\sin^2\theta_W$, (this is discussed in the next chapter), (c) They predict[2] existence of new flavors beyond charm, which seem to be needed experimentally[57], and (d) They offer the possibility[58] that the unifying mass-scale M need be no higher than about 10^4-10^5 GeV to be contrasted with values of M ten orders of magnitudes higher[59] for simple unifying groups such as SU(5), SO(10) or E_7.

In view of these, one is led to pose the question: Assuming that the underlying unifying symmetry has the structure $[SU(4)]^4$, which of the two solutions for quark-charges is indeed chosen through spontaneous symmetry breaking? The major similarities between the two solutions involving properties such as asymptotic freedom, scaling and hadronic spectroscopy have been outlined in the introduction. I shall not enter here into a discussion of how the contributions of gluon-partons to electroproduction-structure functions and to e^-e^+-annihilation above threshold for color-production can help distinguish between the two solutions (integer versus fractional charges). This is reviewed in some detail elsewhere[60]. I present here primarily the signatures of baryon and lepton-number violating quark-decays and proton-decays as well as the signatures for gluon-decays to facilitate their experimental detection.

3.2 The Unconfined Unstable Integer-Charge Quark:

Within quark-lepton unifying symmetries, if quarks acquire integer charges, the exotic X-gauge particles (carrying $B = -L = \pm 1$) mix spontaneously with the flavor W-gauge particles (carrying B=L=0). This induces baryon and lepton-number violations (fermion-number is still conserved). Integer-charge quarks become unstable against decay into leptons. The relevant diagrams, exhibiting the dominant decay mechanisms[13] of quarks are shown in Figures 7 a,b,c. The corresponding dominant decay modes of quarks turn out to be:

$$q_{y,b} \to \nu + \text{mesons } (\pi, K, \eta)$$
$$q^{-}_{red} \to \nu + \text{gluon } (V^{-})$$
$$\bar{q}^{-}_{red} \to \nu + (\pi, \rho, A_1) \qquad (3.1)$$

Other decay modes -- e.g. those involving emission of 3 leptons ($q \to \ell+\ell+\bar{\ell}$) or emission of a single charged lepton (e.g. $q^{o}_{y} \to e^{-} + \pi^{+}$) turn out to be strongly damped[13] compared to the decay modes listed above. (It should be noted that the decay modes $q^{-}_{red} \to \nu$ + mesons (π, ρ, A_1) etc. are expected within the unifying symmetry $[SU(4)]^4$, but not within the restricted basic model $SU(2)_L \times SU(2)_R \times SU(4)'_{L+R}$, for which W_{14} coupling to $(\bar{p}c)$-currents is nonexistent.)

Lifetimes of quarks with physical masses in the range of 2-3 GeV are found to be $\approx 10^{-11} - 10^{-13}$ sec. Such short lived integer-charge quarks would have escaped direct detection by all bubble-chamber as well as emulsion searches made so far, even if they may have been pair-produced in hadronic collisions with reasonable cross-section[61] $\sim 10^{-30} - 10^{-31}$ cm^2 (say) at ISR and Fermilab energies. Emulsion-searches are sensitive to the life-time range $\sim 10^{-11} - 10^{-13}$ sec; however, with production cross-section in the range indicated above, the background-effect in searches made so far would be serious. This, as well as several associated considerations, based on lepto-production experiments (discussed earlier) and proton-stability (see discussions below) obviate the need for the hypothesis of fractionally charged quarks and confinement. Experimental tests of the alternative of unconfined integer-charge quarks are given in the next sections.

3.3 Unstable Proton:

With quarks being unstable, the proton becomes unstable as well against decay into leptons. However, assuming that the physical masses of quarks and diquarks are heavier than those of the proton, the proton - a three quark composite - can decay only provided all three quarks within the proton "simultaneously" convert into leptons.[2] This makes the proton a rather stable object, and accounts for our prolonged existence. Yet the proton must decay. Depending upon the details of quark-decay mechanism their physical masses and the probability of all three quarks within the proton being near the origin ($|\psi(o)|^2$), one obtains[62] a proton lifetime in the range:[63]

$$\tau_{proton} \sim 10^{29} - 10^{32} \text{ years} \tag{3.2}$$

with the dominant decay modes of the proton being of the type

$$p \to 3\nu + (\text{mesons}) \tag{3.3}$$

It is important to note that within the unconfined unstable integer-charge quark-hypothesis, quarks cannot be too longlived ($\tau_{quark} \lesssim 10^{-11}$ sec.), or else they would have been discovered (assuming physical $m_q < 5$ GeV); this in turn implies that the proton, within this hypothesis, can not be too longlived either. A life time of $\approx 10^{32}$ years is in fact a reasonable upper limit for the proton to live within this hypothesis, and thus can

can be used to test the hypothesis. This is unlike alternative models[3] of proton decay, based on confined fractionally charged quarks, in which proton-decay is induced directly by the gauge interactions in the second order. In such models, proton-lifetime depends directly upon the mass of the exotic superheavy gauge particles. Estimating the mass ($\simeq 10^{16}$ GeV) from the unification hypothesis, the proton-lifetime in such models is estimated[64] to be $\approx 10^{38}$ years; in other words the proton is too stable in these models for its decay-possibility to be tested in the forseeable future. This is not so for the case of unconfined unstable integer-charge quarks as mentioned above.

In summary, based on these considerations as well as discussions in Chapter I (involving scaling phenomena, lepto-production experiments and spectroscopy of quark-composites), we see that the hypothesis of liberated integer-charge quarks is a theoretically allowed and at present experimentally viable alternative to the one of confined fractionally charged quarks. Endowed with liberation, the former offers several intriguing observational possibilities for experiments in planning[65]. Some of these are presented below with the assumption of relatively low-mass (~ 1 to few GeV) physical color. A more complete discussion, including those of possible high mass color-threshold ($\gtrsim$ 8 GeV), may be found in Ref. 65.

3.4 Signatures of Physical Color:

In addition to quarks being observable, the octet of spin-1 color-gluons (involving four charged $(V^{\pm}_{\rho}, V^{\pm}_{K*})$ and four neutral members $(\tilde{U}, \tilde{V}, V^{o}_{K*}, \bar{V}^{o}_{K*})$), color-octet $q\bar{q}$-composites in various spin-parity configurations, as well as color-nonsinglet qqq-baryons would be observable entities in the theory, if quarks carry integer-charges. The decay-modes of the lowest colored octet mesons give some of the distinctive signatures for physical color. These are listed below under the assumption that the 1^--gluon octet is the lightest among the color-octet states.

Decay Modes of Neutral Gluons:

There are four neutral members. First consider the ones coupled to diagonal-currents ($\tilde{U}$ and $\tilde{V}$). In the limit of ideal mixing $\tilde{V} = V^o = \frac{1}{2}(\sqrt{3}\ V_8 - V_3)$ and $U \approx [U^o + \mathcal{O}(g/f)W^o_{flavor}]$ are the eigenstates, with $U^o = \frac{1}{2}(\sqrt{3}\ V_3 + V_8)$; in this limit $\tilde{V}$ is decoupled from e^-e^+ and $\mu^-\mu^+$. In general, the physical particles ($\tilde{U}$ and $\tilde{V}$) are mixtures of U and V^o, and both are coupled to e^-e^+ with amplitudes proportional to $\cos\xi$ and $\sin\xi$ respectively, ξ being the mixing angle. The expected decay modes and partial widths are:

	Expected Partial Width (For $m_U \approx$ 1-2 GeV)	
$(\tilde{U},\tilde{V}) \to e^-e^+$ or $\mu^-\mu^+$	6 to 30 KeV ($\cos^2\xi$, $\sin^2\xi$)	
$\to \eta'\gamma$	40 to 1000 KeV ($\cos^2\xi$, $\sin^2\xi$)	
$\to 2\pi\gamma, 4\pi\gamma, K\bar{K}\gamma$	50 to 500 KeV ($\cos^2\xi$, $\sin^2\xi$)	
$\to 3\pi, \rho\pi, 5\pi, K\bar{K}$	10 to 1000 KeV	
$\to 2\pi, 4\pi, K\bar{K}$	10 to 1000 KeV	(3.4)

The existence of such narrow ($\Gamma = \frac{1}{10}$ to few MeV) gluons in the 1.1 to about 1.8 GeV-region is at present compatible with the data involving e^-e^+ and photo-production experiments[66]. There are at present likely candidates for such gluons: the narrow 1498 MeV-resonance, observed at Frascati and reported at the Hamburg Conference, appears to be a promising candidate for the $\tilde{V}$-gluon (with $\sin^2\xi \approx 10^{-2}$). A study of its decay modes, as well as a search for the accompanying narrow $\tilde{U}$-gluon, expected to possess a leptonic partial width larger than that of $\tilde{V}$ (see Eq. (3.4)) in the mass-region within $\pm$ 100 MeV of $\tilde{V}$, should shed further light on such an identification. In general, a search for narrow objects in the region

$\simeq$ 1.1 to 1.8 GeV at Frascati, Orsay and Novosibirsk would be of interest in this regard.

The remaining two neutral members (V^o_{K*} and $\bar{V}^o_{K*}$) possess essentially the same decay-modes and branching ratio as the neutral $\tilde{U}$; their absolute widths are related to that of $\tilde{U}$ by the factor $(m_V^4/m_W^4)(m_V^2/(m_V^2-m_U^2))^2$, where m_V and m_U differ by 10 to 100 MeV, we expect $\Gamma(V^o_{K*}) \sim (\frac{1}{3.6}$ to $\frac{1}{36})\Gamma(\tilde{U})$.

Decay Modes of the Charged Gluons:

	$m_V \approx$ 1-2 GeV
$(V^{\pm}_{\rho}, V^{\pm}_{K*}) \to e^+\nu_e$	$(30 \pm 5)\%$
$\to \mu^+\nu_\mu$	$(30 \pm 5)\%$
$\to 2\pi, 3\pi, K\bar{K}, \ldots$	$(30 \mp 10)\%$
$\to \pi\pi e\nu, K\bar{K}e\nu, \ldots$	(1 to 5)%
$\to \pi e\nu, Ke\nu,$	forbidden by I-spin
$\to \eta e\nu$	forbidden by SU(3) and SU(3)' (3.5)

Note the distinction from charm, for which semileptonic modes involving single Kaon-emission are allowed and frequent. Using $(V^{\pm}_{\rho,K*}$-W-)-mixings[2] their lifetimes are found to be:

$$\tau(V^{\pm}_{\rho}, V^{\pm}_{K*}) \simeq 10^{-13} \text{ sec. } (m_V/\text{ 1 GeV})^{-5} \tag{3.6}$$

Color, in general may be expected to be produced[65] in high energy hadronic collisions, photo-production[66], lepto-production and e^-e^+-annihilation:

$$p + p \to V^+_{\rho} + V^-_{\rho} + X$$
$$\to V^{\pm}_{\rho} + X^{\mp}_{Col} + X$$

Expect $d\sigma_{col} \sim 10^{-29}$-$10^{31}$ cm^2 at Fermilab and ISR energies

$$e^-e^+ \to V^{\pm}_{\rho,K*} + V^{\mp}_{\rho,K*}$$
$$\quad\downarrow \mu\nu \qquad \downarrow e\nu$$

$R_{V\bar{V}} = \frac{1}{8}$
A partial source of (μe)-events, if gluons are light; in this case red quarks are a more dominant source

$e^-e^+ \to X_{col}$ $\qquad$ $R_{Col} = \frac{1}{8}(m_U^2/\mu^2)^2$ (Ref. 22)

μ = mass of gluon-partons, m_U = physical mass of gluon

$\nu_\mu N \to \mu^- + X^+_{col}$, $X^+_{col} \to \mu^+ + \ldots$ $\qquad$ Expect dimuon-rate from color-production $\approx 1/3\%$ (using $m_U \approx 1.4$ GeV, $\bar{\mu}^2 \approx .8$ GeV^2, glue-content $\approx 30\%$).

$\mu^- N \to \mu^- + V^+_\rho + V^-_\rho + X$, $V^+_\rho \to \mu^+\nu$, $V^-_\rho \to \mu^-\bar{\nu}$ $\qquad$ Trimuons

Now consider a possible interpretation of the SPEAR-DORIS μe-events in terms of production and decay of interger-charge quarks.

3.5 SPEAR-DORIS-$\bar{\mu}e$-Events

If (p,n,λ) quarks have a physical mass $\simeq$ 1.8 to 2 GeV, and if gluons are lighter than quarks with $m(q_{red})-m(q_{yel,blue}) < m_\pi$ -- a perfectly feasible possibility -- then one can attribute the $\bar{\mu}e$-events observed[6] at SPEAR and DORIS to pair production and decay of charged red quarks[13]:

$$e^-e^+ \to q^-_{red} + q^+_{red}, \quad q^-_{red} \to V^- + \nu, \; V^- \to e^- + \bar{\nu}_e; \quad q^+_{red} \to V^+ + \nu, \; V^+ \to \mu^+ + \nu_\mu \tag{3.7}$$

(i) Spectrum: The momentum-spectrum[67] of the muon (or the electron) thus arising via a two-step three-body decay of quarks, differs from that arising via a single-step 3-body decay of a heavy lepton in the low and medium momentum region ($p_{e,\mu} \sim$ 200 to 500 MeV/c). But the two spectra coincide in the high-momentum region ($p_{e,\mu} > 600$ MeV/c), especially if charged gluons are relatively light < 1.4 GeV. Experimental measurements are made in the high-momentum-region ($p_\mu > 650$ MeV/c and $p_e > 400$ MeV/c); the data within the errors are compatible with the quark as well as the heavy-lepton hypothesis for the $\bar{\mu}e$-events.

(ii) "Point" Coupling: Quark-electromagnetic form factor, determined by the "temporarily" asymptotically free color-gauge theory, is expected to be nearly unity and slowly varying in the SPEAR energy range. Correspondingly the rate of production of ($\bar{\mu}e$)-events is expected to follow (neglecting small variation) the (1/s)-behaviour characteristic of a point-like heavy lepton. This is observed experimentally. [Taking the branching ratios of $\mu\nu$ and $e\nu$ decays of $V^{\pm}$ to be each $\approx$ 25 to 35% (see Eq. (3.5)), and the asymptotic contribution of pair-production of charged red quarks ($n_r^- n_r^+$) and ($\lambda_r^- \lambda_r^+$) to the R-parameter[68] to be (2/9), the net contribution to R of the μ^+e^--signal arising from $q\bar{q}$-production and decays is given by $R_{q\bar{q}}(\mu^+e^-) \approx (2/9)(1/4 \text{ to } 1/3)^2 \approx (1.6 \text{ to } 2.5)\%$, compatible with the data.]

(iii) Semileptonic signals (e.g. $e^-e^+ \to \bar{\mu}e + \pi^+\pi^-$ + missing momentum): Direct decays of red quarks into $V^{\pm} + \nu$ + (pions) are inhibited by phase-space since gluon is relatively heavy (~ 1.3 to 1.4 GeV) compared to the mass of the quark (~ 1.8 GeV). Semileptonic decays of charged gluons themselves must involve at least a pair of pions, and are also relatively suppressed (see Eq. (3.5)). Combining both sources, the semileptonic ($\bar{\mu}e$)-events are predicted to be $\approx$ 1 to 5% compared to the pure leptonic ($\bar{\mu}e$)-events. This too is borne out by the data.

(iv) Jets: Missing Momentum: The yellow and blue quarks, as well as the neutral red quarks (p_r^o and c_r^o), eventually decay with the emission of neutrinos plus mesons (π,K,η,etc.), but not charged leptons (see Eq. (3.1)). Thus pair-production of these quarks, contributing a value $\approx$ (10/3) - (2/9) = 28/9 to the R-parameter, would lead to jet-like distribution of the decay hadrons (π,K,η) with missing energy and momentum carried away by neutrinos. [The gluonic contribution (1/8) (m_U^2/μ^2) to R may relfect itself in direct production of hadrons through $q\bar{q}$ and gluon-antigluon-recombintion. Such a contribution could lead to an increase in average pion multiplicity $\langle n_{\pi} \rangle$ with increasing energy, as observed experimentally.]

(v) $e^{-}e^{+} \to \bar{\mu}$ + (ρ, or A_1, or π) + missing momentum: Within $[SU(4)]^4$-symmetry charged red quarks may decay into $V^{-} + \nu$ (Fig. 7b) as well as to mesons (ρ, A_1, π, etc.) + ν (Fig. 7c). Thus one would expect semileptonic events of the type ($\bar{\mu}$ + meson + missing momentum) in addition to pure leptonic events. For the quark-hypothesis, the relative frequency of $\rho:A_1:\pi$ semileptonic events depends on the ratios of $\bar{q}q\rho:\bar{q}qA_1:\bar{q}q\pi$ couplings (see Fig. 7c). Thus the suppression of π relative to pure leptonic or semileptonic ρ and A_1-modes, indicated by the recent PLUTO data[69], can be accommodated under the quark-hypothesis for the $\bar{\mu}e$-events. Such an observation, if confirmed, would, however be incompatible with the heavy lepton-interpretation.

We conclude that the present data on $\bar{\mu}e$-leptonic and (μ + meson)-semileptonic events are equally compatible with the quark as well as the heavy lepton-origin of these events (with a possible difficulty for the heavy lepton-interpretation in resoect of the suppression of ($\bar{\mu}\pi$)-events). One needs refined lepton-momentum measurement in the 200-500 MeV/c as well as measurements of rates of semileptonic ($\bar{\mu}e + \pi^{+}\pi^{-}$)-events to help distinguish between these alternatives.

An experimental distinction between the two alternatives would be of major importance inasmuch as that would shed light on a fundamental issue. A few crucial tests, which provide such a distinction are:

(a) The search for the narrow gluons ($\tilde{U}$ and $\tilde{V}$) in $e^{-}e^{+}$-annihilation in the Frascati, Novosibirsk and Orsay regions (As noted above, the mass-region 1.1-1.8 GeV for $\tilde{U}$ and $\tilde{V}$ is suggested on the basis of existing experimental searches ; the same mass region seems to be required at least within the basic model for a consistent interpretation of the ($\bar{\mu}$e)-events in terms of pair production and decays of integer-charge-quarks.),

(b) The search for signals of two step three body decays of parent particles leading to ($\bar{\mu}$e)-events in the lepton-momentum spectrum in the region 200-500 MeV/c at SPEAR and DORIS,

(c) The search for signature semileptonic ($\bar{\mu}$e)-events ($e^-e^+ \rightarrow \bar{\mu}e + \pi^+\pi^-$ + missing momentum) at a 1 to few percent level,

(d) The search for shortlived charged particles ($\tau \sim 10^{-13}$-10^{-15} sec.) via emulsion-bubble-chamber studies, whose semileptonic decay modes always involve a pair of pions or a pair of kaons, but never a single π or K (As discussed above, these particles would be clear signatures for charged gluons).

(e) Last but not least, the search for particles with lifetimes ~ 10^{-12}-10^{-13} sec. via emulsion-bubble-chamber studies, which decay either into ordinary mesons + missing neutrinos like a heavy lepton, but which can scatter relatively strongly against nuclei with cross sections ~ 1 mb (Such particles would be clear signals for decaying integer-charge quarks).

The searches listed above should deserve high priority in that they would help clearly eliminate or establish this intriguing hypothesis of low-mass physical color.

I now finally discuss the question of the unifying mass-scale M.

IV. THE UNITY OF FUNDAMENTAL PARTICLES AND THEIR FORCES

The hypothesis[2,3] that the fundamental particles and their forces are unified at a basic level through a lagrangian characterized by a single coupling constant raises two important questions: (1) At what mass-scale (M) would this complete unification, lost at low energies through spontaneous breaking of the symmetry, be observable? What is the value of the renormalized weak angle $\sin^2\theta_W$?

Over the past, it has generally been claimed[59,70] that the unifying mass-scale M needs to be superheavy ($\gtrsim 10^{15}$ GeV) just in order that "strong" interactions may be strong at low energies. If such a super-heavy mass-scale was inevitable, there would appear to be no hope that "unification" may be manifest even in the foreseeable future.

It is recently realized[58] that such a superheavy mass-scale is however not a general property of the unification-hypothesis, but a consequence of a rather special assumption. The assumption in question is this: the descent of the unifying symmetry G to the low-energy weak interaction-symmetry G_W and the commuting strong interaction-symmetry $SU(3)'_{col}$ is such that the gauge coupling constants associated with G_W and $SU(3)'_{col}$ are equal to each other in the bare symmetric limit.

This assumption is of course satisfied automatically within unifying symmetries such as[3] $SU(5) \supset SU(2)_L \times U(1) \times SU(3)'_{col}$ and $SO(10) \supset SU(2)_L \times SU(2)_R \times SU(3)'_{Col} \times U(1)_{L+R}$. However the flavor-color and left-right symmetric semisimple unifying group[71,72] $[SU(4)]^4 = [SU(4)_A \times SU(4)_B]_{flavor} \times [SU(4)'_C \times SU(4)'_D]_{col}$, and more generally $[SU(n)]^4$ permit a departure from this assumption. For example, corresponding to the descent from SU(4) to GIM-SU(2) for flavor, together with the direct descent of $SU(4)'_{C,D}$

to $SU(3)'_{C,D}$ for color, the coupling constant associated with the GIM-SU(2)-group is smaller by a factor $1/\sqrt{2}$ than that associated with the SU(3)'-color group.

This seemingly small departure involving a factor of only $1/\sqrt{2}$ in the ratio of the bare coupling constants has the far reaching consequence[58] that the unifying mass-scale M is lowered by at least ten orders of magnitudes below all previous estimates. In turn this raises the hope that "unification" may in fact be testable through ongoing cosmic ray-experiments and perhaps also through the next generation of accelerators.

To see the underlying reason behind this dramatic effect, recall first the fermion-gauge boson coupling[71] for the $[SU(4)]^4$-symmetry. Symbolically,

$$\begin{aligned} L = {} & gW_A[\bar{F}_L F_L + \bar{F}^m_R F^m_R]_{flav} + g\, W_B[\bar{F}_R F_R + \bar{F}^m_L F^m_L]_{flav} \\ & + gV_C[\bar{F}_L F_L + \bar{F}^m_R F^m_R]_{col} + gV_D[\bar{F}_R F_R + \bar{F}^m_L F^m_L]_{col} \end{aligned} \tag{4.1}$$

where $F_{L,R}$ denote the "basic" set of four flavor-four color sixteen four component fermions, while $F^m_{L,R}$ denote an analogous set of "mirror" fermions, needed for the cancellation of anomalies.

The symmetry $[SU(4)]^4$ may descend at the primary stage of symmetry breaking through a mass-scale M into either of the following low-energy symmetry structures:

$$[SU(4)]^4 \xrightarrow[M \gg m_W]{\text{Heavy Mass}} \begin{cases} G_W \times SU(3)'_{C+D} & \text{(I)} \\ G_W \times SU(3)'_C \times SU(3)'_D & \text{(II)} \end{cases} \tag{4.2}$$

$SU(3)'_{C+D}$ denotes conventional vectorial color, while $SU(3)'_C \times SU(3)'_D$ denotes chiral color. For either case (I or II), the weak electromagnetic symmetry G_W may be either left-handed or left-right symmetric:

$$G_W = \begin{cases} \text{either} & SU(2)_A \times U(1) \\ \text{or} & SU(2)_A \times SU(2)_B \times U(1)_{C+D} \end{cases} \tag{4.3}$$

$SU(2)_{A,B}$ are the GIM-subgroups of $SU(4)_{A,B}$.

The major distinction between Cases I and II is that for Case II chiral color $SU(3)'_C \times SU(3)'_D$ is preserved as a good low-energy symmetry, it is broken "softly" by a mass-scale $M' \lesssim m_W$ at the secondary stage of symmetry breaking[73]; while for Case I, chiral color breaks down by a heavy mass-scale $M \gg m_W$ into its diagonal sum $SU(3)'_{C+D}$ at the primary stage of symmetry breaking.

Now, let $g_1, g_2, g_3, \bar{g}_3$ and g_4 denote the gauge coupling constants associated with the abelian U(1), weak $SU(2)_{A,B}$, strong $SU(3)'_{C,D}$, the diagonal sum $SU(3)'_{C+D}$ and the unifying group $[SU(4)]^4$ respectively. Noting that diagonal summing (e.g. $SU(3)'_C \times SU(3)'_D \to SU(3)'_{C+D}$) reduces the associated coupling by a factor $1/\sqrt{2}$, we observe the following symmetric limit relationships between the above coupling constants:

Case I: $g_1 = g_2 = \bar{g}_3 = g/\sqrt{2}$ (4.4)

Case II: $g_1 = g_2 = g/\sqrt{2}$ (4.5)

$g_3 = g = \sqrt{2}\, g_{1,2}$

Thus for Case II, the coupling constants $g_{1,2}$ associated with the weak-electromagnetic symmetry G_W are lower by a factor $1/\sqrt{2}$ than that (g_3) associated with the low-energy strong interaction symmetry group.

The full impact of this new feature - the emergence of chiral color as a low energy symmetry (Case II) - may be seen by writing down the renormalisation group-equations for the coupling constants. This may be done under the justifiable assumption that the coupling constants are small ($|g_i| << 1$) in the domain of interest.

Using the standard decoupling theorem arguments[74] and the normalisation conditions ((4.4) and (4.5)), the running coupling constants for $\mu \leq M$ are given by

$$g_i^{-2}(\mu) = 2g^{-2}(M) + 2b_i \ell n(M/\mu) \quad (i = 1,2) \tag{4.6}$$

$$(\bar{g}_3(\mu))^{-2} = 2g^{-2}(M) + 2b_3 \ell n\,(M/\mu) \tag{4.7}$$

$$g_3(\mu)^{-2} = g^{-2}(M) + 2b_3' \, \ell n(M/\mu) \tag{4.8}$$

where $b_2 = -22/(3(4\pi)^2) + b_1$, $b_3 = -11/(4\pi)^2 + b_1$ and $b_3' = b_3 - (b_1)/2$.

For algebraic simplicity assume first that $G_W = SU(2)_A \times U(1)$ and that quarks are fractionally charged. In this case,

$$e^{-2} = g_2^{-2} + (5/3)g_1^{-2} \; ; \quad \sin^2\theta_W = 3g_1^2/(3g_1^2 + 5g_2^2) \tag{4.9}$$

Using Eqs. (4.6) - (4.9) and defining $u_S \equiv \alpha_s(QCD)/\alpha$, we obtain:

<u>Case I</u>:

$$u_S = (\bar{g}_3)^2/e^2 = (8/3)\;[1 - (11\alpha/\pi)\ell n(M/\mu)]^{-1} \tag{4.10}$$

$$\sin^2\theta_W = 1/6 + 5/(9u_s) \tag{4.11}$$

<u>Case II</u>:

$$u_S = g_3^2/e^2 = 2(8/3)\;[1 - (77\alpha/3\pi)\ell n(M/\mu)]^{-1} \tag{4.12}$$

$$\sin^2\theta_W = 2/7 + 10/(21\,u_s) \tag{4.13}$$

In writing (4.12), it is assumed for simplicity that chiral color $SU(3)_C' \times SU(3)_D'$ is broken softly by a mass-scale $M' \sim 1$ GeV, low compared to electroproduction vertex momenta. In this case, electroproduction as well as charmonium physics will still be described[75] by an effective coupling constant $g_3^2/4\pi$, g_3 being the coupling constant associated with chiral $SU(3)'_C \times SU(3)'_D$. Thus one may identify α_s(QCD) with $g_3^2/4\pi$.

Apart from the overall factor of 2, note the crucial distinction between the coefficients of $\ell n(M/\mu)$ for Cases I and II (Eqs. (4.10) and (4.12)). Putting $\mu \simeq 10$ GeV and requiring that $u_s = \alpha_s/\alpha \simeq 15$ (i.e. $\alpha_s \simeq .1$ at $\mu \simeq 10$ GeV), we obtain:

Case I: (Vector-Color)

$$M \approx 10^{15} \text{ GeV}$$
$$\sin^2\theta_W \approx 0.20 \qquad (G_W = SU(2)_A \times U(1)) \tag{4.14}$$

Case II: (Chiral Color)

$$M \simeq 10^5 \text{ GeV}$$
$$\sin^2\theta_W \approx 0.30 \qquad (G_W = SU(2)_A \times U(1)) \tag{4.15}$$

Note the superheavy unifying mass-scale M for Case I and the dramatically lower mass-scale M for Case II, as indicated in the beginning. The variations of the coupling constants g_i as a function of the running momentum μ for the two Cases I and II are shown in Fig. 8.

The results exhibited above for the choice $G_W = SU(2)_A \times U(1)$ are preserved at least qualitatively for the alternative left-right symmetric choice $G_W = SU(2)_A \times SU(2)_B \times U(1)_{C+D}$, except that numerical predictions for $\sin^2\theta_W$ are somewhat altered. The limiting values of $\sin^2\theta_W$ (for $u_s \to \infty$) in this case are found be 1/4 for Case I and 1/3 for Case II.[58, 70] All these predictions are in the experimentally allowed range for $\sin^2\theta_W$ (2.49).

For either choice of G_W, the same qualitative results are found to hold for the case of integer charge quarks as well. The limiting values of $\sin^2\theta_W$ remain unaltered. The main change is that the unifying mass-scale M is found to be lower by about a factor 10 compared to that for the case of fractionally charged quarks. This permits unification at a mass-scale as low as about 10^4-10^5 GeV.[76] Such a mass-scale coincides with the estimates of the exotic X gauge-masses obtained from independent considerations[2] involving experimental upper limits on $K_L \to \bar{\mu}e$-decays as well as shortlived character of integer charge quarks.

CONCLUDING REMARKS

(i) Low-mass-unification-hypothesis suggests that chiral color $SU(3)'_L \times SU(3)'_R$ may be a good low energy symmetry. The QCD-octet of vector color gluons must then be accompanied by a low mass octet of axial color-gluons[71] These axial color-gluons would be neutral and confined (like the vector color gluons) if quarks are fractionally charged, and possess charged members and would be liberated, if quarks are integer-charged. The existence of such light axial gluons (mass ~ 1 to few GeV) seems to be compatible with all known phenomenology such as that encountered in lepto-production. Perhaps their existence may help remove some of the lingering discrepancies between vector QCD and charmonium physics[77].

(ii) The mechanism for low-mass unification involving a symmetric limit dichotomy between "weak" SU(2) and "strong" SU(3)'-color coupling constants, as discussed above, appears to be a special feature of symmetries of the form $[SU(n)]^4$ ($n \geq 4$). It is not available within unifying symmetries such as SU(5) and SO(10). Correspondingly, the unifying mass-mass-scale M needs to exceed $\approx 10^{15}$ GeV within symmetries of the latter type.

(iii) Consistent with natural conservation of quantum numbers, the mirror-theory based on the symmetry-structure $[SU(4)]^4$ does not permit large mixing between the basic (F) and the mirror fermions (F^m). This raises an interesting possibility: if the mixing angles are really tiny $<< 10^{-5}$ (rather than $\mathcal{O}(\alpha)$), the quantum number associated with the lightest mirror quark would be preserved to an extent much better than strangeness and charm; consequently at least some mirror-matter (e.g. composites of the lightest mirror quark and thelightest basic anti-quark ($\bar{p}$)) would be relatively stable possessing life times >> 10^{-10} sec.[78] So also would be the lightest mirror-lepton. In the extreme limit of no F-F^m-mixing, such matter would be absolutely stable. Such semistable or stable objects, if they exist, would provide a new vista for experimental physics.

(iv) Had the unifying mass-scale M been necessarily as large as 10^{15} GeV or higher, the question of a direct test of the hypothesis of unification of the basic forces (weak, electromagnetic, as well as strong) would not even arise, at least within one's own generation. The realization of the low unifying mass-scale $M \simeq 10^4 - 10^6$ GeV serves to remove this gloomy view; it raises the attractive hope that the hypothesis may after all be tested directly through ongoing cosmic ray experiments and also through the next generation of accelerators. Nothing could be more dramatic than the observation that quarks and leptons begin to interact universally in all respects and that simultaneously the so-called "strong" interactions acquire their basic $\mathcal{O}(\alpha)$-strength. That such a possibility could arise at a momentum-scale as low at 10^4-10^5 GeV should provide a strong new incentive to build high energy accelerators approaching this range. The virtual effects of an exotic gauge particle X coupling to quark-lepton-currents can be felt[79] in e^-e^+-annihilation at a centre of mass of energy $\simeq m_X/30$.

(v) The major challenge presently confronting unified theories is a natural understanding of the phenomenologically imposed mass-scales of fermions and gauge-mesons. Mass differences such as that between n-p quarks, ν_e-e, μ-e, or the mass-ratios such as (e/μ), (p/c) and (quark/lepton), as well as the many Cabibbo-like angles and CP-violating phases should hopefully be predictions of the theory.[80] As regards gauge-masses, one would need a natural explanation for the ratio of the superheavy unifying mass-scale M versus the masses of the lighter gauge particles W's and Z's. In this respect, a low unifying mass-scale $M \approx 10^4$-10^6 GeV generates the attractive possibility that perhaps only one super-heavy mass-scale triggers spontaneous symmetry breaking, with radiative corrections of order $\sqrt{\alpha}$ and α providing masses to the lighter gauge mesons (the familiar W's and Z's)[81]. This remains to be actually implemented within a realistic model.

REFERENCES AND FOOTNOTES

1. J. Schwinger, Ann. Phys. (N.Y.) 2, 407 (1957); S. L. Glashow, Nucl. Phys. 22, 57a (1961); Abdus Salam and J. C. Ward, Physics Letters 13, 168 (1964). In the present context, the model based on spontaneous breakdown of the local symmetry $SU(2)_L \times U(1)$ is due to S. Weinberg Rev. Lett. 19, 1264 (1967) and Abdus Salam, in Elementary Particle Theory, Nobel Symposium, Ed. by N. Svartholm (Almqvist, Stockholm, 1968), p. 367. Extension to hadrons is by S. Glashow, J. Iliopoulos and L. Maiani, Phys. Rev. D2, 1285 (1970).

2. J. C. Pati and Abdus Salam, Phys. Rev. D8, 1240 (1973); Phys. Rev. Lett. 31, 661 (1973); Phys. Rev. D10, 275 (1974), and Physics Letters 58B, 333 (1975).

3. H. Georgi and S. L. Glashow, Phys. Rev. Lett. 32, 438 (1974); H. Fritzsch and P. Minkowski, Ann. Phys. (NY) 93, 193 (1975); F. Gürsey and P. Sikivie, Phys. Rev. Lett. 36, 775 (1976); P. Ramond, Nucl. Phys. B110, 214 (1976).

4. These attempts are admittedly incomplete, until gravity is in the picture.

5. L. L. Lewis et al, Phys. Rev. Lett. 39, 795 (1977), P. E. G. et al, Phys. Rev. Lett. 39, 798 (1977). See P. G. Sandars, report at this conference, for an estimate of the shielding effect, and G. Feinberg, report at this conference for a review of theoretical calculations.

6. M. L. Perl et al, Phys. Rev. Lett. 35, 1489 (1975). See M. Perl, report at this conference, for an up-to-date summary.

7. Alternatively one may also pose the question; is left handed neutrino ν_L accompanied by its right handed counterpart ν_R (distinct from $\bar{\nu}_R$), both being two component entities and massless?

8. R. N. Mohapatra, J. C. Pati and A. Salam, Phys. Rev. D13, 1733 (1976); R. N. Mohapatra and J. C. Pati, Phys. Lett. 63B, 204 (1976).

9. See discussions later for a precise definition of left-right symmetry and references thereon.

10. H. Fritzsch and P. Minkowski, Ann. Phys. (NY) 93, 193 (1975), H. Georgi and S. L. Glashow in Particles and Fields - 1974, (APS/DPF Williamsburg), P. 575 (ed. C. E. Carlson).

11. The suggestion of the left $\leftrightarrow$ right symmetric substructure $SU(2)_L \times SU(2)_R \times U(1)_{L+R}$ comprising all matter (quarks as well as leptons) was made by J. C. Pati and Abdus Salam, Phys. Rev. Letters 31, 661 (1973); ibid. Phys. Rev. D10, 275 (1974). The motivations in these work were the realizations of left-right symmetry at the basic level and quantization of electric charge. Motivation based on considerations of CP violation was provided by R. N. Mohapatra and J. C. Pati, Phys. Rev. D11, 566 (1975). The second paper (Phys. Rev. D10, 275 (1974), Ch. IV and VI and Footnote 21) proposes two alternative patterns of spontaneous symmetry breaking; both patterns are consistent with the hypothesis of "natural" left-right symmetry. One of the patterns permits only one, while the other permits two relatively light neutral weak gauge particles. It is this second alternative, which is relevant to present atomic parity experiments (Ref. 5)), and is pursued recently by several authors (See Ref. 12).

12. H. Fritzsch and P. Minkowski, Nucl. Phys. B103, 61 (1976), R. N. Mohapatra and D. P. Sidhu, Phys. Rev. Lett. 38, 667 (1977). A. DeRujula, H. Georgi and S. L. Glashow, Harvard Preprint, 1977. J. C. Pati, S. Rajpoot and Abdus Salam, Phys. Rev. D, 17, 131, 1978 and Physics Lett. 71B, 387 (1977).

H. S. Mani, J. C. Pati and A. Salam "Naturalness of Atomic Parity Conservation within Left-Right Symmetric Unified Theories" - Trieste Preprint IC/77/80, Phys. Rev. D to be published.

R. N. Mohapatra, F. E. Paige and D. P. Sidhu, BNL-Preprint (1977).

13. J. C. Pati, Abdus Salam and S. Sakakibara, Phys. Rev. Lett. 36, 1229 (1976).

14. J. C. Pati and Abdus Salam, "Lepton-Hadron Unification", Proc. 1976 Aachen Neutrino Conference, P. 589 (ed. H. Faissner, H. Reithler and P. Zerwas).

15. Flavor as well as SU(3)'-color-singlet generators Q_o are included in Q_{flav} for convenience.

16. J. C. Pati, S. Rajpoot and A. Salam, "Prequarks and Fractional Charges", Imperial College preprint IC/76/13.

17. H. Fritzsch, M. Gell-Mann, and H. Leutwyler, Physics Letters, B47, 365 (1973), S. Weinberg, Phys. Rev. Letters 31, 494 (1973), and D. J. Gross F. Wilczek, Phys. Rev. D8, 3633 (1973). See also, C. Itoh, et. al. Preprint, 1973, unpublished).

18. D. Gross and F. Wilczek, Phys. Rev. Lett. 30, 1343 (1973); H. D. Politzer, Phys. Rev. Lett. 30, 1346 (1973), and Physics Reports, 14C; No. 4 (1974).

19. See for example H. D. Politzer "Testing of Scaling and QCD with muon Beams", HUTP-77/A038 and references therein; G. Altarelli, G. Parisi and R. Petronzio, Physics Letters 63B, 183 (1976); H. Georgi and H. D. Politzer "Clean Tests of QCD in μp-Scattering (HUTP-77/A063). For tests pertaining to jet-structure, see H. Georgi and M. Machacek, Phys. Rev. Lett. 39, 1237 (1977); G. Steerman and S. Weinberg, Phys. Rev. Lett. 39, 1436 (1977), and E. Farhi, Phys. Rev. Lett. Vol. 39, 1587 (1977).

20. R. N. Mohapatra, J. C. Pati and P. Vinciarelli, Phys. Rev. D8, 3652 (1973); S. Weinberg, Phys. Rev. Lett. 31, 494 (1973).

21. M. K. Gaillard and B. W. Lee, Phys. Rev. Letters 33, 108 (1974); G. Altarelli and L. Maiani, Phys. Letters 52B, 351 (1974).

22. J. C. Pati and A. Salam, Phys. Rev. Lett. 36, 11 (1976); G. Rajasekharan and P. Roy, Pramana 6, 303 (1975); V. Elias, J. C. Pati, A. Salam and J. Strathdee, Pramana 8, 303 (1977).

23. See for example, B. Barish, Particles and Fields, 1974 (APS/DPF-Williamsburg), ed. C. E. Carlson (AIP, New York, 1975).

24. This is the case for example for the unifying symmetry group $G = [SU(4)]^4$ (Ref. 2).

25. M. A. B. Bég and A. Zee, Phys. Rev. Letters 30, 675 (1973). For a list of references on other vector-like models, see R. M. Barnett, Review talk, Brookhaven, APS Meeting, SLAC-PUB 1850 (Sept. 1976).

26. Later $U(1)_{L+R}$ is considered as part of a chiral structure $U(1)_L xU(1)_R$.

27. CP violating phases are suppressed for the present (see later).

28. Left and right handed Cabibbo-angles are restricted to be nearly equal (barring radiative differences) within a unified theory without abelian factors: J. C. Pati, "Unification-Liberation of Quarks and Leptons", 1976 Scottish Univ. Lecture Notes P. 89 (ed. I. M. Barbour and A. T. Davis). In a similar context see, M. A. B. Bég, R. Budny, R. N. Mohapatra and A. Sirlin, Phys. Rev. Lett. 38, 1252 (1977).

29. R. N. Mohapatra and J. C. Pati, Phys. Rev. D11, 2558 (1975). G. Senjanović and R. N. Mohapatra, Phys. Rev. D12, 502 (1975).

30. Mechanisms which help the emergence of the solution b=c "naturally" have been proposed: H. S. Mani, J. C. Pati and A. Salam, Trieste Preprint IC/77/80, Phys. Rev. D (To be published); R. N. Mohapatra,

F. E. Paige and D. P. Sidhu, BNL-Preprint (1977). The mechanisms suggested in these two papers are physically different. The second is an explicit construction making essential use of an abelian component in the gauge-structure. The first depending upon radiative corrections is not subject to this restriction and provides the scope that it can be incorporated within a nonabelian unifying structure.

31. For such a calculation see Q. Shafi and Ch. Wetterich, Univ. Freiburg preprint (1977), and W. J. Marciano and A. I. Sanda, Rockefeller preprint (1978).

32. A. Benvenutti *et al.*, Phys. Rev. Letters 37, 1039 (1976); J. Blietschau *et al.*, Preprint CERN/EP/Phys. 76-55; B. C. Barish, Ca. Tech. preprint CALT-68-544. The clearest distinction is shown by measurements of $\sigma(\bar{\nu}_R p \to \bar{\nu}_R p)$ versus $\sigma(\nu_L p \to \nu_L p)$. See D. Cline *et al.*, Phys. Rev. Letters 37, 252, 648 (1976) and W. Lee *et al.*, Phys. Rev. Letters 37, 186 (1976).

33. H. Fritzsch and P. Minkowski, Nucl. Phys. B103, 61 (1976); J. C. Pati, S. Rajpoot and Abdus Salam, Phys. Rev. D17, 131 (1978); H. Georgi and S. Weinberg, Harvard preprint HUTP77/A052. The equivalence was noted for the special case $x = 0$ (b=c) in the first paper. Its general validity for the group-structure $SU(2)_L \times SU(2)_R \times U(1)_{L+R}$ was proved in the second paper. Its validity for certain extended group-structures is proved in the third paper.

34. J. C. Pati, Proc. Second Orbis Scientae, Coral Gables, Florida, Jan., 1975 (P253-256), ed. by A. Perlmutter and S. Widmayer; J. C. Pati, S. Rajpoot and A. Salam (Ref. 12, Footnote 15).

35. Q. Shafi and Ch. Wetterich, Univ. Freiburg preprint (1977); V. Elias, J. C. Pati and Abdus Salam, Univ. Md. Tech. Rep. 78-043 (1977),

Physics Letters, To be published.

36. A multiplet $A = (2,2, Y=0)$ with Yukawa coupling to fermions gives masses to fermions if $\langle A\rangle \neq 0$; it simultaneously mixes W_L and W_R (Ref. 2).

37. See for example the talks of C. Baltay and J. Steinberger at the European Physical Society meeting, Budapest, Hungary (1977), to appear in the Proceedings. Recent CDHS measurements yield $\sin^2\theta_W = 0.24 \pm .02$ (M. Holder et al., CERN preprint (1977)).

38. R. N. Mohapatra and J. C. Pati, Phys. Rev. D11, 566 (1975).

39. The desirable properties of the milliweak CP violation proposed in Ref. 38 would, of course, be preserved even if there are new quark flavors beyond charm, as long as the new flavors do not mix sensibly with (p,n,λ,c). (See elaboration on this by R. N. Mohapatra and D. P. Sidhu, CCNY, preprint 1977). Finite radiative mixing, if permissible, will be inconsequential in this regard.

40. L. Wolfenstein, Phys. Rev. Lett. 13, 562 (1964); Nucl. Phys. B77, 375 (1974). For a gauge theory version see R. N. Mohapatra, J. C. Pati and L. Wolfenstein, Phys. Rev. D 11, 3319 (1975).

41. The contribution in question is $\sim G_F\,(m_{W_L}/m_{W_R})^2 \sin\delta\,(a^2/c^2)$, which can be limited to be $\lesssim G_F \times 10^{-4}$. ($a^2$ denotes W_L-W_R mixing $(\text{mass})^2$). See also Ref. 38.

42. This lower limit ($\approx$ 30 GeV) on the mass (m_σ) of the relevant left over Higgs-boson corresponds to the experimental constraint that the relation $\eta_{+-} = \eta_{oo}$ is known to hold to better 5% (see Ref. 38). In reference 38, m_σ was allowed to be as high as 10^4 GeV or higher. This led to a prediction for d_n varying between 10^{-24} to 10^{-29} e cm. The constraint imposed here that m_σ should not exceed about 1000 GeV (or else Higgs-fields would begin to interact strongly, see e.g. M. Veltman, Utrecht

preprint, andB. W. Lee, C. Quigg and H. B. Thacker, Phys. Rev. Lett. 38, 883 (1977)) is important: it makes the dipole moment dn to necessarily exceed about 10^{-27} ecm for the isoconjugate model, bringing the same to an experimentally accessible range.

43. W. B. Dress et al., Phys. Rev. D15, 9 (1977).

44. S. Weinberg, Phys. Rev. Letters 29, 1698 (1972). In this work, the symmetric gauge-structure is introduced for quarks for the sake of natural isospin conservation; but leptons are assumed to be singlets of $SU(2)_R$.

45. F. Wilczek and A. Zee, Phys. Lett. 70B, 418 (1977); S. Weinberg, Harvard preprint HUTP-77/A057; H. Fritzsch, Phys. Lett. 70B, 436 (1977); R. N. Mohapatra and G. Senjanovic, City College preprint CCNY-HEP-77/12.

46. The treatment and notations adopted here follow the paper of J. C. Pati, S. Rajpoot and Abdus Salam, Phys. Rev. D17, 131 (1978).

47. For integer charge-quarks, there would be a small mixing (by a mass-scale ~ 1 GeV) of flavor and SU(3)'-color-gauge mesons (See Ref. 2). Such mixing would have no effect however on neutral current-processes below threshold for color-production.

48. This is barring finite $\mathcal{O}(\alpha \ln m_{W_R^\pm}/m_{W_L^\pm})$ radiative correction to $(g_L-g_R)/g_L$.

49. See for example R. Budny, Physics Lett. 55B, 227 (1975); A. McDonald, Nucl. Phys. B75, 343 (1974) and E. Lendavi and G. Pocsik, Physics Lett. 56B, 462 (1975).

50. J. C. Pati, S. Rajpoot and Abdus Salam, Phys. Lett. 71B, 387 (1977).

51. J. Leveille - Imperial College Preprint (1978); V. Barger and R. N. Philips, Univ. of Wisconsin Preprint (1978).

52. Unpublished results of such a preliminary measurement carried out at SPEAR exist, which indicate that the magnitude is less than 3% in

magnitude at a mean center of mass energy ≈ 6.8 GeV (private communications: B. Richter).

53. R. Cahn and F. Gilman, SLAC-PUB (1977).

54. A. Janah, Univ. of Md. preprint (in preparation).

55. H. S. Mani, J. C. Pati, S. Rajpoot and A. Salam, Phys. Lett. (1977).

56. H. S. Mani, J. C. Pati and A. Salam (Ref. 12).

57. S. W. Herb et al., Phys. Rev. Lett. 39, 252 (1977).

58. V. Elias, J. C. Pati and Abdus Salam, "Unification of the Basic Particle Forces at a Mass-Scale of order 1000 m_W", Univ. of Md. Tech. Rep. 78-041 (1977).

59. H. Georgi, H. R. Quinn and S. Weinberg, Phys. Rev. Lett. 33, 451 (1974); F. J. Yndurdin, Nucl. Phys. B115, 293 (1976); M. Chanowitz, J. Ellis and M. K. Gaillard, CERN preprint TH-2312 (1977).

60. See for example Ref. 22, 14 and the review paper: J. C. Pati, "Unification-Liberation of Quarks and Leptons", 1976 Scottish Univ. Summer School Lecture Notes, P. 89 (ed. I. M. Barbour and A. T. Davies).

61. As judged from the pair-production cross-section of J/ψ, it appears that hadrons of 2-3 GeV mass may not be pair produced at ISR-Fermilab energies with cross-sections much bigger than $\approx 10^{-30}$ cm^2.

62. J. C. Pati, S. Sakakibara and Abdus Salam, Trieste preprint IC/75/93 (unpublished).

63. The best "empirical" lower limit on proton-lifetime based on a search sensitive to energetic μ^+'s is at present $\approx 2 \times 10^{30}$ years: F. Reines and M. Crouch, Phys. Rev. Lett. 32, 493 (1974). For a decay-mode-independent search see R. I. Steinberg and J. C. Evans, Science 197, 989 (1977). For an up-to-date review, see M. Goldhaber, Talk at this Conference.

64. See a recent estimate by A. J. Buras, J. Ellis, M. K. Gaillard and D. V. Nanopoulos, CERN-preprint TH2403 (1977).

65. For recent review of the tests of unconfined color, see the paper by J. C. Pati and Abdus Salam, "Design of Future Experiments...", Trieste Preprint IC/77/65.

66. J. C. Pati, J. Sucher and C. H. Woo, Phys. Rev. D15, 147 (1977).

67. S. Sakakibara, B. Kayser, J. C. Pati and G. Zorn (unpublished).

68. This takes into account the photon-gluon-cancellation-effect for lepto-production of color (Ref. 22).

69. Reported at Photon-Lepton Symposium, Hamburg (Aug., 77).

70. V. Elias, Phys. Rev. D14, 1896 (1976); ibid D16, 1586 (1977).

71. J. C. Pati and Abdus Salam, Phys. Lett. 58B, 333 (1975).

72. J. C. Pati, Proc. Second Orbis Scientae Coral Gables, Florida, Jan., 1975 (P. 253) ed. A. Perlmutter and S. Widmayer; H. Fritzsch and P. Minkowski, Ann. Phys. (NY) 93, 193 (1975).

73. At least a soft breaking of SU(3)'-chiral color is needed, since fermions acquire mass.

74. T. Appelquist and J. Carrazone, Phys. Rev. D11, 2856 (1975).

75. With exact chiral color symmetry and massless fermions, the familiar QCD formulae for electroproduction will apply without any change since left and right handed quarks in this case couple to separate octets of gluons with equal coupling constant. In the other extreme limit of infinitely heavy axial color gluons ($m_A >> m_W$) with massless vector gluons, there would be no change in the relevant formulae either except that the coupling constant would be redefined in this case throughout by a normalisation factor $1/\sqrt{2}$.

76. Even such a low unification mass is compatible with the known proton-stability, since proton-decay can occur for symmetries (like $[SU(4)]^4$ only in the third order of effective gauge interactions (Ref. 2).

77. See for example G. Feinberg and J. Sucher, Phys. Rev. Letters 13, 1740. For recent reviews see J. D. Jackson, invited talk at the EPS meeting, Budapest (July 1977), and K. Gottfried, invited talk, Photon-Lepton Symposium; Hamburg (Aug, 1977).

78. Existence of such objects are motivated on different grounds: See for example B. W. Lee and S. Weinberg, Phys. Rev. Lett. 39, 165 (1977); J. C. Pati, VIII GIFT Lectures (June, 1977), Univ. of Md. Tech. Rep. 78-065, To appear in the Proceedings; R. N. Mohapatra and D. P. Sidhu, CCNY Preprint (1977); R. Cahn, Phys. Rev. Lett. (1977).

79. J. C. Pati and Abdus Salam, Phys. Rev. D11, 1137 (1975).

80. For recent attempts in this direction see Ref. 45.

81. S. Coleman and E. Weinberg, Phys. Rev. D7, 1888 (1973); E. Gildener, Phys. Rev. D14, 1667 (1976). For a recent suggestion along these lines see H. S. Mani, J. C. Pati and Abdus Salam (Ref. 12).

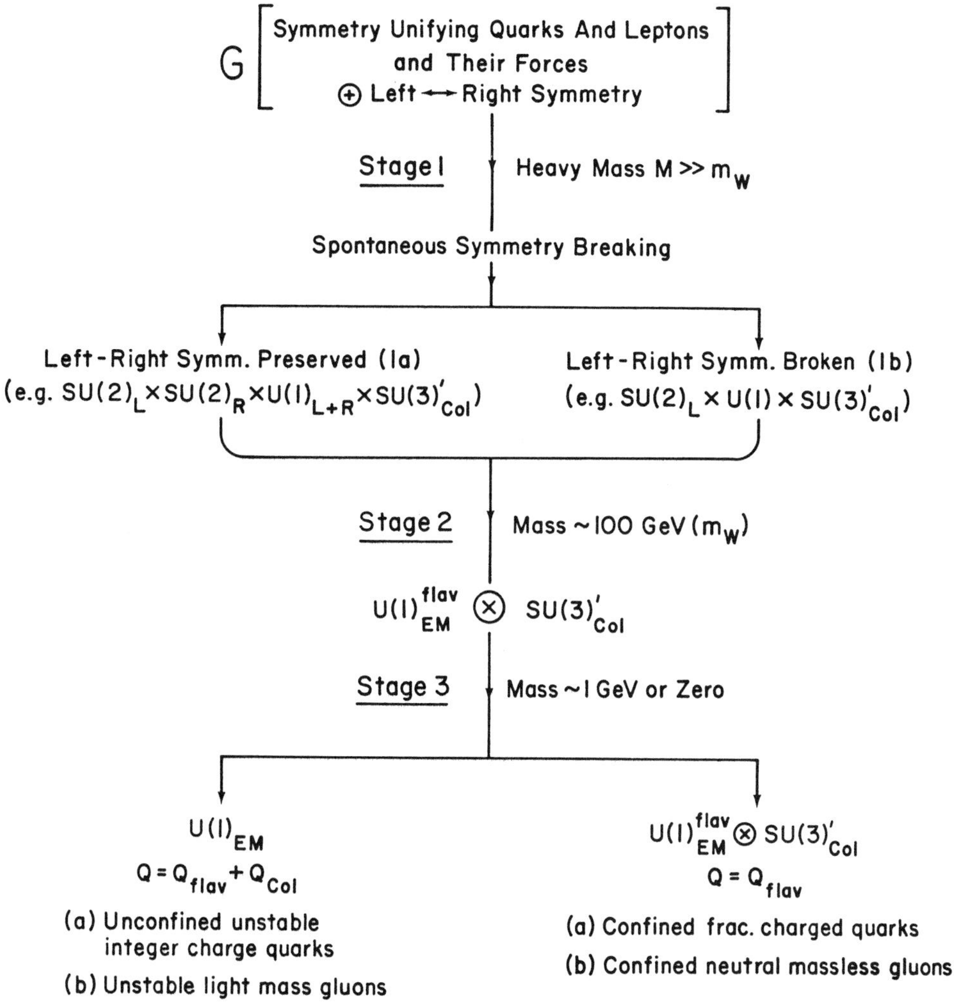

Fig. 1. A chart showing the basic alternatives, which arise within a unified theory, depending upon the nature of spontaneous symmetry breaking. Only the broad features of stage-2 are exhibited. For example, in the transition from stage-1a to stage-2, there is the choice that left-right symmetry may be preserved in the neutral current-sector to an extent much better than in the charged one, or that it may be broken in the charged as well as neutral-current-sectors to nearly the same extent. This determines the extent to which parity is violated in atoms.

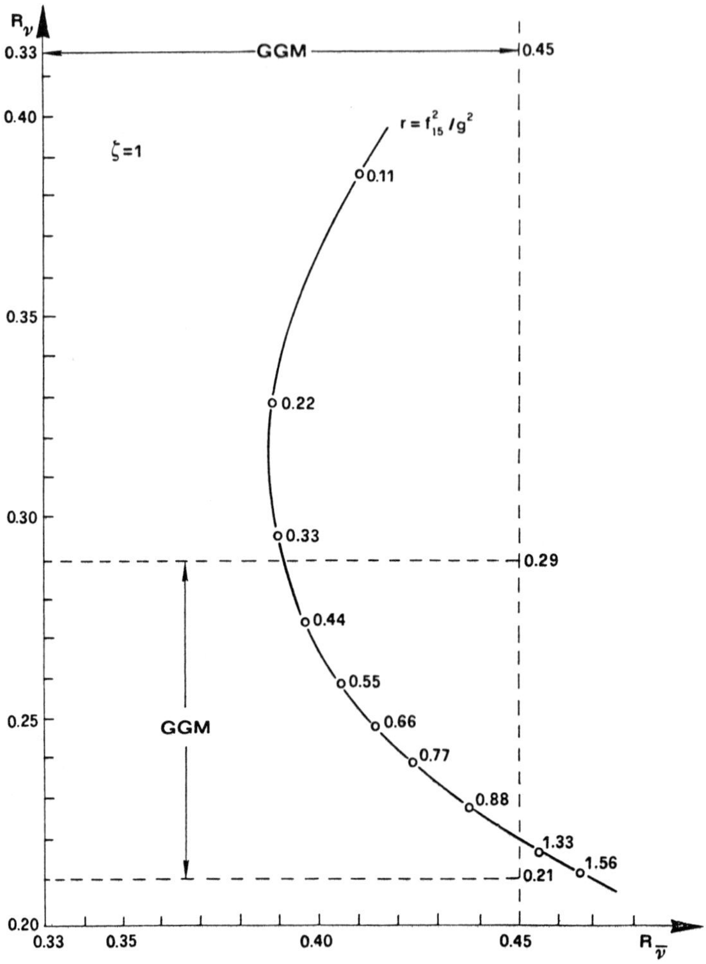

Figure 2

($\sin^2\theta_W = 3r/(6r+2)$, see text)

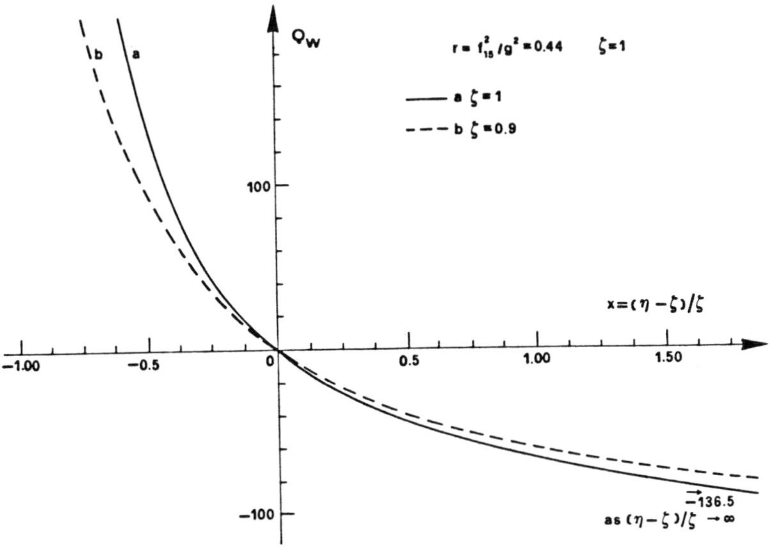

Fig. 3 A plot of the atomic parity violation parameter Q_W for Bismuth vs. the neutral current-parity violating parameter $x \equiv (b^2-c^2)/c^2$ in the absence of radiative corrections to $(g_R-g_L)/g_L$. The radiative corrections begin to be important for $|x| < .2$.

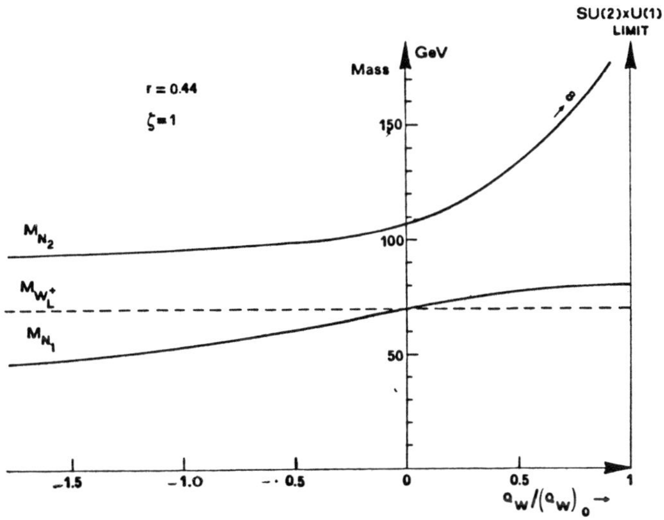

Fig. 4 A plot of the masses of the two neutral weak gauge particles N_1 and N_2 vs. (Q_W/Q_o). Q_W and Q_o denote the atomic parity violation parameters for SU(2) x SU(2) x U(1) and SU(2) x U(1)-theories respectively.

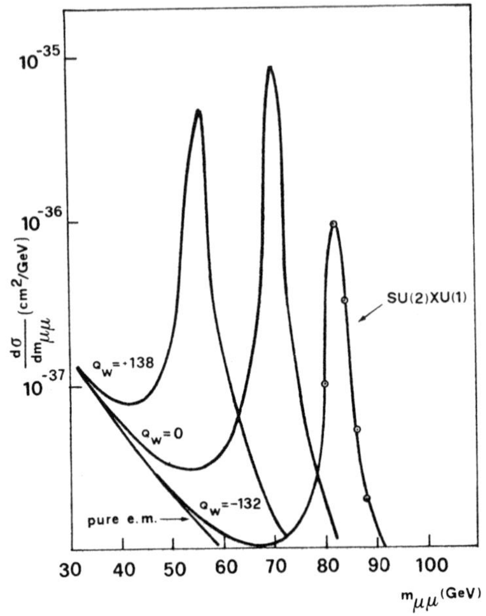

Fig. 5. A plot of $d\sigma/dm_{\mu\mu}$ vs. $m_{\mu\mu}$ for $p\bar{p} \to \mu\bar{\mu}x$ for three different values of atomic parity violation-parameter Q_W (for Bismuth) at $\sqrt{s}$ = 200 GeV.

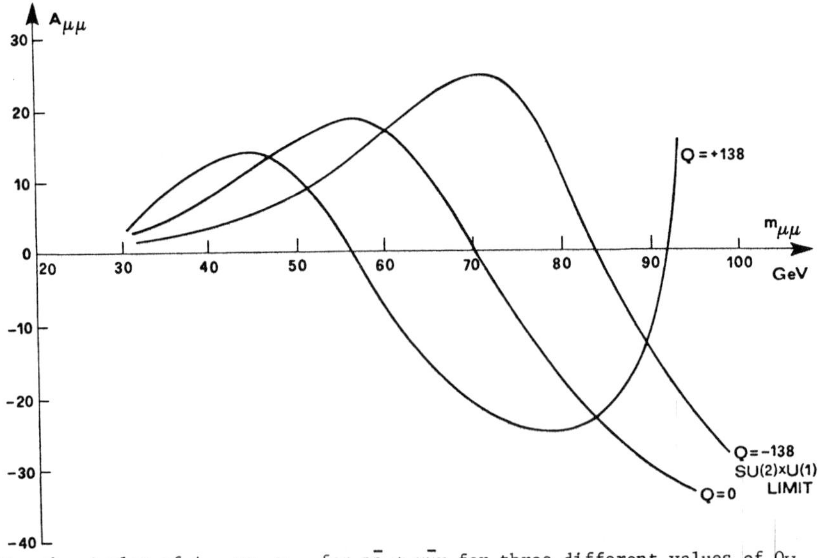

Fig. 6. A plot of $A_{\mu\mu}$ vs. $m_{\mu\mu}$ for $p\bar{p} \to \mu\bar{\mu}x$ for three different values of Q_W at $\sqrt{s}$ = 200 GeV.

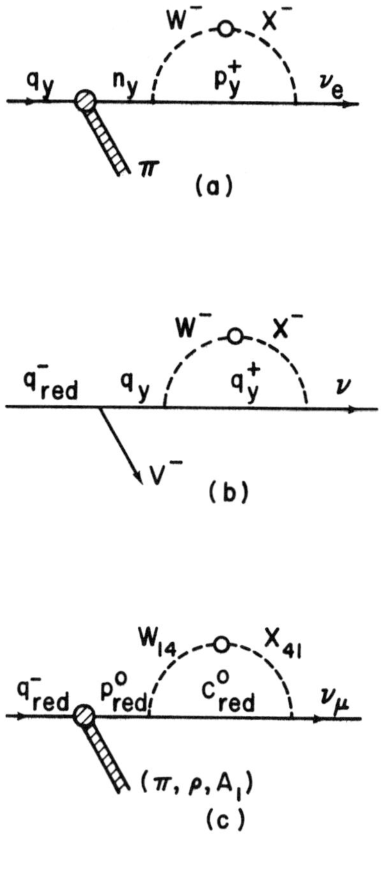

Fig. 7. Dominant Mechanisms for Quark-Decays.

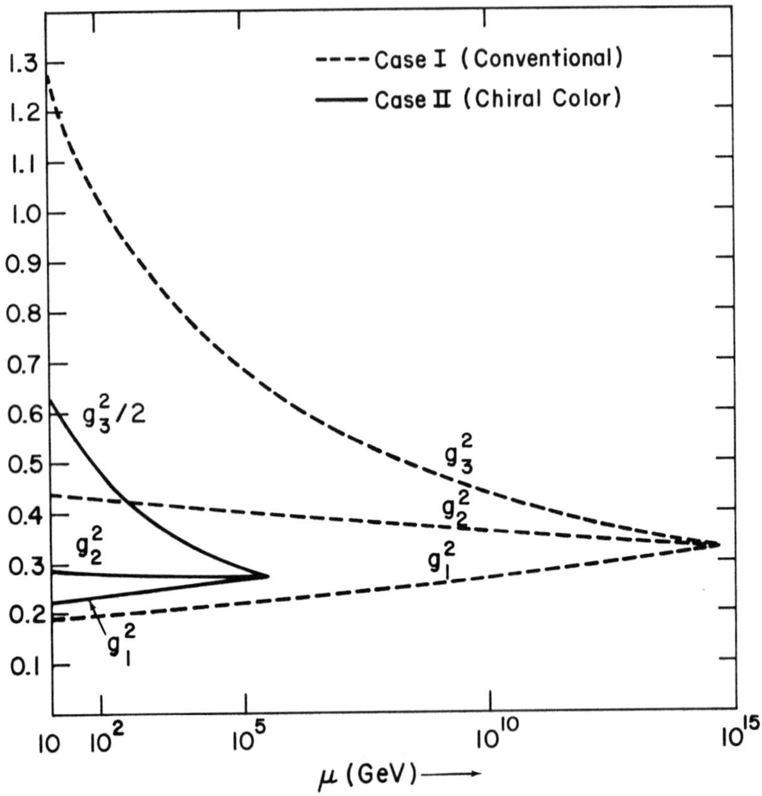

Fig. 8. Effect of the Hierarchy of Symmetry Breaking on the Unifying Mass-Scale M. The variations of g_i's with μ are shown for the two cases of low energy color symmetry I and II (i.e. vector and chiral color) with the choice of $G_W = SU(2)_A \times U(1)$ and of fractional quark charges. For other choice analogous results hold (see text). For either Case I or Case II, it is g_3 which denotes the relevant "strong" interaction coupling constant at low momenta. However, for cases I and II, it is g_3^2 and $g_3^2/2$ respectively, which coincide with the "weak" coupling constants $g_{1,2}$ at large momenta.

ALTERNATIVES TO GAUGE THEORIES

J. D. Bjorken
Stanford Linear Accelerator Center
Stanford University, Stanford, California 94305

To a remarkable degree, gauge principles now permeate the description of almost all of the basic interactions of elementary particles. The proceedings of this Conference make it rather clear why this is so, and why it is hard to find any theorist (including myself) working actively and continuously on theories which lead in a direction contradictory to that of the gauge theories.

This state of affairs is a great tribute to those who created and developed this impressively rich, beautiful, and promising theoretical structure - and this tribute manifestly includes Ben Lee, whom we honor here.

But I think this situation has its dangers. Even the greatest creative endeavors should be made in the face of a strong critical background, in order to be certain that we really proceed in the right direction. To be sure, the absence of criticism of gauge-theory ideology these days is quite understandable. To work on something else is to become a bit of a social outcast, and that is something the younger (untenured) generation may choose not to face. Also, ever since we were taught how to calculate with non-Abelian gauge theories, there have been a lot of calculations - especially for weak processes - to be carried through. Indeed this calculability is a hallmark of the weak-electromagnetic theories: an experimentally accessible observable may be calculated in a definite way. It may be calculated not only in one definite way, but many definite ways with many different definite answers, depending on one's starting point. It is no wonder it is a popular

*Work supported by Department of Energy.

business. Thus far, the alternatives to gauge theories offer nothing much to calculate, but only the supposition that the situation is too complicated to anticipate and to calculate precisely.

In any case it is clearly important to examine the whole gauge-theory enterprise in the most hardheaded and pragmatic way possible: is it conceivable that we can get along without it? And, while raising this question here might seem to be disrespectful to Ben Lee's contributions, this is just not the case. Ben appreciated the need to ask such questions, and in fact was responsible for this subject being included in the program.

What are the arguments for gauge theories? I divide them into two categories: subjective and objective. The subjective arguments are those which persuade even in the absence of data. Objective arguments are those which only persuade in the presence of data. I do not mean to imply that "subjective" and "objective" are euphemisms for "bad" and "good". For me physics without subjective argument isn't worth doing. But however one chooses to weigh subjective and objective arguments, I think that one must make a clear distinction between them. This will be what we try to do here. We shall first try to catalogue objective and subjective arguments for quantum chromodynamics (QCD), and then do the same for weak-electromagnetic gauge theories. Gravity, supergravity, gauge supersymmetry, and superunified theories will not be discussed at all.

I. Arguments for QCD

We begin with objective arguments.

1. Asymptotic Freedom

If one insists upon describing quark interactions by a renormalizable local field theory, then the existence of approximate scaling-behavior is an argument for QCD, as opposed to other renormalizable theories with anomalous dimensions, where scaling-violations might be expected to be bigger. Also, the correct values of the cross section for $e^+e^- \rightarrow$ hadrons, both below and above charm threshold, are rather successfully predicted in QCD but not in other theories. Finally, the scale-breaking effects in electroproduction, while still interpretable in several ways, are in the direction anticipated by QCD.[1]

2. Chiral Structure

The successes of current-algebra, PCAC, and the $(3,\bar{3}) \oplus (\bar{3},3)$ structure of chiral $SU(3) \otimes SU(3)$ symmetry breaking are naturally contained within the QCD Lagrangian. Even the potential trouble with the U(1) problem[2] (there should be a Goldstone mode with mass comparable to the pion) seems to be solved via the rich vacuum structure of QCD.[3]

3. Spectroscopy

Color singlets lie lowest in naive calculations with QCD potentials. Also the splitting of Δ from N can be accounted for by the spin-spin force generated by colored gluon exchange.[4]

4. Charmonium Spectroscopy

There is a certain level of qualitative and semiquantitative agreement with what is expected from QCD, perhaps including some insight into the Zweig-rule. But the hyperfine splittings and some properties of the radiative transitions are not in good condition, so that they tend to cancel out the successes.[5]

None of the above arguments appear to me as especially strong. And I doubt that they are what really motivate theorists to take QCD seriously. Rather there exist strong subjective arguments:

(1) Existence of the color quantum number. I list this here because, while baryon spectroscopy, the $\pi^\circ \rightarrow \gamma\gamma$ decay rate, and the e^+e^- total cross section do provide objective evidence, it is only evidence for global color symmetry, and

not a local gauge principle. One would like to see evidence of the existence (as partons) of the gauge-gluons. But that is still for the future.[1]

(2) Existence of a local gauge-principle is a hypothesis that flows naturally from the previous item. It is a property in common with general covariance and electromagnetic gauge invariance and goes very deep. Perhaps all successful field theories will be gauge theories, and perhaps the key to unification lies in extension of gauge principles.

The idea, embodied in the gauge principle, that existence of color implies existence of strong interaction is also a unifying principle which helps to synthesize lepton-world with quark-world: existence of color is the only feature which distinguishes lepton from quark. But could one accomplish the same end without a gauge principle?

(3) Infrared slavery: Pure QCD is a very sick theory at low momenta and energies. This sickness is turned into radiant health by arguing that this is the key to the problem of apparent confinement of quarks and gluons. There is some support for this from the beautiful work of Wilson[6] and others on lattice QCD. But nothing yet exists that is clearly applicable to the real world.

(4) Renormalizability: We reserve comment on this until we discuss weak-electromagnetic gauge theories.

(5) Dynamics of high energy deep-inelastic processes: The evolution of hadron final states in, say, the process

$$e^+e^- \rightarrow q\bar{q} \rightarrow \text{hadrons}$$

from initial quark pair to final hadron system appears to demand the existence of long-range correlation in rapidity. This in turn suggests the J=1 gluons of QCD as the source of this correlation.[7]

II. Alternatives to QCD

A serious alternative to QCD is the integer-charge (Han-Nambu) version of the quark model, as pursued by Pati and Salam. If color is an exact (ungauged?) symmetry in such a theory, and if only color-singlet states exist at present energies, then the present-day phenomenology is isomorphic to the fractionally charged quark model.[8] On the other hand, if one tries to liberate color at present energies it appears that it can only be done (if at all) within the context of a gauge theory.[9] Therefore that option is not an alternative to gauge theories.

Another alternative is the string model of hadrons. Someday it might be preempted by QCD, but that has not yet happened. So far it does not fit into the mold of standard renormalizable local field-theories. Indeed alternatives may well need to have this property; otherwise the scaling behavior of deep inelastic processes tends to force one back to asymptotically free theory, i.e., QCD.

Could quarks be composites? We generally think that (current) quarks are so fundamental that they should be described in a simple way via a simple and beautiful Lagrangian. Not too long ago there existed such hopes for nucleons and their interactions. And even now a rather pretty - but nonrenormalizable - effective Lagrangian does not do badly in describing low energy interactions of mesons and nucleons. But as energies increase, that description evaporates. Maybe history can repeat.

Continued proliferation in the number of quark flavors might make this notion more attractive, just as nucleon compositeness (or at least "democracy") became increasingly attractive as the continuing discoveries of hadron resonances fattened the Particle Data Group compliations. However, compositeness of quarks is not as easy a notion as compositeness of nucleons. Once one disregards the color assignment, quarks and leptons have remarkably similar properties: the same kind of weak and electromagnetic interactions, and most likely a common origin for their bare mass. It is not much easier to contemplate internal structure for the quark than to contemplate internal structure for the neutrino.

The present situation in strong interactions has been summarized by saying that, while there is little objective evidence in favor of QCD, any known alternative has much less to offer. In other words, QCD is the only known field theory of strong interactions that has a chance of being right.

III. Arguments for Weak-Electromagnetic Gauge Theories

The situation for the weak-electromagnetic interactions is quite different from QCD. Not only are there good subjective arguments for the gauge-theory description, but there is also apparent objective evidence in the remarkable agreement of a great deal of data with the simplest version. Let us begin by listing the subjective arguments:

1. Existence of a gauge-principle: The comments made with regard to QCD again may apply.

2. Unification with electromagnetism: How can one object to that?

3. Renormalizability: Much is made of renormalizability as a requirement or criterion for choosing theories. I have never understood this. If we were to be consistent, we might be obligated to throw out general relativity. It is clear that even if somebody showed that general relatively was incurably nonrenormalizable, the theory would not be abandoned.

I am also unimpressed with related subjective criteria imposed upon weak-electromagnetic gauge theories, such as "anomaly-cancellation" or "naturalness". The anomaly plague only affects observables at a level of precision comparable to that reached in g-2 experiments. Is it obvious the correct theory attains such perfection? The naturalness concept[10] presupposes that the parameters determining the quark and lepton masses and mixings are random variables, and only theories that respect observed properties (such as small strength of $\Delta S = 2$ non-leptonic transitions) for all values of the parameters are to be acceptable. This would imply the masses themselves are random variables. But why, then, the very large ratios such as m_u/m_c? Why the curious degeneracies such as $m_\tau \sim m_c$ and $m_\mu \sim m_s$? It seems that one problem of masses is not that they are randomly distributed, but that they show organization: patterns do exist, but we do not know how to explain them.

I therefore prefer to avoid such criteria, criteria which threaten to exclude the one Correct Theory everyone is searching for. If such a theory really exists, I would expect that when it is found, we will not need such criteria: we will know it is correct.

There is one cogent reason for demanding renormalizability. To presume it, along with QCD, is to presume that we can have a closed, essentially complete description of particle phenomena at all distances down to the Planck distance $\sim 10^{-33}$ cm. If we really can extrapolate our present concepts 18 orders of magnitude we have some reason to ask for perfection. Nevertheless there does remain the possibility that the push from 10^{-15} cm toward 10^{-33} cm may reveal just as much richness as we have found in going from 10^{-8} cm to 10^{-15} cm. If there is such richness we need not demand that the present theoretical description be a closed one. And if it is not closed, then the criterion of renormalizability need not apply.

4. Existence of a mechanism for generation of intermediate-boson and fermion mass: The Higgs-mechanism provides a beautiful way to render an intractable theory into a manageable and calculable form, as well as providing insight into how quark and lepton masses, and perhaps even the Cabibbo-angle and CP violation, might be generated. But when the Higgs sector of a typical theory is studied

in detail, it is rare that output exceeds input. The elegance of the typical fermion and gauge-boson Lagrangian is only surpassed by the clumsiness of the Lagrangian for its Higgs sector. Despite all the effort that has gone into this question, there does not yet exist a detailed picture of mass-generation that persuades.

In addition to subjective arguments there exist objective arguments for the weak-electromagnetic gauge theories. Some of these I consider specious; others however are very impressive.

(1) Existence of neutral currents: Despite the fact that historically the gauge theories were of crucial importance in stimulating the search for $\Delta S = 0$ neutral-currents, this does not imply a logical connection. If only for reasons of symmetry[11], it is easy to motivate the incorporation of $\Delta S = 0$ neutral currents (of strength comparable to charged currents) into theories of weak interactions. It was the absence of $|\Delta S| = 1$ neutral currents which inhibited theorists from doing this. The solution to this problem lay in charm and the GIM mechanism,[12] which leads to the next argument for gauge theories:

(2) Existence of charm and the GIM mechanism: Again the development of gauge theories,[13,14] together with the discovery of neutral currents, was instrumental in stimulating the search for charm.[15] The GIM mechanism neatly solves, within the gauge theory framework, the problem of $\Delta S = 1$ neutral currents. But again it also appears to solve the problem in a more general context. In fact the GIM argument preceded the demonstration that the gauge-theories are renormalizable. We shall also give an example of a more general scheme for which the GIM mechanism applies.

(3) The intermediate boson hypothesis: This suggests introduction of gauge-bosons. It is also the case that at one time the prominence of the ρ- and ω-exchange contribution to the nucleon-nucleon force invited the hypothesis that they

be gauge particles. One may also obtain the V & A structure of the weak-interaction effective Lagrangian by assumptions of chirality conservation.

(4) University of strength of the weak couplings: This is also a predictable consequence of the gauge-theory approach. But similar conclusions have been reached from a phenomenological starting point, assuming only symmetry of the effective Lagrangian under permutation of various fermion degrees of freedom.[16]

Despite the caveats, I would agree that items 3 and 4 do represent objective arguments in favor of gauge theories, although not completely compelling ones. The really strong one is the last:

(5) $SU(2) \otimes U(1)$ agrees with experiment: To the extent that this is true, this not only argues for gauge-theories in general, but also for the specific model at hand - along with those variants which do not affect the predictions for neutrino-induced neutral currents. [Those variants include putting e_R^- into a doublet, as well as making the $SU(2)_L \otimes SU(2)_R \otimes U(1)$ or $SU(2) \otimes U(1) \otimes G$ extensions discussed[17] recently.]

I tried to assess how much objective evidence for the standard model is provided by existing experiments. A good starting point is the data for deep inelastic neutral currents, i. e., the ratios R and $\bar{R}$ of neutral-current to charged-current cross sections. The standard one-parameter model predicts R, given $\bar{R}$, and the predictions work very well. As argued above, even in alternatives to gauge theories we could reasonably expect R and $\bar{R}$ to be nonvanishing and of order 1. To crudely assess the significance of the success of $SU(2) \otimes U(1)$, assume that in a random model $R \leq 1$ and $\bar{R} \leq 1$. What fraction of this piece of R-$\bar{R}$ space does the $SU(2) \otimes U(1)$ prediction cover? Strictly a set of measure zero. But, if for no reason other than theoretical uncertainty, a certain percentage deviation of R and $\bar{R}$ from the prediction should be allowed. Here we take it as $\pm 15\%$. In Fig. 1 is plotted the range of values of R and $\bar{R}$ allowed for the SU(2)⊗U(1) model.[18] The fraction of the space which is acceptable is only about 8%. This is fairly significant. But if one allows other one-parameter gauge theories,[18] then the space

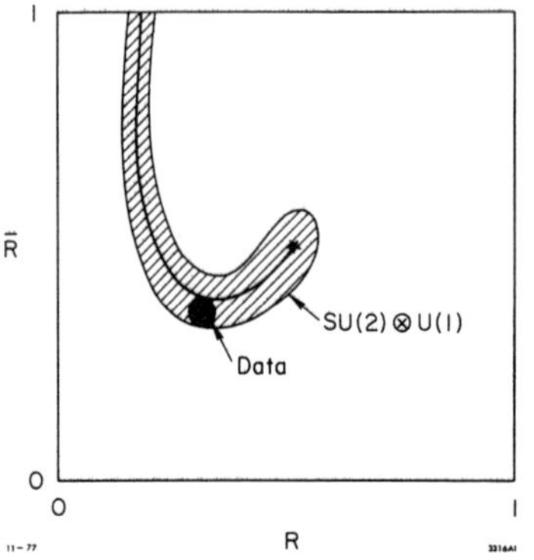

Fig. 1. Region of R-$\bar{R}$ space allowed by the standard SU(2)⊗U(1) model:
$$R = \frac{\sigma(\nu_\mu N \to \nu_\mu \text{ hadrons})}{\sigma(\nu_\mu N \to \mu^- \text{ hadrons})} \quad \text{and} \quad \bar{R} = \frac{\sigma(\bar{\nu}_\mu N \to \bar{\nu}_\mu \text{ hadrons})}{\sigma(\bar{\nu}_\mu N \to \mu^+ \text{ hadrons})}.$$
(The curve is for the most naive quark model and is only schematic.)

quickly begins to fill up (Fig. 2). And just allowing m_Z/m_W to vary from its standard value by a factor κ, with $0.7 \leq \kappa \leq 1.4$, is sufficient (Fig. 3) to fill up half the space. Thus the quantitative success is as much a success of standard SU(2)⊗U(1) as it is a success of general gauge-theory ideology.

This success is further sharpened by the restrictions on the isospin and chiral structure of the neutral-current couplings. In Fig. 4 is shown an analysis of Hung and Sakurai,[19] based on studies by Sehgal[20] of the charge-ratio of leading pions produced in neutral-current processes.[21] The solutions in the first and third quadrants for the neutral-current coupling constants[22] $\epsilon_L(u)$ and $\epsilon_L(d)$ are unacceptable when one includes the information from elastic νp and $\bar{\nu} p$ measurements. The significance of agreement is better than 20%, so that the combined data argue for SU(2)⊗U(1) at better than the 2% level.

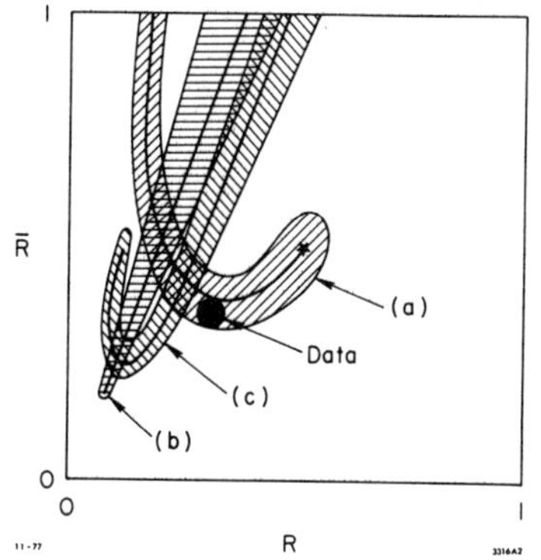

Fig. 2. Region of R-$\bar{R}$ space allowed by various models: (a) standard model, (b) vector-like model, and (c) 5-quark model.

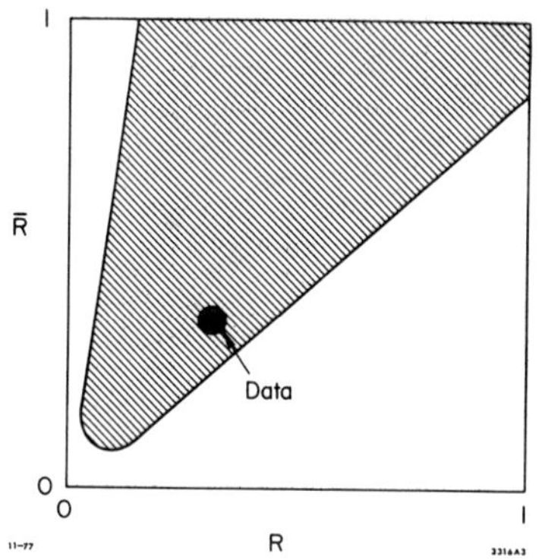

Fig. 3. Region of R-$\bar{R}$ space allowed in the standard model if one varies $\kappa = (m_Z/m_W)\cos\theta_W$ from ~ 0.7 to ~ 1.4.

Sehgal's analysis depends on both the parton model and the assumption that Gargamelle energies are high enough to use that model. So one should still be very careful. But let us cast such doubts to the wind and accept that the agreement of data with the simplest SU(2) ⊗U(1) model is, by any reasonable objective criterion, quite significant, and the chance that it is accidental is at the few percent level or less. This would seem to imply impeccable objective evidence for the weak-electromagnetic gauge theories.

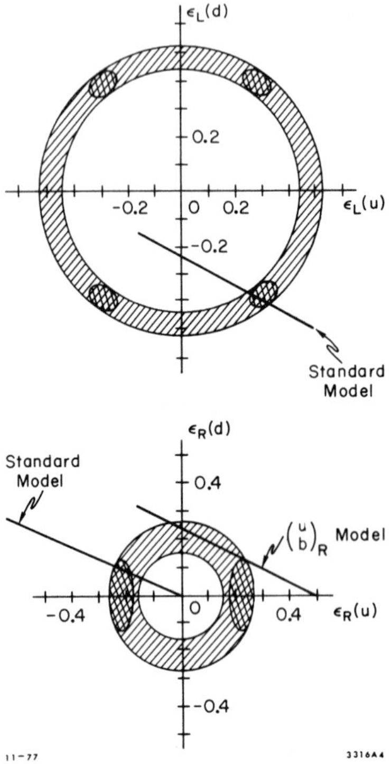

Fig. 4. Region of neutral-current coupling-constant space allowed by deep-inelastic data (hatched) and pion charge-ratio data (cross-hatched). $\epsilon_L(u)$ measures the coupling of ν_μ to a left-handed up-quark; the other ϵ's are defined similarly. The upper right and lower left quadrants in $\epsilon_L(u)$-$\epsilon_L(d)$ space are disfavored by elastic neutral current data.

But before drawing such a conclusion it is necessary to examine whether the Weinberg-Salam effective Lagrangian may be obtained in alternative ways. We now turn to that question.

IV. Alternatives to Weak-Electromagnetic Gauge Theories

Let us accept that the Weinberg-Salam effective Lagrangian for neutrino-induced neutral currents is indeed correct. We shall now argue that it can be obtained in a more general context. To motivate this, we revise history and suppose that both charm and strangeness (including the GIM mechanism) were discovered earliest, that $\Delta S = 0$ neutral-current processes were measured later on, and that the development of renormalizable gauge-theories came latest.

In the first stage, the charged-current effective Lagrangian (with Cabibbo-mixing) would have been determined. And a large number of theorists would find it irresistible not to "complete" the Lagrangian[11] by adding neutral currents in such a way as to give the weak interaction a global SU(2) symmetry:

$$-\mathscr{L}_{\text{eff}} = \frac{G}{\sqrt{2}} \vec{J}_\lambda \cdot \vec{J}^\lambda \tag{4.1}$$

where

$$\vec{J}_\lambda = \sum_i \bar{\psi}_i \gamma_\lambda \left(\frac{1-\gamma_5}{2} \right) \vec{\tau} \psi_i \tag{4.2}$$

with, as usual, the weak doublets

$$\psi_i = \begin{pmatrix} u \\ d\cos\theta_c + s\sin\theta_c \end{pmatrix}, \begin{pmatrix} c \\ s\cos\theta_c - d\sin\theta_c \end{pmatrix}, \begin{pmatrix} \nu_e \\ e^- \end{pmatrix}, \begin{pmatrix} \nu_\mu \\ \mu^- \end{pmatrix}, \cdots \tag{4.3}$$

The GIM mechanism is operative, and the neutral current is strangeness conserving. It has the form

$$-\mathscr{L}_{\text{eff}}^{NC} = \sqrt{2}\, G\, \bar{\nu}_\mu \gamma^\lambda \left(\frac{1-\gamma_5}{2} \right) \nu_\mu J_\lambda^3 \tag{4.4}$$

with

$$J_\lambda^{\ 3} = \bar{u}\gamma_\lambda \left(\frac{1-\gamma_5}{2}\right) u - \bar{d}\gamma_\lambda \left(\frac{1-\gamma_5}{2}\right) d + \cdots \tag{4.5}$$

Skeptics, who might refuse to admit such neutral currents without any objective evidence for them, would still expect an effective neutral-current generated by photon-exchange.[23] The electrically neutral neutrino could have a charge-radius and couple to photon via a vertex function

$$e\Gamma_\lambda(q) = e\bar{\nu}_\mu \gamma_\lambda \left(\frac{1-\gamma_5}{2}\right) \nu_\mu \frac{q^2}{\Lambda^2} \tag{4.6}$$

This leads to a contact-interaction (similar to the low energy neutron-electron interaction) with charged particles of the form

$$-\mathscr{L}^{NC}_{eff} = \frac{-e^2}{\Lambda^2} \bar{\nu}_\mu \gamma_\lambda \left(\frac{1-\gamma_5}{2}\right) \nu_\mu J^\lambda_{em} + \cdots \tag{4.7}$$

with

$$J^\lambda_{(em)} = \frac{2}{3}\bar{u}\gamma^\lambda u - \frac{1}{3}\bar{d}\gamma^\lambda d + \cdots \tag{4.8}$$

What is the verdict of the neutral current experiments? It is simple: one just adds the two contributions, Eqs. (4.4) and (4.7), and the result is precisely the Weinberg-Salam Lagrangian. The single parameter $\sin\theta_W$ of the Weinberg-Salam Lagrangian is equivalent to the single parameter Λ^{-1} (neutrino charge-radius); the connection is

$$2\sqrt{2}\, G_F \sin^2\theta_W = \frac{e^2}{\Lambda^2} \tag{4.9}$$

or

$$|\Lambda| = \frac{53 \text{ GeV}}{\sin\theta_W} \lesssim 100 \text{ GeV} \tag{4.10}$$

So this is the alternative point of view: It is simple enough to be taught to children,[24] and it accounts for all the data on neutrino-induced neutral currents. The assumptions used include:

(a) Existence of a global weak SU(2) symmetry

(b) Existence of a "large" photon exchange contribution

(c) Absence of large electromagnetic corrections to the intrinsic weak couplings via proper diagrams as in Fig. 5. The reason for including this point is that, in this general context, the bare photon coupling might be much larger than e, even at TeV energy scales. (See Section 5.)

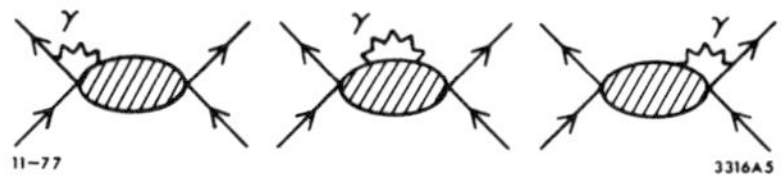

Fig. 5. Proper electromagnetic corrections to intrinsic weak amplitudes, which we neglect.

Assumptions <u>not</u> necessary to make include:

(a) Existence of a local gauge principle

(b) Renormalizability

(c) Intermediate boson hypothesis

(d) Unification of the weak and electromagnetic force

(e) Mass generation via symmetry breakdown of Higgs-bosons.

While I am sure this description must have occurred to others in the past it surprises me that it seems not to be generally known. The brilliance of the gauge theories has blinded us from seeing the simplest of alternatives.

Let us return to our revisionist version of history. The large[25] value for the neutrino charge radius Λ^{-1} (or alternatively $\sin\theta_W$) already implies a "weak cut-off" $\lesssim$ 150 GeV, as we shall describe in the next section. Thus if one assumes existence of $W^{\pm}$, its coupling constant to fermions is necessarily comparable to e. Thus strong motivation for weak-electromagnetic unification would exist, even in the absence of the modern formalism. Development of the modern theory would in turn take us to the present situation, with no doubt the same great enthusiasm for the gauge theories. After all, they provide a comprehensive and calculable framework encompassing all the experimental results and predicting many more. And that is something the simple description does not do.

However, given such a historical background, it seems to me that a critic in search of objective evidence would demand a crucial test to distinguish the gauge theories from other possible options. The simple general description of weak interactions in terms of a global SU(2) symmetry plus an electromagnetic contribution has its own credibility even in the absence of an explicit, fully calculable theory to support it. Nor is it in contradiction with the $SU(2)\otimes U(1)$ gauge theory (or its variants), as we shall elaborate in the next section. The gauge theories are a specific realization of the general picture. But I know no argument why they should be the only possible realization.[26]

The fact that $\sin^2\theta_W$ is positive[27] and of order unity can be considered as objective evidence for the gauge-theories. Beyond that, the crucial tests lie in the future: existence of W and Z with the predicted masses and couplings, evidence for the existence of the Higgs-sector, and (ultimately) evidence that at center-of-mass energies in excess of m_W and m_Z, the weak cross sections are as small as predicted by the renormalizable theories.

V. Additional Consequences

The alternative viewpoint we have sketched does have a few quite strong and specific consequences. One is that the weak cutoff, defined in a way appropriate for charged-current neutrino reactions, must be less than 37 GeV/$\sin^2\theta_W$.

Another is that the yield of neutral weak quanta from e^+e^- colliding beams must be large, as large as in the standard theory[28,29] (where it is concentrated in the Z^0 resonance).

To obtain these results, we only assume that the weak force is generated by some kind of weak quanta (with total J = 1) which are exchanged in the t-channel. The weak quanta may be one or more discrete intermediate bosons, or a more complicated continuum, or both. The weak amplitudes become nontrivial functions of t, analytic in the cut t-plane. The various options are illustrated in Fig. 6. For the standard SU(2)⊗U(1) model the amplitude is dominated by a single pole, with contributions from cuts suppressed by a factor of α. An amplitude with

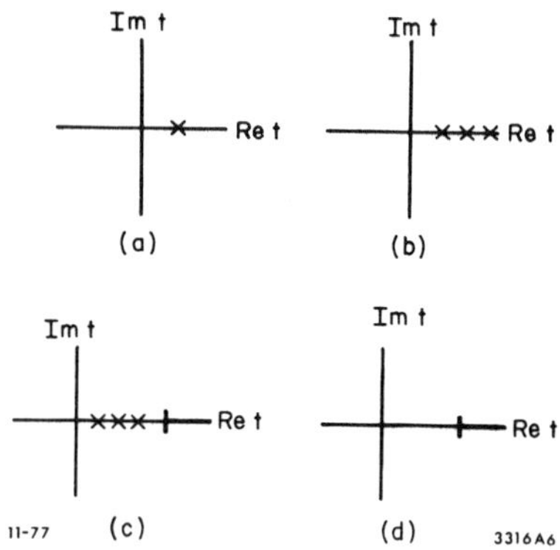

Fig. 6. Options for the analytic properties of the intrinsic weak t-channel amplitude:
(a) Single pole; negligible cut. This is the case of the SU(2)⊗U(1) gauge theory.
(b) Several poles; negligible cut. This could occur for complicated gauge theories.
(c) Some poles + non-negligible cut. This could occur were W's composites of something else, or if they were to interact strongly with each other.
(d) Only cuts. If the weak amplitude has important s and u channel structure, then there could also be a left-hand cut as well. We here disregard this possibility.

poles and cuts of comparable strength could arise were the intermediate bosons composites of something else,[30] or were they to interact strongly with each other.[31] Amplitudes with only cuts might arise were there s, t, and u-channel contributions.[32]

The amplitudes we need consider are illustrated in Fig. 7. Figure 7a shows the intrinsic weak interaction $T_W(t)$. In Fig. 7b, the neutrino charge form factor T(t) is a weak-electromagnetic interference term. In addition, it is important to include the contribution $T_{em}(t)$ to the electromagnetic vacuum-polarization coming from the photonic coupling of neutral weak quanta. The absorptive parts ρ_W, ρ, and ρ_{em} of these amplitudes must satisfy a Schwartz-inequality

$$\rho_W(t)\,\rho_{em}(t) \geq \rho^2(t) \tag{4.11}$$

and thus the electromagnetic contribution is bounded from below. Considerable manipulation of the above inequality leads to a lower bound on the production of weak quanta in e^+e^- colliding beams[34]

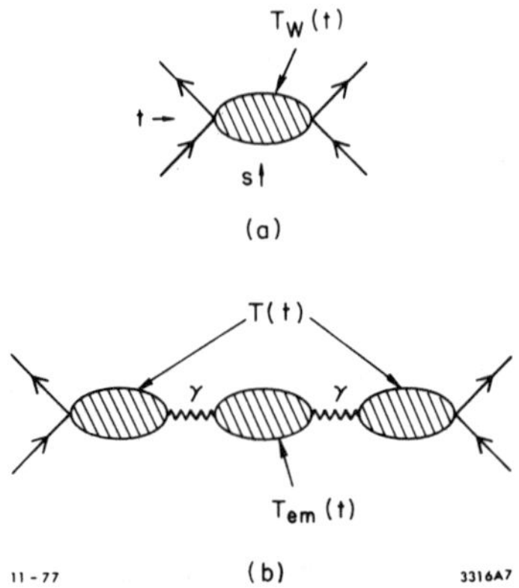

Fig. 7. Schematic of contributions to the neutral current amplitude: (a) Intrinsic weak interaction, and (b) photon exchange, including proper-vertex insertions and photon vacuum polarization contributions.

$$\overline{R} = \int \frac{ds}{s} R(s) \geq \frac{3\pi}{2\alpha} \left\{ \left[1 - \left(\frac{37\ \mathrm{GeV}}{m_W \sin^2\theta_W} \right)^{-2} \right]^{-1} - 1 \right\} \left\{ 1 + \left(\frac{1}{2\sin^2\theta_W} - 1 \right)^2 \right\} \quad (4.12)$$

where

$$R(s) = \frac{\sigma(e^+e^- \to \text{weak quanta})}{4\alpha^2/3\pi s} \quad (4.13)$$

The parameter m_W is not necessarily an intermediate-boson mass, but is what charged-current neutrino experimentalists call an intermediate-boson mass. It is defined in terms of the t-dependence of the Fermi-coupling:

$$G_F(t) = G_F \left(1 + \frac{t}{m_W^2} + \ldots \right) \quad (4.14)$$

or

$$\left.\frac{d\sigma}{dQ^2 d\nu}\right|_{\substack{\text{charged} \\ \text{current}}} = \left.\frac{d\sigma}{dQ^2 d\nu}\right|_{\text{"scaling"}} \cdot \left(1 - \frac{Q^2}{m_W^2} + \ldots \right)^2 \quad (4.15)$$

Eq. (4.12) is plotted in Fig. 8. One sees that the bound

$$m_W < \frac{37\ \mathrm{GeV}}{\sin^2\theta_W} \quad (4.16)$$

is associated with $\overline{R} \to \infty$, i.e., with the electromagnetic charge renormalization $Z_3^{-1} \to \infty$.

The threshold for electromagnetic production of weak quanta need not be the same as m_W (unless the T(t) have no poles). What can be said in general is that the threshold "m_Z" is bounded above as follows:

$$\text{"}m_Z\text{"} \leq \frac{1.5\ \mathrm{GeV}}{\sin^2\theta_W} \sqrt{\overline{R}} \quad (4.17)$$

In other words, it's worth the waiting: For larger "m_Z" one has a larger yield $\overline{R}$.

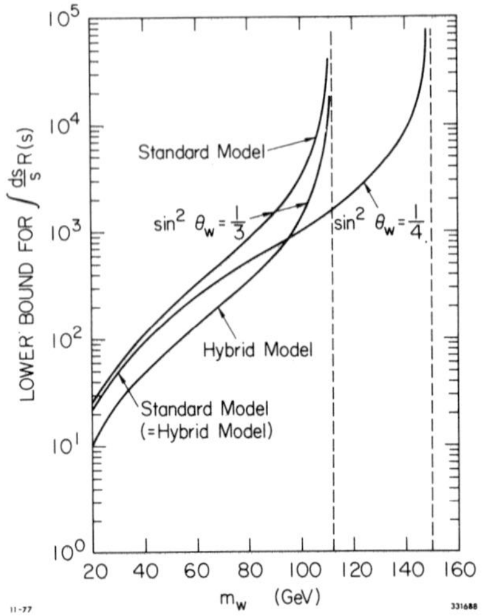

Fig. 8. Lower bound for $\bar{R} = \int \frac{ds}{s} R$, which measures the production of weak quanta by colliding e^+e^- beams. The parameter m_W is defined in Eqs. (4.14) and (4.15); it is the m_W measured in deep-inelastic charged-current neutrino reactions.

It is also enlightening - and simple - to reconstruct the predictions of the standard model. This is accomplished by assuming single-pole dominance for the amplitudes T_W, T, and T_{em}. Let g be the coupling of W's to fermions and $e_0 f$ the direct coupling of W_3 bare photon. The effective neutral current amplitude $\mathcal{M}$ is (including a charge-radius term for the quarks or other fermions):

$$\mathcal{M} = \bar{\nu}_\mu \gamma_\lambda \left(\frac{1 - \gamma_5}{2}\right) \nu_\mu \left\{ \frac{g^2}{(m_W^2 - t)} J_3^\lambda \right. \tag{4.18}$$

$$\left. + \frac{gft}{m_W^2 (m_W^2 - t)} \cdot \frac{(-e_o^2)}{t\left[1 + \frac{e_o^2 f^2}{m_W^2 (m_W^2 - t)}\right]} \left[J_{em}^\lambda + \frac{gft}{m_W^2 (m_W^2 - t)} J_3^\lambda\right] \right\}$$

Identification with the Weinberg-Salam Lagrangian, as $t \to 0$, gives two conditions:

$$\sqrt{2}\, G = \frac{g^2}{m_W^2}$$

$$2\sqrt{2}\, G \sin^2\theta_W = \frac{gfe^2}{m_W^4} \tag{4.19}$$

Definition of the physical charge e and the mass of the Z^o (the latter from the pole in the photon propagator) gives two more:

$$\frac{1}{e^2} - \frac{1}{e_o^2} \equiv \frac{1-Z_3}{e^2} = \frac{f^2}{m_W^4}$$

$$m_Z^2 = m_W^2 + \frac{e_o^2 f^2}{m_W^2} \tag{4.20}$$

Elimination of f and g then gives two equations for m_W and m_Z

$$m_W^2 = \frac{\pi\alpha(1-Z_3)}{2G\sin^4\theta_W} = \left(\frac{37\ \text{GeV}}{\sin^2\theta_W}\right)^2 (1-Z_3)$$

$$m_Z^2 = m_W^2\, Z_3^{-1} \tag{4.21}$$

With only this information we cannot determine m_W and m_Z. What is missing? It is a statement of unification: At short distances ($t \to \infty$), the photon couples only to the hypercharge, not to T_3. If this is the case, the SU(2) symmetry is restored at short distances and the photon (as we have defined it) is transmogrified into the hypercharge generator of unbroken SU(2)$\otimes$U(1). The implication for the amplitude $\mathcal{M}$ is that the proper vertex of the quark or other fermion, as $t \to \infty$, should not depend on J_3^λ. Letting $J_\lambda^{em} = \frac{1}{2} J_\lambda^3 + Y_\lambda$, with Y_λ a hypercharge current which is SU(2) singlet, we find from Eq. (4.18) that we must have

$$\frac{gft}{m_W^2 (m_W^2 - t)} \rightarrow -\frac{1}{2} \qquad (t \rightarrow \infty) \tag{4.22}$$

or

$$2 \frac{gf}{m_W^2} = 1 \tag{4.23}$$

This leads via Eq. (4.19) to

$$m_W = \left(\frac{37 \text{ GeV}}{\sin \theta_W}\right)^2 \tag{4.24}$$

and

$$m_Z^2 = m_W^2 \sec^2 \theta_W \tag{4.25}$$

What do we see from this exercise? It is that the issue of weak-electromagnetic unification is connected with the as-yet-untested short-distance behavior of the theory. Furthermore the predicted W and Z masses (especially m_Z) appear to be sensitive to this hypothesis of unification.

We also see that the simplest $SU(2) \otimes U(1)$ gauge theory fits naturally within the general picture we have discussed. This is in fact true for the known generalizations which preserve the Weinberg-Salam effective Lagrangian for neutrino-induced neutral currents, such as the ambidextrous $SU(2) \otimes SU(2) \otimes U(1)$ models. The most general such version, based on $SU(2) \otimes U(1) \otimes G$, has been discussed by Georgi and Weinberg.[35] It turns out that the neutral-current amplitude in their model breaks up quite naturally into the two pieces (intrinsic SU(2)-invariant weka amplitude and photon-exchange contribution) we discussed in the general context.[36] We shall not go into this in detail.[37] However, Georgi and Weinberg also showed that at least one neutral gauge boson can be no more massive than the standard Z^o: $m_Z \leq m_W \sec\theta_W = (37 \text{ GeV}) \csc\theta_W \sec\theta_W$. With the single-pole approximation for the intrinsic weak amplitude, along with the "unification condition," Eq. (4.22), their result can, not surprisingly, be recovered. However, thus far my attempts to derive this important result from a more general starting point have not been successful, even upon assuming an asymptotic SU(2) symmetry at short distances.

VI. Conclusions

Quantum chromodynamics and the weak-electromagnetic gauge theories provide by far the most profound and promising description of particle interactions we have. Nevertheless, there is as yet not much objective evidence in support of quantum chromodynamics. For the weak interactions, the gauge theories have played a crucial role in guiding weak-interaction experimentation and indeed in correctly predicting many results for neutral-current experiments. However, we have seen that these results are obtainable in terms of a simple and credible framework which, while more phenomenological, is more general than that of the gauge theories. Therefore in assessing the objective evidence for weak-electromagnetic gauge theories I think one should examine those features of the theory which go beyond the level of low-energy phenomenological four-fermion interaction. These include existence of intermediate bosons, weak-electromagnetic unification, renormalizability, and spontaneous symmetry breakdown via the Higgs mechanism. The universal current-current structure of the low-energy effective Lagrangian does provide objective evidence for intermediate bosons. And the large value of $\sin^2\theta_W$ and the consequent low value for "weak cutoff" ($\lesssim$ 150 GeV) provides some objective evidence for weak-electromagnetic unification. But beyond this I find little if any objective evidence for unified, renormalizable gauge theories of weak and electromagnetic interactions.

REFERENCES

1. H. D. Politzer, these proceedings and references cited therein.
2. S. Weinberg, Phys. Rev. D11, 3583 (1975).
3. G. 't Hooft, Phys. Rev. Letters 37, 8 (1976).
4. For example, c.f. L. Susskind, Proceedings of the 14th International Conference on Lepton and Photon Interactions at High Energies, Hamburg (DESY, 1977).
5. For example c.f. K. Gottfried, ibid.
6. K. Wilson, Phys. Rev. D10, 2445 (1974).
7. A. Casher, J. Kogut, and L. Susskind, Phys. Rev. Letters 31, 792 (1973). See also J. Bjorken, Lecture Notes in Physics 56, "Current Induced Reaction," ed. J. Korner, G. Kramer, and D. Schildknecht, Springer-Verlag, Berlin, 1975.
8. H. Lipkin, Phys. Rev. Letters 28, 63 (1972).
9. S. Rajesekharan and P. Roy, Pramana 5, 303 (1975) J. Pati and A. Salam, Phys. Rev. Letters 36, 11 (1976).
10. S. Glashow and S. Weinberg, Phys. Rev. D15, 1958 (1977).
11. S. Bludman, Nuovo Cimento 9, 433 (1958).
12. S. Glashow, J. Iliopoulos, and L. Maiani, Phys. Rev. D2, 1285 (1970).
13. G. 't Hooft, Nucl. Phys. B33, 173 (1971); B35, 167 (1971).
14. S. Weinberg, Phys. Rev. Letters 19, 1264 (1967) A. Salam, "Elementary Particle Theory," ed. N. Svartholm, (Almquist and Wiksell, Stockholm, 1969), p. 367.
15. M. Gaillard, B. Lee, and J. Rosner, Revs. Mod. Phys. 47, 277 (1975).
16. T. Appelquist, J. Bjorken, and M. Chanowitz, Phys. Rev. D7, 2225 (1973).
17. See S. Glashow, these proceedings, for an excellent summary.

18. I have taken these from M. Barnett, Phys. Rev. D13, 671 (1976). The plots are only schematic and not meant to be a serious analysis of present data.

19. P. Hung and J. Sakurai, preprint UCLA/77/TEP/17, Oct. 1977.

20. L. Sehgal, Aachen preprint PITHA-1977-81.

21. K. Kluttig, J. Morfin, and W. VanDoninck, Aachen preprint PITHA-1977-83.

22. The definitions are according to my own preference; c.f. J. Bjorken, "Proceedings of the SLAC Summer Institute on Particle Physics," ed. M. Zipf, SLAC report SLAC-198, Nov. 1976.

23. J. Bernstein and T. D. Lee, Phys. Rev. Letters 11, 512 (1963).

24. A great deal of the gauge theories can also be taught to children. However, I have found that most children are frightened by the Faddeev-Popov ghosts.

25. Actually it is only large compared to the estimate in Ref. 23 provided the anomalous moment κ of the W is set to zero. For nonvanishing κ, their estimate is comparable to what is needed experimentally, but cannot be accurately calculated.

26. There does exist a proof [J. Cornwall, D. Levin, and G. Tiktopoulos, Phys. Rev. D10, 1145 (1974) and references therein to earlier work] that tree-diagram unitarity essentially implies the gauge theories. But if there exist large contributions from unitarity-cuts in weak amplitudes, those arguments do not apply.

27. The general neutrino-charge radius argument does not determine the sign of Λ^2; on the other hand, the gauge theories predict it.

28. B. Richter, Nucl. Insts. and Methods 136, 47 (1976).

29. B. Ioffe and V. Khose, Leningrad preprint LNPI-274.

30. H. Terazawa, K. Akama, and Y. Chikashize, Prog. Theor. Phys. 56, 1935 (1976).

31. G. Feinberg, Phys. Rev. B134, 1295 (1964).

32. This could include schemes along the lines investigated by W. Kummer and G. Segre [Nucl. Phys. 64, 585 (1965)] and by N. Christ [Phys. Rev. 176, 2086 (1968)]; however, we need not insist on renormalizability.

33. J. Bjorken (in preparation).

34. We here assign the right-handed electron as a weak-SU(2) singlet. For the "hybrid" assignment of e_R^- to a weak doublet,[16] the curly bracket on the right-hand side of Eq. (4.12) is replaced by $2\left(\frac{1}{2}\csc^2\theta_W - 1\right)^2$. These formulas are only valid for $\sin^2\theta_W \leq 1/2$.

35. H. Georgi and S. Weinberg, Harvard preprint HUTP-77/A052, Aug. 1977.

36. This was taught to me by Marvin Weinstein, to whom I express my gratitude. It provided important motivation for the entire study.

37. This is carried out in Ref. 33.

CONFERENCE SUMMARY

Steven Weinberg

Lyman Laboratory of Physics
Harvard University
Cambridge, Massachusetts 02138

We are all grateful to our hosts at Fermilab for their warm hospitality, and to the Organizing Committee for arranging a most exciting program. It is clear that this conference has covered far too much ground for any summary, so without attempting apologies, I had better just start.

The neutral current interactions about which we know the most experimentally are those acting between neutrinos (or antineutrinos) and nucleons. It was stated often here, and now seems definitely settled, that the data on these interactions is in excellent agreement with the original simple $SU(2)\otimes U(1)$ gauge theory, in which one makes what I will call the "standard" assignment of right-handed quarks and leptons to $SU(2)$ singlets. The experimental situation as of this summer is summarized here in Table I, and was brought up to date at this conference in the talks of Mann and Haidt. The most precise evidence continues to come from the inclusive deep-inelastic scattering of ν_μ and $\bar{\nu}_\mu$ on nuclei of medium weight, as shown in the first two lines of Table I, but the weight of the other evidence on exclusive [$\nu(\bar{\nu})N \to \nu(\bar{\nu})N$, $\nu(\bar{\nu})N \to \nu(\bar{\nu})\pi N$] and semi-inclusive [$\nu(\bar{\nu})N \to \nu(\bar{\nu})\pi X$] neutrino reactions is already quite substantial.

Purely leptonic neutral current reactions were reviewed here by Rousset. At a given lab energy, the cross sections for neutrino-electron scattering are three orders of magnitude less than for neutrino-nucleon scattering, so it is difficult to make precise measurements of the purely leptonic cross sections. The present data is shown in the last three lines of Table 1. Again, there is agreement with the SU(2) ⊗ U(1) standard model. However, this data requires the values of the $\bar{\nu}_\mu e^-$ couplings g_V, g_A to lie between various ellipses in the g_A-g_V plane, and these ellipses all have a symmetry under the reflection $g_A \leftrightarrow g_V$, so for each fit there is another one with g_A and g_V interchanged. Now the standard model with $\sin^2\theta \simeq 0.27$ fits the data, and this model gives $|g_V| << |g_A|$ (for $\sin^2\theta = 0.25$, $g_V = 0$) so there must be another fit with $|g_A| << |g_V|$. As it happens, this fit corresponds to a modified version of the standard SU(2) ⊗ U(1) model, known as the SU(2) ⊗ U(1) "mixed" or "hybrid" model. (I believe this was first noted by Fayet.) In this model everything is the same as in the standard model for quarks and neutrinos, and under certain conditions on scalar vacuum expectation values there need be no change in anything I said about neutrino-nucleon neutral currents. However, the electron here appears in both left- and right-handed doublets

$$\begin{pmatrix} \nu_e - \varepsilon E^0 \\ e^- \end{pmatrix}_L \qquad \begin{pmatrix} E^0 \\ e^- \end{pmatrix}_R \tag{1}$$

where E^0 is a heavy neutral lepton with mass m_e/ε; also, $(E^0 + \varepsilon\nu_e)_L$ is an SU(2) singlet. (Universality requires $\varepsilon \lesssim 0.07$. Mixing of muonic and electronic leptons is also possible in this model - I leave it out only for simplicity.) Since e^-_L and e^-_R enter symmetrically in the mixed model, the neutral current couplings of the electron in this model would be pure vector, with $g_A = 0$.

A possible method of deciding the assignment of e_R^- was discussed here by Faissner. In the mixed model the $\nu_\mu e^-$ and $\bar{\nu}_\mu e^-$ differential cross sections must be equal. However, the Aachen-Padua group reports that $\sigma(\nu_\mu e^- \to \nu_\mu e^-) = (1.2 \pm 0.6) \times 10^{-42}$ cm$^2 E_\nu$(GeV) and $\sigma(\bar{\nu}_\mu e^- \to \bar{\nu}_\mu e^-) = (2.4 \pm 1.1) \times 10^{-42}$ cm$^2 E_\nu$(GeV). (The Gargamelle results reported by Rousset are consistent with these, but are based on a smaller number of events.) Furthermore, the reported mean electron recoil energy is 0.7±0.17 GeV and 1.10±0.19 GeV for $\nu_\mu e^-$ and $\bar{\nu}_\mu e^-$ scattering, respectively. These results suggest though they do not yet definitely require a difference between $\nu_\mu e^-$ and $\bar{\nu}_\mu e^-$ scattering. Any such difference, if definitely established, would rule out the mixed model. However, if $\sin^2\theta$ is very close to 0.25, the difference is in any case expected to be small. Experimental results on νe and $\bar{\nu} e$ scattering will have to become much more precise before we can use them to draw definite conclusions about the assignment of e_R^- to gauge multiplets. Fortunately, this is a question which has little bearing on the predictions for neutrino-nucleon interactions discussed above.

The data on neutrino-nucleon and neutrino-electron neutral currents is already good enough to provide a valuable input to astrophysicists. The greatest impact here of our knowledge of neutral currents so far has been in the study of supernovae, discussed here by Schramm. The evolution of a massive star prior to supernova depends on neutrino emission processes such as $e^+e^- \to \nu_e\bar{\nu}_e$ and $e^+e^- \to \nu_\mu\bar{\nu}_\mu$, whose rates are substantially affected by neutral current interactions. Then, when the stellar core collapses, most of the energy comes out in the form of neutrinos, at a rate which depends on neutrino scattering processes in the core, and in particular on the neutral-current process of neutrino-nucleon elastic scattering. Finally, the emerging neutrinos interact with the layers of medium-weight elements outside the collapsing core, chiefly through the coherent scattering produced by the isoscalar part of the νN neutral current interaction. As emphasized here by Schramm, there are great uncertainties in the astrophysics itself (such as in the equation of state of the stellar core) which prevent us from reaching definite conclusions so far about whether a given collapsing star will be able to blow off its envelope by neutrino pressure or ordinary blast waves. For the present, it seems unlikely that the supernova calculations are going to feed back into elementary particle physics and tell us anything about neutral currents. I'm afraid that for the foreseeable future, the flow of information in this area will all be in the opposite direction, from particle physics to astrophysics.

One can also study neutral currents in the nucleon-nucleon interaction, as discussed here by Adelberger, Barnes, Nagle, Tadic, and Wilson. It is useful in thinking about this to begin with the old weak-interaction theory, with just charged currents. With a great deal of crossings of fingers, one can perhaps calculate the parity-violating s-wave nucleon-pion coupling from the measured hyperon decay amplitudes, using the octet-dominance formula for s-wave couplings:

$$S(p \to \pi^+ n) = \frac{\sin^2\theta_C}{\sqrt{6}} [4S(\Lambda \to p\pi^-) - 2S(\Xi^- \to n\pi^-)] \qquad (2)$$

a relation derived some years ago by McKellar and Fischbach. However, as Tadic mentioned, there may be large corrections here, perhaps as large as 300%! (Note that both sides vanish in the quark model, and it is not clear that the corrections to the quark model satisfy SU(3) and octet dominance as well as the quark model itself.) It is also not so easy to calculate the effect of a parity-violating pion exchange force on an observable like a γ-ray circular polarization. In addition, despite its long range, the pion exchange force is not the only source of parity violation in nuclear physics; there is also ρ exchange, ω exchange, 2π exchange, etc., etc. Nevertheless, I can imagine getting all this straightened out, and having a definite comparison between the old charged-current theory and experiment for the parity-violating pion exchange force. If theory and experiment do not agree, then one can conclude that there must be some other source of parity violation in nuclear physics, presumably the neutral currents. In my opinion, it is this that makes experiments on parity violation in nuclear physics of direct interest to particle physicists.

On the other hand, it does not seem likely to me that we are going to be able to calculate the nuclear parity violation that would be produced by the neutral currents in any particular gauge model. CP invariance requires the s-wave $\pi\bar{N}N$ coupling to be pure $\Delta I = 1$, so we are chiefly interested here in the $\Delta I = 1$, $\Delta S = 0$ effective weak Hamiltonian produced by $W^{\pm}$ or Z^0

exchange. In the SU(2) ⊗ U(1) gauge model with left-handed quark doublets, these take the form

$$\mathcal{H}^{W}_{\Delta I=1} = \frac{G_F}{2\sqrt{2}} \sin^2\theta_C [L_{84} O_{84} - L_{20} O_{20}] \quad (3)$$

$$\mathcal{H}^{Z}_{\Delta I=1} = \frac{-G_F}{2\sqrt{2}} \left[(1 - 2\sin^2\theta)[L_{84} O_{84} + L_{20} O_{20}] + \tfrac{2}{3}\sin^2\theta L_{15} O_{15}\right] \quad (4)$$

where the O_N are linear combinations of four-fermion operators which cannot be mixed by strong interaction effects; the L_N are corresponding c-number coefficients which would be unity in the absence of strong interactions; and θ is the usual Z^0-γ mixing angle. (Effects of quark masses are not included here; it would be useful to repeat this analysis taking them into account.) Hard-gluon effects tend to reduce L_{84}, so we may perhaps throw away the O_{84} term, as in the derivation of the $\Delta I = \frac{1}{2}$ rule by Gaillard and Ben Lee and by Altarelli and Maiani. The ratio of the O_{20} terms is then a known quantity, $(1-2\sin^2\theta)/\sin^2\theta_C \simeq 8$. However, this still leaves O_{15}. A calculation by Altarelli, Ellis, Maiani and Petronzio shows that L_{20} and L_{15} are both enhanced by hard-gluon effects (L_{15} somewhat more so) but soft-gluon effects make it impossible to calculate either coefficient in detail. And even if we could calculate L_{15}/L_{20}, we still would not know how to calculate the contribution of O_{15} to the $\pi\bar{N}N$ coupling. The only conclusion that can plausibly be reached here is that the neutral currents probably enhance the $\Delta I = 1$ parity violation in the SU(2) ⊗ U(1) gauge theory, because their effect in Eq. (4) is not suppressed by the factor $\sin^2\theta_C$ which appears in Eq. (3), but even so this conclusion could easily be invalidated by a cancellation between O_{20} and O_{15} terms in Eq. (4).

The comparison of the circular polarizations observed in γ transitions in the nuclei F^{19}, K^{41}, Lu^{175} and Ta^{181} with theoretical values has been carried out by Box *et al.* As reported by Tadic, the sign of the pion exchange term needed to fit the data differs from that expected with charged currents alone, and the absolute value is greater by a factor of order 10 - 20. This is roughly what one would expect if the neutral currents violate parity, as in the standard model. [A useful "control" here is provided by results on the parity-violating α decay $O^{16*}(2^-) \rightarrow C^{12}(0^+) + He^4(0^+)$, discussed by Adelberger. The parity violation here arises from mixing of two nuclear levels with $I = 0$, so would not be dramatically affected by an enhancement of the $\Delta I = 1$ parity-violating pion exchange force, which as we have seen is likely to be the chief manifestation of parity-violating neutral currents. The observed width $(1.03 \pm 0.28) \times 10^{-10}$ eV of the $\Delta I = 0$ O^{16} α decay is in reasonable agreement with the value $\approx 10^{-10}$ eV that would be produced by charged currents, in strong contrast with the situation for the parity violation in F^{19}, K^{41}, etc., which can have $\Delta I = 1$ as well as $\Delta I = 0$ terms.] If the observation of an enhanced $\Delta I = 1$ parity violation holds up in future experiments, it might be regarded as evidence for parity violating neutral currents, but as I have indicated, one can not really calculate what enhancement (if any) is to be expected in the $SU(2) \otimes U(1)$ gauge theory.

There is a number of experiments on nuclear parity violation now in the works. I don't have time to go over them all, but four stand out. Measurement of the circular polarization in Ne^{21} γ decay will provide an interesting counterpart to the F^{19} measurement. Measurements on np $\rightarrow$ dγ and pp scattering, discussed here by Wilson and Nagle, will provide direct information on the parity-violating nuclear forces, unobscured by our ignorance of nuclear wave functions. Finally, the parity violation in F^{18} γ decay arises from the mixing of an I = 0 and an I = 1 level, and hence provides the first case (aside from np $\rightarrow$ dγ) of a pure $\Delta I = 1$ parity violation. The calculations of Gari, McGrory, and Offerman indicate that the circular polarization here should be about 0.4×10^{-3} if only charged currents violate parity. The result of the CalTech, CalState, Washington group reported here by Barnes is that the circular polarization is $(-0.6 \pm 2.0) \times 10^{-3}$. Given the uncertainty in the theoretical estimates, this does not allow us to draw any definite conclusions.

So far, I have dealt with the neutral currents in νN, νe, and NN interactions. The remaining area of current experimental effort is the electron-nucleon neutral current interaction, reviewed here by Bouchiat, Feinberg, and Sandars. One looks for parity violating effects due to the exchange of a Z^0 between electrons and nucleus in an atom. As emphasized originally by Bouchiat and Bouchiat, such effects should be enhanced in heavy atoms. This is in part because the interaction between an axial-vector electron current and a vector nucleon current is coherent, giving a parity violation in the standard SU(2) $\otimes$ U(1) model proportional to $Z(2-4\sin^2\theta)-A$. The most significant experiments so far are those which measure the rotation of the plane of polarization of a laser beam passing through bismuth vapor, which can be expressed in terms of the quantity $R \equiv \mathrm{Im}(E1/M1)$ for the relevant atomic transition. As reported here by Sandars, the Oxford group finds $R = (2.7 \pm 4.7) \times 10^{-8}$ at 6476 Å while the Seattle group finds $R = (-0.7 \pm 3.2) \times 10^{-8}$ at 8757 Å. This is very much below the values $R = -25 \times 10^{-8}$ (6476 Å) and $R = -19 \times 10^{-8}$ (8757 Å) originally calculated on the basis of relativistic central field calculations in the $SU(2) \otimes U(1)$ standard model (taking $\sin^2\theta = 0.27$) although perfectly consistent with the $SU(2) \otimes U(1)$ mixed model, where the electron neutral current is pure vector and there is no nuclear coherence.

It has been quite a puzzle what to make of this. To reach conclusions about the weak interactions, one would have to be convinced not only that the measurements are correct, but also that we know how to calculate the value of R predicted by a gauge model in a complicated atom like bismuth. Configuration mixing does not seem to be a problem; both Feinberg and Sandars reported on a recent calculation by Henley, Klapisch, and Wilets, which indicates corrections of at most a few percent. However, it has recently been found that there are large effects involving excitation of the core electrons. Loving and Sandars have calculated these effects to first order in the difference between the true and the effective field Hamiltonian, and find corrections which essentially cancel the leading term in R! They then go on to calculate a certain subset of graphs to all orders, summing the series by solving a differential equation, and find that the parity violation is partially restored, with a new predicted value of $R = -14 \times 10^{-8}$ at 6476 Å. However, now that we know that the original result is subject to these large corrections, we naturally would like to be assured that there are no more large corrections in the larger number of high-order diagrams which are not included in this calculation. I think it has been clear from the tone of the discussion here that most people are reserving judgment on the conclusions to be drawn from the bismuth experiments. As Sandars remarked, more theoretical and experimental work is needed to resolve these problems.

One way that this uncertainty may be eliminated is by experiments on other atoms. These include thallium (experiments at the University of Washington and at Berkeley, reported here by Fortson and Commins), cesium (experiment at ENS, reported by Mme. Bouchiat), and hydrogen and deuterium (experiments at Michigan, University of Washington, and Yale). I list these here in the chronological order in which precise results will probably become available; nature being what it is, this is also the inverse order of ease in understanding the results theoretically. Easiest of all to understand theoretically are of course the experiments in atomic hydrogen and deuterium. I emphasize deuterium here, because by a perversity that doesn't seem to have anything profound about it, the standard model for $\sin^2\theta = 0.25$ gives the electron and proton purely axial-vector couplings to the Z^0, so the parity violation expected in hydrogen is very small, while the neutron has vector as well as axial-vector coupling to the Z^0, and a larger parity violation is expected in deuterium in the standard model. Unfortunately the hydrogen and deuterium experiments are very difficult, because unlike the bismuth experiments, one does not have the help here of large Z^3 and nuclear coherence factors, so that it is not clear that it will be possible to attain the necessary sensitivity.

A quite different experiment on the eN neutral current interaction was mentioned in the discussion session by Hughes. A SLAC-Yale collaboration is studying the deep inelastic scattering of longitudinally polarized electrons on various nuclei at SLAC. The only theoretical uncertainty here is the use of the quark model. As an example of what may be expected, I will quote the helicity dependence of the deep inelastic cross section for electrons on nucleons in the standard model, as taken from a recent paper by Cahn and Gilman:

$$\frac{d\sigma_R - d\sigma_L}{d\sigma_R + d\sigma_L} = -\frac{9G_F q^2}{20\sqrt{2}\pi\alpha}\left\{1 - \frac{20}{9}\sin^2\theta + \text{small y-dependent terms}\right\} . \tag{5}$$

For $q^2 \approx 1\ \text{GeV}^2$, this is in the range of 10^{-5} to 10^{-4}, in comparison with a present upper bound of a few times 10^{-4}. It will be very exciting if these experiments can get down to a sensitivity of 10^{-5}, or even lower.

The bismuth experiments have opened up for us the question, without yet settling it, whether the neutral currents do or do not violate parity conservation. I want to emphasize a well-known but important point about this. If there is only one neutral intermediate vector boson Z^0, as in SU(2) ⊗ U(1) models, then the difference between neutrino-nucleon and antineutrino-nucleon interactions shows immediately that the neutral currents violate parity, because the Z^0 would have to have both vector and axial-vector couplings. However, with more than one Z^0, it would be possible for each to be coupled purely as a vector or as an axial vector, so that parity could be conserved in neutral current interactions. We know that SU(2) ⊗ U(1) must be part of the full weak and electromagnetic gauge group, and our preference for simplicity in elementary particle theories leads us to hope that the standard SU(2) ⊗ U(1) model is the whole story, but it is prudent at least to consider the possibility that SU(2) ⊗ U(1) may be embedded in a larger gauge group of weak and electromagnetic interactions, with more than one Z^0.

If one is going to consider an expanded weak and electromagnetic gauge group, then the question must be considered, why the neutral current interactions of neutrinos with nucleons are so accurately and so comprehensively given by SU(2) ⊗ U(1). A useful point along this line was made by Fritzsch and Minkowski, and Pati, Rajpoot, and Salam, and generalized subsequently by Georgi and myself. There is a certain class of expanded gauge models in which the neutral current interactions of neutrinos are precisely the same as in the usual SU(2) ⊗ U(1) model, although there is more than one Z^0, so that a wide range of results is possible for neutral current interactions of electrons. These models are based on gauge groups of the form SU(2) ⊗ U(1) ⊗ G, with G arbitrary, and with certain limitations on how the scalar fields transform under SU(2) and G. At least within this context, there is a kind of stability to the predictions of the simple SU(2) ⊗ U(1) model for the neutral current interactions of neutrinos, due to the fact that the neutrino is purely left-handed and has zero charge. Thus, however the eN neutral current experiments turn out, we can expect that the predictions of the SU(2) ⊗ U(1) gauge model for the neutral current interactions of neutrinos with nucleons will be a permanent feature of any future theory of weak interactions.

This conclusion was accepted, and taken as a starting point, in Bjorken's talk. As he says, the gauge theory of weak and electromagnetic interactions not only played a crucial role in stimulating the discovery of neutral currents, but also has proved remarkably successful in correctly predicting the details of neutral current neutrino interactions. At past conferences, Bjorken has stressed the need for numerical agreement with accurate experiments to provide support for gauge theories; and that we certainly now have. At this conference he has raised a different issue, whether any other theory could be developed which would yield the same predictions for neutrino interactions as the gauge theory. As he made clear, his comments on what he called the "objective" evidence for the gauge theories of weak interactions were intended not as a reassessment of the agreement between theory and experiment, but only as a statement of his opinions about the possibility of developing such other theories.

No specific theory was proposed by Bjorken; indeed, as far as we know, there does not exist *any* theory with all the properties that he finds would have to be assumed. In brief, he finds that it would be necessary to suppose the existence of a mysteriously large photon exchange associated with a huge neutrino form factor $F(q^2)$ [a reasonable estimate of the quantity $F(q^2)/q^2$ at $q^2 = 0$, as given for example in the 1963 paper of Bernstein and Lee, would be of order 10^{-7} to 10^{-8} GeV^{-2}, while Bjorken finds that a value greater than 10^{-4} GeV^{-2} would be needed]; the existence in some sense of an unbroken SU(2) symmetry [despite the fact that this symmetry would be manifestly violated by quark and lepton masses and by the photon exchange itself]; and the absence of large photon exchange effects where they might interfere with universality. In effect, the large photon exchange term would simply have to be inserted in order to simulate some results of the $SU(2) \otimes U(1)$ gauge theory, and then ignored wherever it would raise troubles. He finds that it would be necessary also to impose a relatively low ultraviolet cut-off of unknown origin, and one would have to assume that the weak currents have the usual properties: V and A form, CVC, PCAC, no derivatives, etc. In contrast, such an effective cut-off and such weak current properties are *consequences* rather than assumptions of renormalizable gauge theories. Bjorken's talk has been helpful, as a fresh demonstration of the validity of the prevailing conclusion, that only the renormalizable gauge theories can furnish a satisfactory framework for the theory of weak interactions.

This leads me to some general remarks about features of gauge theories, which further explain this conclusion. First, about cut-offs. Something must cut off the integrals for higher order weak interactions, and the observed universality of weak couplings gives empirical evidence for an effective ultraviolet cut-off at virtual momenta well below $G_F^{-1/2}$. (In semi-leptonic processes contributions of higher order in the weak interactions are affected by sizable corrections arising from gluon exchange between internal and external quark lines, which do not occur for purely leptonic processes. A cut-off well below order $G_F^{-1/2}$ is needed to suppress these "non-universal" higher order graphs sufficiently.) Now, it is well known that such an effective cut-off is automatically provided in renormalizable unified gauge theories. The integrals in a renormalizable theory begin to converge at momenta of the order of the largest masses in the theory, and m_W and m_Z are predicted in unified gauge theories to be of order $eG_F^{-1/2}$. (As Telegdi remarked, Sirlin's gauge-theoretic calculations show that this effective cut-off is essential in giving the observed values for radiative corrections to universality relations.) The converse also holds; any attempt to arrange for a cut-off would be expected to lead inexorably to a renormalizable gauge theory. Because this has already been tried.

Five years ago a number of people (Cornwall, Levin, and Tiktopoulos; Llewellyn-Smith; Joklegar; Quinn) tried the idea of calculating weak scattering amplitudes in terms of a large number of phenomenological couplings, and then demanding that the amplitudes fall off rapidly enough at high energy to provide a cut-off for higher order effects of weak interactions. (Because trees are simpler than loops, they considered only contributions of pole and not cut singularities, but this seems a technicality.) They found that this demand imposed a large number of algebraic constraints on the coupling constants and masses. Taken together, these constraints required the interactions systematically to take just the form, with just the relations among coupling constants and masses, which would dictate that the theory *is* a renormalizable gauge theory.

Although this may be a minor point, I think that it also might be appropriate to comment here on the relations between various gauge theories and the J/ψ discovery. These 1974 experiments were a turning point, because they unexpectedly provided us with a "hydrogen atom" for exploring the strong interactions. So far, QCD seems to work well. As to weak interactions, it seems to me that the J/ψ discovery should be included, logically as well as historically, among the corroborations of the unified gauge theory. Expectations of a fourth flavor were largely based on the observed absence of strangeness - changing weak interactions of the sort that would be produced either by Z^0 exchange or by higher-order exchange of a W^+W^- pair, and on the need to cancel Adler-Bell-Jackiw anomalies. All these arguments rest logically on the gauge theory. Otherwise there would be no necessity for a Z^0, and no way of predicting its properties. (The real inhibition for theorists against incorporating neutral currents into weak interaction theories was that before the advent in 1967-68 of the unified gauge theory, there was no need for neutral currents.) Further, without a renormalizable theory one could not tell what specific conclusions to draw from W^+W^- exchange [for instance, divergent integrals would have to be cut off at some unknown momentum; a low cut-off would by itself suppress W^+W^- exchange], and there would be no need to cancel ABJ anomalies. For these reasons, there was little interest in a conjectured fourth flavor until 't Hooft and Ben Lee proved the renormalizability of the unified gauge theory in 1971. Indeed, without a renormalizable gauge theory one would have had no way of telling whether an assumed fourth quark would be at an experimentally accessible mass; the first calculation of this mass was carried out by Gaillard and Ben Lee in 1974, using the $SU(2) \otimes U(1)$ gauge theory; and their pre-J/ψ prediction of a quark mass about 1.5 GeV is among the successes of gauge theories. This is not an historical accident; it is only the gauge theories that give us confidence that we know how to deal with weak interactions beyond lowest order perturbation theory.

I would also like to comment more generally on the significance of renormalizability in particle physics. (This is discussed in more detail in an article in the forthcoming Cambridge University Press Einstein Centenary volume.) Renormalizable quantum field theories have a self-consistency and a simplicity which give them an *a priori* attractiveness as possible fundamental theories of elementary particle interactions. Also, the moderate growth of matrix elements with increasing energy in these theories ensures that they are free of unphysical singularities, at least up to exponentially large energies. Renormalizable gauge theories have thus provided a framework for a theory of weak and electromagnetic interactions which is physically and mathematically satisfactory, in the same sense as quantum electrodynamics. It may be that this theory can be extended to a renormalizable theory encompassing QCD and gravitation, perhaps by invoking some sort of supersymmetry. That would be very nice. It is also possible that because of the effects of gravitation on the topology of space-time, its description requires modifications in the fundamental structure of quantum field theory. But even if quantum gravity had to be described by a flat-space quantum field theory which is not renormalizable in the usual sense, calculations indicate that gravity could satisfy a generalized version of the renormalizability condition, which would insure the absence of unphysical singularities, and would lead to a description of the physics of ordinary particles, at energies and masses far below the Planck mass $G_{grav}^{-1/2} = 1.2 \times 10^{19}$ GeV, by a *renormalizable* effective field theory. All non-renormalizable interactions would be quite undetectable, except that the coherence and long range of gravitation allows us to observe its macroscopic effects. Thus, whatever we think of gravity, it makes sense to impose the constraint of renormalizability on our theories of the observable weak, electromagnetic, and strong interactions. And this constraint works. In addition to all the quantitative successes of renormalizable gauge theories discussed at this conference, it is only with a renormalizable theory of weak interactions that we can understand in detail why quantum electrodynamics gives the correct values for the g-2 values of e and μ. Furthermore, the counterpart for the weak interactions of the very well verified absence of a non-renormalizable intrinsic Pauli magnetic moment in QED (which would mess up predictions for g_e-2 and $g_\mu-2$) is the absence of second-class currents. It is a sign of the predictive power of renormalizable gauge theories that for the last few years, even when there appeared to be some experimental indications of second-class terms in weak currents, it was generally acknowledged by theorists that we were unable to fit such terms into any reasonable model which describes strong as well as weak and electromagnetic interactions by a renormalizable gauge theory. There is now quite convincing evidence, summarized here by Wu, that second-class currents are in fact absent.

Let's now return to parity nonconservation and neutral currents. We saw that the bismuth experiments raise the question whether parity may be conserved by the neutral currents in an expanded gauge theory. The only theories that I know of, in which the neutral currents can naturally conserve parity while the charged currents don't, are those in which parity is conserved in the Lagrangian, but is spontaneously broken. So this leads us to the question (and this is in a way the central question of this conference) whether parity is a fundamental symmetry of the Lagrangian.

Within the context of the extensions of $SU(2) \otimes U(1)$ discussed above, the condition that the Lagrangian conserve parity would lead directly to a unique gauge group, $SU(2) \otimes U(1) \otimes SU(2)$, with the two SU(2)'s acting independently on left- and right-handed quarks and (maybe) leptons. Work on this kind of model was reviewed here in the talks by Glashow and Pati. If it turns out that parity is conserved by neutral currents, such models will be important subjects for further experimental and theoretical study.

On the other hand, there is a point that makes me suspect that parity is *not* conserved in the Lagrangian. It is just the fact that the neutrino appears to have zero mass. The only theories of which I know in which the neutrino mass automatically vanishes are the old "two component" theories, i.e., theories in which lepton number is conserved, and in which there are more types of left-handed leptons (as opposed to antileptons) than right-handed leptons. Of course parity cannot then be conserved.

The evidence for lepton (and baryon) conservation was reviewed here by Goldhaber and Reines, and Steigman described the astrophysical limits on the neutrino mass. These limits arise because neutrinos left over from the early universe would be about as numerous as microwave black-body photons, and would therefore provide far too large a cosmic mass density unless their total mass (summed over neutrino flavors) is less than about 40 eV. (As Steigman mentioned, Ben Lee and I have discussed an alternative possibility, that some "neutrinos" may be so heavy that they had already become very rare at the end of the early era of chemical equilibrium. However, such a "neutrino" if stable must be heavier than about 2 GeV, so this clearly does not apply to ν_e, ν_μ, or even ν_τ. This point was also made independently by Hut and by Dolgov, Vysotsky, and Zeldovich. Gunn *et al.* have further raised the lower bound to about 10 GeV by considerations of the dynamics of clusters of galaxies.)

A limit on the neutrino masses which is even better

numerically is set by the apparent absence of neutrino oscillations, discussed here by Reines and Fowler. Suppose we adopt the point of view (as I do) that muon number conservation is unlikely to be a fundamental *a priori* symmetry. The only reason that muon number is so well conserved as it is experimentally is then that most gauge models do not allow large rates for muon number non-conserving processes if the neutrino masses are small. If you take this point of view, then if ν_e and ν_μ have masses, the mixing angle between them should be large, and the observed absence of ν_e-ν_μ oscillations sets an upper bound on $|m(\nu_e) - m(\nu_\mu)|$ of about 0.4 eV. I find it very hard to believe that neutrino mass differences could be as small as 0.4 eV (or even 40 eV) unless neutrinos are massless.

There is another point that suggests that parity may not be a fundamental symmetry of the Lagrangian. It is just that we do not have to assume that parity is conserved on a fundamental level in order to understand why it is conserved by the strong and electromagnetic interactions. In fact, the questions seem to have almost nothing to do with each other! As emphasized in Wilczek's talk, the QCD Lagrangian is extremely simple:

$$\mathcal{L} = -\bar{\psi} Z \gamma^\mu (\partial_\mu - i g_s A^a_\mu t_a)\psi - \bar{\psi} m \psi - \tfrac{1}{4} F^a_{\mu\nu} F^{a\mu\nu} \tag{6}$$

where Z and m are unknown matrices acting on flavor indices of the quark field ψ; g_s is the strong coupling constant; A^a_μ is the gluon field; t_a is the octet of Gell-Mann SU(3) matrices; and $F^a_{\mu\nu}$ is the non-linear Yang-Mills curl of A^a_μ. It might appear that parity could be violated here, because Z and m can contain γ_5 terms; but we can always re-define the quark fields by separate matrix transformations on their left- and right-handed parts in such a way that Z = 1 and m is real, diagonal, and γ_5-free. Thus it appears that not only parity, but also C,T and all flavors are automatically conserved by the strong (and electromagnetic) interactions, despite the fact that these may not be even approximate symmetries of the Lagrangian.

Now, this is all fine as far as strong conservation of flavors and charge conjugation are concerned, but a puzzle has come up as to CP invariance. The new problem here has to do with the existence of solutions of the classical field equations with unusual topologies, treated in Yang's talk. The importance of these solutions in quantum field theories can be seen on the basis of the Feynman path integral formalism. In this formalism, the vacuum expectation value of an operator $O(\psi,A)$ is given by a functional integral

$$\langle O\rangle = \int[d\psi][dA]O(\psi,A)\exp\left(i\int d^4x\mathcal{L}(\psi,A)\right) \quad . \qquad (7)$$

In the functional integral we must include *all* field configurations, including those with unusual topologies, the "non-trivial bundles" discussed here by Yang. A perturbation expansion is generated by looking for some stationary "point" of the action $\int d^4x\mathcal{L}$, i.e., a solution of the classical field equations, and expanding the action in a series of powers of the fluctuations around this solution. In particular, the usual perturbation series is obtained by expanding about the vacuum solution $\psi = A = 0$. However Belavin *et al.* have found that in QCD there are other solutions (named "instantons" by 't Hooft) with unusual topologies characterized by an integer

$$q = \frac{1}{64\pi^2} \int d^4x \, F^a_{\mu\nu} F^a_{\rho\sigma} \varepsilon^{\mu\nu\rho\sigma} \quad . \tag{8}$$

As a result of quantum tunneling effects, expectation values are to be calculated as a sum over q of functional integrals constrained by Eq. (8), with coefficients $\exp(iq\theta)$. However, q changes sign under P and CP, so for $\theta \neq 0$, P and CP are violated. To be a little more precise, the chiral transformations on the quark fields which are needed to eliminate γ_5 terms in m change the effective action by an amount which because of Adler-Bell-Jackiw anomalies is just proportional to q, so that the quantity $\theta + \mathrm{Arg\,Det}\, m_{LR}$ remains invariant. [Here m_{LR} is the coefficient of $\frac{1}{2}(1+\gamma_5)$ in an expansion of m into $\frac{1}{2}(1\pm\gamma_5)$ terms.] The condition for P and CP invariance is that θ should vanish when the quark fields are defined so that m is γ_5-free, i.e., when m_{LR} is real. Hence in general the condition for P and CP invariance in QCD is that $\theta = -\mathrm{Arg\,Det}\, m_{LR}$.

There is no difficulty here in preventing violations of P and CP in the strong interactions. One may simply impose the condition that CP is conserved, so that $\theta = 0$ and m_{LR} is real; the strong interactions will then conserve P as well as CP. The problem is of course that CP is *not* conserved by the weak interactions, and we would like to understand why it should be conserved by the strong interactions, and why the weak interactions do not produce a shift in θ which would introduce a large violation of CP in the strong interactions. The strong interactions could have a value of θ which would just cancel this shift, but why should they?

Peccei and Ouinn have made a very nice suggestion, which was discussed here by Wilczek, of a class of theories in which the CP violation in strong interactions is automatically very small, whatever the value of θ. They proposed that the Lagrangian has a chiral U(1) symmetry, which makes the phase of the mass matrix m_{LR} undetermined in any finite order of perturbation theory. The phase of the mass matrix is then fixed by θ-dependent instanton effects which break this symmetry, and in fact it turns out in detailed calculations that although θ is arbitrary, $\operatorname{Arg}\operatorname{Det} m_{LR}$ is determined to take the value $-\theta$.

At first sight it might appear that the U(1) symmetry of Peccei and Quinn could have no further physical consequences, because after all it *is* strongly broken by instantons. However, Wilczek and I have independently noticed that in fact it would have a dramatic consequence. The dynamics of the scalar fields is governed in tree approximation by the polynomial part $P(\phi)$ of the Lagrangian, which would have to satisfy the Peccei-Quinn symmetry, so the spontaneous breakdown of this symmetry leads to a Goldstone boson, whose mass vanishes in lowest order. Wilczek and I call this the "axion". The Peccei-Quinn symmetry *is* strongly broken by instantons, but the scalar field coupling to quarks is so weak ($\propto G_F^{\frac{1}{2}}$) that the scalar fields know very little about instantons. (Who does?) Thus, the Goldstone boson gets a mass from instanton effects, but a very small mass, with $m_a^2 \approx G_F m^4$, where m is some typical strong interaction mass parameter. For m between 50 MeV and 500 MeV, m_a is between 10 keV and 1 MeV. Leaving aside the massless particles, the axion would be the lightest of all the elementary particles.

A natural question is whether the axion would already have been seen. It is a pseudoscalar electrically neutral boson with semi-weak flavor conserving couplings, so it appears that it would not have been seen in any of the measurements which are usually quoted as evidence against a very light Higgs boson. It is very important to examine whether there is any other evidence against the axion, and if not, to devise experiments which could find it. If axions do not exist, it will be necessary to return to the question of what determines θ.

(Time will not permit me to report on the quantitative side of QCD, discussed here by Politzer, or the theory of black holes, discussed by Perry and Wheeler. I will just mention that the instanton solutions of QCD are playing an increasing role in deriving predictions from QCD, and that they find interesting analogues in general relativity.)

I have emphasized the tasks that lie ahead in an understanding of elementary particle symmetries, but taken all in all, this part of elementary particle theory is in very good shape. (I find it especially satisfying that QCD together with the general framework of renormalizable gauge theories of weak and electromagnetic interactions provides a natural explanation of so many of the successful theoretical ideas of the 1950's and 1960's: CVC, PCAC, current algebra, soft pion theorems, universality, $(3,\overline{3}) + (\overline{3},3)$ symmetry breaking, even conservation of strangeness.) Unfortunately, our mood of optimism shifts when we turn from the symmetries to the particles on which they act. We simply do not have any *a priori* idea of how many flavors of quarks or leptons there ought to be. As far as any fundamental theory is concerned, there might as well be any number of quark and lepton flavors; successive flavors have been pushed on us by experimental data. Still, even without a deep understanding, we might have felt a sense of satisfaction at coming to the end of the story if there were just four flavors of quarks and leptons. Unfortunately, we seem to be faced with a veritable explosion of new quarks and leptons, which reminds us again of how poorly we understand the menu of flavors that nature has to offer.

Perl and Radermacher presented the evidence here for a new lepton, the $\tau(1900)$. Increasingly it seems like another sibling of the e and μ, decaying by the modes $\tau^- \to \nu_\tau + X + \overline{Y}$, with (X,Y) either $(d,u)_{RED}$, $(d,u)_{GREEN}$, $(d,u)_{BLUE}$, (e^-,ν_e), or (μ^-,ν_μ). Most especially, the branching ratios for the e^- and μ^- modes are $(18 \pm 4)\%$, in agreement with the value $\frac{1}{5}$ expected if these five modes have equal rates. It will be very interesting to find if ν_τ is really another neutrino - a stable lepton of small or zero mass.

The evidence for a new quark is less direct but still pretty persuasive. Lederman reported here on the set of new neutral vector hadrons, the $\Upsilon(9410)$, $\Upsilon'(10060)$, and perhaps also a $\Upsilon''(10400)$. It seems most likely that these are the ground state and excited states of a new quark-antiquark bound system, but we do not yet know the charge of the new quark. It is not even ruled out that $\Upsilon(9410)$ and $\Upsilon'(10060)$ are bound states of *two* new quarks.

There are also exotic trimuon events, discussed here by Barger and Reeder. A number of scenarios for these events have been proposed, some of which involve new heavy leptons or new quarks or both. However, the total number of these trimuon events is small, and it is not yet clear that they cannot be explained in more conventional terms.

I have now finished the scientific part of my summary, but with our chairman's forbearance, I want to say a word about this conference and Benjamin Lee. I think that Ben would have enjoyed this conference very much. For one thing, Ben tried throughout his working life to break through the barriers of specialization, and to cover the widest possible range of problems in physics. This conference has brought together a fantastic variety of concerns within physics, ranging from the cosmological and galactic down to the level of atoms, nuclei, elementary particles, and the underlying substratum of quantum fields. Ben would have found all of it stimulating, and would have illumined every part of our discussions.

There is another aspect of this conference that would have pleased Ben Lee. He was a tremendous optimist, always expecting the progress of physics to continue, and always anxious to do everything he could to make it continue. In a way, the most remarkable thing about this conference is the way in which it shows that we have in fact been able to continue our progress. Some of the theory that I have heard about here seems to me to be pretty formidable. And as for the experiments that are being done, if I didn't know that they were being done, I would say that they were impossible. Yet despite the difficulty of these mathematical theories and of these wonderful experiments, we've seen that somehow or other we have been able to keep moving forward. Ben would have liked this very much.

TABLE I

Measured Quantity	SU(2) × U(1) Prediction ($\sin^2\theta = 0.27$)	Observed Value	Source
$\frac{\nu n \to \nu X}{\nu n \to \mu^- X}$	0.28	0.29 ± 0.012	GGM, HPWF, CalT-FNAL, BNL 7' B.C., CDHS, CERN-BEBC
$\frac{\bar{\nu} n \to \bar{\nu} X}{\bar{\nu} n \to \mu^+ X}$	0.38	0.38 ± 0.025	
$\frac{\nu p \to \nu X}{\nu p \to \mu^- X}$	0.50	0.48 ± 0.17	FNAL-IHEP -ITEP-Mich
$\frac{\bar{\nu} p \to \bar{\nu} X}{\bar{\nu} p \to \mu^+ X}$	0.33	0.49 ± 0.14	ANL-CarnMel -Purdue
$\frac{\nu p \to \nu p}{\nu n \to \mu^- p}$	0.10	0.09 ± 0.02	Col-Ill-Rock, HPW-BNL, A-P
$\frac{\bar{\nu} p \to \bar{\nu} p}{\bar{\nu} p \to \mu^+ n}$	0.11	0.2 ± 0.1	HPW-BNL
$\frac{\nu n \to \nu n\pi^0 + \nu p \to \nu p\pi^0}{2\,\nu n \to \mu^- p\pi^0}$	0.24	0.21 ± 0.07	Col-Ill-Rock, GGM, A-P
$\frac{\bar{\nu} n \to \bar{\nu} n\pi^0 + \bar{\nu} p \to \bar{\nu} p\pi^0}{2\,\bar{\nu} p \to \mu^+ n\pi^0}$	0.3	0.37 ± 0.08	Col-Ill-Rock, GGM
$\frac{\nu p \to \nu n\pi^+}{\nu p \to \mu^- p\pi^+}$	0.07	0.13 ± 0.06	ANL 12' B.C.
$\frac{\nu p \to \nu p\pi^0}{\nu p \to \mu^- p\pi^+}$	0.17	0.40 ± 0.22	ANL 12' B.C.
$\frac{\nu n \to \nu p\pi^-}{\nu p \to \mu^- p\pi^+}$	0.07	0.12 ± 0.04	ANL 12' B.C.
$\sigma(\bar{\nu}_e e^- \to \bar{\nu}_e e^-)/E_\nu\,(\mathrm{cm}^2/\mathrm{GeV})$	5.2×10^{-42}	$(5.7 \pm 1.2) \times 10^{-42}$	Reines *et al.*
$\sigma(\nu_\mu e^- \to \nu_\mu e^-)/E_\nu\,(\mathrm{cm}^2/\mathrm{GeV})$	1.4×10^{-42}	$(1.2 \pm 1.0) \times 10^{-42}$	GGM A-P
$\sigma(\bar{\nu}_\mu e^- \to \bar{\nu}_\mu e^-)/E_\nu\,(\mathrm{cm}^2/\mathrm{GeV})$	1.6×10^{-42}	$(1.9 \pm 1.1) \times 10^{-42}$	

Comparison of the SU(2) × U(1) standard model with experiment, for neutral current neutrino reactions. This compilation of experimental and theoretical results was very kindly supplied by C. Baltay. The experimental value given for $\nu p \to \nu p$ has been revised since this table was first prepared, to incorporate recent data from the Aachen-Padua group. I have added recent data on inclusive νp and $\bar{\nu} p$ neutral current cross sections (with y-dependence corrections calculated for $\sin^2\theta = 1/3$) together with theoretical results of Albright, Quigg, Shrock, and Smith for these reactions. Additional data on the semi-inclusive neutral current reactions $\nu n \to \nu X\pi$ and $\bar{\nu} n \to \bar{\nu} X\pi$ is not included here. (Here n denotes a heavy nucleus; X stands for an arbitrary set of hadrons.)

Added Note

The SLAC-Yale group have announced results of the experiment on deep inelastic scattering of polarized electrons, that was mentioned in the Conference Summary. Parity nonconservation in neutral currents is definitely established. The measured asymmetry $A \equiv (\sigma_R - \sigma_L)/(\sigma_R + \sigma_L)$ in deuterium is found to be $A/q^2 = (-9.5 \pm 1.6) \times 10^{-5} \mathrm{GeV}^{-2}$ at $y = 0.21$. This is in excellent agreement with the standard SU(2) ⊗ U(1) gauge model, which gives A/q^2 from $-9.7 \times 10^{-5} \mathrm{GeV}^{-2}$ to $-7.2 \times 10^{-5} \mathrm{GeV}^{-2}$ for the range of $\sin^2\theta$ values (from 0.20 to 0.25) now indicated by neutrino experiments. [In contrast, the assumptions outlined in Bjorken's talk would yield an asymmetry equal to that given by the SU(2) ⊗ U(1) standard model for $\sin^2\theta = 0$, or $A/q^2 = -19.8 \times 10^{-5}\ \mathrm{GeV}^{-2}$, which is more than six standard deviations too large.] Parity nonconservation is also found in hydrogen, and the measured asymmetry is in agreement with the SU(2) ⊗ U(1) standard model, but with larger experimental uncertainties.

LIST OF SPEAKERS

ADELBERGER, E. G.	Physics Department, University of Washington, Seattle
BARGER, V.	Physics Department, University of Wisconsin, Madison
BARNES, C. A.	California Institute of Technology, Pasadena
BJORKEN, J. D.	Stanford Linear Accelerator Center, Stanford University, Stanford
BOUCHIAT, M. A.	Laboratoire de Spectroscopie Hertzienne de l'Ecole Normale Supérieure, Paris
COMMINS, E. D.	University of California, Berkeley
DAVIDSON, J. M.	California Institute of Technology, Pasadena
FAISSNER, Helmut	III. Physikalisches Institut, Technische Hochschule Aachen, Aachen
FEINBERG, G.	Department of Physics, Columbia University, New York
FOWLER, William A.	California Institute of Technology, Pasadena
GLASHOW, S. L.	Harvard University, Cambridge
GOLDHABER, Maurice	Brookhaven National Laboratory, Upton
HAIDT, D.	CERN, Geneva
LEDERMAN, L. M.	Columbia University, New York
LOWRY, M. M.	California Institute of Technology, Pasadena
MANN, A. K.	Department of Physics, University of Pennsylvania, Philadelphia
MARRS, R. E.	Department of Physics, Indiana University, Bloomington
MORINIGO, F. B.	California State University, Los Angeles
NAGLE, D. E.	Los Alamos Scientific Laboratory, Los Alamos
PATI, Jogesh C.	Department of Physics and Astronomy, University of Maryland, College Park
PERL, Martin L.	Stanford Linear Accelerator Center, Stanford University, Stanford
PERRY, M. J.	Department of Applied Mathematics and Theoretical Physics, University of Cambridge, Cambridge
POLITZER, H. David	California Institute of Technology, Pasadena
POTTIER, L.	Laboratoire de Spectroscopie Hertzienne de l'Ecole Normale Supérieure, Paris

RADERMACHER, E.	III. Physikalisches Institut, Technische Hochschule Aachen, Aachen
REEDER, Don D.	University of Wisconsin, Madison
REINES, Frederick	Department of Physics, University of California, Irvine
ROUSSET, A.	DGRST (Delegation Generale a la Recherche), Paris
SANDARS, P. G. H.	Clarendon Laboratory, Oxford University, Oxford
SCHRAMM, David N.	Enrico Fermi Institute and Department of Astronomy and Astrophysics, University of Chicago, Chicago
STEIGMAN, Gary	Yale University Observatory, New Haven
SWANSON, H. E.	University of Washington, Seattle
TADIĆ, Dubravko	University of Zagreb, Zagreb, and "Rudger Bosković" Institute, Zagreb
WEINBERG, Steven	Lyman Laboratory of Physics, Harvard University, Cambridge
WHEELER, John Archibald	Center for Theoretical Physics, The University of Texas at Austin, Austin
WILCZEK, F.	The Institute for Advanced Studies, Princeton
WILSON, Richard	Department of Physics, Harvard University, Cambridge
WU, C. S.	Columbia University, New York
YANG, Chen Ning	Institute for Theoretical Physics, State University of New York, Stony Brook, New York

SUBJECT SUMMARY

AUTHOR INDEX*

**page numbers in italics refer to references*